Electric and Magnetic Fields

Electric and Magnetic Fields

by

STEPHEN S. ATTWOOD

Late Professor of Electrical Engineering
University of Michigan

DOVER PUBLICATIONS, INC., NEW YORK

Published in Canada by General Publishing Company, Ltd., 30 Lesmill Road, Don Mills, Toronto, Ontario.

Published in the United Kingdom by Constable and Company, Ltd., 10 Orange Street, London WC 2.

This Dover edition, first published in 1967, is an unabridged and unaltered republication of the third edition which was published in 1956 by John Wiley and Sons, Inc.

International Standard Book Number: 0-486-61753-X
Library of Congress Catalog Card Number: 66-30008

Manufactured in the United States of America
Dover Publications, Inc.
180 Varick Street
New York, N.Y. 10014

This Dover edition
is dedicated to the memory of
STEPHEN S. ATTWOOD
May 29, 1897 – June 8, 1965

Preface

The electrical engineering curriculum at the University of Michigan for a number of years has included a course in electricity and magnetism. Given in the form of three one-hour recitations and one three-hour computation period per week for one semester, the course serves to coordinate the underclass work in mathematics, mechanics, and physics with the professional work of the last two years. In spite of the time devoted during the first two years to the sciences which are basic for electrical engineering, the student needs further training in the development of the fundamental concepts, formulas, terminology, and units used in electric and magnetic field study. He needs practice under guidance in expressing simple physical ideas in simple mathematical language. Furthermore, the student must learn that there is a considerable gap between the statement of a physical principle and its utilization in the engineering world. It is my opinion, and that of my colleagues, that this course makes the transition from the under- to the upper-class work easier for the student and provides a training in the fundamentals of electricity and magnetism that improves the quality of work in the professional courses given later. The present textbook has been written from the material presented in this course.

In presenting this third edition, it may be well to explain the general arrangement and to indicate some of the points I have wished to emphasize. The book is divided into four parts. The electric field is discussed in the first part. Because of its relative simplicity compared with the magnetic field, the electric field is better adapted to the purpose of developing the field concepts of potential, field intensity, flux density, etc. Moreover, such a large part of the electrical engineering curriculum deals with magnetic relations that the electric field is often neglected. The electric field deserves the emphasis that it receives in being discussed first.

The magnetic field is treated in the second part, where the field formulas are developed under the assumption of constant permeability. The first chapter (Chapter 9) of this part has been completely rewritten to emphasize current, rather than magnetic poles, as the fundamental source of the magnetic effect. The magnetic pole concept has not been abandoned entirely since its value as a pedagogical tool is too well established.

The ferromagnetic field, in the third part, introduces the modifications of the basic magnetic relations necessary to account for the use of magnetic material in which permeability cannot be considered constant. It is believed that this division in the study of magnetism aids the student to a considerable degree in obtaining a clearer interpretation of the fundamental relations. The sections on magnetic materials have been improved by the inclusion of magnetization curves for a wider range of materials, and by the addition of elementary discussions of the basic physical relations determining dia-, para-, and ferromagnetism.

The fourth part, "Combined Electric and Magnetic Fields," treats in an elementary manner the interactions between electric and magnetic fields that simultaneously occupy the same space. This part is concluded by the presentation of the electromagnetic (Maxwell's) equations, which summarize concisely the basic relations developed independently earlier in the text.

Though circuits have not been neglected, the author has striven to emphasize the *field* point of view. The curriculum in electrical engineering of a few years back was developed largely on the basis of the electric circuit; but with the demand for closer limits in design, with the growth of electronics and radio communication, and with the tendency of schools to expand the curriculum along more theoretical lines has come a need for including the study of fields in addition to the study of circuits.

A feature of this text is the use of a large number of drawings and sketches of both electric and magnetic fields of various geometrical forms. Much time and effort have been spent in giving the correct shapes to the lines of flux and the equipotential surfaces. In addition, the solution of a considerable number of problems has been given symbolically.

Another feature worthy of explicit statement is the use of the methods of "field mapping" which have come into prominence in the last few years. Field-mapping methods have greatly increased the number of problems which can be treated quantitatively without the use of the higher forms of mathematics.

A collateral, but by no means unimportant, benefit that should accrue to the student from the presence of many carefully developed field drawings is the stimulation of his imaginative sense. Perhaps the greatest weakness of the average student is his inability to form a satisfactory mental image of geometrical forms. It is hoped that the training received in this course will help to develop this important sense.

During and since World War II a definite and unmistakable swing toward the use of the official mks system of units developed. Consequently, these are the units adopted for use in this edition. Furthermore, there appears to be an increasing trend toward the use of the rationalized form of the mks system. In line with this development, the formulas have been recast in the rationalized forms (from the cgs forms used in the earlier editions). Appendix I has been added to facilitate transfer of magnitudes between the mks and cgs systems.

I wish to take this opportunity to acknowledge my sincere appreciation of the labor of my colleagues, Professors W. G. Dow and H. S. Bull, in reading and criticizing the manuscript, and of Dr. W. H. Bixby, who spent many hours sketching certain of the field maps.

STEPHEN S. ATTWOOD

Electrical Engineering Department
University of Michigan
March 1949

Contents

PART I. THE ELECTRIC FIELD

PART II. THE MAGNETIC FIELD

List of Tables

PART I

THE ELECTRIC FIELD

CHAPTER 1

THE ELECTROSTATIC FIELD

A. UNITS AND ELECTRONIC BACKGROUND

The modern electron theory states that all matter is electrical in constitution. We do not know what electricity is, but we can learn its behavior under certain conditions and thus predict its probable behavior under other circumstances. The purpose of this book is, first, to examine the methods by which electricity is made available; second, to state the experimental laws governing electrical interactions; and, lastly, to develop a theory which encompasses these relations, couched in terminology that is common in the electrical engineering profession.

In carrying out this program it is convenient to divide electrical engineering problems into two classes:

1. Circuit problems, in which attention is focused primarily upon the distributions of voltage, current, and power occurring in various arrays of batteries, generators, resistances, inductances, condensers, and tubes under conditions where the current is confined to the array elements and to the conducting paths connecting them together.

2. Field problems, in which the relevant quantities to be studied are the strengths of the electric fields associated with differences of electric potential, the strengths of magnetic fields surrounding currents, the force effects of the fields, the storage of energy in electric and magnetic field form, and the propagation of radiant energy resulting from the interaction of electric and magnetic fields. This book is concerned primarily, though not exclusively, with problems of this type.

The first portion of the book deals with electrostatics, which is the simplest form of field study since the field depends only on the static distributions of electrical charges. Magnetostatics, in the second part, is somewhat more involved since the magnetic field effects are related to currents or the movements of charge. In the third part, consideration is given to the special techniques required in dealing with the properties of the ferromagnetic materials. The last part discusses the interactions of electric and magnetic fields varying in time, which lead to the propagation of energy in the form of electromagnetic waves. Problems involving low-frequency voltages and currents often may be solved by methods which are essentially circuital, but with increasing frequency the field approach is used more and more, and at the higher frequencies it is indispensable.

1.1. Units and Dimensions. Quantities defined in subsequent chapters (such as current, voltage, energy density, and field strength) are specified quantitatively in terms of certain *units* agreed upon by the International Committee on Weights and Measures (ICWM) and the International Electrotechnical Commission (IEC) composed of representatives of technical organizations in many countries. These quantities may also be expressed in terms of *five* fundamental units and thus form a system of units. Quantities so defined will, in general, be expressed in two or more of the fundamental units and are said to have the *dimensions* of these units.

Mechanical quantities, such as force, power, and energy, can be expressed in not more than three fundamental units. The units ordinarily chosen (and approved by the ICWM) as fundamental for the mechanical relations are length, mass, and time, whose dimensions are L, M, and T. For example, force is mass times acceleration or mass times length divided by time squared. Consequently the dimensional formula for force is

$$F = L^1M^1T^{-2} \tag{1.1}$$

and for energy, the product of force and length, the dimensional formula is

$$W = FL = L^2M^1T^{-2} \tag{1.2}$$

Extension of this system to include electric and magnetic quantities requires a fourth unit. For this the ICWM has chosen μ_0, the magnetic permeability of free space, but they might equally well have chosen any other electric or magnetic quantity, for instance, the volt, the ohm, the coulomb, the ampere, or the henry. The assignment of a particular value for μ_0, however, determines the sizes of the units of these alternative quantities. For the purpose of writing dimensional formulas for electric and magnetic quantities it is more satisfactory to adopt a basic physical quantity as fundamental. Consequently we shall choose the coulomb as the fourth fundamental unit with dimensions represented by Q. The dimensional formulas of electric and magnetic quantities then take the form

$$\text{Quantity} = L^aM^bT^cQ^d \tag{1.3}$$

with a, b, c, and d each assuming one of the integral values ± 0, 1, 2, 3, as prescribed by the basic physical entities entering into the particular quantity.

The values of the exponents *a*, *b*, *c*, and *d* for the physical quantities discussed are given in Appendix I.

To complete the system of units to include heat in addition to mechanics, electricity, and magnetism, it is necessary to add a fifth fundamental unit. This last is usually taken as temperature. We shall not, however, need this unit since the problems discussed subsequently present, at most, only elementary thermal aspects.

1.2. Cgs Units. For many years the units used for scientific purposes were based on the cgs system, in which the units of length, mass, and time are the centimeter, the gram, and the second, respectively. This system, though satisfactory for use with mechanical quantities, proved unduly complicated when electric and magnetic quantities had to be measured. As developed the cgs system was divided into two sub-systems, the electrostatic sub-system (esu) based on Coulomb's law of force between charges and the electromagnetic sub-system (emu) based on Coulomb's law of force between magnetic poles. The latter was referred to also as the "absolute" system. As a result each electric or magnetic quantity requires three different sizes of unit, one for each sub-system and a third for the so-called "absolute practical" system in which, for convenience, the units are taken as multiples or sub-multiples of the electromagnetic units. By way of illustration we may note that, for current, 1 statampere (esu) is approximately equal to $1/(3 \times 10^{10})$ abampere (emu) and the abampere is equal to 10 "absolute practical" amperes; for voltage, 1 statvolt (esu) is approximately equal to 3×10^{10} abvolts (emu) and the abvolt is $1/10^8$ "absolute practical" volt. To complicate the matter further we may note that laboratory voltmeters and ammeters are calibrated in the International System of Units in which the units differ from the absolute practical values by small amounts which, for measurements requiring high accuracy, are not negligible.

Appendix I is a table giving the number of units in the various systems required to describe a particular amount of any of the more common electric and magnetic quantities.

1.3. Mks Rationalized Units.[1] Coulomb's Law. In 1904, Giorgi proposed that the meter replace the centimeter as the unit of length, that the kilogram replace the gram as the unit of mass, but that the second be retained as the unit of time. He showed that, if a fourth fundamental unit of electric or magnetic character is chosen of correct size, the whole forms a *single* comprehensive system of units including

[1] For a more extended presentation of the development of the mks units, consult *Electrical Engineering*, December 1935, p. 1373; February 1939, p. 78; July 1947, p. 737.

the practical electrical units such as the volt, ampere, ohm, watt, and joule.

At its meeting in 1935 the IEC adopted the absolute system of mechanical units (mks system) proposed by Giorgi and decided that the substitution of this system in place of the cgs system should take effect Jan. 1, 1940. In 1938 the IEC adopted μ_0, the magnetic permeability of free space, as the fourth fundamental unit, thus completing the system far enough to include electric and magnetic quantities.

The fundamental unit of length is the meter, defined as the distance between two specified lines on a bar of platinum-iridium kept at the International Bureau near Paris, when the bar is at the temperature of 0° C, at standard atmospheric pressure, and is supported at the two neutral points 0.285 meter from the center of the bar. Two copies are deposited at the U. S. National Bureau of Standards.

The fundamental unit of mass is the kilogram, defined as the mass of a cylinder of platinum-iridium kept at the International Bureau. Two copies are deposited at the U. S. National Bureau of Standards.

The fundamental unit of time is the second, defined as one 86,400th part of a mean solar day. Radio and telegraph signals are sent out each second by the Naval Observatory near Washington.

The fourth fundamental unit, μ_0, the magnetic permeability of free space, is taken as $4\pi/10^7$ henrys per meter in the "rationalized" mks system used in this book. (In the unrationalized mks system, preferred by some writers, μ_0 is taken as 10^{-7} henry per meter.) The value chosen is prescribed by the desire to relate the mechanical units to electric and magnetic units through basic physical formulas without introducing multiplying factors not based on physical laws.

The question of "rationalization" has not yet been settled by the IEC or the ICWM. The phrase "rationalization" refers merely to the problem of whether the factor 4π should be introduced in certain formulas and omitted from others or not. Those authors who wish to rationalize the formulas arbitrarily introduce the factor 4π into formulas dealing with point quantities, whence it will appear as 2π in line problems and disappear entirely from parallel plate or uniform field problems. The present author feels that rationalization of the formulas is a desirable step to take and that it generally leads to simpler forms. Consequently this book uses the rationalized system of units and formulas.

For the study of fields we shall need also ϵ_0, the dielectric constant of free space (called also the permittivity or specific capacitance of free space). The value of ϵ_0 depends upon the value chosen for μ_0 and on c, the velocity of electromagnetic waves (or light) in free space.

According to wave theory

$$\frac{1}{\sqrt{\mu_0 \epsilon_0}} = c \tag{1.4}$$

or

$$\epsilon_0 = \frac{1}{\mu_0 c^2} \tag{1.5}$$

For $c = 2.998 \times 10^8$ meters per second and $\mu_0 = 4\pi 10^{-7}$

$$\epsilon_0 = 8.854 \times 10^{-12} \quad \text{farad per meter} \tag{1.6}$$

For $c \cong 3 \times 10^8$, the convenient approximate value,

$$\epsilon_0 \cong \frac{1}{36\pi 10^9} \quad \text{farad per meter} \tag{1.7}$$

The remaining units required to complete the mks system may be defined in terms of the preceding fundamental units with the aid of basic physical formulas. For instance, the unit of force is the newton, defined as that force which will give a 1-kilogram mass an acceleration of 1 meter per second per second.

$$F = ma \quad \text{newtons} \tag{1.8}$$

(The newton is equal, approximately, to 0.225 pound.)

The unit of energy or work is the joule or watt-second, given by a force of 1 newton acting through a distance of 1 meter.

$$W = FL \quad \text{joules or watt-seconds} \tag{1.9}$$

The unit of power is the joule per second or the watt

$$P = \frac{W}{T} \quad \frac{\text{joules}}{\text{second}} \text{ or watts} \tag{1.10}$$

The basic law of electrostatics is Coulomb's law which states that two electric charges of like sign repel each other (and unlike charges attract each other) with a force which is proportional to the product of the charges and varies inversely as the square of the distance between them. Quantitatively, for the mks rationalized units, this law takes the form

$$F = \frac{Q_1 Q_2}{4\pi\epsilon_0 r^2} \quad \text{newtons} \tag{1.11}$$

for two charges in a vacuum, separated by distance r.

Consequently, unit charge (the coulomb) is one that repels a like charge in a vacuum at a distance of 1 meter with a force of

$$\frac{1}{4\pi\epsilon_0} = 8.99095 \times 10^9 \cong 9 \times 10^9 \quad \text{newtons}$$

This is an enormous force, equal, very approximately, to 1,000,000 tons. We shall usually expect to find concentrations of electricity which are only small fractions of a coulomb, and forces that are correspondingly much lower.

Three of the subatomic electrical particles entering into the construction of all atoms are the electron, the proton, and neutron. The electron and the proton have identical values of charge, the former negative and the latter positive, equal to

$$e = 1.602 \times 10^{-19} \quad \text{coulomb} \tag{1.12}$$

Thus there are 0.6242×10^{19} electrons or protons in 1 coulomb. The neutron, as its name implies, is a particle with zero charge. Further information on these particles is given in Table 1.3.

1.4. The International Units. Laboratory measuring devices such as voltmeters, ammeters, and wattmeters were calibrated in terms of the international units which were intended to be the precise equivalents of the absolute values given by the mks system. There are, however, differences in the sizes of the units in the two systems amounting in certain instances to 0.05 per cent. In making measurements of high accuracy errors of this magnitude may have to be accounted for.

The cause of this discrepancy lies in the fact that the international units are based on physical standards whose specifications are slightly in error. The specifications for the international standards are as follows:

The International Ohm is the resistance offered to an unvarying electric current by a column of mercury at the temperature of melting ice, 14.4521 grams in mass, of a constant cross-sectional area, and of a length of 106.300 centimeters.

The International Ampere is the unvarying electric current which, when passed through a solution of silver nitrate in water in accordance with standard specifications, deposits silver at the rate of 0.00111800 gram per second.

The International Volt is the electric difference of potential which, when steadily applied to a conductor whose resistance is one international ohm, will produce a current of one international ampere.

For practical purposes, it is agreed that the Weston normal (saturated) cadmium cell has a potential difference of 1.0183 international volts at 20° C.

Experimental work has repeatedly been carried out by physical laboratories in various countries, at the request of the International Committee on Weights and Measures, (ICWM), to determine as precisely as possible the magnitudes of the international units in terms of the absolute mks units. The values obtained by the National Bureau of Standards, for use in the United States after January 1, 1948, are given in Table 1.1.

TABLE 1.1

Relation of International Units and Mks Units

1 international ohm	(U. S.)	=	1.000495 absolute ohms
1 international volt	(U. S.)	=	1.000330 absolute volts
1 international ampere	(U. S.)	=	0.999835 absolute ampere
1 international coulomb	(U. S.)	=	0.999835 absolute coulomb
1 international watt	(U. S.)	=	1.000165 absolute watts
1 international joule	(U. S.)	=	1.000165 absolute joules
1 international henry	(U. S.)	=	1.000495 absolute henrys
1 international farad	(U. S.)	=	0.999505 absolute farad

As a matter of practice, measuring instruments made before January 1, 1948, and calibrated in the international units may require the use of the correction factors of Table 1.1 when the accuracy of the measurements is sufficiently high. Instruments made subsequently are calibrated in absolute units.

1.5. Electrical Character of the Atom. All matter is divisible into small clusters called molecules which are relatively stable under many conditions. Chemical reactions indicate that the molecules are composed of smaller units, or atoms, and modern physical methods of investigation have shown that the atom is still further divisible into two units, positively charged nuclei, and electrons (negative electricity). The total positive charge of the nucleus, in the *normal* or *neutral* atom, is equal to the total negative charge of all the electrons which surround the nucleus. Nearly all the atom's mass is associated with the nucleus. The 96 elements of the chemical periodic table differ from one another in the amount of mass and charge in each nucleus and in the number of electrons. If the elements are numbered consecutively from 1 to 96, the value for any particular atom, known as the "atomic number," gives the number of electrons surrounding the nucleus of that atom. The "atomic weight," based on 16 for oxygen, is a measure of the relative amount of mass of the atom. These values, for the first 36 atoms, are exhibited in Table 1.2.

In this table the column of valency gives the relative chemical combining proportions for the formation of molecules. For instance, phosphorus with valencies 3,5 joins with chlorine, valency 1, to form phosphorus trichloride PCl_3 or phosphorus pentachloride PCl_5. The elements are divided into groups, and the upper end element in each

TABLE 1.2

Electron Data for Normal Atomic States

Atomic No.	Name	Symbol	Atomic* Weight	Valency	Atomic* First Ionization Potential, Electron-Volts†	Number of Electrons — Electron Symbols n, l: 1,0 1s	2,0 2s	2,1 2p	3,0 3s	3,1 3p	3,2 3d	4,0 4s	4,1 4p
1	Hydrogen	H	1.0079	1	13.53	1							
2	Helium	He	4.003	0	24.46	2							
3	Lithium	Li	6.94	1	5.37	2	1						
4	Beryllium	Be	9.02	2	9.28	2	2						
5	Boron	B	10.82	3	8.28	2	2	1					
6	Carbon	C	12.006	2,4	11.22	2	2	2					
7	Nitrogen	N	14.008	3,5	14.48	2	2	3					
8	Oxygen	O	16.	2	13.55	2	2	4					
9	Fluorine	F	19.	1	18.6	2	2	5					
10	Neon	Ne	20.18	0	21.47	2	2	6					
11	Sodium	Na	22.997	1	5.12	2	2	6	1				
12	Magnesium	Mg	24.32	2	7.61	2	2	6	2				
13	Aluminum	Al	26.97	3	5.96	2	2	6	2	1			
14	Silicon	Si	28.06	4	8.12	2	2	6	2	2			
15	Phosphorus	P	31.02	3,5	.....	2	2	6	2	3			
16	Sulfur	S	32.065	2,4	10.3	2	2	6	2	4			
17	Chlorine	Cl	35.457	1	12.96	2	2	6	2	5			
18	Argon	A	39.94	0	15.69	2	2	6	2	6			
19	Potassium	K	39.10	1	4.32	2	2	6	2	6	..	1	
20	Calcium	Ca	40.08	2	6.09	2	2	6	2	6	..	2	
21	Scandium	Sc	45.10	3	6.7	2	2	6	2	6	1	2	
22	Titanium	Ti	47.9	4	6.81	2	2	6	2	6	2	2	
23	Vanadium	V	50.95	3,5	6.76	2	2	6	2	6	3	2	
24	Chromium	Cr	52.01	3,6	6.74	2	2	6	2	6	5	1	
25	Manganese	Mn	54.93	2,3	7.41	2	2	6	2	6	5	2	
26	Iron	Fe	55.84	2,3	7.83	2	2	6	2	6	6	2	
27	Cobalt	Co	58.94	2,3	8.5	2	2	6	2	6	7	2	
28	Nickel	Ni	58.69	2,3	7.61	2	2	6	2	6	8	2	
29	Copper	Cu	63.57	1,2	7.68	2	2	6	2	6	10	1	
30	Zinc	Zn	65.38	2	9.36	2	2	6	2	6	10	2	
31	Gallium	Ga	69.72	3	5.97	2	2	6	2	6	10	2	1
32	Germanium	Ge	72.60	4	8.09	2	2	6	2	6	10	2	2
33	Arsenic	As	74.93	3,5	10.0	2	2	6	2	6	10	2	3
34	Selenium	Se	78.96	2	9.5	2	2	6	2	6	10	2	4
35	Bromine	Br	79.916	1	11.8	2	2	6	2	6	10	2	5
36	Krypton	Kr	83.7	0	13.94	2	2	6	2	6	10	2	6

* These are relative atomic weights based on 16 for the oxygen atom. Multiply these values by 10^{-3} to obtain atomic weights in kilograms per mole.

† 1 electron-volt = 1.602×10^{-19} joule. (See Art. 1.29.)

group has a valency of unity and is very active in chemical combination. In the middle of each group the valency number increases and frequently is multiple, indicating multiple methods of combination into molecular form. Elements at the lower end of each group have zero valency and are notably inactive chemically. Problems involving chemical combinations of atoms, ionization, and excitation depend primarily on the arrangements of the electrons in each atom, as indicated in the columns on the right in Table 1.2. Discussion of the electronic structure of the atom has been placed in Appendix II.

1.6. The Hydrogen Atom—Energy Levels. The simplest atomic nucleus is that of hydrogen, called the proton, with a mass which is 1836.5 times the mass of the electron. These masses are, respectively, $m_p = 1.6725 \times 10^{-27}$ kilogram and $m_0 = 9.1066 \times 10^{-31}$ kilogram. The sum of these quantities gives 1.6734×10^{-27} kilogram for the mass of the hydrogen atom, which is 1837.5 times the electronic mass, and the relative atomic weight is 1.0079. Consequently the mass of a particle of unit atomic weight (no such particle actually exists) is $1837.5/1.0079 = 1823.3$ times the electronic mass.

The electric charge is $e = 1.602 \times 10^{-19}$ coulomb (defined in Art. 1.3), positive for the proton and negative for the electron. Values of electric charge have been determined by the Millikan oil-drop experiment, while ratios of charge to mass have been calculated from experiments in which beams of electrical particles are deflected as they cross a magnetic field.

The size of the particles, electron and proton, can be computed from electromagnetic theory if it is assumed that the charges are small spherical balls. From this theory, given in Chapter 16, the radii of the particles, inversely proportional to their masses, are

$$R_0 = \frac{e^2}{6\pi\epsilon_0 m_0 c^2} \quad \text{and} \quad R_p = \frac{e^2}{6\pi\epsilon_0 m_p c^2} \tag{1.13}$$

in which c is the velocity of light, 2.998×10^8 meters per second. These values are given in Table 1.3. Experiments of various kinds agree as to the order of magnitude. In addition to the electron and proton, several additional particles have been discovered in the last few years but we shall mention only the neutron. Atomic nuclei are composed of protons and neutrons, which are particles with mass but without charge. (See Table 1.3.) In the neutral atom the nucleus must therefore contain as many protons as there are orbital electrons.

The simplest picture of the hydrogen atom is given by the Bohr theory, which, in its simplest form, assumes that an electron rotates in free space at radius "a" in a circular orbit about the nucleus. The

TABLE 1.3

PROPERTIES OF ELEMENTARY PARTICLES

	Charge in Coulombs	Mass in Kilograms	Radius R in Meters	Relative Atomic Weight	Ratios
Electron	$e = -1.602 \times 10^{-19}$	$m_0 = 9.1066 \times 10^{-31}$	$R_0 = 1.885 \times 10^{-15}$	....	$\frac{e}{m_0} = 1.759 \times 10^{11}$
Proton (Hydrogen Atom Nucleus)	$e = +1.602 \times 10^{-19}$	$m_p = 1.6725 \times 10^{-27}$	$R_p = 1.025 \times 10^{-18}$	....	$\frac{m_p}{m_0} = 1836.5$
Neutron	0	$m_n = 1.6747 \times 10^{-27}$	....	1.00866	$\frac{m_n}{m_0} = 1838.9$
Hydrogen Atom	....	$m_H = 1.6734 \times 10^{-27}$	0.53×10^{-10} (1st orbit)	1.0079	$\frac{m_H}{m_0} = 1837.5$
Atom of Unit Atomic Weight (none exists)	....	$m_1 = 1.6603 \times 10^{-27}$	....	1.0	$\frac{m_1}{m_0} = 1823.3$

attractive force between the positive nucleus and negative electron is equal to the product of the charges divided by the square of their separation distance (see Coulomb's law, expressed in Art. 1.3), and it is balanced by the centrifugal force on the electron. Then

$$\frac{e^2}{4\pi\epsilon_0 a^2} = \frac{m_0 v^2}{a} \tag{1.14}$$

In this equation the electron orbital velocity v and the orbital radius a are unknowns. An additional relation is obtained by using Planck's unit of "action," $h = 6.624 \times 10^{-34}$ joule-second. According to Planck, energy multiplied by time, energy divided by frequency, or momentum multiplied by distance always occur in integral multiples of h, say nh. The momentum of the electron multiplied by the circumference of the orbit gives the action and is

$$m_0 v \cdot 2\pi a = nh \tag{1.15}$$

The integer n is called the total quantum number.

From equations 1.14 and 1.15, the electron velocity and orbital radius are

$$v = \frac{e^2}{2\epsilon_0 nh} \quad \frac{\text{meter}}{\text{second}} \tag{1.16}$$

$$a = \frac{n^2 h^2 \epsilon_0}{\pi m_0 e^2} \quad \text{meters} \tag{1.17}$$

and the number of revolutions per second is

$$f = \frac{v}{2\pi a} = \frac{m_0 e^4}{4\epsilon_0{}^2 n^3 h^3} \tag{1.18}$$

The potential energy of the electron in the presence of the positive charge is equal to the integral of the force $e^2/(4\pi\epsilon_0 r^2)$ between the proton and the electron, separated distance r, over the distance from infinity to the final position at radius a. This is equal to

$$W_p = \int_{r=\infty}^{r=a} \frac{e^2}{4\pi\epsilon_0 r^2}\,dr = -\frac{e^2}{4\pi\epsilon_0 a} = -\frac{m_0 e^4}{4\epsilon_0{}^2 n^2 h^2} \quad \text{joules} \tag{1.19}$$

This quantity is reckoned from the zero state of energy (arbitrarily chosen as a reference) at infinite separation.

The kinetic energy is

$$W_k = \tfrac{1}{2} m_0 v^2 = +\frac{m_0 e^4}{8\epsilon_0{}^2 n^2 h^2} = -\tfrac{1}{2} W_p \quad \text{joules} \tag{1.20}$$

The difference between the potential energy and the kinetic energy is supposed to have radiated away in bringing the electron from infinity to its orbital position. The total energy of the electron is

$$W_T = W_p + W_k = \tfrac{1}{2}W_p = -W_k = -\frac{m_0 e^4}{8\epsilon_0{}^2 n^2 h^2} \text{ joules} \quad (1.21)$$

Taking $n = 1$, in the foregoing equations, gives the values for the *first* circular orbit of the hydrogen atom:

$$\begin{aligned} a_1 &= 0.53 \times 10^{-10} \text{ meter} & W_{p_1} &= -4.3 \times 10^{-18} \text{ joule} \\ v_1 &= 2.19 \times 10^{6} \text{ meters/sec} & W_{k_1} &= +2.15 \times 10^{-18} \text{ joule} \\ f_1 &= 0.65 \times 10^{16} \text{ rps} & W_{T_1} &= -2.15 \times 10^{-18} \text{ joule} \end{aligned} \quad (1.22)$$

For the second circular orbit $n = 2$, giving the radius four times as great, the velocity half, the frequency one-eighth, and the energies one-quarter as large.

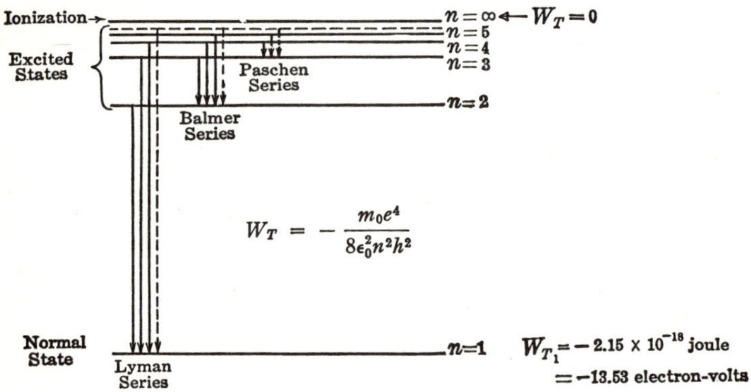

FIG. 1.1. Electronic energy levels for the hydrogen atom with circular orbits.

The normal state of the atom corresponds to $n = 1$. Higher values of n, with larger orbits and higher total energies (smaller negative values), correspond to "excited" states. In the limit $n = \infty$, the electron is removed completely from the nuclear influence, the total energy is zero (by assumption), and the atom is "ionized." The ionization energy for the hydrogen atom is equal to the total energy of the electron in the normal state, $n = 1$, that is 2.15×10^{-18} joule, which is equivalent to 13.53 electron-volts since one electron-volt is equal to 1.602×10^{-19} joule. (See Art. 1.29.)

The possible energy states or levels for the hydrogen atom with circular orbits are shown in Fig. 1.1. If the electron is raised to a higher energy (excited) state by any means, such as a collision by the atom with a second free electron, it will soon drop down to one of the lower states, and in so doing it will emit the difference in energy in the form of radiation with a frequency

$$f_r = \frac{W_n - W_{n'}}{h} = \frac{m_0 e^4}{8\epsilon_0^2 h^3}\left(\frac{1}{n'^2} - \frac{1}{n^2}\right) \tag{1.23}$$

The free space wave length λ_0 of the radiation is equal to the free space velocity of light divided by the frequency.

$$\lambda_0 = \frac{c}{f} \quad \frac{\text{meters}}{\text{second}} \tag{1.24}$$

Only certain frequencies or wave lengths can be radiated (or absorbed); in fact, only those values are possible for which the quantum numbers n and n' are integer. Each possible frequency or wave length corresponds to a line in the hydrogen spectrum. The spectral lines are grouped into series as shown in Fig. 1.1.

Atomic structure in general is not so simple as this discussion of the hydrogen atom would imply. Here we have given only the simple Bohr picture of the simplest atom of all. Attempts to extend this picture to the remaining 95 atoms and to molecules of all kinds have failed. Predicted values of wave length for the spectral lines have not been confirmed experimentally. It would seem that the Bohr type of atom is one in which too many arbitrary assumptions have been made. See Appendix II for a continuation of this discussion.

1.7. States of Matter—Ideal Gas Law. Any substance may be classed as a gas, liquid, or solid, according to the freedom of motion attained by its individual molecules. With this motion we associate the idea of temperature. A higher temperature indicates more violent motion and the acquisition of energy in the heat form. As the temperature is lowered toward absolute zero the molecules approach a stationary condition. The movement of a gaseous molecule is restricted by its collisions with other molecules which change its direction repeatedly. The result is a haphazard motion of any one molecule, but the average motion of the whole gas is zero if it is contained in a closed vessel, motion to the right being balanced by equal motion to the left. Liquid molecules possess less freedom of motion, as they are restricted to a larger extent by the relatively close packing of the molecules. The motion of a molecule in a solid is primarily vibratory, so that any given molecule retains its average position, and the substance as a whole maintains its shape without the aid of confining walls.

In liquids and solids the atoms or molecules are packed together so closely that the average distance between two particles is approximately the same as the size of the particle, about 10^{-10} meter. To a certain extent the particles overlap and the motion of any one is severely restricted by its neighbors, and we should expect to find the external properties of liquids and solids relatively complicated. In gases the spacing between molecules is about 10 times as large even at normal temperature pressure (0° C, 0.76 meter of mercury). At lower pressures the spacing increases proportionately. The gaseous molecule spends a large portion of its life between collisions in regions remote from even its nearest neighbors, and the laws governing its behavior are correspondingly simplified.

The ideal gas law which actual gases follow quite closely, taken from the kinetic theory of gases, is

$$pV = RT = NkT = \tfrac{1}{3}Nmv_{\text{rms}}^2 \qquad (1.25)$$

in which p = pressure in newtons per square meter.
V = volume of a kilogram-molecule (mole).
R = gas constant of a kilogram-molecule = 8.3144 joules per degree Kelvin.
T = temperature in degrees absolute (Kelvin).
N = Avogadro's number = 6.023×10^{23} molecules in a kilogram-molecule (mole).
k = Boltzmann's constant = 1.380×10^{-23} joule per degree Kelvin.
m = mass of a gas molecule in kilograms.
v_{rms} = root mean square speed of the particles in meters per second.
= the square root of the average of the squares of all the velocities, taken at any instant, irrespective of direction.

In terms of properties per cubic meter equation 1.25 may be written

$$p = \frac{R}{V}T = \frac{N}{V}kT = nkT = \tfrac{1}{3}nmv_{\text{rms}}^2$$
$$= \tfrac{1}{3}\rho v_{\text{rms}}^2 \quad \frac{\text{newtons}}{\text{square meter}} \text{ or } \frac{\text{joules}}{\text{cubic meter}} \qquad (1.26)$$

in which $n = \dfrac{N}{V}$ = number of molecules per cubic meter.
$\rho = nm$ = density of gas in kilograms per cubic meter.

Since $\frac{1}{2}mv_{\text{rms}}^2$ is the average kinetic energy of one molecule, the average kinetic energy per cubic meter is

$$\omega_k = \tfrac{3}{2}p = \tfrac{3}{2}nkT = \tfrac{1}{2}nmv_{\text{rms}}^2 = \tfrac{1}{2}\rho v_{\text{rms}}^2 \quad \frac{\text{joules}}{\text{cubic meter}} \qquad (1.27)$$

Under normal-temperature-pressure conditions, $T_0 = 273.18°$ K (0° C), $V_0 = 0.022414$ cubic meter per mole, $p_0 = 1.01328 \times 10^5$ newtons per square meter, and the number of molecules per cubic meter, Loschmidt's number, is $n_0 = N/V_0 = 2.687 \times 10^{25}$. The average atomic spacing under these conditions is $(2.687 \times 10^{25})^{-1/3} = 33.39 \times 10^{-10}$ meter. The density $\rho_0 = n_0 m$ kilograms per cubic meter and the root-mean-square velocity $v_{\text{rms }0}$ depend upon the mass m of the molecule. Per cubic meter the kinetic energy equals 1.52×10^5 joules, equivalent to 5.65×10^{-21} joule per molecule or 0.0353 electron-volt per molecule. Table 1.4 gives the values of ρ_0 and $v_{\text{rms }0}$ for a number of gases.

According to the kinetic theory a gas is in a continuous state of thermal agitation, and each molecule collides repeatedly with its neighbors. At each collision a particular molecule changes its direction of flight abruptly, interchanging energy, momentum, and velocity with its colliding neighbor. Any one molecule pursues therefore a series of zigzag paths of various directions and lengths, with various values of velocity between each successive pair of collisions. If, in passing through a collision, a particular molecule has its velocity increased, its colliding neighbor must have had its velocity decreased.

Any finite volume contains a tremendously large number of molecules, which encounter a correspondingly large number of collisions per second. From an external point of view the kinetic energy of the gas appears to be constant, and we may reasonably expect to find some law which governs the "distribution" of *total* velocities among the molecules, that is, some law which states that, at any instant, we shall find a definite fraction of the total number of molecules which have total velocities in some velocity range, say between v and $v + dv$, *irrespective of direction.*

Maxwell's distribution law of total velocities is the one followed by ordinary gases in equilibrium, and under certain conditions by electrons in a vacuum. According to this law relatively few molecules lie in a total velocity range dv at either low or high values of velocity (irrespective of direction); the greatest number lie in a range dv about a particular velocity known as the "most probable velocity" v_{mp}. Above and below this particular velocity the relative number of molecular velocities is smaller. This is illustrated by Curve II in Fig. 1.2. If R is the ratio of any total velocity to the most probable velocity, the equation of Curve II is

$$\frac{4}{\sqrt{\pi}} R^2 e^{-R^2} \quad \text{with} \quad R = \frac{v}{v_{\text{mp}}} \tag{1.28}$$

and

$$\frac{4}{\sqrt{\pi}} R^2 e^{-R^2}\, dR \tag{1.29}$$

is the fraction of the molecules whose total velocities lie in the range dv, between v and $v + dv$, or between R and $R + dR$.

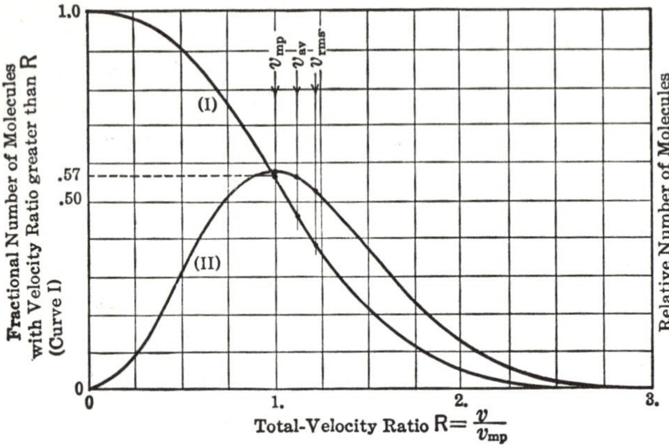

FIG. 1.2. Maxwellian total-velocity distribution curves.

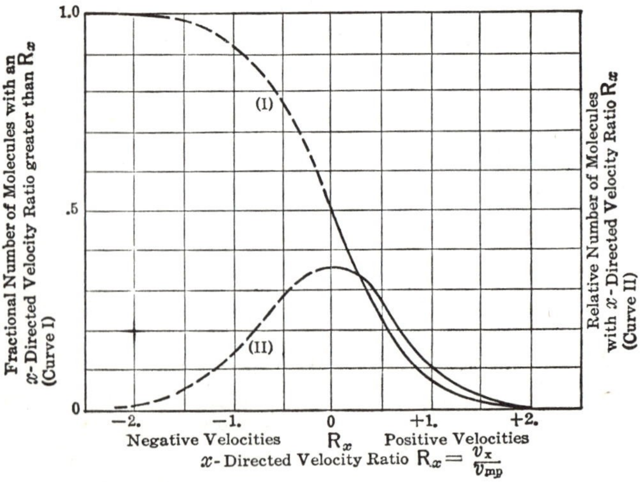

FIG. 1.3. Maxwellian x-directed velocity distribution curves.

The integral of equation 1.29, between $v = v$ and $v = \infty$, or between $R = R$ and $R = \infty$, gives the fractional number of particles whose velocities *exceed* v, or whose velocity ratio *exceeds* R. This is plotted

TABLE 1.4

Molecular Properties of Gases, Liquids, Solids

Gas	Molecular Mass in Kilograms	Atomic or Molecular Weight kg/mole	Density ρ_0 kg/m³ NTP*	$v_{rms\ 0} = \sqrt{\frac{3p_0}{\rho_0}}$ m/sec NTP*	$v_{av\ 0} = \frac{v_{rms\ 0}}{1.086}$ m/sec NTP*	Mean-Free-Path L_0 Meters NTP*	Molecular Diameter σ Meters	Collisions per Second NTP* $= \frac{v_{av\ 0}}{L_0}$
Unit Atomic Weight Gas	1.6603×10^{-27}	1.0×10^{-3}	0.0446	2.61×10^{3}	2.4×10^{3}			
Atomic H	1.6734×10^{-27}	1.0079×10^{-3}	0.0449	2.6×10^{3}	2.39×10^{3}		1.06×10^{-10}	
He	6.64×10^{-27}	4.003×10^{-3}	0.1793	1.302×10^{3}	1.2×10^{3}	18.45×10^{-8}	2.12×10^{-10}	0.65×10^{10}
H_2	3.34×10^{-27}	2.0156×10^{-3}	0.0898	1.84×10^{3}	1.69×10^{3}	11.48×10^{-8}	2.69×10^{-10}	1.473×10^{10}
N_2	46.5×10^{-27}	28.0166×10^{-3}	1.249	0.494×10^{3}	0.454×10^{3}	6.19×10^{-8}	3.66×10^{-10}	0.734×10^{10}
O_2	53.2×10^{-27}	32×10^{-3}	1.425	0.462×10^{3}	0.425×10^{3}	6.68×10^{-8}	3.53×10^{-10}	0.636×10^{10}
Air	48.1×10^{-27}	28.952×10^{-3}	1.293	0.484×10^{3}	0.445×10^{3}	6.3×10^{-8}	3.64×10^{-10}	0.707×10^{10}
H_2O Gas	29.95×10^{-27}	18.0156×10^{-3}	0.801	0.616×10^{3}	0.567×10^{3}	7.2×10^{-8}	3.41×10^{-10}	0.787×10^{10}
Electron Gas	9.1066×10^{-31}	5.49×10^{-7}	2.445×10^{-5}	1.115×10^{5}	1.025×10^{5}			
Liquids and Solids	**Molecular Mass in Kilograms**	**Molecular Weight kg/mole**	**Density kg/m³**	**Molecules/m³**	**Molecular Spacing in Meters**			
Water	29.9×10^{-27}	18.016×10^{-3}	1.0×10^{3}	3.37×10^{28}	3.1×10^{-10}			
Hg (0° C)	333×10^{-27}	200.61×10^{-3}	13.595×10^{3}	4.09×10^{28}	2.9×10^{-10}			
Al (20° C)	45×10^{-27}	27.1×10^{-3}	2.7×10^{3}	6.07×10^{28}	2.545×10^{-10}			
Fe (20° C)	92.7×10^{-27}	55.84×10^{-3}	7.87×10^{3}	8.55×10^{28}	2.27×10^{-10}			
Cu (20° C)	104.7×10^{-27}	63.57×10^{-3}	8.89×10^{3}	8.48×10^{28}	2.28×10^{-10}			
Co	97.8×10^{-27}	58.94×10^{-3}	8.9×10^{3}	9.18×10^{28}				
Ni	97.5×10^{-27}	58.69×10^{-3}	8.85×10^{3}	9.15×10^{28}				
Gd	261.3×10^{-27}	157.3×10^{-3}						

*** NTP (normal temperature pressure—0° C, 0.76 meter of mercury.)**

as Curve I in Fig. 1.2.

$$\text{Fraction exceeding } v \text{ or } R = \frac{4}{\sqrt{\pi}} \int_{R=R}^{\infty} R^2 e^{-R^2}\, dR$$
$$= \frac{2}{\sqrt{\pi}} R e^{-R^2} + 1 - \operatorname{erf} R \qquad (1.30)$$

in which erf R is the error function or the probability integral. Tabular values of the probability integral are given in both Peirce's and Dwight's integral tables.

Three velocities of special interest are shown in Fig. 1.2. The most probable velocity, the average velocity of all the molecules, and the root-mean-square velocity, stand to each other in fixed ratios.

$$\frac{v_{mp}}{v_{av}} = \frac{1}{1.128} \qquad \frac{v_{av}}{v_{rms}} = \frac{1.128}{1.225} = \frac{1}{1.086} \qquad (1.31)$$

Let us now consider a particular direction in the gas, say the x direction, and consider only the x components of all the velocities, v_x, at any instant. For these component velocities we have Maxwell's distribution law of x-directed velocities. This law is required if it is desired to compute the number of molecules or electrons which flow across an area taken normal to the x direction. It is necessary to consider both positive and negative velocities in the x direction. For a gas in equilibrium there are like laws for the distributions in the y and z directions.

Let R_x be the ratio of any x-directed component of velocity v_x to the most probable (total) velocity v_{mp}. Then the equation of Curve II, Fig. 1.3, is

$$\frac{1}{\sqrt{\pi}} e^{-R^2} \quad \text{with} \quad R_x = \frac{v_x}{v_{mp}} \qquad (1.32)$$

and

$$\frac{1}{\sqrt{\pi}} e^{-R_x^2}\, dR_x \qquad (1.33)$$

is the fraction of the molecules whose x-directed velocity components lie in the range dv_x, between v_x and $v_x + dv_x$, or between R_x and $R_x + dR_x$.

The integral of equation 1.33, between $v_x = v_x$ and $v_x = \infty$, or between $R_x = R_x$ and $R_x = \infty$, gives the fractional number of molecules whose x-directed velocity components *exceed* v_x, or whose x-directed velocity ratio *exceeds* R_x. v_x and R_x may be positive or negative. The integral is plotted as Curve I in Fig. 1.3.

$$\text{Fraction exceeding } v_x \text{ or } R_x = \frac{1}{\sqrt{\pi}} \int_{R_x = R_s}^{\infty} e^{-R_x^2}\, dR_x$$
$$= \tfrac{1}{2}(1 - \operatorname{erf} R_x) \qquad (1.34)$$

By virtue of its thermal motion a gas is a viscous substance which supports shearing forces. Consider two solid material planes parallel to each other in the gas. If one plane is moved in a direction parallel to the other, the intervening parallel layers of gas are moved in a parallel direction with varying degrees of velocity. The gaseous motion is transmitted through the mechanism of collisions. From the kinetic gas theory, the coefficient of viscosity η is

$$\eta = 0.5\rho v_{av} L \qquad (1.35)$$

where L is the average distance between collisions and is called the "mean-free-path." From measurements of η, with ρ and v_{av} known, L may be computed. Values of mean-free-path for normal-temperature-pressure conditions are given in Table 1.4.

The mean-free-path, by the kinetic theory, is related to the size of the molecule and to the number of molecules per cubic meter, by the rule

$$L = \frac{1}{\sqrt{2}\,\pi\sigma^2 n} \qquad (1.36)$$

in which σ is the diameter of the molecule, which is assumed to be spherical. Since pressure varies directly with n, L varies inversely with pressure. Molecular diameters computed by this formula are given in Table 1.4.

For a mixture of gas molecules and a small percentage of electrons, the electron mean-free-path is $4\sqrt{2}$ times as large.

$$L_e = \frac{4}{\pi\sigma^2 n} \qquad (1.36a)$$

For comparative purposes certain molecular values for solids and liquids have been given also in Table 1.4.

1.8. Conductors and Insulators. All materials contain at least a few units of electric charge which are free to move in the presence of an electric field. In a gas the free charge, which is a mixture of a few positive ions, negative ions, and electrons among the neutral atoms, is the result of the passing of ionizing radiations, such as ultraviolet, radioactive, and cosmic rays, which originate outside the body of the gas itself. An additional source of free charge may be present in liquids and solids, due in large measure to the close packing of the molecules. In some liquids we may expect to find electrolytic dissociation into positive and negative ions. In solids some of the electrons are freed from molecular binding, but the positive ions commonly are fixed in position. So it is that when an electric field is imposed on a material, free charge is present which can and does move under its influence, and the electric charge in motion constitutes a current.

All materials therefore conduct electricity to some extent, and the conductivity, or current density per unit field strength, is a measure of their conducting power. The range in conductivity is tremendous. At the high conductivity end of the range we find silver with a conductivity value which is about 10^{23} times as large as that possessed by mica at the low end. Materials with relatively high values of conductivity are called *conductors* and those with relatively low values are called *insulators*. See Tables 1.5 and 4.1 where values of conductivity are listed.

The explanation of the enormous variation in conductivity lies in the fact that the electric charges, under the influence of an electric field, move with varying degrees of freedom which depend upon the character of the molecular world that they inhabit.

In the metallic conductors, which have high conductivities, we find that each atom contributes one or more electrons to form an electronic cloud of about 10^{29} electrons per cubic meter. Individual electrons have velocities which range from zero up to 10^6 meters per second, and move only short distances (order of 10^{-8} meter) between collisions with the positive ion residues of the atoms. The collisions restrict the motion of the electrons. In the presence of an electric field the cloud as a whole may drift forward slowly (order of 1 meter per second) and this drift motion constitutes the electric current. Energy passes from the field to the electrons, and on to atoms through the mechanism of collisions. The material and its atoms therefore offer *resistance* to the passage of current.

In *insulators*, the conductivities are relatively low, and the resistivity values correspondingly high. In these materials the amount of free charge per cubic meter is probably very small, and it is believed that the charge in moving induces characteristic vibrations of the molecules. This requires that the electrons deliver the energy, which they have received from the field, to the vibrating molecules. Apparently this is a mechanism which is very effective in restricting the motion of the electric charges.

B. ELEMENTARY FACTS AND DEFINITIONS

1.9. Definitions and Experimental Facts. Any physical theory is based primarily on a series of facts obtained by experiment. As an aid to the correlation and explanation of these facts it is usually necessary to develop a series of concepts or ideas by definition. The facts and definitions together form the theory, which is considerably clarified by recognizing the distinction between its factual and conceptual components.

1.10. Vectors and Scalars. Quantities requiring magnitude and direction for their specification are called vector quantities. Force is an excellent example. Vector addition by the parallelogram law is required for the summation of vector quantities.

Quantities needing only magnitude for their description are called scalar quantities. Work is an example. If work done upon a body is called positive, work done by the body is necessarily negative. For scalar quantities, scalar addition only is required.

1.11. Static Electricity. Electrical particles are never quiescent, but if the distribution as a whole remains fixed, the particles may be regarded as stationary from the observer's point of view. In this sense we may use the term static electricity.

1.12. Separation of Electricity. Neutral atoms containing equal numbers of positive charges and electrons produce no external electrical effect. In some manner a separation in space between the two kinds of electricity must be effected in order to cause electrical phenomena.

The separation is effected in a number of ways:

(1) *By friction between unlike substances, followed by their separation.* The surface atoms of one material become positively charged by giving up electrons to the second material, which then appears negatively charged.

(2) *By use of batteries.* The poles of a battery are charged with equal amounts of opposite types of electricity.

(3) *By movement of a conductor across a magnetic field.* By and during the movement of a conductor across a magnetic field, equal quantities of opposite kinds of electricity appear toward the ends of the conductor. Upon cessation of the movement the electricities recombine and the conductor again appears neutral. This is the fundamental phenomenon upon which the action of a generator depends.

(4) *By induction.* If a conductor is brought into the neighborhood of electricities already separated, some of its free electrons move to one

part of the surface and the remaining part of the surface appears positively charged. The electricities of the conductor return to their normal neutral distribution upon removal of the inducing cause.

(5) *By polarization.* When an insulator is brought into the neighborhood of separated electricities, the atoms are stretched or polarized, but they return to their normal state when withdrawn from the field.

(6) *By ionization.* Ions and electrons accelerated by an electric field and colliding with neutral atoms and molecules may, when sufficient energy has been attained, cause the atoms and molecules to divide into positive and negative elements, usually positive ions and electrons.

(7) *By photoelectric action.* If light rays of short wave length, usually visible or ultraviolet, are directed upon a conductor, electrons may be released from the conductor and passed on to a second conductor which can collect them. The performance of the photoelectric tube depends upon this phenomenon.

(8) *By thermionic action.* When a conductor, such as tungsten, is raised to a high temperature, some of its electrons have their energies raised to the point where they may escape from the conductor and may be collected by a second conductor. Vacuum tubes of many types utilize this method.

(9) *By action of radioactive and cosmic rays.* Emanations of these types passing through material substances ionize some of the molecules into positive ions and electrons. A condition of stability is usually reached in which the rate of generation of the ions is balanced by their removal, that is, by either recombination or actual removal from the region.

1.13. Electric Field (Definition). An electric field is said to exist in any portion of space or matter if an electric charge suspended therein has a force exerted upon it.

Whenever electricities are separated, an electric field exists somewhere in their vicinity. This definition is decidedly qualitative, so we turn to Coulomb's law to find a quantitative statement of the forces between electrical particles, which may then be used to define the strength of the electric field.

1.14. Coulomb's Law (Experiment). **Dielectric Constant.** Two quantities of separated electricity, each confined to a region of dimensions small compared with the distance between their centers, located in a dielectric medium that is extensive and isotropic, exert equal and opposite forces upon each other along the line joining their centers; the forces are proportional directly to the quantities of electricity, and inversely to the dielectric constant of the medium and the square of the distance between centers. For like kinds of electricity the forces are repulsive; for unlike kinds, attractive. The medium in which this experiment is performed must be extensive so that the boundaries do not affect the force. Furthermore, the medium must be isotropic, or possessed of the same properties in all directions.

Coulomb's law for point charges, in mks rationalized units, takes the form

$$F = \frac{Q_1Q_2}{4\pi\epsilon r^2} \quad \text{newtons} \tag{1.37}$$

with force F given in newtons, charge Q in coulombs, distance r in meters. In a medium which is free space (vacuum)

$$\epsilon_0 = 8.854 \times 10^{-12} \cong \frac{1}{36\pi 10^9} \quad \frac{\text{farad}}{\text{meter}} \tag{1.38}$$

In any other dielectric medium

$$\epsilon = \epsilon_0\epsilon_r \quad \frac{\text{farads}}{\text{meter}} \tag{1.39}$$

ϵ is the dielectric constant of the medium, ϵ_0 is the dielectric constant of free space, both with the units farads per meter, and

$$\epsilon_r = \frac{\epsilon}{\epsilon_0} \tag{1.40}$$

a pure numeric, is called the relative dielectric constant. For most dielectric materials used as insulators, values of ϵ_r lie between 1 and 80. See Table 1.5. A few materials with unusual properties may have values of ϵ_r of several hundred or several thousand.

Some authors prefer a different terminology for ϵ, calling ϵ the permittivity of the medium, ϵ_0 the permittivity of free space, and reserving the term dielectric constant for the numeric ϵ_r.

According to equation 1.37 the Coulomb force between two charges is lower in any dielectric medium than in free space in proportion to the value of ϵ_r for the medium. The cause of this decrease in force action is found in the polarization or stretching of the molecules of the dielectric, a matter which is discussed more fully in Chapter 3.

1.15. Electric Charges—Bound and Free. By bound charges we shall mean those charges, occurring in pairs of equal and opposite sign, which are held in positions that are approximately fixed by forces of atomic and molecular origin. Bound charges, in the presence of an electric field, shift slight distances from their fixed positions, positive charges with the field and negative charges opposite to the field direction. The atoms are then stretched and the material is said to be polarized. A neutral atom and a perfect insulator contain only bound charges. All other charges are free and they, unless restrained, will move under the influence of an electric field. Free charges appear on the surfaces of conductors and are the ones which take part in a general movement of electricity from place to place.

In using Coulomb's law (equation 1.37) the separated charges Q_1 and Q_2 are free charges only. The force-reducing effect of the bound charges of the intervening dielectric is described quite sufficiently by the use of the dielectric constant ϵ. The dielectric constant, in a sense, reduces the influence of a dielectric composed of myriads of atoms to a single number.

1.16. Electric Field Definitions. The electric field is described quantitatively by the two vector quantities, electric field intensity E and electric flux density D, and the two scalar quantities, electric flux ψ and electric potential V. They are all derived ultimately from Coulomb's law, and the electric field may be described in terms of any one of these quantities. These four methods of describing the electric field find adequate justification for existence in practice.

1.17. Electric Intensity at a Point (Definition). The electric intensity or field strength at any point in an electric field is a vector quantity E, whose magnitude is the ratio of the force to the charge $dF/+dQ$ upon a positive charge placed at the point, and whose direction is identical with the force.

In the form of an equation:

$$\text{Electric intensity } E = \frac{dF}{+dQ} \quad \frac{\text{newtons}}{\text{coulomb}} \quad \text{or} \quad \frac{\text{volts}}{\text{meter}} \qquad \text{(Vector)} \tag{1.41}$$

(The units, volts per meter, will be clear following the definition of potential gradient.)

The dimensional formula is

$$E = \frac{\text{force}}{\text{charge}} = L^1M^1T^{-2}Q^{-1} \tag{1.42}$$

This form may be compared with the dimensional table in Appendix I.

The intensity possesses several properties worthy of explicit statement:

(1) It is directly proportional to the *free* charges in the field.

(2) Being measured by the force, it is proportional to $1/\epsilon$ and therefore its magnitude *depends* upon the medium.

(3) Its magnitude varies continuously from point to point throughout any *one* medium but may change abruptly across the interface between two different media.

If the intensity is obtained for every point in an insulator of known dielectric constant, the field is completely determined. In general, the intensity varies from point to point both in magnitude and direction, becoming larger whenever a free charge is approached owing to the influence of the inverse square distance relation.

Intensity is often referred to as the force per unit positive charge placed at the point under investigation.

$$E = \frac{\text{force}}{\text{unit positive charge}} \quad \text{or} \quad \frac{\text{newtons}}{\text{coulomb}} \qquad \text{(Vector)} \qquad (1.43)$$

This statement is satisfactory only with the condition that the unit charge shall not, by induction, cause any redistribution of the charges creating the field.

Intensity is not a force; rather it is an electrical property of the field, a distinction which is emphasized by noting the difference in their dimensional formulas. Let us say that intensity is *measured* by a force. The intensity at a field point may be represented by placing the tail of a vector at the point. The vector is given a length proportional to the magnitude of the intensity, and the same direction. Later it will be shown that the vector intensity is simply the negative of another vector (reversed vector) called the electric potential gradient.

1.18. Dielectric Strength—Breakdown (Definition). The dielectric strength of an insulator is the maximum permissible electric intensity or electric potential gradient. See Table 1.5.

If the dielectric strength is exceeded at any point in the insulation, the latter *may* break down. Consider air as an insulator. With a uniform field (intensity vectors everywhere parallel), which is obtained between two parallel plate electrodes and is approximately attained in the sphere gap with large spheres at small spacing, any increase of the intensity above the value of the dielectric strength leads almost instantaneously to the passage of a spark through the air. During the passage of the spark the voltage across the gap drops close to zero, and breakdown has occurred.

But with diverging or converging fields a "self-sustaining" discharge such as a corona may be set up, with the emission of the familiar corona light from regions adjacent to the conductor. See Art. 2.14.

The dielectric strength is not a fixed constant for a material but depends on many other factors such as thickness, rate of voltage application, temperature, impurities, condition of electrode surfaces, radii of electrode surfaces, etc.

1.19. Electric Flux Density (Definition). The electric flux density D at a point in the electric field is a vector quantity which has the same direction as the intensity at the point, but whose magnitude is given by the product of the intensity and the dielectric constant.

Electric flux density

$$D = \epsilon E \quad \frac{\text{lines of electric flux}}{\text{square meter}} \quad \text{or} \quad \frac{\text{coulombs}}{\text{square meter}} \qquad \text{(Vector)} \quad (1.44)$$

The dimensional formula is

$$D = \epsilon E = L^{-2}Q \qquad (1.45)$$

The properties of the flux density are:

(1) It is directly proportional to the *free* charges in the field.

(2) Inasmuch as intensity E is proportional to $1/\epsilon$, $D = \epsilon E$ must be *independent* of the medium.

(3) It varies continuously from point to point throughout any *one* medium but may change abruptly across the interface between two different media.

The primary function of the vector D is to give a point-by-point vector description of the field that is independent of the character of the medium. Inasmuch as E is also a vector point quantity, it is desirable to find a method of representing D other than by the use of vectors.

Following Fig. 1.4, proceed in the following manner. At any field point P describe a small area dA_n normal to the intensity, and draw a series of lines $d\psi$, of density $D = d\psi/dA_n$ lines per square meter, normal to the area and consequently parallel to the intensity. Give the lines the same direction as that of the intensity. The lines so drawn are called lines of electric flux.

Electric flux density $D = \epsilon E = \dfrac{d\psi}{dA_n}$

$$\frac{\text{lines of electric flux}}{\text{square meter}} \quad \text{or} \quad \frac{\text{coulombs}}{\text{square meter}} \qquad \text{(Vector)} \quad (1.46)$$

The units, coulombs per square meter, will be clear after reading Art. 1.21.

1.20. Electric Flux Lines (Definition). The number of lines of electric flux ψ crossing any area A is a scalar quantity obtained by integrating the normal component of the vector flux density D over the area.

$$\text{Electric flux lines } \psi = \iint_A D_n \, dA$$

lines of electric flux or coulombs (Scalar) (1.47)

Figure 1.5 illustrates the method of carrying out the process indicated by equation 1.47. Forming the area integral of a vector function is a general mathematical process applicable to any vector field.

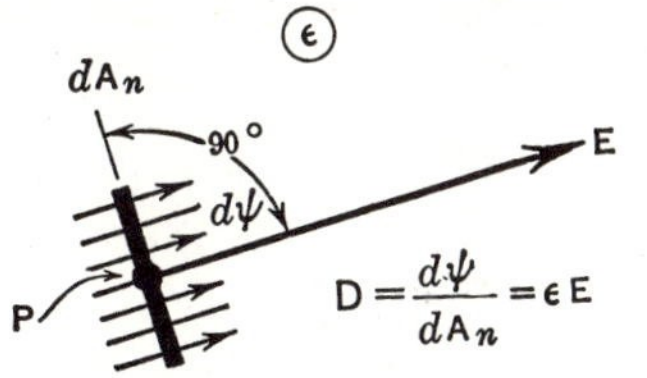

FIG. 1.4. Method of drawing electric flux lines.

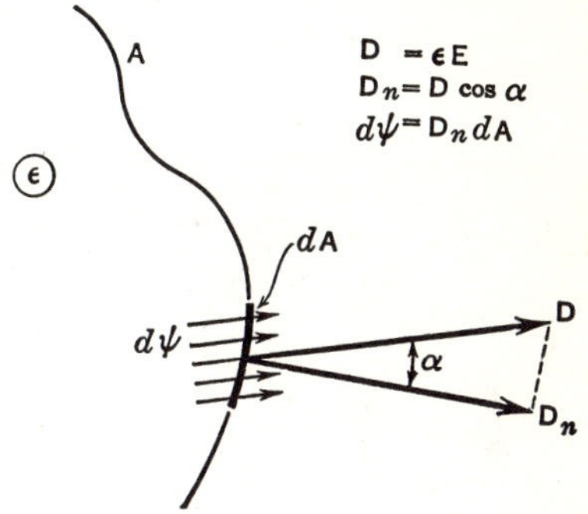

FIG. 1.5. Method of calculating the flux crossing area A.

The dimensional formula gives the same fundamental units for electric flux as for electric charge.

$$\psi = DA = (L^{-2}Q)(L^2) = Q \tag{1.48}$$

The significance of this lies in the fact that the electric flux is merely a method of describing the space effects of separated electricities. The density of the flux at any field point divided by the dielectric constant at once gives the intensity at the point, which is merely the force effect of the electric charge.

1.21. Field of the Single Point Charge. The concepts just developed will be clarified by studying their application to the field of a single point charge. By point charge we mean a charge placed in a region of small finite dimensions so that viewed from the outside it may be treated mathematically as a point.

When electricities are separated it is possible either to remove one of them (say the negative charge) or to confine the attention to the immediate neighborhood of the other. In either case, for mathematical purposes, we must consider that this negative electricity is infinitely far away and concentrate our attention upon the positive electricity which remains nearby.

A single point charge of $+Q$ coulombs placed in a medium of dielectric constant ϵ creates an electric field in its vicinity that possesses spherical symmetry. The intensity or field strength E_1 at any point P_1 (see Fig. 1.6), distant r_1 from the charge $+Q$, is obtained by

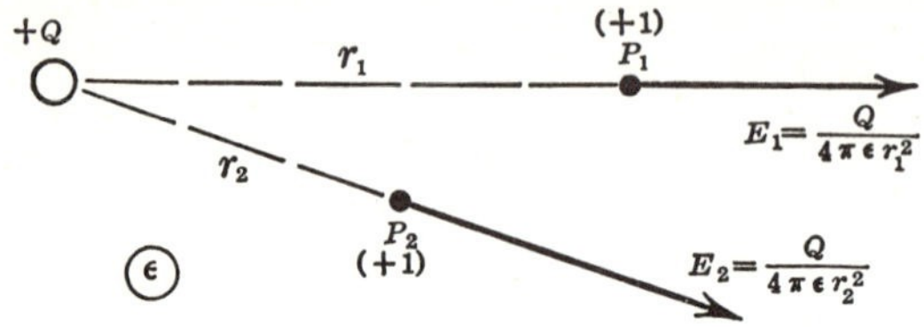

FIG. 1.6. Field strength about a point source.

placing unit positive charge at P_1 and applying Coulomb's law (equation 1.37) and the general definition of intensity (equation 1.41).

$$E_1 = \frac{\text{force on unit positive charge at } P_1}{\text{unit positive charge}} = \frac{\left[\dfrac{Q(+1 \text{ charge})}{4\pi\epsilon r_1{}^2}\right]}{(+1 \text{ charge})}$$

$$= \frac{Q}{4\pi\epsilon r_1{}^2} \; \frac{\text{newtons}}{\text{coulomb}} \;\text{ or }\; \frac{\text{volts}}{\text{meter}} \quad \text{(Vector radially outward)} \qquad (1.49)$$

Over any given sphere about $+Q$ the intensity is constant in magnitude, radial and outward in direction, as $+Q$ and $+1$ repel each other. The intensity increases rapidly as the size of the sphere decreases.

The flux density possesses the same direction as the intensity but is of ϵ times the magnitude. It has the same value at a given point whatever the medium may be.

$$D_1 = \epsilon E_1 = \frac{Q}{4\pi r_1{}^2} \; \frac{\text{lines or coulombs}}{\text{square meter}} \qquad \text{(Vector)} \qquad (1.50)$$

Figure 1.7 shows the variation of E and D with the radial distance r.

The number of lines of electric flux ψ crossing any sphere of radius r is, by equation 1.47,

$$\text{Flux } \psi = \text{Flux Density} \cdot \text{Area} = D \cdot A_{\text{sphere}}$$

$$\psi = \frac{Q}{4\pi r^2} \cdot 4\pi r^2 = Q$$

$$\text{lines of electric flux or coulombs} \qquad \text{(Scalar)} \qquad (1.51)$$

Straight multiplication of D by A suffices in this case because the density is constant and normal over the sphere area. Equation 1.51 states that there are Q lines of flux outward over every sphere about the point charge. The area need not be a sphere; any closed area about the point would pass the same amount of flux.

For a charge $-Q$, the directions of the vectors E and D are inward and the $-Q$ lines of flux are directed in toward the charge. Figures

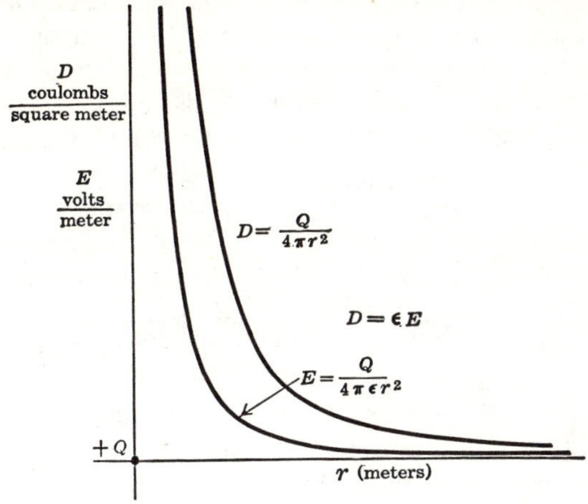

FIG. 1.7. Single point charge. Magnitude of vectors D and E.

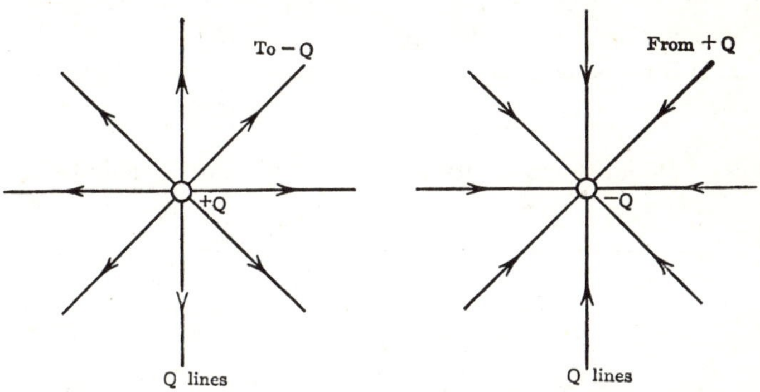

(a) Positive charge (b) Negative charge

FIG. 1.8. Fields of point charges.

1.8a and 1.8b show the flux fields about $+Q$ and $-Q$ charges, respectively. No attempt is made to draw exactly Q lines. It is sufficient if enough lines are drawn to indicate clearly the general field shape, emphasizing the idea that the lines lie closer together where the flux density and intensity magnitudes have larger values. It must not be assumed from these special cases that the flux lines are generally straight. In fact, they are generally curved.

Let us turn for a moment to the purely geometrical concept of solid angle, and then express any portion of the flux from a point charge in terms of this concept.

1.22. Solid Angle. The solid angle subtended at a point by any area is a pure number equal to the projection of that area upon a sphere of unit radius about the point. The unit of solid angle is called the steradian. There can be a maximum of 4π steradians about the point, occurring only when a closed area surrounds the point. See Fig. 1.9.

1.23. Solid Angle of a Right Circular Cone. Following Fig. 1.10, let a right circular cone AOB form a solid angle Ω, and a plane angle α with its axis Ox. Pass a sphere of radius R and a tangent cylinder of axis Ox about the point O. The cone then cuts off sphere cap CDE, whose projection on the cylinder is $FGHI$. Geometry shows that these two are of equal area, each equal to $2\pi R(R - r_\alpha) = 2\pi R^2(1 - \cos\alpha)$. Consequently the projection of the sphere cap CDE upon unit sphere about O is $2\pi(1 - \cos\alpha)$, or the solid angle Ω at point O is

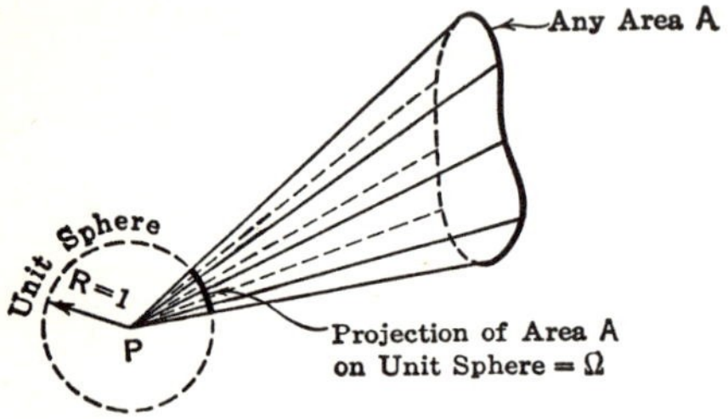

FIG. 1.9. Solid angle subtended at point P by area A.

$$\Omega = 2\pi(1 - \cos\alpha) \quad \text{steradians} \tag{1.52}$$

By choosing equal increments for $\cos\alpha$, or choosing equal distances along diameter JD, and passing normal planes, equal areas are cut off on the sphere surface.

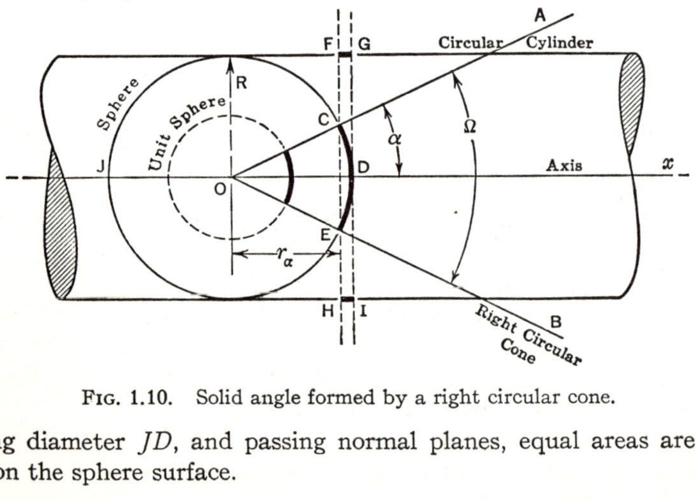

FIG. 1.10. Solid angle formed by a right circular cone.

1.24. Flux from a Point Source. The electric flux from a point source $+Q$ is Q lines emerging equally in all directions. To divide this flux into equal conal groups about some axis, it is necessary merely to pass a sphere about the charge, divide the diameter into as many parts as there are groups required, and pass normal planes at the division points. Equal spherical annular areas are cut off. See Fig. **1.11** where four of the eight equal flux groups are shown.

The percentage of flux crossing any sphere cap of angle α is:

$$\text{Per cent flux} = \frac{2\pi R^2(1 - \cos \alpha)}{4\pi R^2} 100 = \tfrac{1}{2}(1 - \cos \alpha)100 \qquad (1.53)$$

The graph of Fig. 1.12 gives the result in simple form. It is essentially a displaced cosine curve.

1.25. Sources and Sinks. Each free unit positive charge (coulomb) is the starting point or source of one line of electric flux. Similarly each free unit negative charge is the ending point or sink of one line of electric flux. Flux lines, in electrostatics, begin and end only at points where *free* charges are located. They must therefore be continuous throughout intervening space irrespective of the dielectric media present.

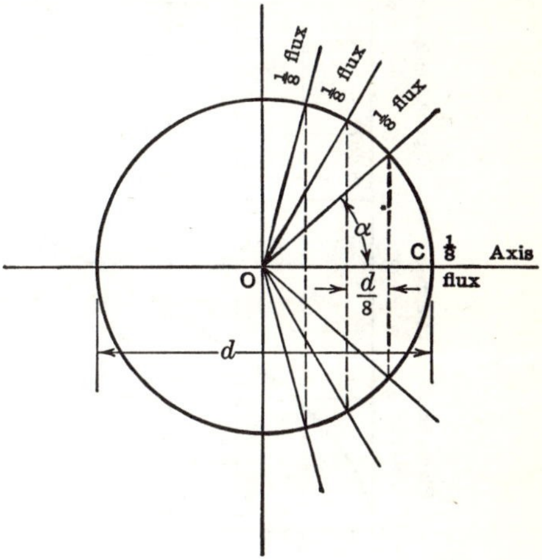

FIG. 1.11. Equal flux groups about a point source.

A positive charge placed in the field and permitted to move slowly will follow the flux line through the point from source to sink.

1.26. Gauss Theorem. If any closed area is passed about a region containing several point charges, the *net* outward flux across the closed area is equal to the *net* positive charge inside. In determining the net quantities, positive charge and outward flux are reckoned $+$, the others $-$. The statement follows merely as an extension of the properties of the single point charge.

By way of illustration, consider a gas-filled vacuum tube. A closed area so shaped as just to enclose the cathode surrounds only negative charge and is crossed only by inwardly directed flux, or the net outward flux is counted as negative. A closed area enclosing the adjacent cathode-fall space will enclose more negative charge than positive; hence the net outward flux is negative. Farther out still there is the plasma region where the concentrations of positive and negative space charge are approximately equal, and the net outward flux is zero. Around the anode, with its positive charge, the net outward flux is positive. Obviously, in free space devoid of any charge at all, the net outward flux is zero, which merely means that all the lines which enter the region continue on through and leave.

1.27. Electric Potential Difference. So far, three means of describing the electric field have been found—two vector quantities, intensity

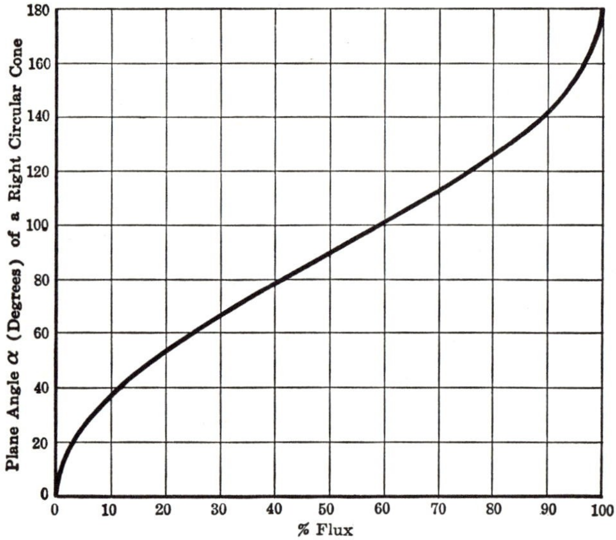

FIG. 1.12. Percentage of flux about a point source.

and flux density, and the scalar quantity flux. Electric potential difference is an additional scalar concept serving the same purpose. These quantities are, of course, all related, but frequently one is better adapted to the purpose in hand than the others.

Let us suppose that in a certain region we have a medium of dielectric constant ϵ throughout which there is an electric field created by positive and negative electricities located either outside or inside. To make it as general as possible let the intensity and flux density vary in magnitude and direction from point to point, giving flux lines which are curved, as shown in Fig. 1.13.

1.28. Electric Potential Difference (Definition). The electric potential difference $V_2 - V_1$ between two points P_2 and P_1 in an electric field is the work per unit positive charge performed in moving the unit positive charge from P_1 to P_2 against the field intensity.

At any point P along the path taken by the unit positive charge, the intensity of the field is E newtons per coulomb. $(-E)$ newtons per coulomb is then the force per unit positive charge which must be applied externally to overcome the effect of the field itself. Then $(-E)\ ds \cos \alpha$ is the *external* work done upon the unit charge to force

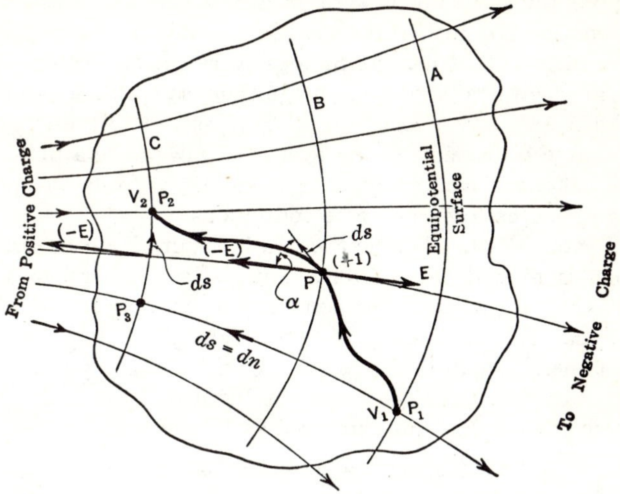

FIG. 1.13. Potential difference between points P_2 and P_1.

it through a small distance ds along the path. Cos α enters because work is done along only that part of the motion that is parallel to the force. α then is the angle between $-E$ at a point and the path element ds through it. Integrating this over the path chosen from P_1 to P_2 gives

Electric potential difference

$$V_2 - V_1 = \int_{P_1}^{P_2} (-E)\ ds \cos \alpha = \frac{\text{newton-meters}}{\text{coulomb}}$$

$$= \frac{\text{joules}}{\text{coulomb}} = \text{volts} \quad \text{(Scalar)} \tag{1.54}$$

The dimensional formula is

$$\text{Electric potential difference} = \frac{\text{work}}{\text{charge}} = \frac{\text{force} \times \text{distance}}{\text{charge}}$$

$$= \frac{L^1M^1T^{-2} \cdot L}{Q} = L^2M^1T^{-2}Q^{-1} \tag{1.55}$$

The integration indicated in equation 1.54 is known as the line integral of the vector E and the result is a scalar quantity.

Electric potential difference must not be thought of as work—it is an electrical quantity, work per unit charge. but is *measured* by work. A similar distinction was made between intensity and force. The unit of potential is the joule per coulomb and is called the volt in the mks system.

Since potential is the line integral of E, it depends upon the medium (proportional to $1/\epsilon$) just as does E itself. Another important property of potential is that it is continuous everywhere. Practically, the intensity must always be finite in magnitude, so that the difference of potential between any two points in space given by moving a unit positive charge from one to the other must always be a finite result. Thus the potential can never change suddenly from one point to another and is therefore said to be continuous.

1.29. Electron-Volt. The *electron-volt* is a unit of *energy*, not potential, which is related to the volt. An electron (charge = 1.602×10^{-19} coulomb) moved from one point to another between which there is a potential difference of 1 volt or 1 joule/coulomb acquires an amount of energy equal to the product of the charge by the potential difference. In mks units the energy is equal to 1.602×10^{-19} joule, which is equal to an amount expressed as 1.0 electron-volt.

$$1 \text{ electron-volt} = 1.602 \times 10^{-19} \text{ joule} \tag{1.56}$$

1.30. Electric Equipotential Surface (Definition). An equipotential surface in the electric field is one over which a unit positive charge may be moved without doing work. Consequently, the electric intensity is normal to such a surface.

In Fig. 1.13 three of the many possible equipotential surfaces are drawn (A, B, and C). In proceeding from A to B to C, external work must be done upon a unit positive charge. In the region shown, the same difference of potential exists between any two points on two different equipotential surfaces irrespective of the path followed by the unit positive charge. Hence it is often easier to compute the difference of potential between P_2 and P_1 by following another path, P_1 to P_3 along a flux line ($\alpha = 0$, $\cos \alpha = 1.0$, and $ds = dn$, parallel everywhere to $-E$) where work is done, and then proceeding from P_3 to P_2 along an equipotential surface where no work is done (as ds is perpendicular to $-E$, $\alpha = 90°$, and $\cos \alpha = 0$). By choosing a flux line path we may restate equation 1.54 as

$$V_2 - V_1 = \int_{P_1}^{P_3} (-E)\, dn \quad \text{volts} \tag{1.57}$$

where $ds \cos \alpha = dn$ for a movement along a flux line, i.e., normal to equipotential surfaces.

1.31. Zero Electric Potential Surface or Region (Definition). Any surface or region lying outside of an electric field may be defined as a zero electric potential surface or region. This does not preclude the possibility of having zero potential surfaces inside the field as well.

1.32. Absolute Electric Potential of a Point (Definition). The absolute electric potential at any point in an electric field is the external work done upon a unit positive charge in moving it from a surface or region of zero electric potential to the point in the field.

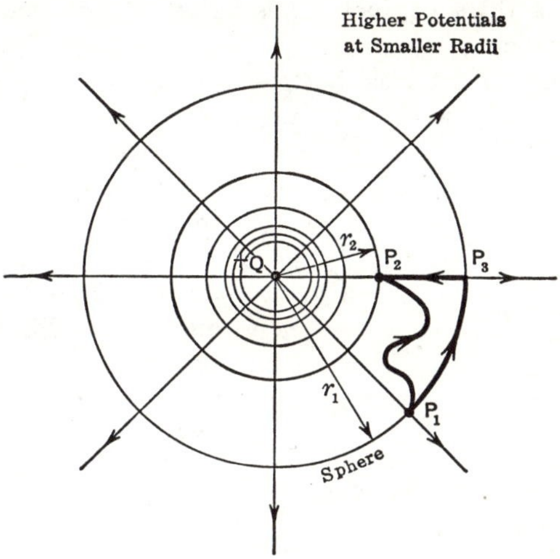

FIG. 1.14. Potential difference in a point charge field (showing equal increments of potential).

1.33. Electric Potential of a Single Point Charge ($+Q$). Let us return to the case of the single point charge $+Q$ considered in Art. 1.21 and calculate the potential. We have seen that this field has a vector intensity $E = Q/(4\pi\epsilon r^2)$, radially outward, and a vector flux density $D = Q/(4\pi r^2)$, and that Q lines of flux emerge from the charge.

Consider two points P_1 and P_2 in Fig. 1.14 as radii r_1 and r_2 respectively, through which two spheres are passed centering at the charge. The spheres are equipotential surfaces. Suppose we move a unit positive charge along the sphere surface from P_1 to P_3 and then along a flux line from P_3 to P_2, adding up or integrating the work per unit charge over the journey. From P_1 to P_3 over an equipotential surface no work is done. From P_3 to P_2 the external work done upon the unit positive charge gives the potential difference between P_2 and P_1:

$$V_2 - V_1 = \int_{P_1}^{P_2} (-E)\, ds \cos \alpha \qquad \text{(General definition)}$$

$$= \int_{r=r_1}^{r=r_2} \left(-\frac{Q}{4\pi\epsilon r^2}\right) dr \cdot 1.0 = +\frac{Q}{4\pi\epsilon r}\Bigg]_{r=r_1}^{r=r_2}$$

$$= \frac{Q}{4\pi\epsilon}\left(\frac{1}{r_2} - \frac{1}{r_1}\right) \frac{\text{joules}}{\text{coulomb}} \text{ or volts} \qquad \text{(Scalar)} \qquad (1.58)$$

Figure 1.14 shows the equipotential surfaces for equal increments of potential. Note that, as one approaches $+Q$, the intensity grows rapidly larger and the equipotentials for equal increments of potential are increasingly close together.

The difference of potential between any two points P_2 and P_1 in an electric field is independent of the path chosen. This is generally true even when many charges are present, because we may treat any static electric field as the sum of fields of single point charges. As a consequence, the difference in potential around any closed path in a static electric field must be zero.

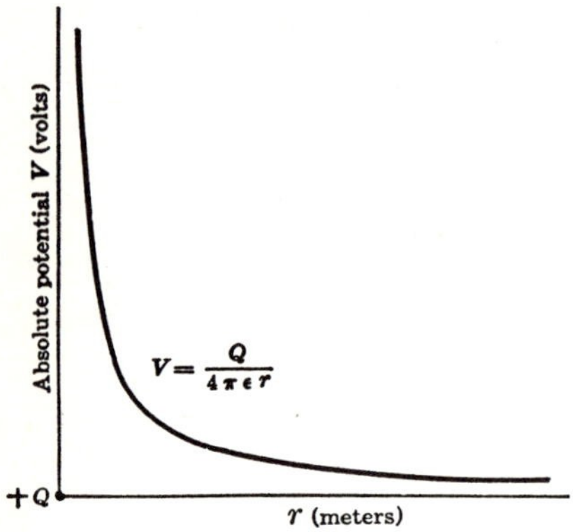

FIG. 1.15. Absolute potential in a point charge field.

The zero potential surface of this field is found only at infinite radius because there only is the intensity zero. The absolute potential of any point P_2 in the field is given by letting r_1 equal infinity in the equation for the difference of potential (equation 1.58). Hence

$$\text{Absolute potential } V_2 = \frac{Q}{4\pi\epsilon r_2} \quad \text{volts} \qquad \text{(Scalar)} \quad (1.59)$$

which is positive if Q is positive and depends upon the medium. For smaller radii, the values of the absolute potential grow larger, as shown in Fig. 1.14, varying in a hyperbolic manner, as shown in Fig. 1.15.

1.34. Electric Potential Gradient. Conversely to the process for determining the potential difference as the line integral of the negative intensity, if the equations for the equipotential surfaces are known, it may be desirable to work back to the intensities. This reverse process is to obtain the electric potential gradient or directional derivative of the potential.

In Fig. 1.16, at point P on equipotential surface B, let the intensity be E. Let the surface B' be another equipotential surface very close to B, normally distant $dn = ds \cos \alpha$. Then

$$\Delta(V_{B'} - V_B) = dV = -E\, ds \cos \alpha = -E\, dn$$

is the difference of potential between B' and B.

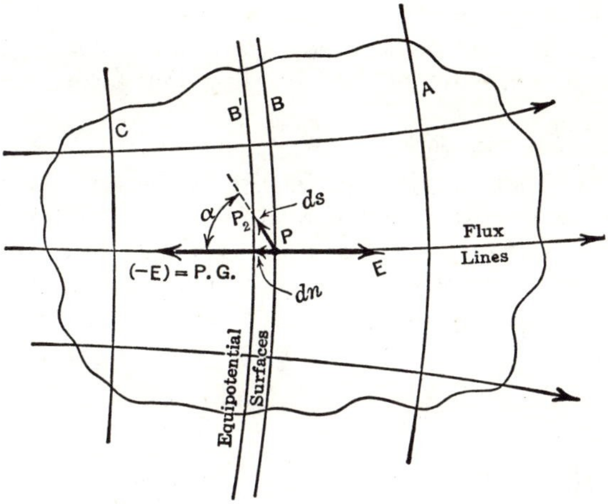

FIG. 1.16. Method of obtaining the potential gradient.

The electric potential gradient at P is

$$\text{P.G.} = \frac{dV}{dn} = -E \quad \frac{\text{newtons}}{\text{coulomb}} \text{ or } \frac{\text{joules}}{\text{coulomb}} \cdot \frac{1}{\text{meter}}$$

$$= \frac{\text{volts}}{\text{meter}} \qquad \text{(Vector)} \tag{1.60}$$

The electric potential gradient at a point in an electric field is a vector quantity given by the difference in potential dV taken over a small distance dn along the flux line through the point, divided by the distance, both measured in a direction opposite to that of the flux line.

Another way of stating it is: The potential gradient at a point is a vector quantity given by the maximum space rate of change of potential at that point.

The gradient at a point has a magnitude equal to that of the intensity, but it is opposite in direction. Consequently the units are the same—newtons per coulomb or volts per meter.

Referring again to Fig. 1.16, if we depart from point P on surface B to P_2 on B' along a line ds, which is not normal to B, the same difference in potential dV is encountered. Dividing dV by ds we get a component of the vector gradient in the ds direction, which is smaller than the gradient itself.

Gradient component

$$= \frac{dV}{ds} = -E \cos \alpha = -E_s \quad \frac{\text{volts}}{\text{meter}} \qquad \text{(Vector)} \tag{1.61}$$

Obviously the gradient is the largest space rate of change of potential (directional derivative) and has its direction normal to the equipotential surface.

If the equipotential surfaces are given in the form of equations in Cartesian coordinates, it is possible to take the three partial derivatives of the potential function V with respect to the three directions ∂x, ∂y, and ∂z, and obtain three vector components of the gradient in the x, y, and z directions. By adding these three vectors at 90° to each other the total vector gradient is obtained.

$$\left.\begin{aligned} \text{P.G.}_x &= -E_x = \frac{\partial V}{\partial x} && x\text{-direction vector} \\ \text{P.G.}_y &= -E_y = \frac{\partial V}{\partial y} && y\text{-direction vector} \\ \text{P.G.}_z &= -E_z = \frac{\partial V}{\partial z} && z\text{-direction vector} \end{aligned}\right\} \tag{1.62}$$

$$\begin{aligned} \text{P.G.} &= -E = -(E_x \oplus E_y \oplus E_z) \qquad \text{(Vector sum)} \\ &= \text{gradient } V \end{aligned} \tag{1.63}$$

This process is possible only when a potential function exists. The necessary condition is that the line integral of the vector function over a closed path is zero. In this event the potential difference

$$\oint_{\text{closed path}} (-E)\, ds \cos \alpha = \Delta V = 0 \tag{1.64}$$

over a closed path is zero and the potential difference between two points is the same irrespective of the path followed between the points. This condition is fulfilled in the *static* electric field, and again in the *static* magnetic field when the path does not enclose any current.

1.35. Principle of Linear Superposition. If several causes, each *linearly* related to its own effect, combine to produce a total effect, the latter is the sum of the separate effects and the principle of linear superposition is said to apply. Conversely, when the relation between cause and effect is *nonlinear*, superposition does not hold.

Illustration of Superposition. If a certain force (cause) produces a certain acceleration (effect) upon a given mass, doubling the force doubles the acceleration because cause and effect are directly proportional to one another (linearly related).

Illustration of Nonsuperposition. A certain torque (cause) drives a fan at a certain speed (effect). Doubling the torque will not produce twice the speed, as the torque is proportional to the square of the speed (nonlinearly related).

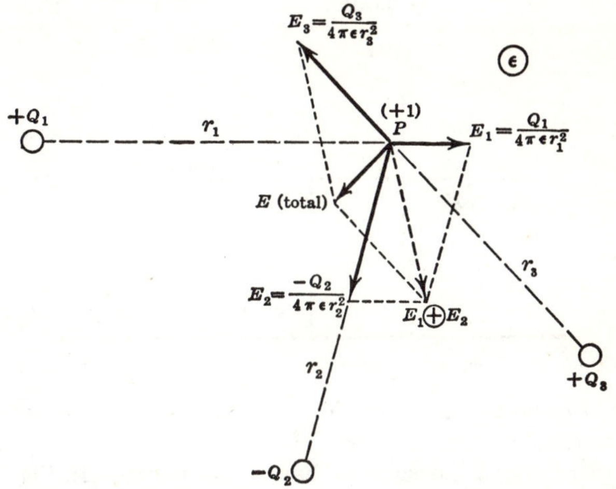

FIG. 1.17. Vector addition of intensities.

Effects that are superposable must be summed by vector or scalar addition according to whether they are vector or scalar quantities, respectively. Whether effects are superposable or not is a matter that must ordinarily be left to experiment to decide.

C. FIELDS OF DISCRETE CHARGE DISTRIBUTIONS

1.36. Field of Several Point Charges. Point charges, either positive or negative or both, may be distributed in space to suit the pleasure of the experimenter, as shown in Fig. 1.17. It is desired to find the electric intensity or potential gradient and the absolute potential at any field point P.

Both the intensity and the potential are superposable, by vector and scalar addition, respectively, inasmuch as both of these field effects are linearly related to the causative charges. Experiment supports this statement.

Intensity. Figure 1.17 illustrates the vector addition of separate intensities at point P where the unit positive charge is placed. The separate intensities are (in this case) $Q_1/(4\pi\epsilon r_1^2)$ and $Q_3/(4\pi\epsilon r_3^2)$ repulsion, and $-Q_2/(4\pi\epsilon r_2^2)$ attraction, as given by equation 1.49.

The process may be written as an equation for general use:

$$E_p = \sum \frac{Q}{4\pi\epsilon r^2} = \underbrace{\frac{Q_1}{4\pi\epsilon r_1^2} \oplus \frac{Q_2}{4\pi\epsilon r_2^2} \oplus \frac{Q_3}{4\pi\epsilon r_3^2} \oplus \oplus}_{\text{(vector addition)}}$$

$$\frac{\text{newtons}}{\text{coulomb}} \quad \text{or} \quad \frac{\text{volts}}{\text{meter}} \qquad \text{(Vector)} \qquad (1.65)$$

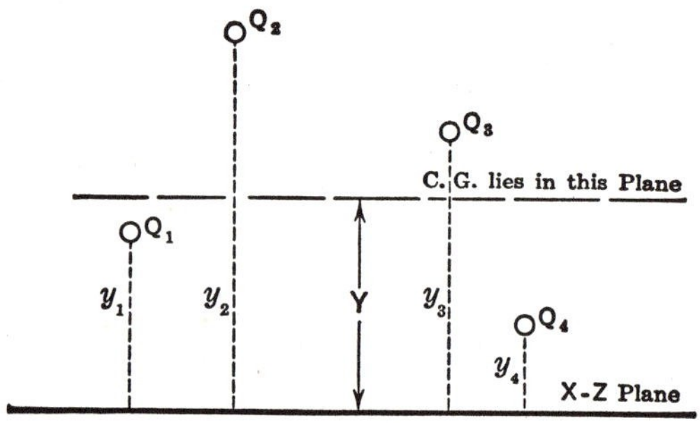

FIG. 1.18. Center of gravity—distance from the X-Z plane.

Absolute Potential. By equation 1.59 the absolute potential at a point P distant r from charge Q is $V_p = Q/(4\pi\epsilon r)$ volts, a scalar quantity with its sign the same as that of the charge. In Fig. 1.17, the absolute potential of point P is

$$V_p = \frac{+Q_1}{4\pi\epsilon r_1} + \frac{-Q_2}{4\pi\epsilon r_2} + \frac{+Q_3}{4\pi\epsilon r_3} \quad \text{volts}$$

In general, we may write:

$$V_p = \sum \frac{Q}{4\pi\epsilon r} = \underbrace{\frac{\pm Q_1}{4\pi\epsilon r_1} + \frac{\pm Q_2}{4\pi\epsilon r_2} + + +}_{\text{(scalar addition)}} \quad \text{volts} \qquad \text{(Scalar)} \qquad (1.66)$$

1.37. Center of Gravity of Charge Distribution. Any distribution of charges has a center of gravity. Choose any point in space as origin and pass the three planes of coordinates through it. By taking the product of the magnitudes of the charges (signs must be considered) by the perpendicular distances from the X–Z plane, and, dividing by the sum of the charges (signs considered), the distance Y from the X–Z plant to the center of gravity is obtained. See Fig. 1.18.

$$Y = \frac{\pm Q_1 y_1 \pm Q_2 y_2 + + \pm Q_n y_n}{\pm Q_1 + \pm Q_2 + + \pm Q_n} \tag{1.67}$$

Proceed similarly to obtain the distances from the other two planes.

1.38. Shape of Field Far Distant from Charges. At field points which are quite distant from the charges compared with the distance between the charges themselves, the electric equipotential surfaces appear spherical, and the intensity and flux lines appear radial, being asymptotic to straight lines drawn through the center of gravity of the charge system.

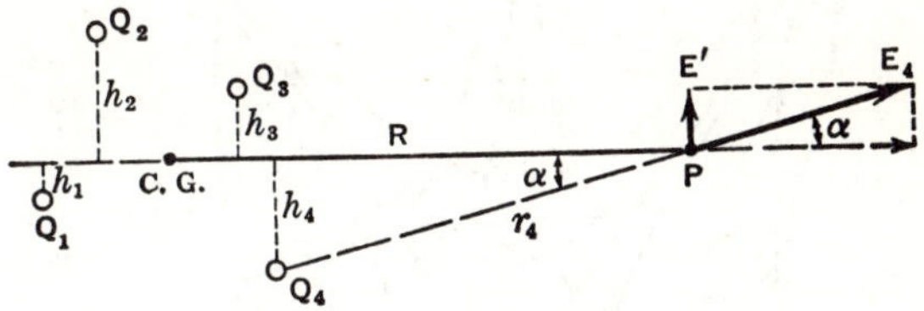

FIG. 1.19. Intensity far from charges is directed through the center of gravity.

Assume, as shown in Fig. 1.19, that charges Q_1, Q_2, Q_3, and Q_4 have their center of gravity at C.G. Pass a straight line of length R from C.G. to any distant field point P where the intensity is to be found. Consider one of the charges (say Q_4), whose distance from P is r_4 and from the line R normally is h_4.

The intensity at P due to Q_4 is $E_4 = Q_4/(4\pi\epsilon r_4^2)$. The component perpendicular to R is $E' = E_4 \sin\alpha = \dfrac{Q_4}{4\pi\epsilon r_4^2}\cdot\dfrac{h_4}{r_4}$, which is approximately equal to $Q_4h_4/(4\pi\epsilon R^3)$ as R is approximately equal to r_4 under the assumed conditions.

Now considering all the charges, the sum of the components of intensity at P, normal to R, is

$$E' = \Sigma E \sin\alpha = \frac{Q_1h_1}{4\pi\epsilon R^3} + \frac{Q_2h_2}{4\pi\epsilon R^3} + \frac{Q_3h_3}{4\pi\epsilon R^3} + \frac{Q_4h_4}{4\pi\epsilon R^3}$$

$$= \frac{1}{4\pi\epsilon R^3}(Q_1h_1 + Q_2h_2 + Q_3h_3 + Q_4h_4) \tag{1.68}$$

However, the value in parentheses is zero, as it is merely the numerator of equation 1.67 in which Y is zero because the line R is taken through the center of gravity C.G. Hence the intensity component E' for all the charges is zero. Consequently the intensity is radial from C.G. and the equipotential surfaces are spheres, at large distances from the charges.

1.39. Singular Points and Lines. Points or lines in the electric field where the intensity or gradient is zero are called singular points or lines. At such points or lines the electric equipotential surface passing through the points or lines crosses itself at least twice.

Two special cases of singular points and lines are proved in Arts. 2.16 and 2.17, but the general proof for axially symmetric cases may be found in a number of texts on electrical theory.[1]

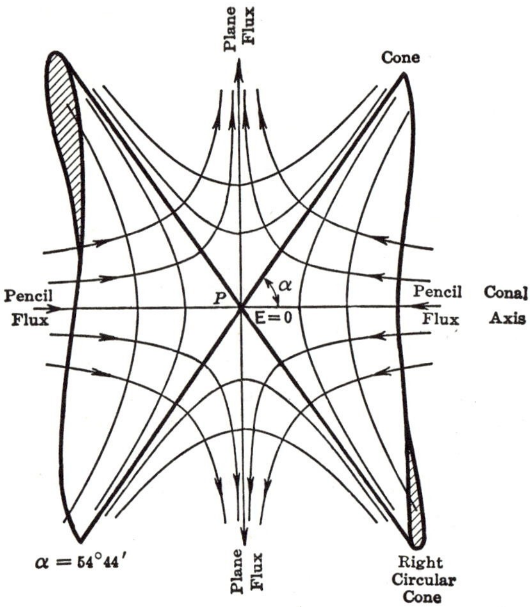

FIG. 1.20. Field about a singular point. (See Art. 2.16.)

Points. Through the singular point P there is a line, as shown in Fig. 1.20, along which the flux approaches the point from both directions as a pencil of flux, changing direction by 90° at the singular point, to spread out into a plane of flux. The pencil flux is also the conal axis of the equipotential surface passing through the point. The surface for axially symmetric cases is a right circular cone forming plane angle $\alpha = 54° \, 44'$ with the pencil flux. Other surfaces and flux lines are shaped as shown. They are, in fact, hyperbolas and hyperboloids of revolution.

[1] Livens, *The Theory of Electricity*, First Edition only, 1918, p. 98. Jeans, *Electricity and Magnetism*, Fifth Edition, Columbia University Press, 1925, p. 59. If the geometry is not axially symmetric the cone will not be right circular in cross section.

Lines. At a line singularity, the equipotential surface through the line may cross itself two or more times, depending upon the character of the charge distribution responsible, and form planes in the neighborhood of the line. If the equipotential crosses itself n times, the angle β between planes is $180°/n$. Halfway between the equipotential planes at $\alpha = \beta/2 = 180°/(2n)$, we find planes of flux that enter in one sector, change direction abruptly by angle β at the singular line, and depart in another plane of flux. Figure 1.21 gives

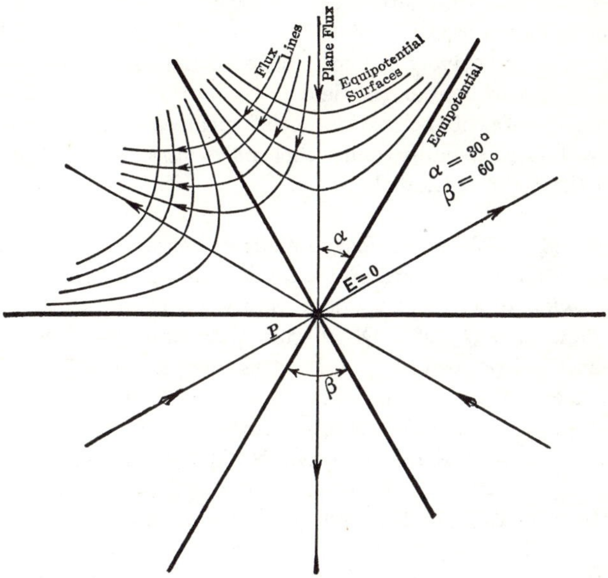

FIG. 1.21. Field about a line singularity. (Equipotential surface crosses three times. See also Fig. 2.17.)

the picture of an equipotential surface crossing three times at the singular line. In the most common case the equipotential surface crosses only twice, as shown in Fig. 2.17.

1.40. Field Sketching. Field sketching is the process of representing an electric (or other) field in a cross-section sketch by drawing representative flux lines and equipotential surfaces in that cross section.

It is not necessary to draw precisely Q lines of flux; any proportional number is satisfactory. Nor need many of the equipotential surfaces be drawn. A sufficient number of lines and surfaces should be drawn to indicate the general geometrical character of the field. A well-drawn sketch will show and emphasize the following points, so far as they may apply to a given case:

(1) Boundaries—surfaces which occur wherever the medium is changed: (*a*) Conductor-insulator boundaries, where free charge is located. (*b*) Insulator-insulator boundaries, where the flux lines and equipotential surfaces may be refracted.

(2) Singular points or lines.

(3) Center of gravity of charge system.

(4) Shape of the field at large distances from the charges, as well as close thereto.

(5) Geometrical symmetries of any kind.

(6) Distribution of the lines of flux so that by direction and proximity they give a relatively correct picture of the variation of the vector flux density D throughout the field.

(7) Distribution of the equipotential surfaces so that by proximity they represent reasonably truly the space variation of the absolute potential throughout the field.

Six cases of simple point charge fields are shown for the purpose of applying numerous principles of electric field study and of gaining a better knowledge of the methods of representation.

1.41. Field of $+Q$ and $-Q$. Figure 1.22 shows the flux field of two equal charges of opposite kinds, in cross section through the line containing the charges. The flux lines are shown in the upper part. In the lower part the method of obtaining the directions and relative densities of flux by the aid of Coulomb's law is illustrated.

The charges may be considered to be distributed over conducting spheres (boundaries) small enough compared with the distance between $+Q$ and $-Q$ that they may be treated as point charges. Very close to each small sphere, the lines of flux have a spherically radial distribution as the opposite charge is relatively far away. Farther out, both charges influence the intensity at any point, bending the flux and altering the flux density. At any given field point the flux direction is tangent to the intensity. All the Q lines leaving $+Q$ must enter $-Q$. Where the intensity and flux density are lower, the lines are spaced farther apart. If a proportional number of lines are drawn, their spacing should represent the relative densities of the field.

The line of charges is an axial symmetry containing straight flux lines, about which the entire flux field is symmetrical. The perpendicular plane to the line of charges bisecting the distance between them is a plane symmetry. Except for the directions of the flux lines, the sketch is symmetrical about this plane. There are no singular points or lines in this case, and the center of gravity is infinitely far away.

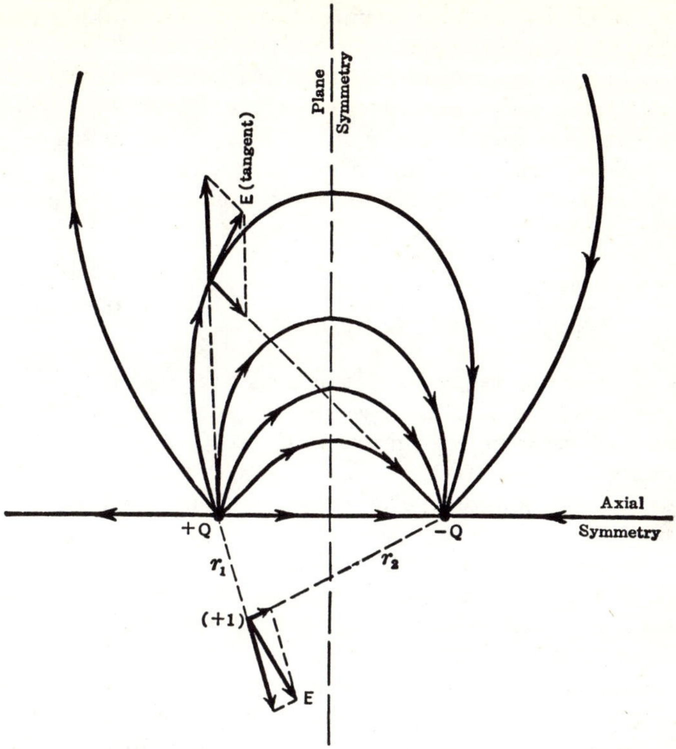

FIG. 1.22. The flux field of $+Q$ and $-Q$ for equal increments of flux.

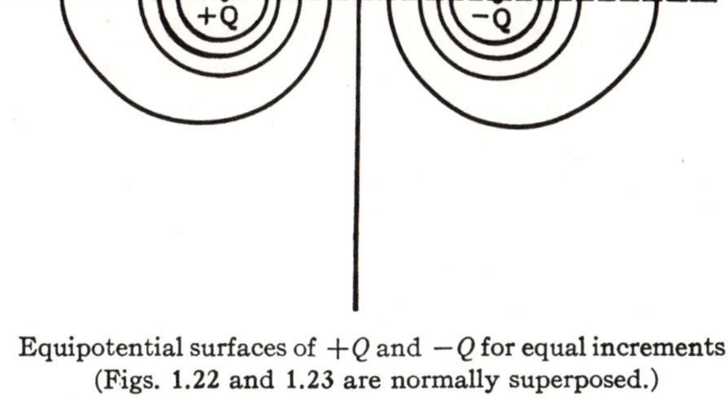

FIG. 1.23. Equipotential surfaces of $+Q$ and $-Q$ for equal increments of potential. (Figs. 1.22 and 1.23 are normally superposed.)

Figure 1.23 shows several equipotential surfaces which are always closed and are everywhere normal to the flux lines. + and − marks are used to distinguish the positive and negative absolute potentials. The higher potentials are marked with several +'s, the lower potentials with several −'s. A smaller number of + or −marks indicates potentials of intermediate value. In this particular case the absolute potential equation is given by the algebraic sum of the absolute potentials of the two charges taken separately:

$$V_p = \frac{+Q}{4\pi\epsilon r_1} + \frac{-Q}{4\pi\epsilon r_2} = \frac{Q}{4\pi\epsilon}\left(\frac{1}{r_1} - \frac{1}{r_2}\right) \text{ volts} \qquad \text{(Scalar)} \qquad (1.69)$$

where r_1 and r_2 are the respective distances to point P from $+Q$ and $-Q$. If V_p in equation 1.69 is assigned any definite value, positive or negative, letting r_1 and r_2 vary so as to satisfy it, we may trace out the position of that particular potential. In this case the equipotential surfaces are not spheres. The zero equipotential surface ($V_p = 0$) is satisfied by $r_1 = r_2$. It is therefore the bisecting plane normal to the line of charges. If the sketcher draws surfaces for equal increments of potential, the surfaces will be spaced more closely where the intensity or potential gradient is higher.

1.42. Field of $+Q$ and $+Q$. Figure 1.24 shows the flux field of $2Q$ lines and equipotential surfaces drawn together for two equal charges of like kinds, separated a distance s. This field has a singular point and a center of gravity both of which lie at the midpoint between the charges. Close in, the lines of flux are bent rather sharply, but, far out, they become approximately straight and radial, becoming asymptotic to straight lines through the center of gravity at O.

By using the solid angle relations given in Art. 1.24, the flux lines may be drawn, if desired, to represent equal increments of flux. Flux line DEF in Fig. 1.24 must emerge from $+Q$ at 54° (see Fig. 1.12) with the line of charges, to enclose one fifth of the flux of $+Q$. Far out, DEF becomes asymptotic to a line through O, forming an angle with the line of charges of 78.5°, on the assumption that the charge system is $+2Q$ located at the center of gravity. There are two symmetries—the axial symmetry through the charges and the plane symmetry of the normal plane through O.

Far out, the equipotential surfaces are positive and concentrically spherical about the center of gravity. As the charges are approached, the surfaces assume higher values of positive potential and become somewhat dumbbell-shaped. Surface B through the singular point crosses itself to form a tangent cone of $\alpha = 54° \ 44'$ at the singular

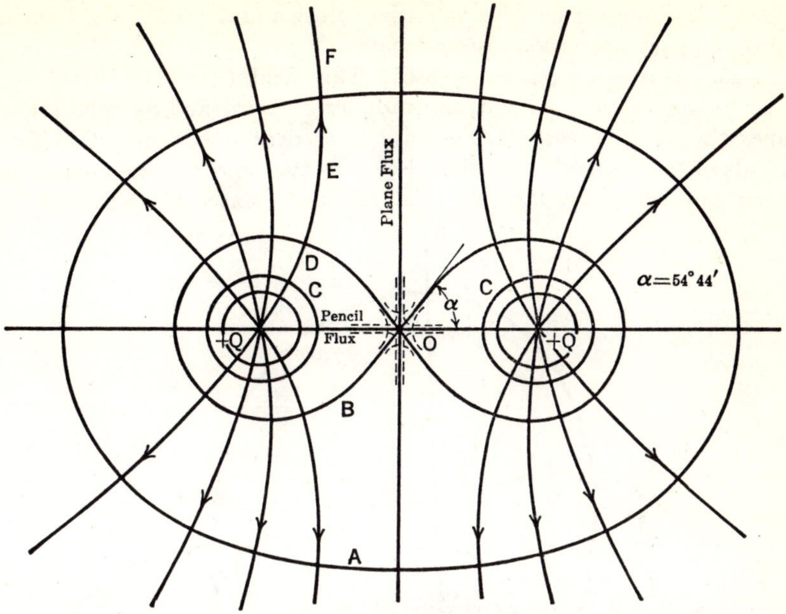

FIG. 1.24. Field of $+Q$ and $+Q$.

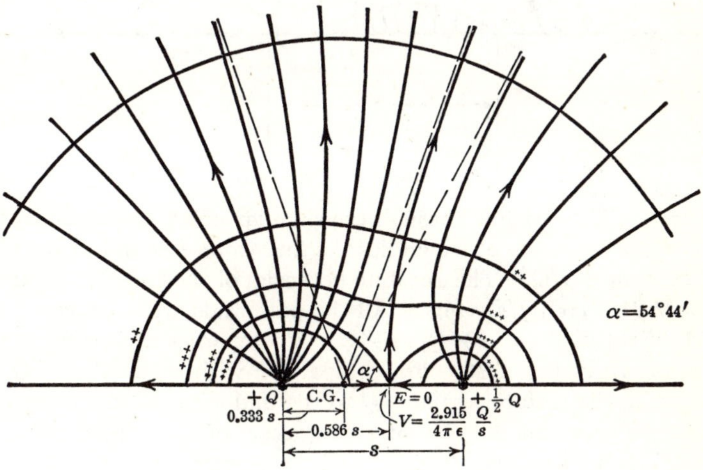

FIG. 1.25. Field of $+Q$ and $+\frac{1}{2}Q$.

point. Still higher positive potential values find the surface C divided into two parts, one around each charge.

1.43. Field of $+Q$ and $+\frac{1}{2}Q$. This field (Fig. 1.25) has $\frac{3}{2}Q$ lines of flux which, far out, are radially outward and asymptotic to lines through the center of gravity located at $s/3$ from $+Q$. Two thirds of the lines or those from $+Q$ cross two thirds of the area of an equipotential sphere far out. The singular point is placed at $0.586s$

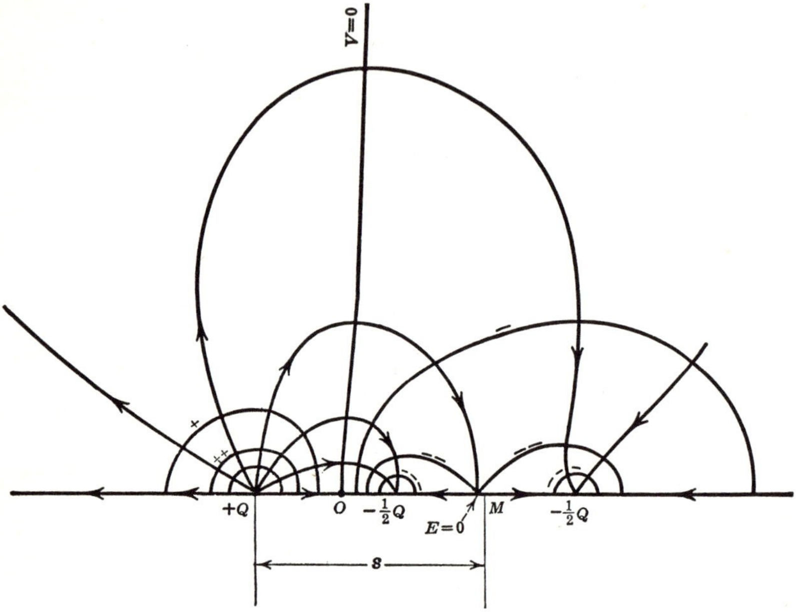

FIG. 1.26. Field of $+Q$, $-\frac{1}{2}Q$, $-\frac{1}{2}Q$.

from $+Q$. The flux line through the singular point divides the region into two parts, in one of which lies the flux from $+Q$ and in the other the flux from $+\frac{1}{2}Q$. The absolute potential of the singular point surface is $(2.915/4\pi\epsilon)(Q/s)$ volts. All the potentials are positive.

1.44. Field of $+Q$, $-\frac{1}{2}Q$, $-\frac{1}{2}Q$. The charges are arranged in a line as shown in Fig. 1.26. For comparison with the field of $+Q$ and $-Q$ let s be the distance from $+Q$ to the midpoint M of the negative charges. This point is the center of gravity of the negative charges alone. The field as a whole has its center of gravity at infinity. The singular point $E = 0$ is somewhat to the left of M. The flux line from $+Q$ to $E = 0$ divides the flux from $+Q$ into two equal parts, each going to $-\frac{1}{2}Q$. Close in, the positive potentials show no peculiarities. The zero potential is warped as shown. The negative potentials of greater negative value than that passing through the singular point are double surfaces.

1.45. Field of $+Q$ and $-\frac{1}{2}Q$. Figure 1.27. Half of $+Q$'s flux reaches the negative charge; the rest goes to infinity. Far out, the lines are radial and the potentials are concentric spheres about the center of gravity, which lies to the left of $+Q$. Coming into higher positive potentials, the surfaces become heart shaped in general form

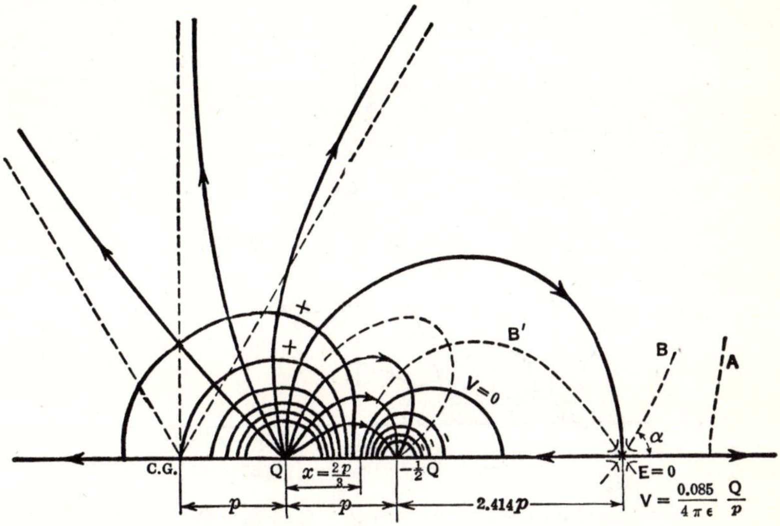

FIG. 1.27. Field of $+Q$ and $-\frac{1}{2}Q$. (See also Fig. 5.9.)

A, until the double surface BB' through the singular point is reached. Inside the closed surface B' the potentials decline to zero and on to increasing negative values about the negative charge.

The zero potential surface is a sphere. No others are. This is shown in the following manner. Let r_1 and r_2 be the respective distances from $+Q$ and $-q$ (in this case $-q = -\frac{1}{2}Q$). Then we have

$$V_p = \frac{+Q}{4\pi\epsilon r_1} + \frac{-q}{4\pi\epsilon r_2} = 0 \qquad \text{(See equation 5.35.)}$$

or

$$\frac{r_1}{r_2} = \frac{Q}{q} = 2 \qquad \text{(in this case)} \tag{1.70}$$

which is a sphere, as a sphere is the locus of points whose ratio of distances to two fixed points is constant.

1.46. Field of $+Q$, $-Q$, $+Q$, $-Q$. If charges are arranged in a line in the order shown in Fig. 1.28, a new point of interest appears. At O is found a singular line which is a circle symmetrical about the line of charges. Here the flux lines change direction sharply at 90°, and the equipotential surface (which is zero in this case) crosses itself at 90°. One branch is the center bisecting plane and the other is a

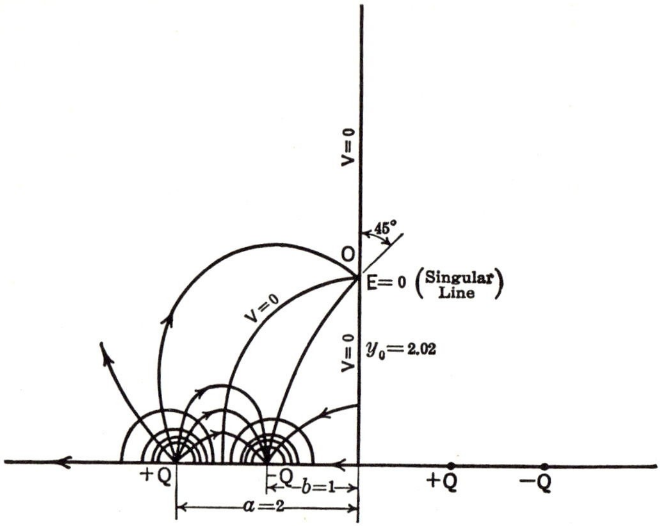

FIG. 1.28. Field of $+Q$, $-Q$, $+Q$, $-Q$.

surface shaped like a thick discus crossing the axis between $+Q$ and $-Q$.

The height of the singular line above the axis is given by

$$y_0 = (ab)^{1/3}(a^{2/3} + b^{2/3})^{1/2} \tag{1.71}$$

D. FIELDS ABOUT CHARGED CONDUCTORS

1.47. Densities of Electricity. Electricity may be distributed throughout a volume, as in a vacuum tube between the cathode and plate, or spread over the surface of a charged conductor, and, theoretically at least, placed charge to charge forming a line distribution. It is convenient to define these densities as:

ρ = number of coulombs per cubic meter (volume density).
ρ_s = number of coulombs per square meter (surface density).
ρ_l = number of coulombs per meter of length (linear density).

1.48. An Isolated Charged Conductor. Electric charges of one sign placed inside the volume of a conducting body flow out to the surface under the action of their mutually repulsive forces. Both the volume and the surface of the conductor must become equipotential, for, if they did not, a difference of potential would exist between parts of the conductor, creating an electric field which would cause the charges to redistribute themselves until the equipotential situation prevailed. Inside the volume of the conductor, the intensity is zero.

The distribution of charges (variation of the surface density) depends upon the shape of the conductor. About all that can be said is that, in general, from experiments we know that the surface density increases rapidly where the convex curvature is sharpest and decreases markedly with concave curvature. Figure 1.29 illustrates this point.

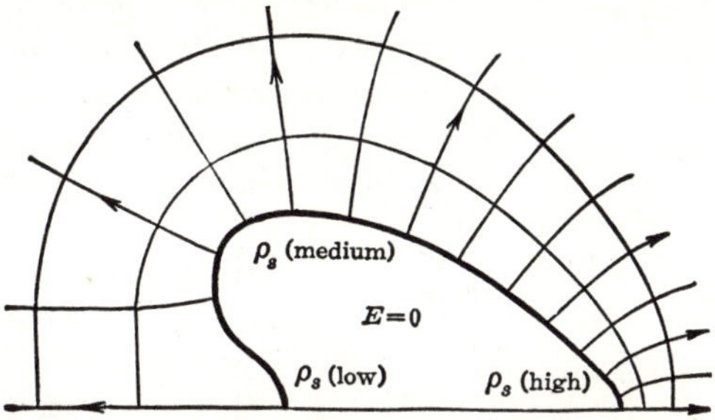

FIG. 1.29. Field about an isolated conductor.

The lines of flux leave the surface normally as it is equipotential. Then curving, they become spherically radial at large distances from the body. The body has an absolute potential that is obtained by computing the external work done upon a unit positive charge in moving it from infinity to any point on or inside the body. The potential is positive if the body possesses a positive charge. Equipotential surfaces in the field close to the body are shaped much like the body but approach the surface more closely where the density of charge is large. Far out from the body where the flux lines are radial, the equipotential surfaces approximate spheres.

Intensities. Just off the Surface. Since the flux lines leave the surface normally, where the charge density is ρ_s charges per square meter, we may say that the flux density D (lines per square meter) just off the surface in the dielectric is $D = \rho_s$. Dividing by the dielectric constant ϵ the intensity is

$$E_{\text{off surface}} = \frac{\rho_s}{\epsilon} \quad \frac{\text{newtons}}{\text{coulomb}} \quad \text{or} \quad \frac{\text{volts}}{\text{meter}} \tag{1.72}$$

normal to the surface, outward if ρ_s is positive, inward if ρ_s is negative.

Inside the Surface. Inside the volume of the conductor

$$E_{\text{inside}} = 0 \tag{1.73}$$

On the Surface. On the surface, the intensity is one half the value just off the surface, outward for positive ρ_s, inward if ρ_s is negative.

$$E_{\text{surface}} = \frac{\rho_s}{2\epsilon} \tag{1.74}$$

This is easily shown to be true with the aid of Fig. 1.30. Consider the surface of the conductor, shown to be divided into two parts, a small disk area A_d upon which there is a surface charge density ρ_s, and the remainder of the area A_r. Points 1, 2, and 3 are respectively just inside, on, and just outside the disk area. Temporarily, for the purposes of this discussion, assume that ρ_s is positive. Due to the disk charge alone, the intensities at points 1 and 3 are equal but oppositely directed.

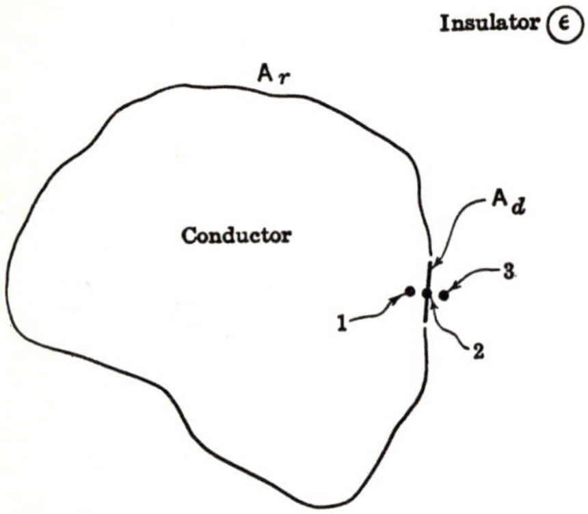

FIG. 1.30. Conductor surface intensity.

At point 2 the disk charges can exert no force upon a unit positive charge placed thereon, as they are in the same plane. The following schedule illustrates the matter:

$$\text{Intensity due to disk } E_D = \begin{Bmatrix} \text{at 1} \longleftarrow \\ \text{at 2} \quad 0 \\ \text{at 3} \longrightarrow \end{Bmatrix} (\text{for } +\rho_s)$$

Due to the rest of the charges over A_r, the intensities at points 1, 2, and 3 are equal and similarly directed, as the points are considered to be placed very close together. Their schedule follows:

$$\text{Intensity due to } A_r \text{ is } E_r = \begin{Bmatrix} \text{at 1} \longrightarrow \\ \text{at 2} \longrightarrow \\ \text{at 3} \longrightarrow \end{Bmatrix} (\text{for } +\rho_s)$$

At point 1, the total intensity is zero (equation 1.73), as it is inside the conductor, or

At point 1, total $E_1 = E_r - E_D = 0$

At point 3, the total intensity is ρ_s/ϵ (equation 1.72) or

$$\text{At point 3, total } E_3 = E_r + E_D = \frac{\rho_s}{\epsilon}$$

From these two equations we obtain:

$$E_r = E_D = \frac{\rho_s}{2\epsilon}$$

Hence, at point 2, the total intensity on the surface is

$$E_r = E_{\text{surface}} = \frac{\rho_s}{2\epsilon}$$

If ρ_s is negative, the intensities are reversed in direction.

Mechanical Pressure on a Surface. To say that the intensity on the surface is $\rho_s/(2\epsilon)$ means that a unit positive charge placed thereon experiences that much force per unit positive charge. Then the ρ_s charges per square meter experience a mechanical pressure or force per unit area of

$$p = \frac{\rho_s^2}{2\epsilon} \quad \frac{\text{newtons}}{\text{square meter}} \qquad \text{(Normal outward)} \tag{1.75}$$

The pressure is always outward, tending to expand the conductor surface into the dielectric, for if ρ_s is positive, the intensity is outward and so is the pressure; if ρ_s is negative, the intensity is inward, but negative charge tends to move against the intensity or outward.

Further on it is shown that $D^2/(2\epsilon)$ in joules per cubic meter is equal to the energy density ω_e in the field, whence it follows that the normal mechanical pressure on the surface is *numerically* equal to the energy density just off the surface.

$$p = \frac{\rho_s^2}{2\epsilon} = \frac{D_{\text{off}}^2}{2\epsilon} = \omega_e \quad \frac{\text{newtons}}{\text{square meter}} \text{ or } \frac{\text{joules}}{\text{cubic meter}} \tag{1.76}$$

1.49. Conductor Point—Corona. A sharp point on a conductor is merely a surface of very high curvature. Consequently the charge density ρ_s is very high locally and produces a very large potential gradient just off the surface in the dielectric. With a sufficiently high gradient (3×10^6 volts per meter for normal air—see Table 1.5), free electrons that may be present in this region acquire enough energy in falling against the field to ionize neutral atoms into new electrons and positive ions, which are now propelled in opposite directions by the field. The new electrons as well as the original group continue to generate still more electrons and positive ions. The process is known as "cumulative ionization."

Some of the neutral gas atoms, in colliding with electrons, may be raised to "excited" energy states below the values needed for ionization, and later in returning to their normal states they emit this energy in the form of radiation. Some of the radiation lies in the visible portion of the spectrum. This visual display is known as corona. It is to be found regularly in air with high voltages, and with conductors having sharp points or high curvature.

Corona is found only in fields where the flux diverges or converges, where the intensity varies in magnitude from place to place, and in these fields the corona occurs only in those portions where the intensity is greater than the dielectric strength. The corona light appears to come from a limited region about the conductor point. At the edge of the corona the intensity is approximately equal to the dielectric strength. At the conductor surface itself the intensity may be much higher.

If the voltage is raised sufficiently above the value producing corona, a sharp discharge called a "spark" results and the dielectric is broken down. Gases and liquids may recover their dielectric strengths after the spark has passed.

1.50. Two Charged Conductors. The distribution of electricity ρ_s over one conductor is such as to make the surface an equipotential, resulting in a greater charge density ρ_s on parts of the surface of sharper convex curvature. When two charged conductors are brought into proximity, each conductor and its surface must still be an equipotential. The electric distribution on one conductor is now not only due to that conductor's shape but is also dependent on the distribution on the other conductor. For instance, two bodies both positively (or negatively) charged and brought into the same neighborhood redistribute their charges owing to the mutual repulsion of the two sets of charges. The charge density on the near sides of the conductors is sparser, and on the far sides denser, than would be the case if the two bodies were completely isolated.

1.51. Two Conductors Equally and Oppositely Charged. The Condenser. Two conductors brought into proximity, but separated by a dielectric or insulator, and charged to an equal degree with electricities of opposite kinds constitute a simple type of condenser. Acting under the mutual attraction of the two sets of electricity, the near sides of the two conductors are more densely charged and the far sides more sparsely. A difference of electric potential exists between the two charged bodies, because work must be done upon a unit positive charge in order to move it from the negatively charged body to the positive.

More general and more complicated types of condenser occur when there are more than two conductors involved, especially if there are several sources of voltage connected to various pairs of conductors. Special cases are considered in Chapter 5.

E. CAPACITANCE AND ENERGY STORAGE

1.52. Capacitance (Definition). The capacitance (or capacity) of a condenser, charged to $+Q$ and $-Q$, is the ratio of the charge magnitude on either conductor Q to the difference in potential between the conductors ΔV. The unit of capacitance is the farad.

$$\text{Capacitance } C = \frac{Q}{\Delta V} \quad \frac{\text{coulombs}}{\text{volt}} = \text{farads} \qquad \text{(Scalar)} \qquad (1.77)$$

The dimensional formula is

$$\text{Capacitance} = \frac{\text{electric charge}}{\text{electric potential}}$$

$$= \frac{Q}{L^2M^1T^{-2}Q^{-1}} = L^{-2}M^{-1}T^2Q^2 \qquad (1.78)$$

It is found by experiment that increasing the charge on the condenser by any given percentage increases the potential difference in the same degree, so long as the potential gradient or intensity does not exceed the dielectric strength of the insulation at any point. Consequently, the capacitance which is the ratio between these two is a constant, independent of either. Capacitance is purely a function of the geometry of the system and of the dielectric material used. The capacitance is always directly proportional to the dielectric constant ϵ, inasmuch as the potential difference introduced in the denominator of equation 1.77 is itself inversely proportional to ϵ.

1.53. Energy Stored in a Condenser. Total Energy. The fact that forces are required to separate positive and negative electricity indicates that energy is stored in a charged condenser. The condenser is charged by connecting it to a battery or generator, and its conductors become extensions of the battery terminals. Momentarily there is an electric current or charge movement through the wire and battery, until the condenser by becoming charged builds up a potential difference V_c equal to that of the battery V_b but oppositely directed in the wire circuit. The total amount of charge that flows onto the condenser is $Q = CV$, $+Q$ on one conductor connected to the positive terminal of the battery and $-Q$ on the other.

At any stage during the charging process, let the charge on the condenser be q causing a proportional condenser potential V_c.

$$q = CV_c$$

and

$$dq = C\,dV_c$$

Remembering that the potential V_c means that V_c units of work (joules) must be done to move one unit positive charge from the negative to the positive conductor, $V_c\,dq$ is the number of joules needed to move dq charges against V_c potential. Hence the total work required completely to charge the condenser is

$$W_e = \int V_c\,dq = \int_{q=0}^{q=Q} \frac{q}{C}\,dq = \frac{1}{2}\frac{Q^2}{C} = \frac{1}{2}QV = \frac{1}{2}CV^2 \text{ joules} \quad (1.79)$$

This energy is stored in the field of the condenser and is recoverable under appropriate conditions.

1.54. Energy Density of the Electric Field. It is believed that the energy stored in a condenser is located in the electric field about the charges and not in the charges themselves. It is pertinent then to investigate the energy density or joules per cubic meter stored in the electric field. Consider a volume in the electric field bounded by lines of flux and two equipotential surfaces, as shown in Fig. 1.31, sufficiently small so that the intensity and flux density may be considered as constant throughout. The energy stored in this volume is one half the product of the boundary charge and the difference of potential.

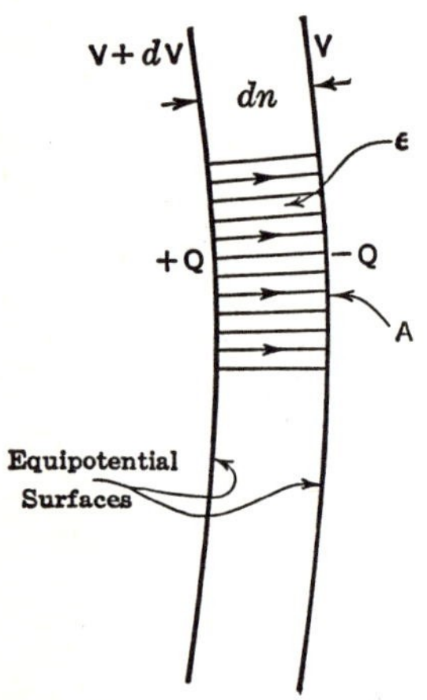

FIG. 1.31. Uniform field in a small volume.

$$dW_e = \tfrac{1}{2}Q\,dV$$

but

$$Q = \rho_s A = DA$$

The difference in potential dV is equal to the product of the intensity and the distance dn, as the intensity is constant throughout the volume.

$$dV = E\,dn = \frac{D}{\epsilon}\,dn$$

Hence

$$dW_e = \frac{1}{2}(DA)\left(\frac{D}{\epsilon}\,dn\right) = \frac{D^2}{2\epsilon}(A\,dn) = \frac{D^2}{2\epsilon}\,dv \quad \text{joules}$$

as the volume $dv = A\,dn$. The energy density is

$$\omega_e = \frac{dWe}{dv} = \frac{D^2}{2\epsilon} = \frac{DE}{2} = \frac{E^2\epsilon}{2} \quad \frac{\text{joules}}{\text{cubic meter}} \tag{1.80}$$

Conversely the volume integral of the energy density gives the total energy stored in the volume considered.

$$W_e = \iiint_v \omega_e\,dv = \iiint_v \frac{D^2}{2\epsilon}\,dv = \tfrac{1}{2}QV \quad \text{joules} \tag{1.81}$$

remembering that the flux density in general varies from point to point in the field.

F. BOUNDARY CONDITIONS

1.55. Boundary Conditions. A boundary is a closed surface which separates two media having different electrical properties. These surfaces may be divided into two classes—conductor–insulator boundaries, and insulator–insulator boundaries—according to the character of the media on both sides of the surface. It is assumed in this discussion that no current is flowing either in the conductor or in the insulator.

Conductor–Insulator Boundary. The conditions given here are merely a restatement of Art. 1.48, which discusses the electric field of a charged conductor. The surface is an equipotential, carrying free charge density ρ_s that varies according to the shape of the conductor and the proximity of other conductors. Inside the conductor, the intensity is zero; just outside in the insulating region, the intensity is ρ_s/ϵ normal to the surface, outward or inward as ρ_s is positive or negative, respectively. On the surface the intensity is one half the outside value.

$$\begin{aligned} E_{\text{off surface}} &= \frac{\rho_s}{\epsilon} \\ E_{\text{on surface}} &= \frac{\rho_s}{2\epsilon} \\ E_{\text{inside}} &= 0 \end{aligned} \tag{1.82}$$

See Art. 4.19 for the case of current flow in the conductor.

Insulator–Insulator Boundary with Zero Surface Charge. The common surface between two *perfect* insulators is assumed to be devoid of any free charge that might be present due to friction. The presence of the bound charges of polarization is ignored because the properties of the two insulators are adequately treated by the use of their respective dielectric constants.

Following Fig. 1.32*a*, let us describe a small closed rectangular path 1–2–3–4–1 about the surface, making 2–3 and 4–1 infinitesimal in length and 1–2 and 3–4 each *s* meters. The work done in moving a unit positive charge around this or any other closed path in the electrostatic field is zero. Let E_1 be the intensity in medium one, divisible into E_{T1} and E_{N1}, the tangential and normal components, respectively. E_2, E_{T2} and E_{N2} represent similar quantities in medium two. Then the work from point 1 to point 2 plus that from 3 to 4 must total zero, as the other two parts of the path are infinitesimal, or

$$+E_{T1}s + E_{T2}(-s) = 0$$

whence

$$E_{T1} = E_{T2} \tag{1.83}$$

That is, the tangential components of the intensities are equal in the two media at nearby points across the boundary. The normal components are not equal.

Referring to Fig. 1.32*b*, describe a small right cylinder of area A and length dn with its axis normal to the surface. Since there is no free charge inside the cylinder, as much flux must leave one flat circular face as enters the other one. The curved cylindrical area which is infinitesimal need not be considered. Under these conditions the flux entering one circular face is equal to that leaving the other face, or

$$\psi = D_{N1}A = D_{N2}A$$

and

$$D_{N1} = D_{N2} \tag{1.84}$$

The normal components of the flux density are equal in the two media at nearby points across the boundary. The tangential components are not equal.

Let α_1 and α_2 be the angles which the flux density D (or intensity E) vectors make with the surface normal. Then take the ratio

$$\frac{\tan \alpha_1}{\tan \alpha_2} = \frac{\left(\dfrac{D_{T1}}{D_{N1}}\right)}{\left(\dfrac{D_{T2}}{D_{N2}}\right)} = \frac{D_{T1}}{D_{T2}} = \frac{\epsilon_1 E_{T1}}{\epsilon_2 E_{T2}} = \frac{\epsilon_1}{\epsilon_2} \tag{1.85}$$

using the two results previously obtained.

Summing the results, at an insulator-insulator boundary:

(1) Tangential intensities are continuous. $E_{T1} = E_{T2}$ (1.86)

(2) Normal flux densities are continuous. $D_{N1} = D_{N2}$ (1.87)

(3) E and D make angles with the surface normal, whose ratio of tangents is equal to the ratio of dielectric constants.

$$\frac{\tan \alpha_1}{\tan \alpha_2} = \frac{\epsilon_1}{\epsilon_2} \tag{1.88}$$

These results do not determine the angles but merely specify what the relations in one medium must be if those in the other medium are

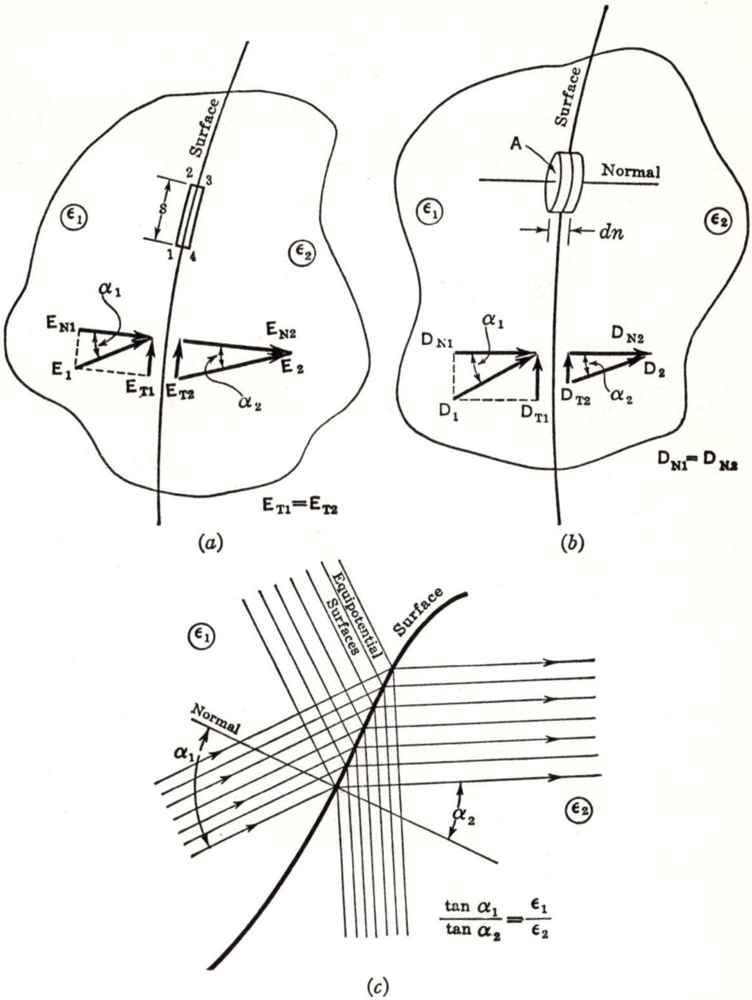

FIG. 1.32. Insulator-insulator boundary with zero surface charge. (a) Tangential intensities are equal. (b) Normal flux densities are equal. (c) Flux and equipotential surface refraction for $\epsilon_1 > \epsilon_2$.

known. Figure 1.32c shows the refracted flux lines and equipotential surfaces in the neighborhood of the boundary upon the assumption that $\epsilon_1 > \epsilon_2$.

Two special cases are of particular importance:

(a) *Boundary Normal to the Flux.* If the boundary surface is placed in an electric field everywhere normal to the flux, the flux continues on across the boundary without refraction. Then

$$\alpha_1 = \alpha_2 = 0$$

$$E_{T1} = E_{T2} = D_{T1} = D_{T2} = 0$$

$$D = D_{N1} = D_{N2} = \epsilon_1 E_1 = \epsilon_2 E_2 \tag{1.89}$$

See Fig. 1.33a. The boundary is an equipotential surface. The flux densities are equal, but the intensities are inversely proportional to

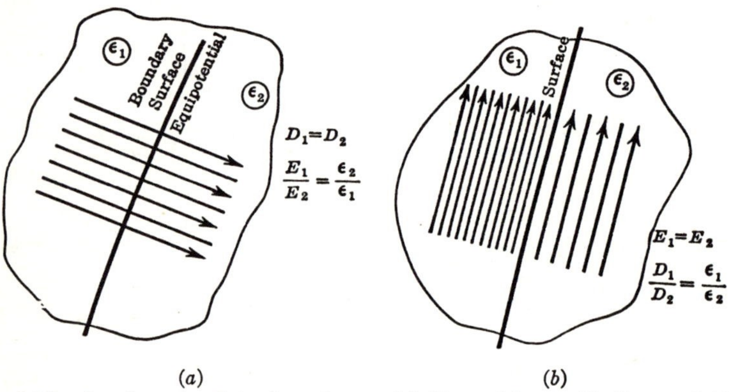

FIG. 1.33. Insulator-insulator boundary. (a) Normal flux. (b) Tangential flux.

the dielectric constants. The shape of the field is not altered under these conditions.

(b) *Boundary Parallel to the Flux.* If the boundary surface is introduced into the field everywhere parallel to flux lines, the flux lines remain parallel to the surface. Then

$$\alpha_1 = \alpha_2 = 90°$$

$$D_{N1} = D_{N2} = 0$$

$$E = E_1 = E_2 = \frac{D_1}{\epsilon_1} = \frac{D_2}{\epsilon_2} \tag{1.90}$$

See Fig. 1.33b. The intensities are now equal and the flux densities are directly proportional to the dielectric constants.

TABLE 1.5

PROPERTIES OF INSULATORS

RELATIVE DIELECTRIC CONSTANT, DIELECTRIC STRENGTH, VOLUME CONDUCTIVITY

Material	Relative Dielectric Constant ϵ_r	Dielectric* Strength, $\frac{\text{volts}}{\text{meter}}$	Volume Conductivity, $\sigma\ \frac{\text{mhos}}{\text{meter}}$
Vacuum	1.0	....	....
GASES			
Air (Atmosphere Pressure)	1.00059	3×10^6	....
CO_2	1.00098	....	....
H_2	1.00026	....	....
LIQUIDS			
Abietic Acid	4	21×10^6	10^{-4}
Alcohol (Ethyl)	25	....	10^{-3}
Carbon Tetrachloride	2.21	66×10^6	......
Ether (Ethyl)	4.8	....	10^{-3}
Paraffin Oil	....	18×10^6	10^{-14}
Petroleum	2.1	....	10^{-14}
Transformer Oil (Mineral)	2.5	15×10^6	10^{-12}
Turpentine	2.2	....	....
Water (Distilled)	80	....	10^{-4}
(Fresh)	80	....	5×10^{-3}
(Sea)	80	....	4
SOLIDS			
Asbestos	2.7	3×10^6	....
Asphalt	2.5	....	....
"Bakelite"	4.5	21×10^6	10^{-14}
Beeswax	2	....	10^{-13}
Cotton	....	4×10^6	....
Ebonite	2.5	60×10^6	10^{-14}
Fiber	2.5	2.1×10^6	10^{-8}
Glass (Light Flint)	7	30×10^6	10^{-13}
(Hard Crown)	7	30×10^6	10^{-13}
(Plate)	5	30×10^6	10^{-13}
("Pyrex")	4.5	90×10^6	10^{-12}
Gutta-Percha	4	15×10^6	10^{-13}
Marble	8	6×10^6	10^{-9}
Mica	6	90×10^6	10^{-15}
"Micarta"	4.1	....	10^{-9}
Paper (Dry Manila)	2	2.4×10^6	10^{-7}
(Impregnated)	3.5	15×10^6	....
Paraffin	2.3	30×10^6	10^{-14}

*Usually obtained from test in a uniform field.

Table 1.5 continued on page 64.

TABLE 1.5. (*Continued*)

PROPERTIES OF INSULATORS

RELATIVE DIELECTRIC CONSTANT, DIELECTRIC STRENGTH, VOLUME CONDUCTIVITY

Material	Relative Dielectric Constant ϵ_r	Dielectric* Strength, $\frac{\text{volts}}{\text{meter}}$	Volume Conductivity, $\sigma \frac{\text{mhos}}{\text{meter}}$
SOLIDS			
Polystyrene	2.6	20×10^6	10^{-14}
Porcelain	4.5	13.5×10^6	10^{-12}
Quartz (Fused)	4	30×10^6	10^{-17}
Resin	2.5	30×10^6	10^{-15}
Rubber (Pure)	2.5	21×10^6	10^{-14}
(Hard)	3	21×10^6	10^{-14}
Shellac	3	12×10^6	10^{-14}
Slate	7	3×10^6	10^{-6}
Soil (Moist)	30	....	2×10^{-2}
(Fertile)	15	....	5×10^{-3}
(Rocky)	7	....	10^{-3}
(Dry)	4	....	10^{-2}
(Very dry)	4	....	10^{-3}
Sulphur	4	....	10^{-15}
Wood	4	6×10^6	10^{-8}

* Usually obtained from test in a uniform field.

Insulator–Insulator Boundary with Surface Charge ρ_{s12}. A surface charge density ρ_{s12} may be present on the interface between two insulators. Whether it is originated by external agents such as friction or is due to the passage of the small currents that inevitably flow through the insulation, it changes the boundary conditions. In place of equation 1.84, we must use

$$D_{N2} - D_{N1} = \rho_{s12} \tag{1.91}$$

Equation 1.83 still is valid, but equation 1.85 is no longer true.

The boundary conditions outlined in this article are most nearly approached with high frequencies when the dielectric constants control the distribution of flux. At low frequencies and for d-c conditions, the currents and fields may be controlled by the conductivity of the insulating material. See Art. 4.19.

1.56. Consequences of Boundary Conditions. An electric field lying entirely in one dielectric medium may be altered as follows:

(1) By introducing any conducting body, the field shape is altered in such a way as to make the volumes and surfaces of all conductors equipotentials. Free charges must exist on the surfaces.

(2) By introducing a conducting body so that its surface lies entirely along an equipotential surface of the original field, the field shape is preserved. Free charges exist on the conductor surface.

(3) By introducing a second dielectric of different dielectric constant, the flux shape is altered, being refracted at the boundary surface, unless the new dielectric is so shaped as to have its surface (*a*) lie everywhere along an equipotential, or (*b*) lie everywhere along flux lines.

1.57. Properties of Insulators. Table 1.5 gives the relative dielectric constant, dielectric strength, and volume conductivity for a number of insulators. The values given are for d-c or low-frequency conditions. The effect of frequency is discussed in Chapter 3.

PROBLEMS

1. From consideration of the valencies given in Table 1.2, how would you expect carbon and oxygen to unite in forming a molecule? Sodium and chlorine?

2. A ball with a mass of 100 grams is tossed straight upward 1000 cm. Compute the kinetic energy at the highest and lowest positions. What is its velocity when it reaches the bottom again? Compute the potential and total energies at any position if the potential energy is assumed zero at the highest position. Compare with the hydrogen atom.

3. Find the force of repulsion between two electrons spaced 10^{-10} meter (which is the order of magnitude of an atomic distance). How much *energy* is required to bring them to this spacing from an infinite distance?

4. For the hydrogen atom, compute the frequencies and wave lengths of the two extreme lines in the Lyman series shown in Fig. 1.1. Compare these values of wave length with those in the visible range, which lie between 0.4×10^{-4} and 0.8×10^{-4} cm.

5. At ground level what changes occur in the pressure, density, molecules per cubic meter, rms velocity, mean-free-path, collisions per second, in air if the temperature is raised from 0° C to 300° K?

6. Compute the v_{mp} and v_{av} for 300° K, Prob. 5. What fraction of the molecules have total velocities greater than v_{mp}, than $0.5\ v_{mp}$, than $1.5\ v_{mp}$, than zero? What fraction of the particles have x-directed velocities greater than v_{mp}, than $0.5\ v_{mp}$, than $1.5\ v_{mp}$, than zero?

7. Name several physical quantities that are scalars; several that are vectors.

8. How are the fundamental units of mass, length, and time obtained?

9. Give some methods other than those mentioned in the text by means of which positive and negative electricity may be separated.

10. What is the maximum number of units of solid angle possible? Find the solid angle subtended at a point on the axis of a right circular cone of 8-cm altitude and base of 6-cm radius.

11. In a plane geometric surface placed 8 cm from a point charge (+0.06 microcoulomb) describe a circle of 6-cm radius whose center is at the foot of the perpendicular from the charge. Medium dielectric constant $\epsilon_r = 3$. Compute the number of flux lines crossing the circular area.

12. Consider two right circular coaxial cones, whose surfaces make angles of 60° and 30°, respectively, with the axis. Compute the solid angle between the conal surfaces.

13. Following Fig. P13:

(*a*) Find the electric field intensity at point P_1. (As it is a vector quantity, magnitude and direction must be specified.)

(*b*) What is the potential difference between P_1 and P_2? Which has the higher absolute potential?

14. Two point charge systems ($+0.1 \times 10^{-6}$ and -0.1×10^{-6} coulomb) are located 8 cm apart. Find the absolute potential and the electric field intensity at a point 5 cm from both point charge systems. Specify both of these quantities completely.

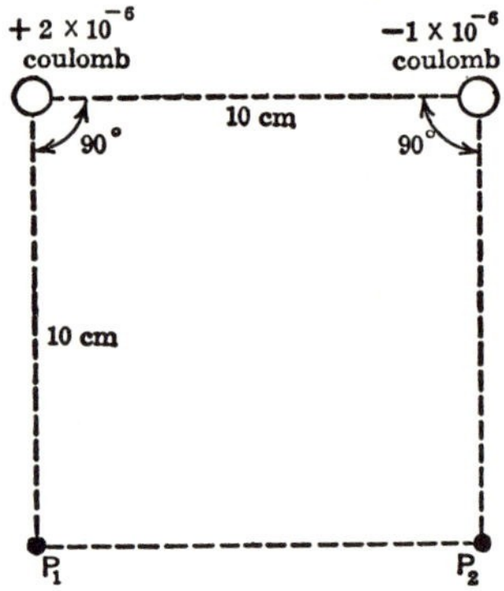

FIG. P13.

15. Three point charge systems (two of them are $+0.1 \times 10^{-6}$ and one is -0.1×10^{-6} coulomb) are placed 10 cm from each other. Find the absolute potential and the electric field intensity at the midpoint of the triangle. Specify both of these quantities completely. Locate the center of gravity. Sketch the flux lines and equipotential surfaces in the plane of the charges, using the properties of the singular point.

16. Where is the center of gravity of two equal charges having opposite signs?

17. Find the singular point in Prob. 13, also the center of gravity.

18. Under what conditions would a singular line be found?

19. Sketch the field of three parallel line charges lying in the same plane, $\rho_{l1} = +10 \times 10^{-6}$ coulomb, $\rho_{l2} = +5 \times 10^{-6}$ coulomb, $\rho_{l3} = +5 \times 10^{-6}$ coulomb. The middle line is equidistant from the two outer lines.

20. Sketch the flux lines and equipotential surfaces of four charges arranged in a line at equal distances, for relative values of $+1, +1, -1, -1$. Locate singular point, center of gravity, $V = 0$ surface.

21. Sketch the field about three line charges arranged in equilateral formation, for relative values of $\rho_{l1} = +2$, $\rho_{l2} = -1$, $\rho_{l3} = -1$. The spacing is 20 cm. Locate the two singular lines and one point on the zero potential surface. Sketch variation in potential along the center line.

22. A charged conductor is surrounded by air. Compute:

(*a*) The maximum charge density ρ_s which the conductor can hold.

(*b*) The conductor surface mechanical pressure at that ρ_s.

(*c*) If the air pressure is increased to 10 atmospheres, estimate the maximum charge density possible.

23. A metallic sphere of 10-cm radius is charged with 0.1×10^{-6} coulomb (+), spread uniformly over the surface, and is surrounded by a medium having a relative dielectric constant ($\epsilon_r = 4$). Find the intensities just off the surface, and on the surface. Find the mechanical pressure tending to swell the sphere. What is the intensity inside the sphere?

24. At the center of a 10-cm cube, place $+Q$ coulombs. The relative dielectric constant of the medium is $\epsilon_r = 2$. Find the number of flux lines crossing one cube face.

25. The lines of flux of an electric field pass from air into glass, making an angle of 30° with the normal to the plane surface separating the air and glass, on the air side of the surface. The value of the relative dielectric constant for the glass is 5.0. The field strength in the air is 100 volts per meter. What is the flux density in the glass, and what angle do the lines make with the normal on the glass side?

26. How many joules are stored in a 0.25-microfarad condenser charged to 24,000 volts? What are the charge and force on each plate, for a dielectric with $\epsilon_r = 4$, if the spacing between the plates is 0.5 cm?

CHAPTER 2

ELECTRIC FIELDS OF SIMPLE GEOMETRIES

In this chapter we shall confine our attention to electric fields created by distributing electricity in simple geometric patterns. Some of the cases considered are primarily of academic interest, serving to illustrate the fundamental principles established in the last chapter. Other cases lead to important practical applications.

2.1. Spherical Shell of Charge. A spherical conductor, solid or hollow, of external radius R charged with Q coulombs, due to its

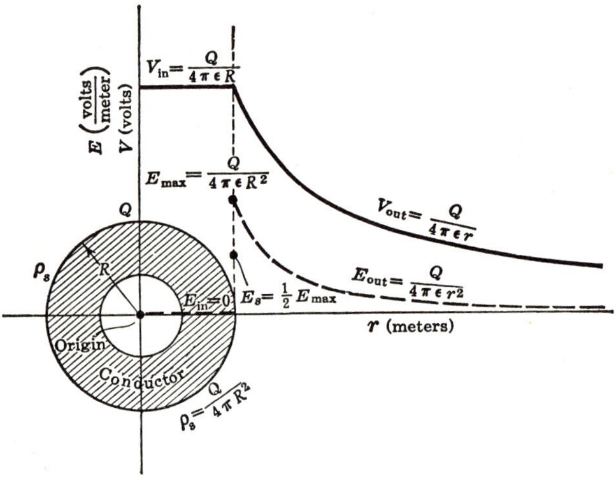

FIG. 2.1. Charged sphere—intensity and potential.

symmetry distributes the electricity uniformly over the external surface with a density $\rho_s = Q/(4\pi R^2)$ coulombs per square meter. From symmetrical considerations we see that the external flux field is spherically radial and that the equipotential surfaces are concentric spheres, both about the center of the sphere. Clearly then the external field is identical with that of the single point charge placed at the sphere center. We may then carry over the results of Arts. 1.21 and 1.33 at once. Inside the outer radius R there is no electric field. (See Art. 1.48.) All parts of the volume and surface of the sphere must therefore be equipotential.

Let r be the distance from the sphere center to any point where the intensity or potential gradient and the absolute potential are sought, as shown in Fig. 2.1. Inside the sphere the gradient is zero; outside, it is a spherically radial vector of magnitude $E = Q/(4\pi\epsilon r^2)$, rising to a maximum value just off the surface of $Q/(4\pi\epsilon R^2)$ newtons per coulomb or volts per meter. The surface value of the gradient is $Q/(8\pi\epsilon R^2)$, one half of the adjacent external value. The flux density vector D lines per square meter is ϵ times the gradient. Outside the sphere, the absolute potential is $V = Q/(4\pi\epsilon r)$ volts, rising to a maximum of $Q/(4\pi\epsilon R)$ at the surface. Throughout the volume of the sphere the potential remains constant at the surface value, as no difference of potential can exist between points lying in a region of no electric field.

Ordinarily a condenser is presumed to be composed of two conductors and an intervening dielectric. In this case, let the sphere conductor have a charge $+Q$. Then $-Q$ is to be found upon a concentric sphere of infinite radius. The difference of potential by equation 1.58 is

$$\Delta V = V_R - V_\infty = \frac{Q}{4\pi\epsilon}\left(\frac{1}{R} - \frac{1}{\infty}\right) = \frac{Q}{4\pi\epsilon R} \quad \text{volts} \tag{2.1}$$

Hence the capacitance of the sphere is

$$\text{Capacitance } C = \frac{Q}{\Delta V} = 4\pi\epsilon R \quad \text{farads} \tag{2.2}$$

The capacitance of a sphere is the product of 4π, the dielectric constant of the external medium, and the sphere radius.

The energy stored in the electric field is given by equation 1.79.

$$\text{Energy } W_e = \tfrac{1}{2}Q \cdot \Delta V = \frac{Q^2}{8\pi\epsilon R} \quad \text{joules} \tag{2.3}$$

Charge Q placed on a sphere of smaller radius would be at a higher potential and would store more energy in the surrounding field.

2.2. Concentric Conducting Spheres Oppositely Charged. If the terminals of a battery of potential difference ΔV are connected to two concentric spheres of radii r_1 and r_2 separated by dielectric ϵ (see Fig. 2.2), $+Q$ and $-Q$ coulombs will flow onto the two conductors. The exact relation is given by $Q = C\,\Delta V$, where C is the capacitance of the electric system. Assuming a given charge Q, let us compute the potential difference and capacitance.

Owing to symmetry, the equipotential surfaces are concentric spheres and the flux lines are spherically radial, extending from the outside surface of the inner sphere to the inside surface of the outer sphere. In this region (2) we may apply the intensity and potential equations of the single point charge.

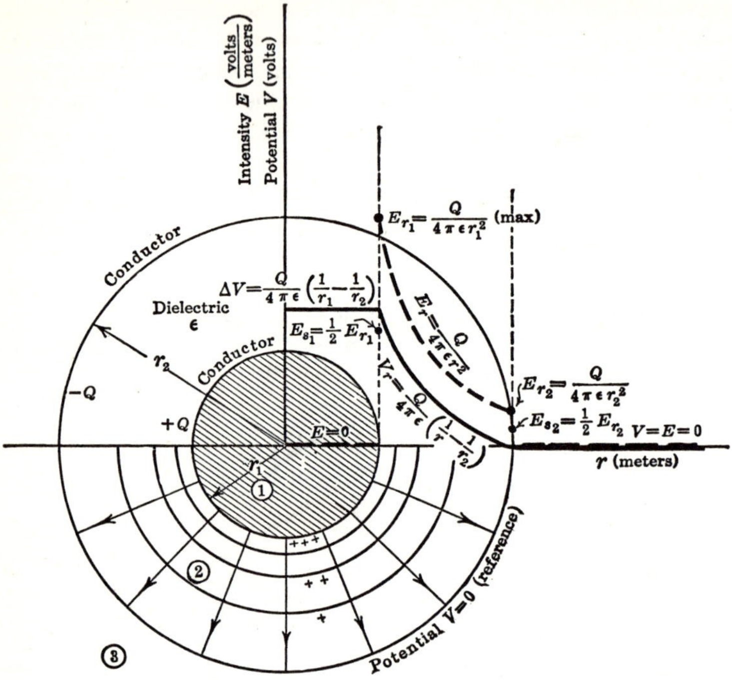

FIG. 2.2. Concentric spheres—intensity and potential.

There are five intensities to be obtained:

$E = 0$ inside of the outer surface of the inner sphere. (2.4)

$E = \dfrac{Q}{8\pi\epsilon r_1^2}$ on the outer surface of the inner sphere.

$E = \dfrac{Q}{4\pi\epsilon r^2}$ between spheres, varying from a maximum of $E = \dfrac{Q}{4\pi\epsilon r_1^2}$ at inner edge of dielectric, to a minimum of $E = \dfrac{Q}{4\pi\epsilon r_2^2}$ at outer edge of dielectric.

$E = \dfrac{Q}{8\pi\epsilon r_2^2}$ on the inner surface of the outer sphere.

$E = 0$ outside of the inner surface of the outer sphere.

The maximum gradient in the dielectric occurs at the inner dielectric edge and is equal to $E = Q/(4\pi\epsilon r_1^2)$. This value must be not greater than the dielectric strength given in Table 1.5 for the dielectric in question. Since there is no field in region 3, we may choose to call the potential of the outer sphere zero as a reference surface. The potential at any point (radius r) in region 2 is then the difference of potential between that point and the outer sphere as given by equation 1.58 for a single point charge field.

$$V_r - 0 = \frac{Q}{4\pi\epsilon}\left(\frac{1}{r} - \frac{1}{r_2}\right) \quad \text{volts} \tag{2.5}$$

The potential difference between the two spheres is

$$\Delta V = V_1 - V_2 = V_1 - 0 = \frac{Q}{4\pi\epsilon}\left(\frac{1}{r_1} - \frac{1}{r_2}\right) \quad \text{volts} \tag{2.6}$$

The capacitance of this condenser then is

$$C = \frac{Q}{\Delta V} = \frac{4\pi\epsilon}{\left(\dfrac{1}{r_1} - \dfrac{1}{r_2}\right)} \quad \text{farads} \tag{2.7}$$

2.3. Circular Ring of Charge. Arcing Rings. Let us arrange Q charges uniformly spaced in the form of a circular ring, as shown in Figs. 2.3*a* and 2.3*b*. The field possesses axial as well as plane symmetry. The center of the ring O is a singular point in addition to serving as the center of gravity. The lower part of Fig. 2.3*b* is a sketch of the field and equipotentials. Far out, the field is radial about the center of gravity, and the equipotentials are spheres.

The intensity may be computed only on the axis by the simple mathematical tools at hand. The upper part of Fig. 2.3*b* shows the variation in the axial intensity. The magnitude is computed with the aid of Fig. 2.3*a*. A small part of the ring charge dQ causes an intensity at point P of $dE = dQ/(4\pi\epsilon s^2)$, directed along line s. We need consider only the axial component $dE_a = dQ \cos\alpha/(4\pi\epsilon s^2)$, inasmuch as the sum of the components normal to the axis dE_R is zero when all the elementary charges of the ring are taken into account. The result may be stated at once:

$$\text{Axial intensity } E_a = \frac{Q}{4\pi\epsilon s^2}\cos\alpha = \frac{Q}{4\pi\epsilon}\frac{r}{s^3}$$

$$= \frac{Q}{4\pi\epsilon}\frac{r}{(r^2 + R^2)^{3/2}} \quad \frac{\text{volts}}{\text{meter}} \tag{2.8}$$

The difference in potential between two points on the axis is

$$(V_A - V_B)_{\text{axis}} = \frac{Q}{4\pi\epsilon}\left(\frac{1}{\sqrt{r_A^2 + R^2}} - \frac{1}{\sqrt{r_B^2 + R^2}}\right) \tag{2.9}$$

High-tension power-line string insulators frequently are protected by the addition of "arcing rings" so arranged that abnormally high

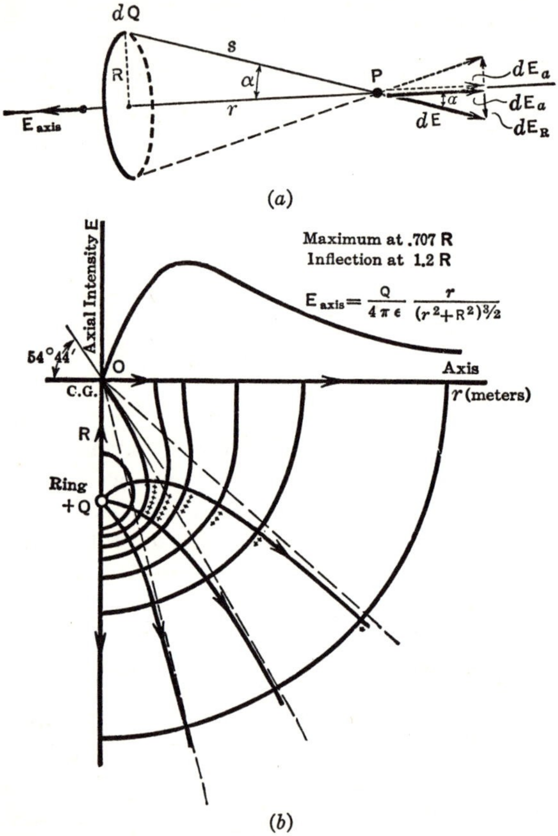

FIG. 2.3. Circular ring of charge. (*a*) Axial intensity. (*b*) Shape of field.

potentials, caused usually by lightning transients, may be relieved by flash-overs in the air from ring to ring rather than over the surface of the porcelain insulating elements. A discharge over the latter path may well break the porcelain and cause an expensive shutdown of the line for repairs. The arcing rings frequently are metallic coaxial circles placed at the ends of each insulator string.

2.4. Circular Disk of Charge. It is assumed that $+Q$ charges are spread *uniformly* over a circular disk of radius R with a surface density $\rho_s = Q/(\pi R^2)$ coulombs per square meter. The lower part of Fig. 2.4*b* is a sketch of the field drawn so that equal quantities of flux emerge from equal areas on the disk and arrive at large distances at the proper solid angle. Point O is the center of gravity. The field has axial and plane symmetry.

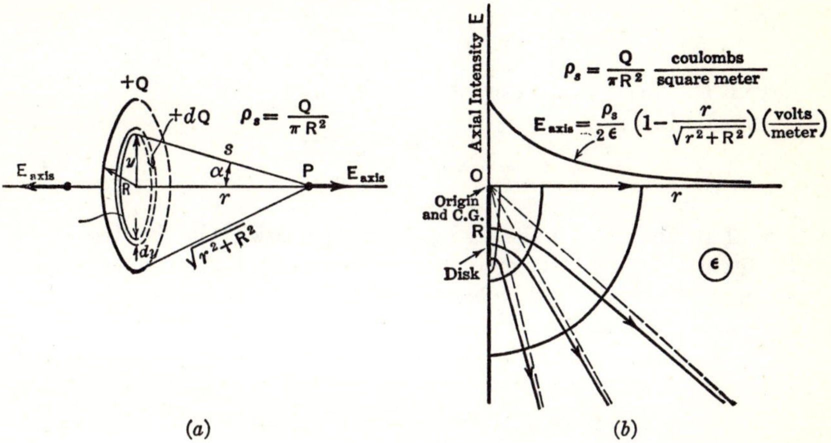

FIG. 2.4. Circular disk. (*a*) Axial intensity. (*b*) Shape of field.

We shall obtain only the axial intensity as we shall use the results of the ring considered in the preceding article. Consider an elementary circular ring of radius y and thickness dy, charged to surface density $\rho_s = Q/(\pi R^2)$, as shown in Fig. 2.4*a*. Its axial intensity at P is given by equation 2.8:

$$dE_P = \frac{dQ}{4\pi\epsilon s^2}\cos\alpha$$

but

$$dQ = 2\pi y\,dy\,\rho_s$$

Also

$$\frac{y}{s} = \sin\alpha$$

$$\frac{r}{s} = \cos\alpha$$

$$y = r\tan\alpha$$

$$dy = r\sec^2\alpha\,d\alpha = \frac{r\,d\alpha}{\cos^2\alpha}$$

Hence

$$dE_P = \frac{2\pi y\, dy \rho_s}{4\pi\epsilon s^2}\frac{r}{s} = \frac{\rho_s}{2\epsilon}\sin\alpha\, d\alpha$$

$$E_{\text{axial}} = \frac{\rho_s}{2\epsilon}\int \sin\alpha\, d\alpha = \frac{\rho_s}{2\epsilon}\Big[-\cos\alpha\Big]_{\alpha=0}^{\alpha=\cos^{-1}\frac{r}{\sqrt{r^2+R^2}}}$$

$$= \frac{\rho_s}{2\epsilon}\left(1 - \frac{r}{\sqrt{r^2+R^2}}\right) \quad \frac{\text{volts}}{\text{meter}} \tag{2.10}$$

Just off the disk surface the intensity is

$$E_{\text{off disk}} = \frac{\rho_s}{2\epsilon} \quad \frac{\text{volts}}{\text{meter}} \tag{2.11}$$

oppositely directed on the two sides of the disk. On the disk itself the intensity is zero. It should be noted here that this is not the case of a charged metallic disk, which has charge over both sides that is not uniformly distributed.

The absolute potential along the axis is

$$V_{\text{axis}} = \frac{\rho_s}{2\epsilon}\left(\sqrt{r^2+R^2} - r\right) \tag{2.12}$$

2.5. Field of Two Parallel Conducting Plates. Let us bring close together two conducting plates whose inner sides are flat and parallel and whose inner areas (A each) are very large compared with the separation s (Fig. 2.5). If the plates are charged to $+Q$ and $-Q$ by connection to a battery, practically all the charge is distributed over the inner surfaces as a result of induction, leaving the outer areas very sparsely charged.

In the region between the plates the equipotential surfaces must be planes. Consequently the flux lines which are normal run parallel from plate to plate throughout this region, and a *uniform* field is said to exist. In a uniform field the flux lines are parallel, the flux density or intensity has the same magnitude and direction throughout, and the equipotentials are parallel planes equally spaced for equal increments of potential. Since the flux lines are uniformly dense in the inner region, the charge density ρ_s must be uniformly dense over the inner conductor surfaces. To a high degree of accuracy we may say that

$$\rho_s = \frac{Q}{A} = D = \epsilon E \quad \frac{\text{coulombs}}{\text{square meter}} \tag{2.13}$$

where E is the intensity in the dielectric. On the surface the intensity is as usual of half value. See Fig. 2.5.

The absolute potential of the mid-plane is zero. The difference of potential between the two conductors is obtained by taking the product of the intensity in the dielectric by the separation s, inasmuch

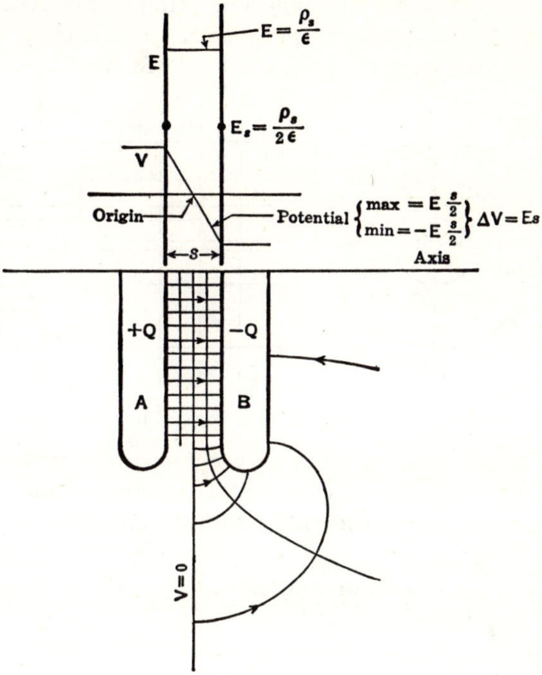

FIG. 2.5. Field of two parallel plates.

as the former is constant. Hence the potential varies as a straight line function between the plates, as shown in Fig. 2.5.

$$\Delta V = V_A - V_B = Es$$

$$= \frac{\rho_s s}{\epsilon} = \frac{Ds}{\epsilon}$$

$$= \frac{Q}{A\epsilon} s \quad \text{volts} \tag{2.14}$$

The capacitance of this condenser is

$$C = \frac{Q}{\Delta V} = \frac{A\epsilon}{s} \quad \text{farads} \tag{2.15}$$

The preceding discussion has neglected the effects of the edges and the backs of the plates where the so-called fringing flux exists. The influence of these effects is reduced practically to negligibility by the assumption originally made that A is very large compared with s.

2.6. Field of an Infinite Straight Line of Charge. Suppose electric charges are placed in an infinitely long straight line with a *uniform* linear density of $+\rho_l$ coulombs per meter. There are $\psi = \rho_l$ lines of flux emerging from each meter of the line. Symmetry demands that

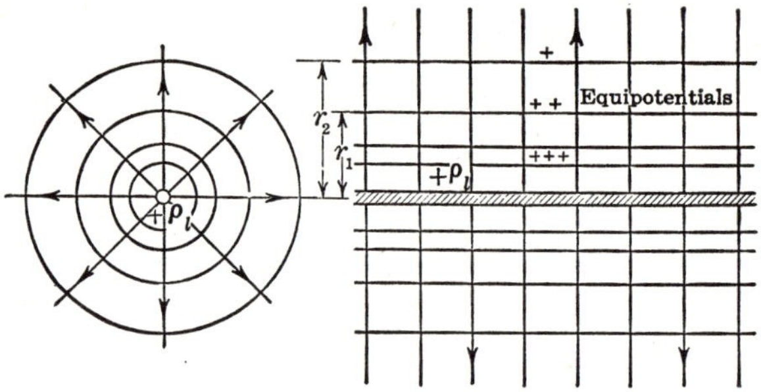

FIG. 2.6. Field of an infinite straight line of charge.

the lines of flux extend outward in a cylindrically radial manner as shown in Fig. 2.6. The lines are straight and of uniform normal density over any cylinder concentric with the charged line. At radius r the density is

$$D = \frac{\rho_l}{2\pi r} \quad \frac{\text{lines or coulombs}}{\text{square meter}} \qquad \text{(Radial vector)} \tag{2.16}$$

and the intensity is

$$E = \frac{\rho_l}{2\pi\epsilon r} \quad \frac{\text{newtons}}{\text{coulomb}} \text{ or } \frac{\text{volts}}{\text{meter}} \qquad \text{(Radial vector)} \tag{2.17}$$

The equipotential surfaces are concentric cylinders. Applying the general defining equation for potential difference (equation 1.54), the potential difference between any two cylinders of radii r_1 and r_2 is

$$V_1 - V_2 = \int (-E)\, ds \cos \alpha = \int (-E)\, dr$$

$$= \int_{r=r_2}^{r=r_1} \left(-\frac{\rho_l}{2\pi\epsilon r} \right) dr = +\frac{\rho_l}{2\pi\epsilon} \ln \frac{r_2}{r_1} \quad \text{volts} \tag{2.18}$$

in which the form $\ln x$ is equivalent to $\log_e x$. The absolute potential V_1 at any point r_1 proves to be infinite, as may be seen by choosing

$V_2 = 0$ at $r_2 = \infty$. Therefore it is possible to speak only of potential differences in this field.

This problem is the first of those that are called *two dimensional.* In fields of this sort one direction (here the axial) finds no change in the field, either in magnitude or direction.

2.7. Concentric Metallic Cylinders. The basis for the solution of this case is the infinitely long straight uniformly charged line of ρ

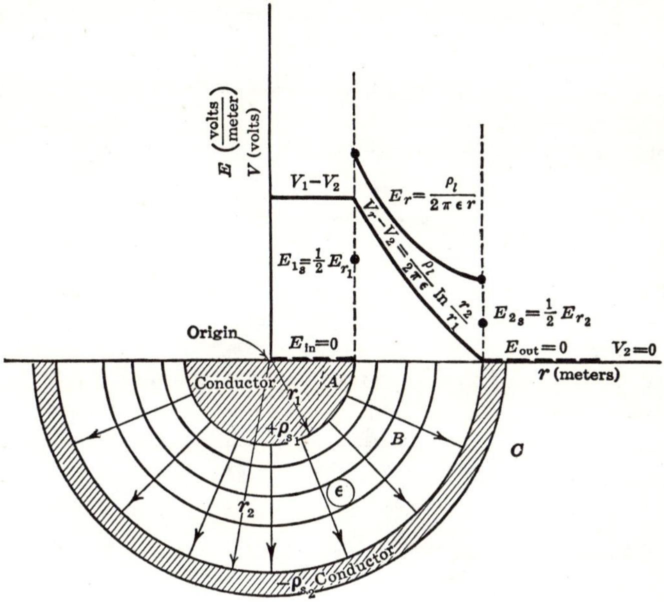

FIG. 2.7. Field of concentric cylinders.

charges per meter given in Art. 2.6. There it was shown that the equipotential surfaces are concentric cylinders. Concentric cylinders which are metallic necessarily form equipotential surfaces, and the field between them is the same as that formed by an infinite line placed at the center. The charges, however, distribute themselves over the outer surface of A and the inner surface of C (Fig. 2.7) with a surface density ρ_s coulombs per square meter sufficient to form an equivalent line density of ρ_l charges per meter.

$$\rho_l = 2\pi r_1 \rho_{s1} = 2\pi r_2 \rho_{s2}$$

Then the intensity or gradient in region B is, by equation 2.17,

$$E_p = \frac{\rho_l}{2\pi\epsilon r} \tag{2.19}$$

reaching a maximum in region B just off surface A of

$$E_{p\,\max} = \frac{\rho_l}{2\pi\epsilon r_1} \tag{2.20}$$

Regions A and C are free from any field. The surface intensities are just one half of the adjacent medium values.

Using surface 2 as a potential reference ($V_2 = 0$), the difference in potential between it and any field point inside is, by equation 2.18,

$$V_p - V_2 = \frac{\rho_l}{2\pi\epsilon} \ln \frac{r_2}{r} \tag{2.21}$$

The total potential difference between the two cylinders is

$$V_1 - V_2 = \frac{\rho_l}{2\pi\epsilon} \ln \frac{r_2}{r_1} \tag{2.22}$$

Any two points, taken either in A or C, can have no potential difference in the absence of any field there. Figure 2.7 gives the results in graphic form.

The capacitance of this condenser for s meters of length is

$$C = \frac{Q}{\Delta V} = \frac{\rho_l s}{\dfrac{\rho_l}{2\pi\epsilon} \ln \dfrac{r_2}{r_1}} = \frac{2\pi\epsilon s}{\ln \dfrac{r_2}{r_1}} \quad \text{farads} \tag{2.23}$$

2.8. Concentric Conductor Cable. A concentric conductor cable is built of a cylindrical copper core of solid or stranded wire, surrounded by a dielectric over which a concentric lead sheath is placed. For low voltages, the dielectric is often made of rubber; for higher voltages, the dielectric is made of a high-grade manila paper wrapped on tightly in a spiral to sufficient thickness. The paper is impregnated with viscous oils and oil-resin mixtures, which are supposed to fill up any air holes and improve the dielectric strength of the cable.

Cables for 66,000 and 132,000 volts use a cored-center conductor, in which oxygen-free oil is maintained under pressure, thus helping to eliminate air bubbles and at the same time reduce the tendency to form oxidation products.

Cables are primarily for use in cities where local ordinances require that all wiring be placed underground in concrete ducts. Inasmuch as the length of the cable is very great compared with its radial dimensions, the infinite cylinder formulas of the preceding section may be used.

2.9. Concentric Conductor Cable with Graded Insulation. Inasmuch as the inner layers of the dielectric are subjected to potential gradients considerably higher than those existing in the outer layers, it is occasionally desirable to divide the dielectric into concentric cylinders, as shown in Fig. 2.8, composed of different materials. Again assuming $\pm\rho_l$ coulombs per meter of length, we note from the symmetry of the geometry that the flux must be radial. Hence the division between the two dielectric regions B and C is an equipotential surface.

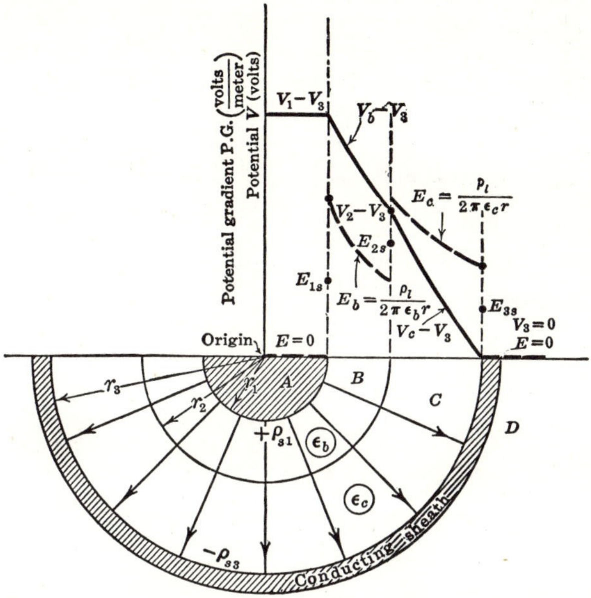

FIG. 2.8. Concentric conductor cable with graded insulation.

The gradients on the two sides of this surface are normal and have magnitudes that vary inversely as the dielectric constants. (See equation 1.89.) Anywhere in the two media the gradients are

$$E_b = \frac{\rho_l}{2\pi\epsilon_b r} \quad (r_1 < r < r_2) \tag{2.24}$$

and

$$E_c = \frac{\rho_l}{2\pi\epsilon_c r} \quad (r_2 < r < r_3) \tag{2.25}$$

These values at the maximum points must not exceed the permissible gradients. (See Table 1.5.)

The potential expression must now be given in two sections. Using the sheath as a reference ($V_3 = 0$), the difference of potential between any point in C and the sheath is

$$V_C - V_3 = \frac{\rho_l}{2\pi\epsilon_c} \ln \frac{r_3}{r} \tag{2.26}$$

The total potential difference between radii r_2 and r_3 is

$$V_2 - V_3 = \frac{\rho_l}{2\pi\epsilon_c} \ln \frac{r_3}{r_2} \tag{2.27}$$

Between any point in B and radius r_2, we have

$$V_B - V_2 = \frac{\rho_l}{2\pi\epsilon_b} \ln \frac{r_2}{r} \tag{2.28}$$

Between radii r_1 and r_2 we find

$$V_1 - V_2 = \frac{\rho_l}{2\pi\epsilon_b} \ln \frac{r_2}{r_1} \tag{2.29}$$

Hence the total potential drop between the conductor and sheath is

$$\begin{aligned} V_1 - V_3 &= (V_1 - V_2) + (V_2 - V_3) \\ &= \frac{\rho_l}{2\pi}\left(\frac{1}{\epsilon_b} \ln \frac{r_2}{r_1} + \frac{1}{\epsilon_c} \ln \frac{r_3}{r_2}\right) \quad \text{volts} \end{aligned} \tag{2.30}$$

By properly proportioning the radii and the dielectric constants, the maximum gradients in the B and C regions may be made the same percentage (if desired) of their own dielectric strengths or permissible gradients. Or it may be desirable to arrange the media so that the largest possible potential difference between the conductor and the sheath may be used. In any case, it ordinarily proves desirable to use the medium with the larger ϵ nearer to the inner conductor.

2.10. Condenser-Type Bushings. Where a high-voltage wire is brought out from a transformer, it is necessary to insulate it from the transformer case by means of a bushing. In particular, the condenser-type bushing is composed of several metallic cylinders placed concentrically about the wire and separated from each other by insulating layers. The successive metallic cylinders are given the same cylindrical area, so that the outer ones are necessarily shorter. See Fig. 2.9*a*.

Any given positive conductor charge induces an equal negative charge on the inside of the adjacent cylinder, sending an equal positive charge to the exterior radius. The process continues through the succeeding layers. Each layer outside the next then has a greater concentration of charges per meter of length but the same value of total charge. In this way the maximum gradients in the successive layers of insulator may be made approximately equal, thus permitting the greatest possible potential difference from conductor to case in a limited space. Figure 2.9*b* shows rather roughly the great increase in

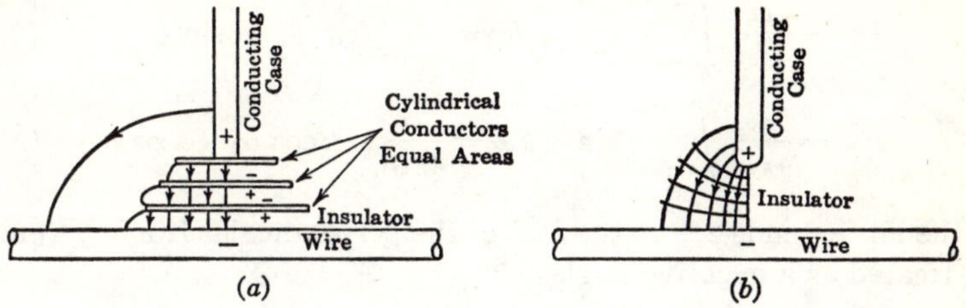

FIG. 2.9. Shape of field. (*a*) With condenser-type bushing. (*b*) Without bushing.

flux density occurring around the conductor case if a condenser bushing is not used. The greater intensity might cause a dielectric breakdown that would be avoided by using the bushing.

2.11. Field of a Finite Straight Line Uniformly Charged—Intensity. Let $+Q$ unit charges be arranged in a straight line of length s with constant line density ρ_l charges per meter, as shown in Fig. 2.10. The intensity at P is perhaps most easily obtained by adding up separately the vertical and horizontal components resulting from all the small elements dx.

$\rho_l\, dx$ is the small charge on that part of the wire distant p from point P. Its intensity at P is $dE = \rho_l\, dx/(4\pi\epsilon p^2)$. The two components are

$$dE_V = \frac{\rho_l\, dx}{4\pi\epsilon p^2} \cos \alpha \tag{2.31}$$

$$dE_H = \frac{\rho_l\, dx}{4\pi\epsilon p^2} \sin \alpha \tag{2.32}$$

But

$$x = r \tan \alpha$$

$$dx = r \sec^2 \alpha \, d\alpha$$

Also

$$p = r \sec \alpha$$

Substituting these values into equations 2.31 and 2.32, we have

$$dE_V = \frac{\rho_l}{4\pi\epsilon r} \cos \alpha \, d\alpha$$

$$dE_H = \frac{\rho_l}{4\pi\epsilon r} \sin \alpha \, d\alpha$$

Integrating:

$$E_V = \frac{\rho_l}{4\pi\epsilon r} \int_{\alpha=\alpha_1}^{\alpha=\alpha_2} \cos \alpha \, d\alpha = \frac{\rho_l}{4\pi\epsilon r} (\sin \alpha_2 - \sin \alpha_1) \tag{2.33}$$

$$E_H = \frac{\rho_l}{4\pi\epsilon r} \int_{\alpha=\alpha_1}^{\alpha=\alpha_2} \sin \alpha \, d\alpha = \frac{\rho_l}{4\pi\epsilon r} (-\cos \alpha_2 + \cos \alpha_1) \tag{2.34}$$

If the line extends across the foot of the perpendicular from P, α_1 must be treated as a negative angle.

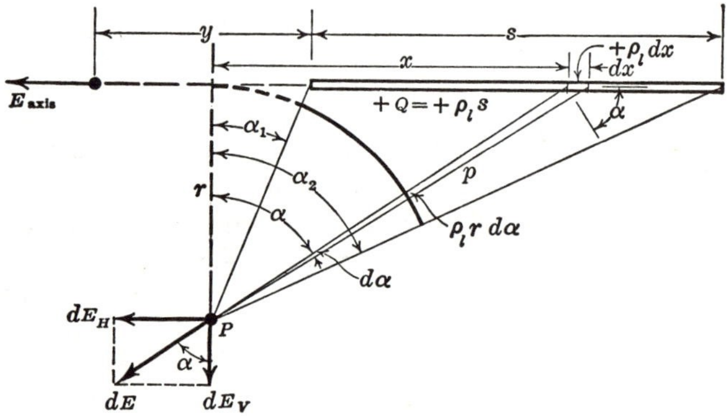

FIG. 2.10. Finite line of charge—intensity.

The infinitesimal intensity at P is

$$dE = \frac{\rho_l}{4\pi\epsilon r} d\alpha = \frac{\rho_l}{4\pi\epsilon r^2} (r \, d\alpha) \tag{2.35}$$

which means that the charge $\rho_l r \, d\alpha$ placed on the circle of radius r in the angle $d\alpha$ produces the same intensity at P as the actual charge $\rho_l \, dx$ on the line. Hence the total charge along the circle in the range $(\alpha_2 - \alpha_1)$ is the equivalent of the whole line charge, and obviously the total intensity has a direction that bisects $(\alpha_2 - \alpha_1)$.

On the axis ($r = 0$), equation 2.33 reduces to zero and equation 2.34 becomes indeterminate. At some axial point beyond the end of the wire, let y equal the distance to the near end of the wire and $y + s$ the distance to the far end. Then

$$\cos \alpha_2 \rightarrow \frac{r}{y+s} \quad \text{and} \quad \cos \alpha_1 \rightarrow \frac{r}{y}$$

and

$$E_{\text{axis}} = E_{H\text{axis}} = \frac{\rho_l}{4\pi\epsilon}\left(\frac{1}{y} - \frac{1}{y+s}\right) \tag{2.36}$$

The Equipotential Surfaces. Let the length of the line be s, and let ρ_l be the number of coulombs per unit length. Then, if we take the

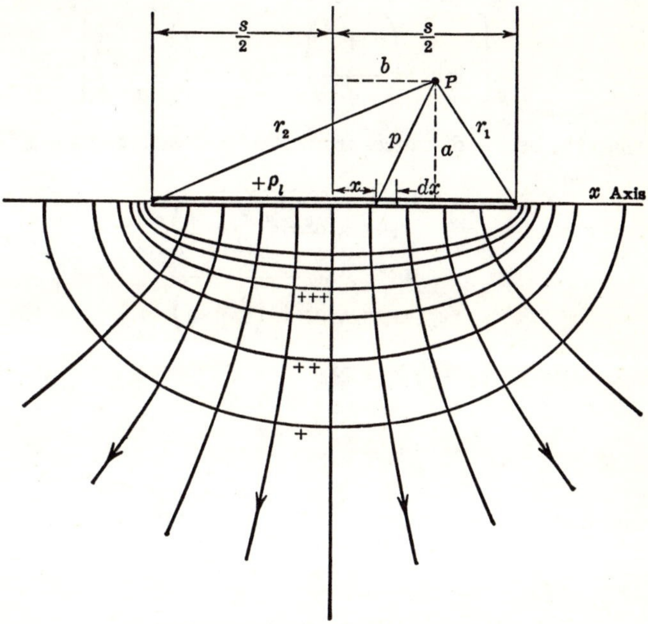

FIG. 2.11. Finite line of charge, shape of field.

line along the x axis in Fig. 2.11 with its center at the origin, the potential at any point $P(b, a)$ due to an element of the line dx is

$$dV = \frac{\rho_l\, dx}{4\pi\epsilon p}$$

and

$$p = \sqrt{a^2 + (b-x)^2}$$

and the absolute potential at P due to the whole line is

$$V = \frac{\rho_l}{4\pi\epsilon}\int_{-s/2}^{s/2} + \frac{dx}{\sqrt{a^2 + (b - x)^2}}$$

$$= -\frac{\rho_l}{4\pi\epsilon}\ln\left[(b - x) + \sqrt{a^2 + (b - x)^2}\right]_{-s/2}^{s/2}$$

$$= \frac{\rho_l}{4\pi\epsilon}\ln\left[\frac{b + \frac{s}{2} + \sqrt{a^2 + \left(b + \frac{s}{2}\right)^2}}{b - \frac{s}{2} + \sqrt{a^2 + \left(b - \frac{s}{2}\right)^2}}\right] \tag{2.37}$$

$$= \frac{\rho_l}{4\pi\epsilon}\left[\sinh^{-1}\left(\frac{b + \frac{s}{2}}{a}\right) - \sinh^{-1}\left(\frac{b - \frac{s}{2}}{a}\right)\right] \tag{2.38}$$

Let r_1 and r_2 be the distances from P to the ends of the line, respectively. Then

$$r_1 = \sqrt{a^2 + \left(\frac{s}{2} - b\right)^2}$$

$$r_2 = \sqrt{a^2 + \left(\frac{s}{2} + b\right)^2}$$

$$r_2^2 - r_1^2 = 2bs$$

Hence

$$b = \frac{r_2^2 - r_1^2}{2s}$$

Substituting the foregoing values in equation 2.37, we obtain

$$V = \frac{\rho_l}{4\pi\epsilon}\ln\left(\frac{\frac{r_2^2 - r_1^2}{2s} + \frac{s}{2} + r_2}{\frac{r_2^2 - r_1^2}{2s} - \frac{s}{2} + r_1}\right)$$

If we multiply by $2s$, and divide the numerator and denominator each by $(r_2 - r_1 + s)$, we get

$$V = \frac{\rho_l}{4\pi\epsilon}\ln\left(\frac{r_1 + r_2 + s}{r_1 + r_2 - s}\right) \tag{2.39}$$

Equation 2.39 shows that V is constant when $(r_1 + r_2)$ is constant, and therefore it determines the equipotential surfaces, which are evidently confocal ellipsoids with their foci at the ends of the line.

Since the lines of flux are always normal to the equipotential surfaces, they form in this case a system of confocal hyperbolas. Furthermore, the direction of the intensity at any point P bisects the angle between r_1 and r_2, inasmuch as this direction is always normal to the ellipsoid through the point.

2.12. Two Parallel Infinite Straight Lines Oppositely Charged. Let ρ_l be the linear charge density, positive on one line and equally negative on the other line, separated a distance $2h$. As shown in Fig. 2.12,

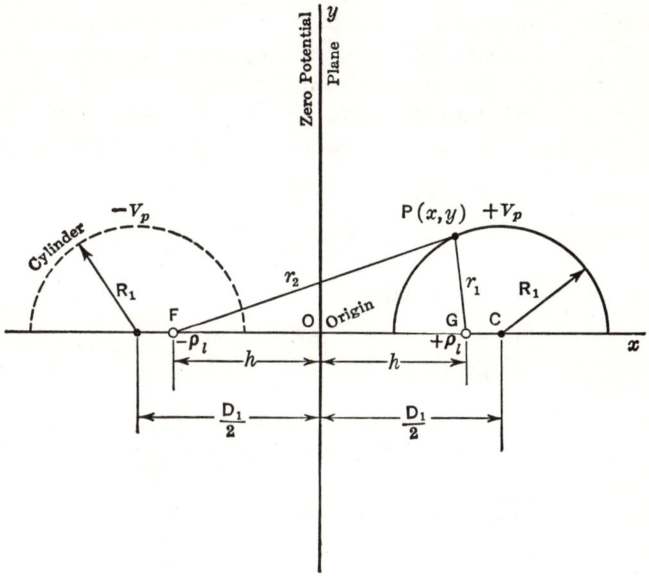

FIG. 2.12. Two line charges—equipotential cylinders.

the plane including the lines is a symmetry and contains straight parallel flux lines, while the normal bisecting plane is a symmetry of zero potential.

The absolute potential V_p at any field point P is given by the difference of potential between P and origin O of potential zero, using equation 2.18.

$$V_p = \frac{(+\rho_l)}{2\pi\epsilon} \ln \frac{h}{r_1} + \frac{(-\rho_l)}{2\pi\epsilon} \ln \frac{h}{r_2}$$

$$V_p = \frac{\rho_l}{2\pi\epsilon} \ln \frac{r_2}{r_1} \quad \text{volts} \qquad \text{(Absolute potential)} \qquad (2.40)$$

Remembering that ρ_l and ϵ are constants, we obtain the equation of an equipotential surface by assigning some value to V_p (either positive or negative). Rewriting equation 2.40:

$$\frac{r_2}{r_1} = e^{\left(\frac{2\pi\epsilon V_p}{\rho_l}\right)} = \text{constant } A, \text{ for any fixed value of } V_p \quad (2.41)$$

Since a constant ratio of the distances from a point to two lines describes a cylinder, we know at once that the equipotential surfaces in this field are cylinders, one series about each line of charge. The cylinders are, however, not concentric. Furthermore, an equal negative potential V_p gives the same-sized cylinder, but surrounding the negative instead of the positive line of charge.

This may be put into the more familiar Cartesian form. Let the coordinates of point P be (x, y) measured from the origin O. Then

$$r_2^2 = (h + x)^2 + y^2$$

$$r_1^2 = (h - x)^2 + y^2$$

But since $r_2^2 = A^2 r_1^2$ from equation 2.41

$$(h + x)^2 + y^2 = A^2[(h - x)^2 + y^2]$$

Rearranging in quadratic form,

$$\left(x - h\frac{A^2 + 1}{A^2 - 1}\right)^2 + (y)^2 = \left(h\frac{2A}{A^2 - 1}\right)^2 \quad (2.42)$$

which is the equation of a circle with

$$\text{Center coordinates } \left(+h\frac{A^2 + 1}{A^2 - 1}, 0\right) \quad \text{or} \quad \frac{D_1}{2} = h\frac{A^2 + 1}{A^2 - 1} \quad (2.43)$$

and

$$\text{Radius equal to } h\frac{2A}{A^2 - 1} = R_1 \quad (2.44)$$

From these last two equations we may easily obtain h in terms of D_1 and R_1 as

$$R_1^2 = h^2\left[\left(\frac{A^2 + 1}{A^2 - 1}\right)^2 - 1\right] = h^2\left[\left(\frac{D_1}{2h}\right)^2 - 1\right] = \left(\frac{D_1}{2}\right)^2 - h^2$$

whence

$$h = \sqrt{\left(\frac{D_1}{2}\right)^2 - R_1^2} \quad (2.45)$$

Also we may see that

$$A = \frac{D_1}{2R_1} + \sqrt{\left(\frac{D_1}{2R_1}\right)^2 - 1} = \frac{r_2}{r_1} \tag{2.46}$$

as from equations 2.43 and 2.44

$$\frac{D_1}{2R_1} = \frac{A^2 + 1}{2A}$$

then

$$\sqrt{\left(\frac{D_1}{2R_1}\right)^2 - 1} = \sqrt{\left(\frac{A^2 + 1}{2A}\right)^2 - 1} = \frac{A^2 - 1}{2A}$$

and

$$\frac{D_1}{2R_1} + \sqrt{\left(\frac{D_1}{2R_1}\right)^2 - 1} = \frac{A^2 + 1}{2A} + \frac{A^2 - 1}{2A} = A = \frac{r_2}{r_1}$$

Finally then, the absolute potential of any point P is

$$V_p = \frac{\rho_l}{2\pi\epsilon} \ln \frac{r_2}{r_1} = \frac{\rho_l}{2\pi\epsilon} \ln \left[\frac{D_1}{2R_1} + \sqrt{\left(\frac{D_1}{2R_1}\right)^2 - 1}\right] \quad \text{volts} \tag{2.47}$$

The difference of potential between equal-sized positive and negative cylinders is

$$\Delta V = 2V_p = \frac{\rho_l}{\pi\epsilon} \ln \left[\frac{D_1}{2R_1} + \sqrt{\left(\frac{D_1}{2R_1}\right)^2 - 1}\right] \quad \text{volts} \tag{2.48}$$

$$= \frac{\rho_l}{\pi\epsilon} \cosh^{-1} \left(\frac{D_1}{2R_1}\right) \tag{2.49}$$

Flux Line Shape. The flux lines, which are of course orthogonal to the equipotential cylinders, are arcs of circles beginning at $+\rho_l$ and ending at $-\rho_l$. In Fig. 2.13 let P be any field point. Pass radial planes through P and the two lines of charge making angles α_1 and α_2 radians as shown. Pass any surface from P to P_0, a point in the plane of the lines of charge. Now for a moment consider the real flux field divided into its two component fields, each associated with only one line of charge, for 1 meter depth along the lines. The flux crossing surface PP_0 to the left is

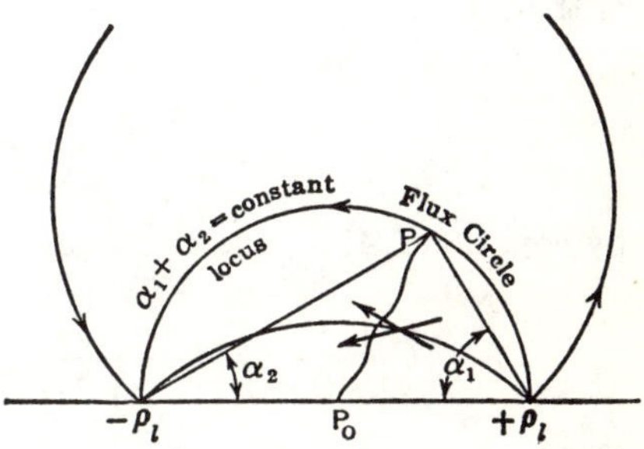

FIG. 2.13. Two line charges—flux circles.

$$\psi = \rho_l\left(\frac{\alpha_1}{2\pi}\right) + \rho_l\left(\frac{\alpha_2}{2\pi}\right) \quad \text{lines of flux} \tag{2.50}$$

Holding this sum constant as P is moved from point to point in space causes P to describe a locus of constant flux crossing surface PP_0. In this case

$$\alpha_1 + \alpha_2 = \text{constant}$$

But this is merely the condition for P to describe a circular arc including the lines of charge. Consequently the real flux lines are circular arcs through the lines of charge.

2.13. Two Equally Charged Parallel Cylindrical Infinite Wires. Let their center distance be D and their radii R. If a battery or generator of potential difference ΔV is connected to these wires, a current flows until the wires are charged oppositely with ρ_l coulombs per meter spread over the surface and the potential difference between the wires is equal to that of the battery. This case fits exactly the geometry of the preceding article, so that we may carry over that solution. Then

$$\Delta V = V_1 - V_2 = \frac{\rho_l}{\pi\epsilon}\ln\left[\frac{D}{2R} + \sqrt{\left(\frac{D}{2R}\right)^2 - 1}\right] \quad \text{volts} \tag{2.51}$$

The charges are spread over the cylindrical surfaces in such a way as to make them equipotential. They are distributed more densely over the nearer parts of the conductors. However, at field points external to the conductors, the field appears exactly as though the flux lines had emerged from pure geometric lines of charge spaced $2h$ meters apart. See Fig. 2.14. These lines may be called the electrical centers of the conductors and are closer together than the wire centers.

$$h = \sqrt{\left(\frac{D}{2}\right)^2 - R^2} \tag{2.52}$$

by equation 2.45. The capacitance of this condenser, for s meters, is

$$C = \frac{Q}{\Delta V} = \frac{\rho_l s}{\dfrac{\rho_l}{\pi\epsilon}\ln\left[\dfrac{D}{2R} + \sqrt{\left(\dfrac{D}{2R}\right)^2 - 1}\right]}$$

$$= \frac{\pi\epsilon s}{\ln\left[\dfrac{D}{2R} + \sqrt{\left(\dfrac{D}{2R}\right)^2 - 1}\right]} \quad \text{farads} \tag{2.53}$$

If $D/(2R) > 10$, the capacitance, to 0.1 per cent accuracy or better, is

$$C \cong \frac{\pi \epsilon s}{\ln\left(\dfrac{D}{R}\right)} \tag{2.54}$$

The potential gradient has its largest values along the plane of the wire centers. In this plane only we may write $r_1 + r_2 = 2h$. The

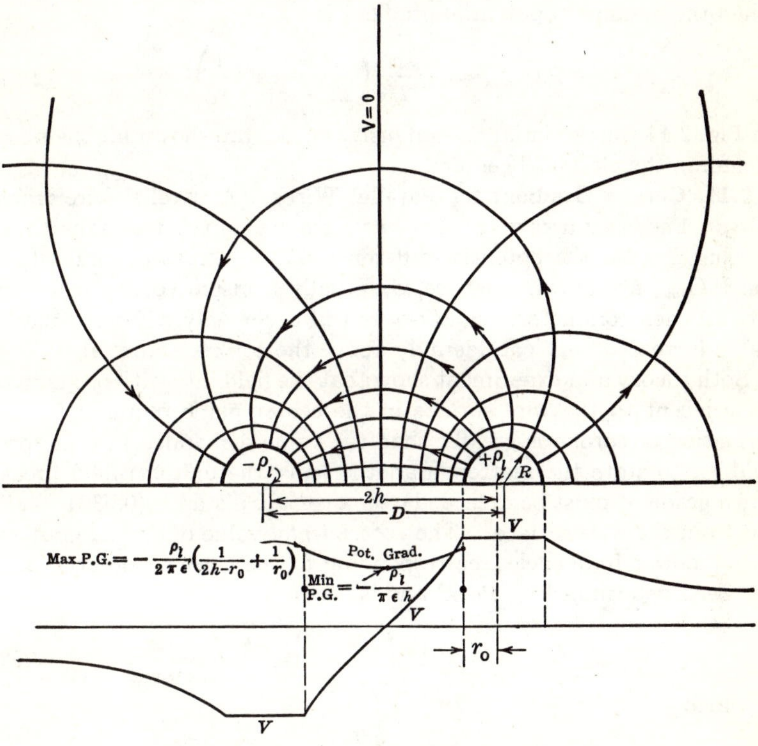

FIG. 2.14. Electric field of two parallel wires.

absolute potential in this plane at any point $r = r_1$ from the positive electrical center is

$$V_p = \frac{\rho_l}{2\pi\epsilon} \ln \frac{2h - r}{r} \tag{2.55}$$

Then the potential gradient is

$$-E_r = \text{P.G.} = \frac{dV_p}{dr} = -\frac{\rho_l}{2\pi\epsilon}\left(\frac{1}{2h - r} + \frac{1}{r}\right) \tag{2.56}$$

a vector in the plane of electrical centers and normal to them. This has a minimum value $P.G._{min} = -\rho_l/(\pi\epsilon h)$ when $h = r$ at the midpoint and reaches its maximum value in the medium just off the surface of the conductor. $D - 2R$ is the distance between the conductor surfaces. Hence at the point of maximum gradient,

$$r = r_0 = h - \frac{D - 2R}{2}$$

Then the maximum potential gradient is

$$P.G._{max} = -\frac{\rho_l}{2\pi\epsilon}\left(\frac{1}{2h - r_0} + \frac{1}{r_0}\right) \tag{2.57}$$

In Fig. 2.14 the potential and intensity curves are shown for the plane including the electrical centers.

2.14. Corona Gradient for Parallel Wires. A parallel wire transmission line in air uses wires of radius R small compared with the center spacing D. As the potential difference ΔV is increased, and with it the $P.G._{max}$ at the wire surface, eventually a visual corona will form around each conductor. Spark-over will occur only when the potential is increased very considerably above the corona potential.

Both theory and experiment show that the field intensity or gradient E_c adjacent to the wire surface in the center plane required for the formation of corona is greater than the value demanded in a uniform field, or to state the matter in another way the uniform field breakdown gradient must be exceeded over a certain distance $(0.0301\sqrt{\delta R})$ out from the wire surface. The *experimental* value of corona gradient E_c for power frequencies, corresponding to the corona potential ΔV_c, has been determined by Peek[1] and is

$$E_c = 2.98 \times 10^6\,\delta\left(1 + \frac{0.0301}{\sqrt{\delta R}}\right)\ \frac{\text{volts}}{\text{meter}} \tag{2.58}$$

in which

$$\delta = \frac{3.92b}{273 + T}$$

is the relative air density at barometer reading b in centimeters of mercury and at temperature $T°$ centigrade. At $T = 25°\ C$ and $b = 76$ centimeters of mercury, $\delta = 1.0$. E_c is plotted against $1/\sqrt{R}$ in Fig. 2.15 for $\delta = 1.0$.

With E_c set equal to $P.G._{max}$, the corona potential ΔV_c can be computed.

[1] Peek, *Dielectric Phenomena in High Voltage Engineering*, McGraw-Hill, 1929, p. 64.

2.15. Corona Gradient for Concentric Cylinders. The experimental value of corona gradient adjacent to the inner cylinder with air dielectric is given by Peek as

$$E_c = 3.1 \times 10^6 \delta \left(1 + \frac{0.0308}{\sqrt{\delta R}}\right) \frac{\text{volts}}{\text{meter}} \tag{2.59}$$

For this geometry the thickness of the corona layer is $0.0308 \sqrt{\delta R}$. E_c is plotted against $1/\sqrt{R}$ in Fig. 2.15.

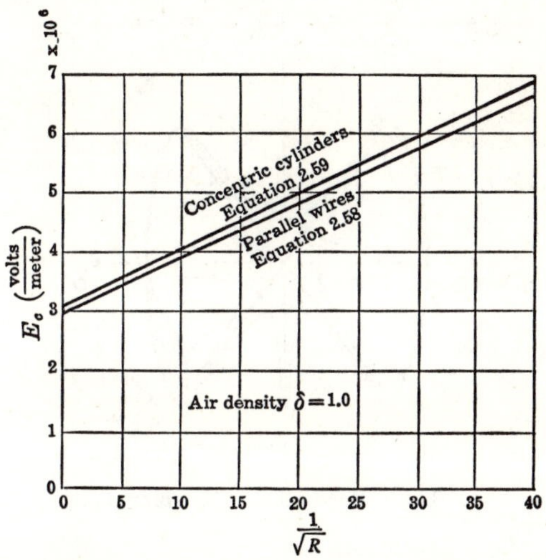

FIG. 2.15. Off-surface corona gradient.

2.16. The Singular Point of $+Q$ and $+Q$. The field in the neighborhood of the singular point, O in Fig. 2.16, between two equal positive point charges is most readily obtained by studying the potential expression. Taking the origin at O, let P be any neighboring point, distant r_1 and r_2 from the two charges. The difference of potential between P and O is

$$V_P - V_0 = \frac{Q}{4\pi\epsilon}\left(\frac{1}{r_2} - \frac{1}{h}\right) + \frac{Q}{4\pi\epsilon}\left(\frac{1}{r_1} - \frac{1}{h}\right) = \frac{Q}{4\pi\epsilon}\left(\frac{1}{r_2} + \frac{1}{r_1} - \frac{2}{h}\right)$$

Let

$$\frac{4\pi\epsilon(V_P - V_0)}{Q} = A = \frac{1}{r_2} + \frac{1}{r_1} - \frac{2}{h}$$

$$= [(h + x)^2 + y^2]^{-\frac{1}{2}} + [(h - x)^2 + y^2]^{-\frac{1}{2}} - \frac{2}{h} \tag{2.60}$$

For points not far from the singular point

$$h + x \gg y$$

$$h - x \gg y$$

and we may expand equation 2.60 in powers of y by use of the binomial theorem.

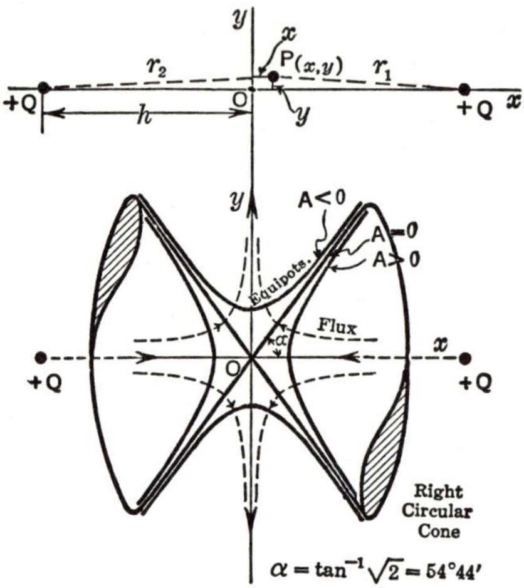

FIG. 2.16. Field about a singular point.

For $y < a$,

$$(a + y)^n = a^n + na^{n-1}y + \frac{n(n-1)}{\underline{\lfloor 2}} a^{n-2}y^2 + \frac{n(n-1)(n-2)}{\underline{\lfloor 3}} a^{n-3}y^3 +++$$

Hence

$$[(h + x)^2 + y^2]^{-\frac{1}{2}} = (h + x)^{-1} - \tfrac{1}{2}(h + x)^{-3}y^2 +-+$$

$$[(h - x)^2 + y^2]^{-\frac{1}{2}} = (h - x)^{-1} - \tfrac{1}{2}(h - x)^{-3}y^2 +-+$$

Furthermore

$$(h + x)^{-1} = h^{-1} - h^{-2}x + h^{-3}x^2 -+-$$

$$(h - x)^{-1} = h^{-1} + h^{-2}x + h^{-3}x^2 +++$$

$$(h + x)^{-3} = h^{-3} - 3h^{-4}x +-+$$

$$(h - x)^{-3} = h^{-3} + 3h^{-4}x +++$$

Then

$$(h + x)^{-1} + (h - x)^{-1} = 2h^{-1} + 2h^{-3}x^2 ++$$

$$-\frac{y^2}{2}\,[(h+x)^{-3}+(h-x)^{-3}] = -\frac{y^2}{2}\,(2h^{-3}) = -y^2h^{-3} ++$$

$$A = 2h^{-3}x^2 - h^{-3}y^2$$

$$Ah^3 = 2x^2 - y^2 \tag{2.61}$$

When $V_p - V_0 > 0$, $A > 0$. Hyperboloid of revolution of two sheets about line of charges.

$V_p - V_0 = 0$, $A = 0$. Right circular cone.

$V_p - V_0 < 0$, $A < 0$. Hyperboloid of revolution of one sheet about line of charges.

For the cone

$$A = 0$$

and

$$y^2 = 2x^2$$

$$y = \pm\sqrt{2}x$$

$$\tan\alpha = \frac{y}{x} = \pm\sqrt{2}$$

$$\alpha = 54^\circ\,44' \tag{2.62}$$

These equipotential surfaces are shown in Fig. 2.16, and the flux lines are shown dotted.

2.17. The Singular Line of $+\rho_l$ and $+\rho_l$. The field about a singular line lying halfway between two equal line charges is easily obtained. See Fig. 2.17. The potential difference between P and O is

$$V_p - V_0 = \frac{\rho_l}{2\pi\epsilon}\ln\frac{h}{r_2} + \frac{\rho_l}{2\pi\epsilon}\ln\frac{h}{r_1} = \frac{\rho_l}{2\pi\epsilon}\ln\frac{h^2}{r_1r_2} = \frac{\rho_l}{\pi\epsilon}\ln\frac{h^4}{r_1{}^2r_2{}^2}$$

$$r_1{}^2 = (h-x)^2 + y^2$$

$$r_2{}^2 = (h+x)^2 + y^2$$

$$\frac{h^4}{r_1{}^2r_2{}^2} = A = e^{\frac{4\pi\epsilon(V_p - V_0)}{\rho_l}}$$

$$h^4 = A[(h-x)^2 + y^2][(h+x)^2 + y^2]$$

$$h^4 = A[y^4 + 2h^2y^2 + h^4 + x^4 - 2h^2x^2 + 2x^2y^2]$$

or

$$1 = A\left[\underset{①}{\frac{y^4}{h^4}} + \underset{②}{\frac{2y^2}{h^2}} + \underset{③}{1} + \underset{④}{\frac{x^4}{h^4}} - \underset{⑤}{\frac{2x^2}{h^2}} + \underset{⑥}{\frac{2x^2y^2}{h^4}}\right]$$

In the neighborhood of the singular line x and y are of the same order of magnitude, but each is small compared with h. Consequently terms ①, ④, and ⑥ may be dropped as negligible compared with the others. Then

$$h^2 = A(h^2 - 2x^2 + 2y^2)$$

$$x^2 - y^2 = \frac{h^2}{2}\frac{(A-1)}{A} \tag{2.63}$$

When $V_p - V_0 > 0$, $A > 1$. Hyperbolic cylinder.

$V_p - V_0 = 0$, $A = 1$. Two planes at 90°.

$V_p - V_0 < 0$, $A < 1$. Hyperbolic cylinder.

When $A = 1$, $y = \pm x$

$$\tan\alpha = \frac{y}{x} = \pm 1$$

$$\alpha = 45° \tag{2.64}$$

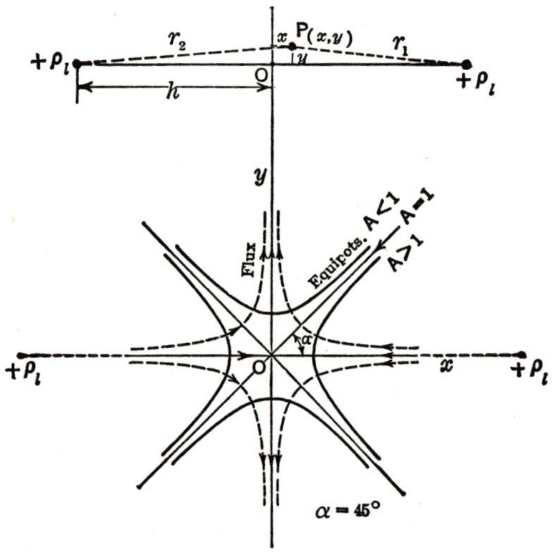

FIG. 2.17. Field about a singular line. (Equipotential surface crosses twice. See also Fig. 1.21.)

These equipotential surfaces are shown in Fig. 2.17, and the flux lines are shown dotted.

PROBLEMS

1. Assume that the metallic sphere in Fig. 2.1 has a radius of 5 cm, is charged with 10^{-7} coulomb, and is surrounded by an insulating medium having a relative dielectric constant equal to 4. Find:

(*a*) The charge density ρ_s.
(*b*) The electric intensities inside, on, and just off the surface.

(*c*) The flux density at a radius of 20 cm.
(*d*) The absolute potential of the sphere.
(*e*) The difference in potential between the center of the sphere and a point at a radius of 20 cm.
(*f*) The mechanical pressure on the surface.
(*g*) The capacitance of the sphere.
(*h*) The energy stored in the electric field.
(*i*) The energy density at a point 20 cm from the center.

2. Integrate the energy density in Prob. 1 throughout the field volume, and compare with the value obtained in Prob. 1*h*.

3. The earth has a radius of 6.37×10^6 meters. What is its capacitance in microfarads? If the maximum permissible electric gradient is 30,000 volts per centimeter, find the earth's maximum possible absolute potential. (The actual gradient at the earth's surface is about 1 per cent of the above value.)

4. Two concentric metallic spheres of radii $r_1 = 1$ cm and $r_2 = 5$ cm are charged to a potential difference of 6000 volts. The inner sphere has the higher potential. The relative dielectric constant is equal to 4. Find:

(*a*) The five values of potential gradient mentioned in Art. 2.2.
(*b*) The flux density.
(*c*) The capacitance and energy stored.
(*d*) The equipotential surface having the value +3000 volts.

5. A hollow conducting sphere of radius 10 cm carries a positive charge of 0.1 microcoulomb. Then another conducting sphere of radius 5 cm carrying 0.04 microcoulomb negative charge is placed concentrically. The specific inductive capacity of the medium between the spheres is 3. The charge on the inner sphere must induce an equal but opposite charge on the inside surface of the outer sphere. The medium outside the outer sphere is air. Find:

(*a*) Potential difference between the two spheres.
(*b*) The intensity 20 cm from the center of the system.
(*c*) Flux density midway between the two spheres.
(*d*) The absolute potential of the inner sphere.

6. Move the inner sphere in Prob. 5 to an eccentric position without permitting it to touch the outer sphere. Draw a neat sketch of the lines of induction and equipotential surface both inside and outside the outer sphere.

7. Assuming that the force of attraction of a spherical mass shell of radius R is M/R^2 at a point immediately outside the shell, prove that the attraction of a solid homogeneous mass sphere for unit mass is proportional to the distance from the center for all points inside the mass. Draw the curve of attraction for points inside and outside the mass.

8. Two thin concentric metallic spheres of radii 4 and 5 cm are separated by a dielectric of specific inductive capacity ($\epsilon_r = 2$) and of "dielectric strength" of 150,000 volts per centimeter. Compute the maximum permissible difference of potential in volts. Draw the curves of intensity, flux density, and potential.

9. Two concentric spherical metal shells of radii r_1 and r_2 = constant are separated by a dielectric. The shells are held at a constant difference of potential = ΔV. Find the value for the radius r_1 of the inner shell such that the greatest potential gradient anywhere in the dielectric will be a minimum.

10. Consider a concentric spherical shell of rubber ($\epsilon_r = 3$) with radii 1 and 2 cm. Place $Q = +10^{-7}$ coulomb at the center. Find the potential difference between the inner rubber surface and an external point at radial distance 10 cm.

11. Following Fig. 2.3, compute the difference in potential between two points on the axis distant 5 and 10 cm from the singular point. The charge on the ring is 0.1 microcoulomb at a radius of 10 cm. The dielectric is air. Find the axial distance to the maximum intensity.

12. For Fig. P12, find the difference of potential between points A and B, and find the potential gradient at A.

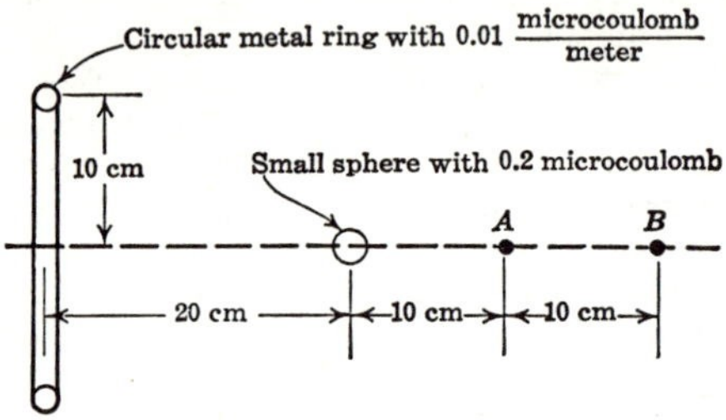

FIG. P12

13. A circular ring, 20-cm radius, possesses a total charge of +1 microcoulomb. At the center a small sphere, radius 0.5 cm, is charged to −0.2 microcoulomb. Find the difference of potential between the singular point and the sphere surface. Make a sketch of the field in a plane through the axis.

14. Two parallel metallic plates, each of 100 sq cm, are separated 0.1 cm by an insulator with a relative dielectric constant of 3 and a "dielectric strength" of 10^5 volts per centimeter. Find the charge density on the plates, the potential difference, the flux density, and the energy stored for the maximum permissible potential gradient. Compute the force on each plate.

15. Two circular plates, 40 cm in diameter, are placed parallel to each other in a vertical position and 0.5 cm apart. A potential difference of 6000 volts is maintained between the plates. One plate is grounded; hence its potential may be taken to be zero. A disk of rubber, 0.25 cm thick, is placed between the plates next to the grounded plate. The relative dielectric constant for rubber is 3. Assume that the field is uniform between the plates. Compute:

(*a*) The difference of potential across the air and across the rubber.
(*b*) The potential gradient in the air and in the rubber.
(*c*) The dielectric flux density in the air and in the rubber.
(*d*) The total number of charges on each plate.
(*e*) The force of attraction on the plates in newtons.
(*f*) The capacitance.
(*g*) Draw curves of potential and potential gradient for the space between the plates.

Repeat for the case in which the rubber occupies half the volume but extends completely from one plate to the other.

16. A 500,000 circular mil cable (0.814-in. diameter) is covered with $^{22}/_{64}$ in. (0.343 in.) of impregnated paper ($\epsilon_r = 3.5$) and a lead sheath. The difference of potential between the wire and the sheath is 12,000 volts. Plot curves of potential and potential gradient between the wire and the sheath. What is the value of the maximum potential gradient, and where does it occur? What is the maximum voltage which may be impressed on this cable? At 12,000 volts how much energy is stored per meter of length? Compute the capacitance per meter.

17. Consider a cable with a fixed sheath radius (r_2 = constant) and a fixed applied potential difference ΔV. The maximum potential gradient occurs in the dielectric just off the surface of the inner conductor (radius r_1). Derive the expression for the maximum gradient as the conductor radius r_1 is varied. For what particular radius will the maximum gradient become a minimum?

18. Apply the results of Prob. 17 to the data of Prob. 16. Plot E_{max} against r_1. What influence should the shape of this curve have on the design of a single conductor cable?

19. Two very long concentric brass tubes of radii 0.1 and 5.0 cm have a potential difference of 1200 volts. The intervening medium has permittivity $\epsilon_r = 3$. If the above potential difference is the maximum permissible:

(*a*) Find the "dielectric strength" of the medium, naming the units.

(*b*) If the tubes should lose their concentricity, sketch neatly the resulting field and equipotential surfaces.

20. A single conductor cable consists of a copper conductor of 1.0-cm diameter surrounded by a layer of impregnated paper, 0.25 cm thick, and a lead sheath. The paper has a relative dielectric constant of 3.0 and a dielectric strength of 120,000 volts per centimeter. What is the maximum difference of potential between the conductor and the sheath which may safely be applied to this cable?

21. A single conductor cable, 1 mile long, is tested at a potential of 120,000 volts between conductor and sheath. The conductor has a diameter of 2.0 cm and the paper insulation is 1.0 cm thick; $\epsilon_r = 3.5$ for paper. If the insulation breaks down at this voltage, how much energy is dissipated at the point of rupture?

22. Two very long concentric cylinders have radii of 2 and 10 cm, respectively. They are separated by fiber of $\epsilon_r = 2.5$. The difference of potential between the two cylinders is 3000 volts. Find the values of the intensity and the difference of potential between the outer cylinder and the point at:

(*a*) Halfway between the two cylinders.

(*b*) The center of the inner cylinder which is hollow.

23. A cable has a wire, radius $r_1 = 0.2$ cm, and sheath, radius $r_3 = 2.0$ cm, separated by two layers of insulation. The inner layer has $\epsilon_{ra} = 6$, the outer $\epsilon_{rb} = 3$. The dielectric strength of the inner layer is 300,000 volts per centimeter, the outer 240,000. Find:

(*a*) The radius of division between the two layers so that their respective fields both reach their dielectric strengths simultaneously.

(*b*) The maximum permissible potential difference, wire to sheath.

24. A 1,000,000 circular mil cable (1.155-in. diameter) is covered by 0.280 in. of rubber and a lead sheath. The difference in potential between the wire and the sheath is 45 kv ($\epsilon_r = 3$ for rubber). Grounded sheath.

(*a*) Plot curves of potential and potential gradient for all points between the wire and the sheath.

(*b*) What is the maximum potential gradient and where does it occur?

(*c*) With this sheath, for what size cable will this gradient be a minimum?
(*d*) Compute the capacitance per kilometer of length.

25. Refer to Prob. 24. Replace rubber of Prob. 24 with 0.140 in. of varnished cambric ($\epsilon_{rc} = 5$) next to the wire and 0.140-in. paper ($\epsilon_{rp} = 2.5$) next to the sheath.

(*a*) Plot curves of potential and potential gradient.

The cambric dielectric strength = 150 kv/cm; that of the paper is 100 kv/cm.

(*b*) What is the maximum permissible voltage?
(*c*) Compute the capacitance per kilometer of length.

26. Consider a 1,000,000 circular mil (1.155-in. diameter) cable with a 0.001-in. uniform air film next to the conductor, then 0.279-in. rubber, then the sheath. ϵ_r for rubber = 3, the potential difference is 45 kv between conductor and sheath. Plot the curves for potential and potential gradient.

27. Assume a long straight pipe, 4-in. outside diameter, carrying steam at 400° F, and that the outside pipe temperature is that of the steam. The covering is 3 in. thick of 85 per cent magnesia. Temperature at surface of covering is 100° F. Coefficient of heat conductivity for covering is 0.45 Btu per square foot of normal area, per inch thick, per degree (F) temperature difference per hour.

(*a*) Find the temperature throughout the covering.
(*b*) Find Btu per hour lost in 1 ft of pipe.
(*c*) Repeat (*a*) and (*b*) for a 2-in. covering—other things remaining fixed.
(*d*) Draw curves of temperature.

28. Assume a long straight steam pipe of 4-in. diameter and 400° F covered by equal volumes of hair felt ($k_H = 0.32$) and Asbestocel ($k_A = 0.64$) so that the total thickness of the double covering is 3 in. Temperature of outside covering is 100° F. k = heat conductivity = Btu per square foot of normal area, per inch thick, per degree (F) temperature difference per hour.

(*a*) For Asbestocel inside, find the temperature at the division between hair felt and Asbestocel, also the temperature at midpoints of hair felt and Asbestocel.
(*b*) Find heat dissipation per hour per foot of pipe.
(*c*) Repeat (*a*) and (*b*) for hair felt inside.
(*d*) Draw curves of temperature against radius.

29. Following Fig. P29, compute the field intensity at point P_1 and the potential difference between points P_1 and P_2. Locate the singular point.

30. Consider a 20-cm finite straight line of charge ($\rho_l = 1$ microcoulomb per meter). Find the intensity at a point on the line of charge produced, distant 25 cm from the charge center. Find the absolute potential.

31. Consider two long parallel wires, (4/0) diameter = 0.460 in., with 3000 volts between them. One inch from center to center. Draw to three times full scale:

(*a*) The equipotential surfaces for 250-volt changes.
(*b*) The curve of potential for points on line from center to center.
(*c*) The potential gradient along the line joining the centers.
(*d*) The flux lines for 10 per cent increments of flux.

Compute the capacitance per foot of length.

32. Consider two long parallel wires, each of radius equal to 3 cm, separated center distance 10 cm, with a potential difference of 20,000 volts. Draw:

(*a*) The equipotential surfaces $V_p = \pm 5{,}000$ volts.
(*b*) The potential gradient curve for points along the line of centers.
(*c*) The flux lines for 10 per cent increments of flux.

Compute the capacitance per meter of length.

33. A transmission line has its 3 wires arranged in equilateral formation, with a spacing of 50 cm. One wire is charged to $+10^{-8}$ coulomb per centimeter, the others to -0.5×10^{-8} coulomb per centimeter each. Find the location of one point where $V = 0$, and locate the singular line. Sketch field, both flux lines and equipotential surfaces, according to principles we have considered.

34. A certain high-voltage transmission line operates at 287,000 volts rms. The instantaneous peak voltage is $\sqrt{2}$ times as much. The spacing between conductors is 750 cm. With an air dielectric, compute the minimum radius of wire required if the field strength at the wire surface is not to exceed 30,000 volts per

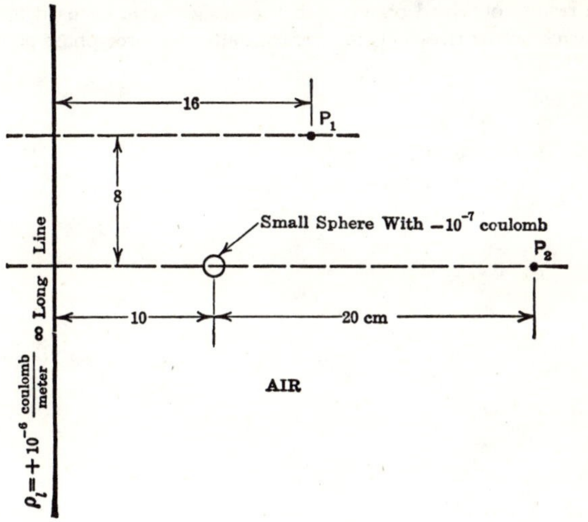

FIG. P29

centimeter. Compare this value with the radius actually used, 1.75 cm. With the latter radius used, what is the field strength at the wire surface? Compute the corona voltage for a radius of 1.75 cm. Plot a curve of field strength at the wire surface against wire radius as variable, for fixed values of spacing and potential difference. See *Trans. A.I.E.E.*, **54**, May 1935, p. 507.

35. Given a metallic prolate spheroid (ellipse rotated about its major axis) carrying 0.2×10^{-6} coulomb, find the surface density of charge at the ends of the axes. Major axis = 10 cm. Minor axis = 6 cm. Air medium. Find the capacitance.

36. Compute the capacitance of a condenser with air dielectric between conductors whose surfaces form confocal ellipsoids of revolution about their major axes. For one conductor, the major axis is 10 cm and the minor axis 6 cm; for the other conductor the major axis is 12 cm.

37. Derive expressions for the potential and the potential gradient for points outside and inside a sphere of radius R, throughout which electricity is distributed with a uniform volume density ρ.

38. Derive expressions for the potential and the potential gradient for points outside and inside a circular cylinder of radius R, throughout which electricity is distributed with a uniform volume density ρ.

39. Three-phase 60-cycle power is carried by three cables of the type illustrated in Fig. 2.7, except that the inner conductor has a cored center as stated in Art. 2.8. The inner radius of the lead sheath is $r_2 = 3.6$ cm. The insulation is manila paper oil-immersed for which $\epsilon_r = 3.5$ and the dielectric strength equals 150,000 volts per centimeter. The operating voltage, line to line, is 132,000 volts rms and the wire to sheath peak potential is $\sqrt{2} / \sqrt{3}$ times as much.

Compute the inner wire radius r_1 which requires the minimum value of gradient in the adjacent insulation. What is the value of the minimum gradient and the capacitance per meter? Plot the value of gradient in the adjacent insulation as the inner wire size r_1 is varied between values of zero and r_2. Between what limits may r_1 be used without exceeding the dielectric strength of the insulation?

Compare results obtained above with the design actually used for which $r_1 =$ 1.7 cm. If each wire carries 200 amperes compute the three-phase power delivered.

CHAPTER 3

ELECTRIC POLARIZATION AND INDUCTION

The purpose of this chapter is to make a closer study of a dielectric under the influence of an electric field, and to observe the effect on the electric field of introducing additional insulators and conductors.

A. POLARIZATION

3.1. Condenser with a Homogeneous Dielectric. As the basis for this discussion let us choose a simple condenser composed of two conductors separated by a homogeneous and isotropic dielectric which is characterized electrically by its dielectric constant $\epsilon = \epsilon_0\epsilon_r$. If the conductors are given surface free charge density $+\rho_s$ and $-\rho_s$, respec-

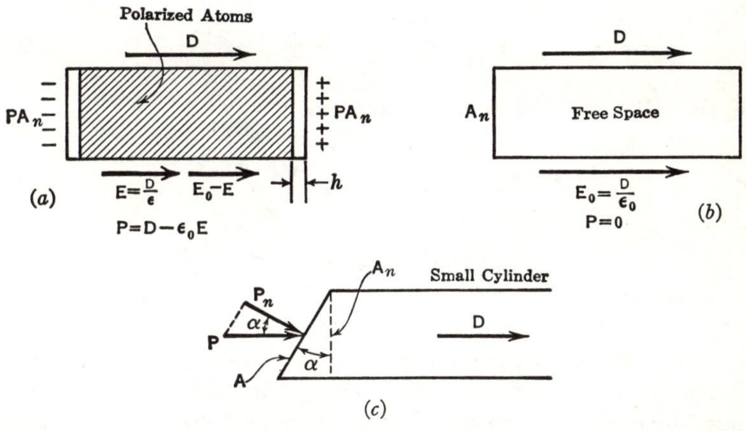

FIG. 3.1. Polarization. (*a*) In a dielectric. (*b*) In a vacuum. (*c*) On a slant area A.

tively, the dielectric becomes the seat of an electric field that at any point may be described by the vectors of flux density D and intensity E. At any field point let us describe a small cylinder with its axis chosen parallel to the flux and of normal cross-section area A_n (Fig. 3.1). If the dielectric is a vacuum, the intensity is equal to $E_0 = D/\epsilon_0$. With any other dielectric the flux density has the same value for the same free charge density $\pm\rho_s$, but the intensity is reduced to $E = D/\epsilon$, as given by the Coulomb experiment. To what influences must we ascribe the reduction in intensity $E_0 - E = D/\epsilon_0 - D/\epsilon$?

3.2. Polarization. The atoms of a perfect insulator yield no free charges in the presence of an electric field but are elastically stretched or polarized to an amount depending on the magnitude of the intensity. The positive charges are urged in the direction of the intensity, the negative charges against it.

The volume of the small cylinder under consideration in the dielectric is still electrically neutral, but at one end there is a positive charge and at the other a negative charge, as shown in Fig. 3.1*a*. These are bound charges, securely fastened to atoms which are merely stretched. Figure 3.1*b* shows the conditions in a vacuum. Let $P = \pm\rho_s'$ charges per square meter be the bound charge density over either normal cylindrical end surface A_n. At any internal point in the dielectric (inside the cylinder) the reduction in intensity must be caused by this charge shift, which tends to set up a uniform electric field oppositely directed to the intensity caused by the free charge density $\pm\rho_s$. The reduction in intensity then is equal to

$$E_0 - E = \frac{D}{\epsilon_0} - \frac{D}{\epsilon} \tag{3.1}$$

and, since P and D are dimensionally equal,

$$\epsilon_0 E_0 - \epsilon_0 E = D - \frac{D}{\epsilon_r} = P = \rho_s' \tag{3.2}$$

Hence

$$P = D\left(1 - \frac{1}{\epsilon_r}\right) = D - \epsilon_0 E = \rho_s' \tag{3.3}$$

P is known as the polarization vector. It has the same direction as the vectors D and E and a magnitude equal to $D - \epsilon_0 E = \rho_s'$.

Let us define $\rho_\pm'$ as the *volume* charge density of either plus or minus *bound* charge. In the presence of a field the charge densities of opposite sign are shifted a small distance apart, h in Figure 3.1*a*, in the direction of the field. The amount of charge in the end layer per square meter cross section of cylinder is then ρ_{s+}' or ρ_{s-}', equal numerically to the value of P, and also equal numerically to $\rho_\pm' h$. Hence

$$|P| = |\pm\rho_s'| = |D - \epsilon_0 E| = |\rho_\pm' h| \tag{3.4}$$

The quantity $\rho_+' h$ is equal to the moment per cubic meter of bound charge, a vector with the same direction as D or E.

In summary, the polarization P at a point is a vector quantity having the same direction as D or E at the point, equal to $(D - \epsilon_0 E)$, and equal also to the bound charge moment per cubic meter and to the bound surface charge density on a small area taken normal to the field.

3.3. Polarization at a Slant Surface. As shown in Fig. 3.1*c*, if the small cylinder end A is chosen not perpendicular to the field, the bound charge density upon the area A is

$$P_n = P \cos \alpha \tag{3.5}$$

where α is the angle between the area A and the area A_n normal to the field.

3.4. Dielectric Constant as a Function of Frequency. We have referred in Art. 3.2 to polarization by the action of a field in *stretching* the atoms. This is attributable to the action of the field in distorting the electron orbits, with the result that the center of gravity of the positive nucleus and the center of gravity of the electron shells or orbits are shifted a very small distance h (far less than the atomic radius) from coincidence, which produces a polarized atom. If h is proportional to the field E applied, so also is P and thus D, and the dielectric constant ϵ is independent of the field strength. Experimentally this is found to be true. Furthermore, since the electrons have so little mass, we should expect, in the presence of an alternating field, that the polarization would respond practically instantaneously, and we should expect the dielectric constant to be independent of frequency. Actually, however, the dielectric constant does depend on the frequency, and several distinct phenomena may operate simultaneously to determine its value.

The influence of these phenomena may be brought out in the following manner. Suppose that the dielectric is exposed to alternating electric fields with a very wide range of frequencies. At the higher frequencies optical methods are required. Light waves are electromagnetic waves, so that passing light through a dielectric is equivalent to exposing the latter to an electric field.

Frequencies in the ultraviolet range are so high and the reversals of the electric field so rapid that the electrons in atoms, though of small mass, cannot follow the field. The atom cannot then be polarized and the relative dielectric constant is unity. In a somewhat lower frequency range the electrons follow the field reversals almost instantaneously, the atom is polarized by the field, and the relative dielectric constant has a value greater than unity. This is illustrated in Fig. 3.2 by the change $\epsilon_{r5} = 1$ to ϵ_{r4}. Between the two ranges there is an absorption band in which the dielectric constant changes very rapidly, and in which the losses per cycle are relatively high since the light is highly absorbed by the dielectric. For frequencies below the band the relative dielectric constant is $\epsilon_{r4} = 1 + \epsilon_{rE}$. In this, ϵ_{rE} is the contribution of "electronic polarization." All atoms can be polarized or stretched electronically, and the relative dielectric constant of any dielectric must therefore be greater than unity below the absorption band.

In the infrared range a similar change occurs. Again in passing through this band there is marked increase in the losses and a fairly

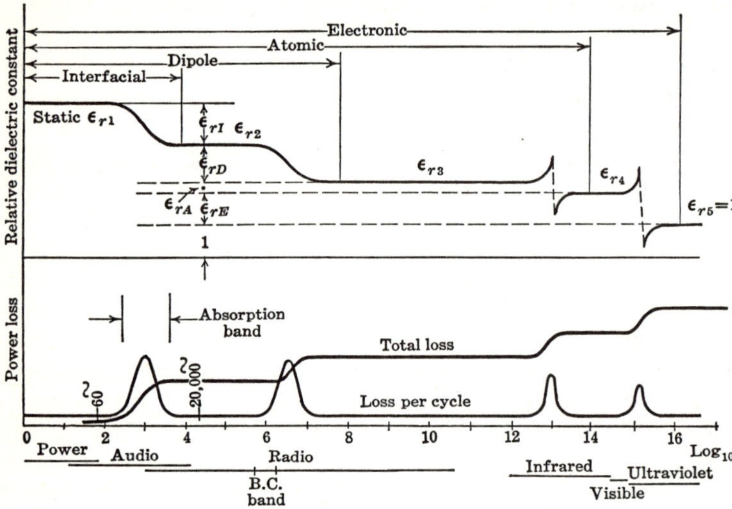

FIG. 3.2. Dependence of relative dielectric constant and losses on frequency.

abrupt increase in the dielectric constant to $\epsilon_{r3} = 1 + \epsilon_{rE} + \epsilon_{rA}$. The mechanism responsible for adding ϵ_{rA} is known as "atomic polarization." For illustration, a sodium chloride (NaCl) crystal is composed of alternate ions, Na^+ and Cl^-. The electric field of a light wave with frequency below that of the infrared absorption band causes the ions of opposite sign to oscillate in opposite directions simultaneously. ϵ_{rA} is usually small compared with the other components of ϵ_r.

Dropping down into the radio range of frequencies, where electrical methods may be used, that is, controllable frequency applied to the material placed between electrodes, another increase of loss and dielectric constant is found, provided that the material is composed of *dipolar molecules*. This is a molecular and not an atomic phenomenon. If a molecule is so constructed that the center of gravity of its positive charge does not coincide with the center of gravity of the negative charge, even with no electric field applied, the molecule is a "permanent dipole." Under an applied field the dipolar molecules *tend* to orient their dipole axes toward the field direction, but orientation is limited by thermal agitation which tends to knock the molecule out of alignment with the field and set up a random orientation. Between these two influences the dipole axes "on the average" lean slightly in the field direction and thus add to the polarization of the material as a whole, and the dielectric constant is increased by ϵ_{rD}. This is, of course, additional to the electronic or stretch polarization to which every atom contributes. Since thermal action tends to destroy or at least decrease the alignment with the field, it is to be expected that an increase in temperature will decrease the contribution to ϵ_{rD} of the polar molecules. Water is a highly polar molecule; carbon tetrachloride (CCl_4) is nonpolar. Though the absorption band is frequently found in the radio range, it may occur for certain materials in the audio range or even in the power range of frequencies.

The last absorption band due to "interfacial polarization" is likely to be found in the power or audio ranges, but for some materials it may be found in the radio range. Interfacial polarization is present whenever the dielectric is a composition of two or more dissimilar dielectrics with different conductivities. In this event charge flows more freely under an applied field through one dielectric component than through the others, and free charge is built up on the interfaces between the components, as well as on the electrodes. Thus free charge is to be found "absorbed" throughout the dielectric, and since it increases the total charge of the condenser the capacitance and dielectric constant is increased by ϵ_{rI} to the value $\epsilon_{r1} = 1 + \epsilon_{rE} + \epsilon_{rA} + \epsilon_{rD} + \epsilon_{rI}$.

At *very low frequencies* or with a *constant potential* applied, the dielectric constant reaches its highest value ϵ_{r1}, but not instantaneously. Time is required for the "absorbed" charge to "soak into" the dielectric. When a battery is connected to a condenser through a small resistance the electrodes are charged almost instantaneously (perhaps a thousandth of a second) to a value $Q_g = C_gV$, in which C_g is the geometric capacitance determined entirely by the dielectric constants of the several components in the dielectric. In Fig. 3.3, i_N is the corresponding *normal* charging current to the condenser. As time passes, more and more charge is absorbed *slowly* inside the dielectric, an absorption current i_a flows during this period, and the capacitance rises to a final value C_0 determined by the conductivities of the components of the dielectric. i_c is the final and continuing current related to the conductivity of the dielectric. The absorption current may continue to flow, in some cases, for a considerable period—hours, days, weeks. Any material which does not have an absorption current with d-c potential does nothave a change of dielectric constant ϵ_{r2} to ϵ_{r1} with a-c potential applied. See Art. 4.24.

The lower part of Fig. 3.2 illustrates both the power loss per cycle, which is relatively high only in an absorption band, and the total power loss. The latter curve is the integral of the loss per cycle curve—its shape therefore depends on the precise characteristics of the former. Certain dielectrics at least tend to follow the step and plateau

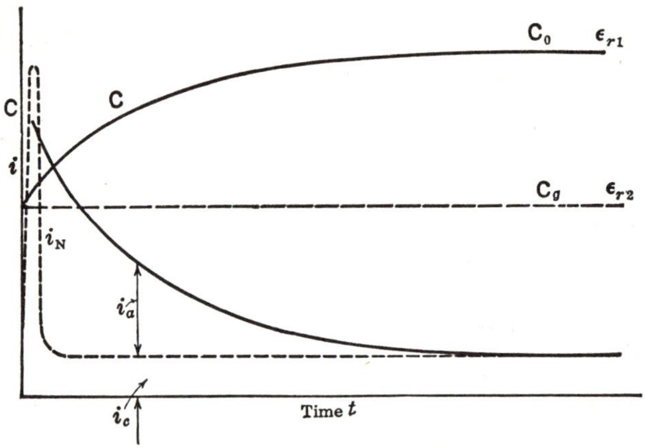

FIG. 3.3. Variation of C and i in a condenser, with constant voltage applied, resulting from absorption.

form of curve for total loss. Application of these principles is found in the industrial process known as "dielectric heating," used, for instance, in drying wood and setting the glue in plywood. For the type of power loss curve shown it is advantageous to use a frequency high enough to utilize the properties of the various absorptive mechanisms available in the dielectric.

3.5. Condenser with Two or More Dielectrics. The problem of determining the field becomes considerably more complicated when two or more dielectrics are present, because then different degrees of polarization must be taken into consideration. When polarization takes place in any portion of a dielectric, the accompanying change in intensity occurs throughout the entire insulating region and redistributes the free charge on the conductors, so that there is a still further change in the intensities in the dielectrics. But the intensity change affects the magnitude of the polarization. This apparently endless chain of influences finally creates an electric field that is determined by the boundary conditions.

At a conductor–insulator boundary the free charges are redistributed in such a manner that the conductor surface is an equipotential, while at an insulator–insulator boundary the field is refracted according to the following conditions:

1. Normal flux densities are continuous. $D_{N1} = D_{N2}$
2. Tangential intensities are continuous. $E_{T1} = E_{T2}$

3.6. Field of Two Dielectrics. Figure 3.4. This problem can be solved mathematically in only a few simple geometrical cases. If dielectric ϵ_2 is a sphere, a cylinder, or a prolate ellipsoid of revolution

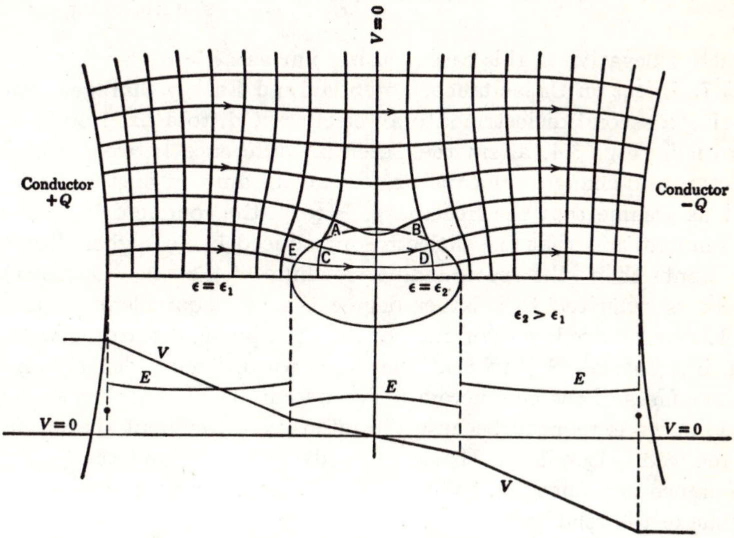

FIG. 3.4. Field with two dielectrics. $\epsilon_2 > \epsilon_1$.

immersed in an original uniform field, the solution is known. Here we shall content ourselves with a few general remarks.

Consider two equipotential surfaces AC and BD. A and B are points just outside the surface in dielectric ϵ_1: C and D are just inside. The intensity along CD is lower than along AB, as it is inside the more intensely polarized region ϵ_2. Hence points C and D must be farther apart than points A and B in order to give an equal potential change. Consequently, the equipotentials must be bent as shown, and the flux must be directed so that an increased flux density occurs in the region having the higher dielectric constant.

At any insulator–insulator surface boundary point, as E in Fig. 3.4, the surface layer of polarization or bound charge is given in two parts, the negative layer from one dielectric and the positive layer from the other dielectric:

(1) The negative layer $P_{N2} = D_{N2}\left(1 - \frac{1}{\epsilon_{r2}}\right)$ (3.6)

(2) The positive layer $P_{N1} = D_{N1}\left(1 - \frac{1}{\epsilon_{r1}}\right)$ (3.7)

Hence the total bound charge at the interface is

$$P_N = +P_{N1} - P_{N2} = D_N\left(\frac{1}{\epsilon_{r2}} - \frac{1}{\epsilon_{r1}}\right) \frac{\text{coulombs}}{\text{square meter}} \tag{3.8}$$

which is negative in this case at this point as $\epsilon_{r2} > \epsilon_{r1}$.

3.7. Effect on Capacitance, Potential, and Energy Storage. Introducing a second dielectric into an electric field, to a fixed position as shown in Fig. 3.4, alters the potential difference between the conductors and changes the capacitance and the amount of stored energy. Let us assume that the free charge $\pm Q$ on the conductor is constant in amount and that the dielectric introduced has a higher dielectric constant than the surrounding medium ($\epsilon_2 > \epsilon_1$). Dielectric 2 becomes polarized to a larger degree than an equivalent volume of dielectric 1 placed at the same position, producing a rearrangement of the free charges $\pm Q$ so that they are crowded more densely on the nearer faces of the conductors. The potential difference between the conductors is reduced because the intensity throughout the insulator is reduced. But if the charge is fixed in amount and the potential difference is reduced, it follows that the capacitance is increased by virtue of the relation

$$\underline{Q} = \underset{\uparrow}{C}\,\underset{\downarrow}{\Delta V} \tag{3.9}$$

A more direct way of arguing the matter follows. The capacitance is a function of the geometry and of the dielectric constant. Replacing a part of the dielectric with one of higher dielectric constant necessarily increases the capacitance.

With a fixed charge and a decreased potential the energy stored is smaller.

$$\underset{\downarrow}{W_e} = \frac{1}{2}\underline{Q}\underset{\downarrow}{V} \quad \text{or} \quad \frac{1}{2}\frac{\underline{Q}^2}{\underset{\uparrow}{C}} \tag{3.10}$$

If the dielectric 2 is now moved from its position to a new position where the original intensity (before introducing 2) had a larger value, the capacitance is increased to an even greater degree than before, and the potential difference and energy storage are reduced still further.

Should the dielectric introduced be of smaller dielectric constant, the converse relations hold—the capacitance is decreased, and for a fixed charge the energy stored is increased. Moving the dielectric to an originally more intense part of the field magnifies these changes in capacitance, potential, and energy storage.

3.8. Point of View—*P* versus ϵ. There are two ways of considering any problem involving polarized media.

(1) *Physical Point of View.* Treat all charges as free, including the multitudinous bound charges of the polarized atoms. The relative dielectric constant no longer has any meaning and may be considered as unity. This is the point of view chosen for an exposition of the electron theory.

(2) *Engineering Point of View.* The charges are divided into free and bound. The multitudinous atomic charges of polarization are now treated as a unit by the introduction of the relative dielectric constant, experimentally obtained. The forces are expressed in terms of the free charges only, modified by the use of the relative dielectric constant. This treatment is much simpler and is adaptable to the solution of engineering problems.

B. INDUCTION

3.9. Condenser with Three or More Conductors. Into the field of a simple condenser composed of two conductors and a single dielectric, let us introduce a third conductor (or more) and study the modifications required in the field. In Fig. 3.5 conductors A and B are charged to $+Q$ and $-Q$, respectively, and neutral conductor C is brought to some arbitrary position as shown. Conductor C must be an equipotential and possess zero intensity internally. But C can eliminate its internal field only by having a portion of its free electronic cloud appear upon its surface, leaving another part of its surface positively charged to an equal degree. Hence we find $+q$ and $-q$ on the surface of conductor C. These charges in turn cause a redistribution of $+Q$ and $-Q$ so as to keep the conductor surfaces of A and B equipotential. The distribution or surface density of the charges depends upon the inducing field as well as upon the conductor shape. Furthermore, q is less than Q. Otherwise C, owing to its greater charge, would send lines out to infinity to charges that do not exist in this problem.

The electric field in the dielectric is altered by the introduction of C due to the induced charge $\pm q$ and the redistribution of $\pm Q$. Conductor C assumes an absolute potential that depends upon the position assigned to it. In this figure it is positive. Curve $V_AV_CV_B$ gives the distribution of absolute potential through the three conductors and the field.

Line GG around the surface of conductor C, separating the positively and negatively charged regions, is called a *line-of-no-electrification.* Along GG the surface charge density must be zero, as it is neither

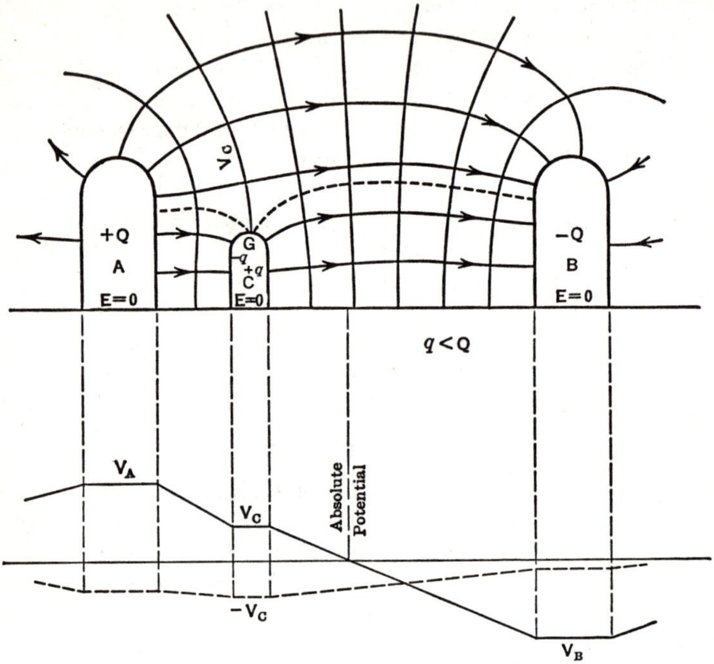

FIG. 3.5. Electrostatic induction on a third conductor (C).

positive nor negative. But this is the condition that line GG shall be a singular line, at which an equipotential surface crosses itself at 90° and the flux lines come in at 45° to the surface. Hence the field equipotential surface of the same absolute value as that of conductor C leaves that conductor normally along the line-of-no-electrification.

Some of the flux lines from A go straight through to B. Others go to C, and an equal number of lines start at C and go to B. Conductor C is at an intermediate potential between A and B, which have the highest and lowest potentials, respectively, in the field.

3.10. Effect on Capacitance, Potential, and Energy Storage. Introducing a third conductor into an electric field alters the capacitance between the two original conductors, as well as their potential difference and the energy storage. For present purposes we may regard the third conductor with its induced charges $\pm q$ as a dielectric possessed of an extreme degree of polarization. That is to say, a high degree of polarization may cause electrons to leave the atoms in a dielectric and flow to the boundaries. In this case the permissible potential gradient has been exceeded, the dielectric has become a conductor, and induction has replaced polarization. It follows that introducing a conductor into the field increases the capacitance, decreases the potential difference between the conductors of fixed charge $\pm Q$, and decreases the energy storage.

3.11. Shielding. A closed conductor shields its interior from *steady* external electrostatic influences, as it tends to keep this region free from electric fields. During *changing* external electric influences, the shielding becomes only partially effective, as it takes time for the induced charges (see Fig. 3.5) to arrive at positions necessary to make the surface equipotential.

3.12. Grounding. Suppose conductor C with positive potential in Fig. 3.5 is now grounded by connecting it through a conducting wire to some point outside the field. Consequently the potential of C now becomes zero. Practically, a conductor is grounded by connecting it to the earth or any other conducting body of large dimensions (hence large capacitance), compared with conductors A, B, and C, that lies outside their field. From a mathematical point of view the ground may be regarded as zero potential located at infinity.

Figure 3.6 shows the field under this new condition. The only way in which the potential of conductor C can become zero is to permit the conductor to acquire a negative charge $(-q_1)$ from the grounding body (infinity). The broken curve in Fig. 3.5 gives the negative potential distribution among the conductors and field due to the presence of $-q_1$ on C alone. The potential curve of Fig. 3.6 is the combined potential due to both distributions. Conductor A is at a lowered positive potential, C at zero, and B at an increased negative value. The increased negative charge on C $(-q')$ is tied by flux lines to an increased portion of $+Q$ on A. The remaining positive charge of A sends its lines of flux partly to ground, which must be regarded as a fourth conducting body in the field, and the remainder directly to a portion of $-Q$. The ground, which has a net positive charge $+q_1$, now sends q_1 more flux lines to conductor B than under the condition of conductor C ungrounded $(-q_1 = -q' + q'')$. Proceeding leftward from conductor A one finds the $V = 0$ surface crossing the center plane and farther on the singular line. From the $V = 0$ surface on to the left the potential along the center plane is negative, reaching a minimum at the singular line.

Comparing Figs. 3.5 and 3.6, it may be noted, in relative terms, that the difference of potential in the former is 8 units and in the latter is only 7 units. Consequently, grounding conductor C has increased the capacitance in the ratio $8/7$. If Q is taken, relatively, as 8 units, $\pm q$ in Fig. 3.5 is 2.5 units, compared with $q' = -3.5$ and $q'' = +1.5$ units

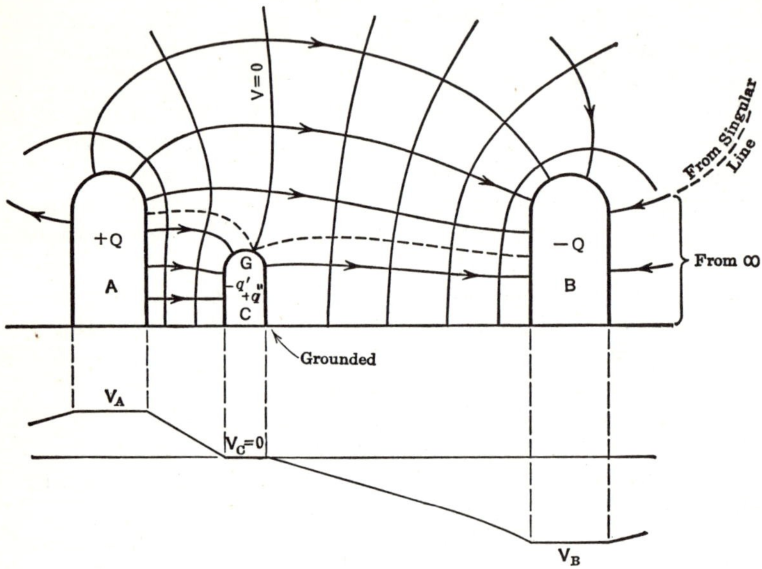

FIG. 3.6. Induction on a grounded conductor (C).

in Fig. 3.6. Consequently -2 units must have come from earth when conductor C was grounded.

3.13. Induction on a Conductor Surrounding Others. Another type of induction situation occurs when one conductor completely surrounds others. As shown in Fig. 3.7, let us assume that conductors A and B with charges $+Q_1$ and $-Q_2$ ($Q_1 > Q_2$) are placed inside conductor C, assumed originally electrically neutral.

There can be no field inside C's metallic volume. Hence there must be a net induced charge on the inner surface of C equal to the net inducing charge on conductors A and B. Conductor A sends part of its flux lines directly to B; others go to C inducing a negative charge $-q_1$ over part of the C inner surface. B with its negative charge induced $+q_2$ on the remaining inner C surface. Finally, since

C was originally neutral, a positive charge $+Q_3$ must be found upon the outer surface of C.

$$Q_1 > Q_2$$

$$+Q_1 - Q_2 = -(-q_1 + q_2) = +Q_3 \tag{3.11}$$

Between $-q_1$ and $+q_2$ is found a singular line GG. The singular line potential is that of conductor C and appears as a diaphragm

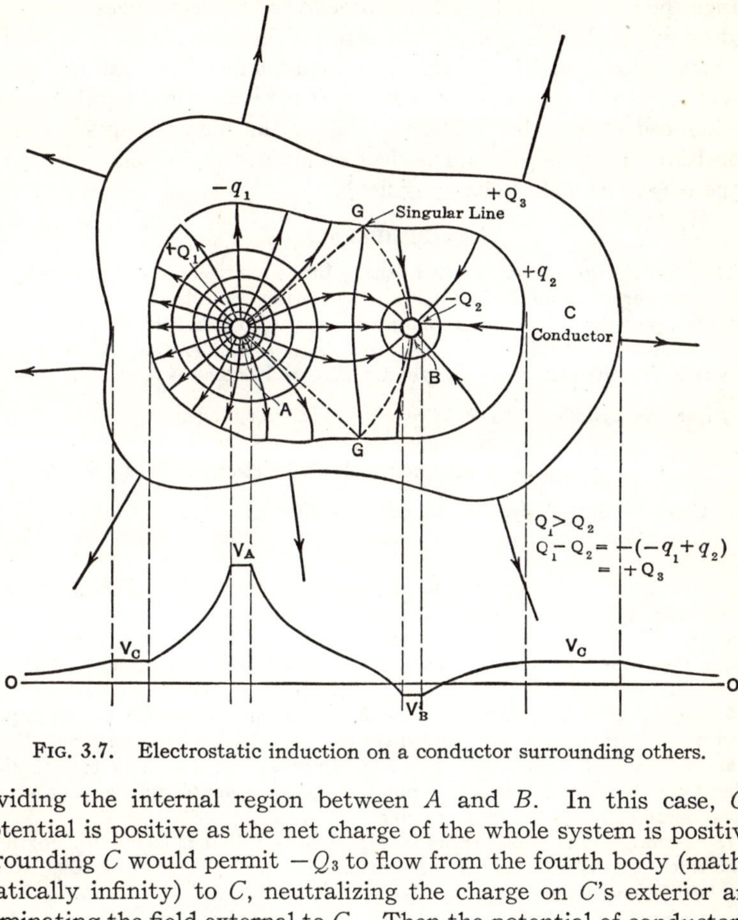

FIG. 3.7. Electrostatic induction on a conductor surrounding others.

dividing the internal region between A and B. In this case, C's potential is positive as the net charge of the whole system is positive. Grounding C would permit $-Q_3$ to flow from the fourth body (mathematically infinity) to C, neutralizing the charge on C's exterior and eliminating the field external to C. Then the potential of conductor C would become zero.

Within certain limits, conductor C shields the internal and external regions from each other. Charge Q_3 is not changed in magnitude or distribution even though charges Q_1 and Q_2 are moved to new positions, or are touched together to form a single positive charge $+(Q_1 - Q_2)$, or are touched to the inside surface of the conductor, eliminating the internal charge and field system entirely. Charge Q_3 and the external field are altered only by a change in the net charge inside the conductor. Conversely, a new charge brought into the external field creates an induced charge distribution on the outer surface of the conductor, which is superposed on Q_3 but does not change the internal field. It is assumed that these changes have been made slowly. In the presence of rapidly changing electric fields, the conductor does not shield each region completely. As the conductor falls short of being a perfect conductor, time is required for the changing induced charge distribution to follow the field changes. Consequently, nothing less than a perfect conductor can shield completely in the neighborhood of changing fields.

PROBLEMS

1. It can be proved that the flux density inside a circular cylindrical dielectric (ϵ_2), placed normal to an original uniform field (ϵ_1) (without the dielectric), is a uniform field, provided that the free charges giving the original field are relatively far away. Sketch the field for the two cases, $\epsilon_2 > \epsilon_1$ and $\epsilon_2 < \epsilon_1$. Compute the polarization on the surface of the cylinder, if $\epsilon_{r1} = 1$ and $\epsilon_{r2} = 5$, and $E_1 = 10^4$ far away from the cylinder. $E_2 = E_1 \dfrac{2\epsilon_1}{\epsilon_1 + \epsilon_2}$ inside the cylinder.

The same statement may be made for a spherical dielectric, with $E_2 = E_1 \dfrac{3\epsilon_1}{2\epsilon_1 + \epsilon_2}$.

2. Compute the polarization charge in the dielectric adjacent to the conductor in Prob. 1 at the end of Chapter 2. Repeat for the rubber-air interface in Prob. 15 of Chapter 2.

3. Sketch the field after a circular cylinder of metal has been placed normal to an original uniform field. Repeat for a metallic sphere.

4. In high-voltage equipment, metallic parts that may be touched by the operator are usually solidly grounded. Explain the benefits derived, in terms of electrostatic principles.

5. High-frequency bridges and other high-frequency apparatus are often provided with several shields, one around the other. Why is this necessary?

6. An insulator with $\epsilon_r = 5$ has 10^{29} molecules per cubic meter. If each molecule has 50 orbital electrons, compute the polarization stretch distance h for a field strength of 3×10^6 volts per meter.

7. Assume that the conductors in Fig. 3.5 are cylinders of infinite length of capacitance $C_{AB} = 10^{-10}$ farad per meter and that $V_A - V_B = 8000$ volts. Calculate Q, q, and V_C.

Following Fig. 3.6, ground conductor C. Assume Q remains constant and calculate V_A, V_B, $V_A - V_B$, q', q'', and the charge per meter $(-q_1)$ which flows from the earth.

8. In Fig. 3.7, let $Q_1 = +20$, $Q_2 = -12$, $q_2 = +5$, in relative terms, and compute q_1 and Q_3.

CHAPTER 4

ELECTRIC CURRENT

RESISTANCE—LINEAR AND NONLINEAR

The movement of electric charges from place to place constitutes one type of current. Another type is the displacement current which is to be found wherever there is an electric field changing in time. With the current come new forces and new phenomena. This chapter is devoted to the study of the electric current and its influence upon the electric field. The study of the *magnetic* effect of the electric current is reserved for Part II, which is entirely devoted to this important subject.

A. ELECTRIC CURRENT

4.1. Current Caused by Moving Charge. An electric current is said to flow through a portion of space when electric charges are moving in the region. This is a qualitative statement which fails to tell how much current is flowing. The quantity of current is determined by setting up an area in the region and counting the number of unit charges or coulombs crossing the area per unit of time in a specified direction.

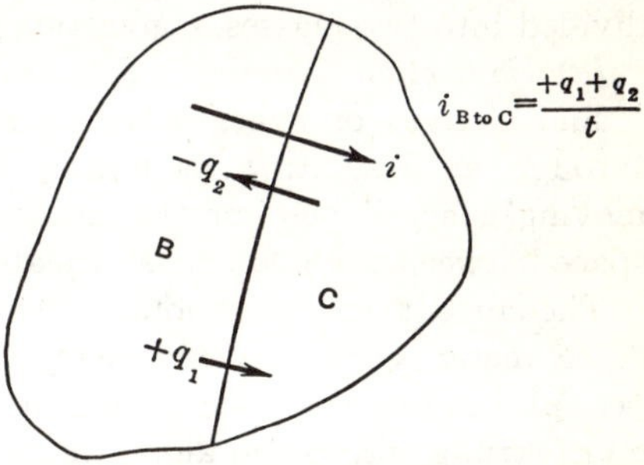

FIG. 4.1. Electric current from *B* to *C*.

Suppose that, in time t, $+q_1$ coulombs of electricity flow from B to C and $-q_2$ coulombs from C to B (Fig. 4.1). Then the average current from B to C is

$$i = \frac{q_1 + q_2}{t} \quad \frac{\text{coulombs}}{\text{second}} \text{ or amperes} \qquad \text{(Scalar)} \qquad (4.1)$$

It is to be noted that a negative charge movement against the specified direction is reckoned as a positive current. If the current is changing in time, the count of charge movement must be made over a small period of time. Formally, this is summed up:

Electric Current (Definition). An electric current across a surface is the sum of the time rates of flow of positive and negative charges across the surface.

$$i = \frac{dQ}{dt} \quad \frac{\text{coulombs}}{\text{second}} \text{ or amperes} \qquad \text{(Scalar)} \tag{4.2}$$

Current is a *scalar* quantity as it involves merely the *number* of charges partaking in the movement per unit of time.

The dimensional formula is

$$\text{Electric current } i = \frac{\text{electric charge}}{\text{time}} = Q^1T^{-1} \tag{4.3}$$

4.2. Current Density (Definition). The electric current density at a point is a *vector* quantity whose direction is that of the positive charge movement at the point and whose magnitude is the number of coulombs per second per square meter which cross a small area placed normal to the current flow.

$$\text{Current density } J = \frac{di}{dA} \quad \frac{\text{amperes}}{\text{square meter}} \qquad \text{(Vector)} \tag{4.4}$$

Currents resulting from charge movement may conveniently be divided into two classes, convection and conduction types.

(*a*) *Convection Currents.* A convection current occurs when individual charges or charged bodies move from one place to another through an insulating medium. Illustrations of this type are: a moving charged belt, or the electronic movement in the evacuated space between the filament and plate in a vacuum tube.

The current density is obtained easily. Let there be n particles per cubic meter, each carrying charge q, with velocity v. The charge per cubic meter is $\rho = nq$. Consider a number of boxes, each 1 meter cube, strung end to end along the line of flow. With velocity v there will be v boxes of charge crossing 1 square meter, taken normal to the flow, per second. Hence convection the current density is

$$J_{\text{convection}} = nqv = \rho v \tag{4.5}$$

For charges of both signs

$$J_{\text{convection}} = n_+q_+v_+ + n_-q_-v_- = \rho_+v_+ + \rho_-v_- \tag{4.6}$$

(*b*) *Conduction Currents.* An electric current flowing by conduction from one point to another requires a material conductor connection between these points. The current flows along the conductor, whatever its shape. With this type of current we concentrate our attention not on the individual electron or charge but rather on a *cloud* of electric charges forcing its way forward as a group with a relatively low drift velocity among a cloud of obstructing atoms. Drift velocities of the cloud in a *metallic* conductor are of the order of 1 meter per second, while the random motion of the individual particles comprising the cloud are frequently of the order of 10^6 meters per second.

The metals are good conductors and form the best media for this type of current. Dielectrics, or insulators, which always have the conducting property to a certain extent, pass small conduction currents.

In Art. 4.13 Ohm's law, usually written $I = V/R$, is applied to a conductor of any length or cross section and reduces, for a meter cube, to

$$J_{\text{conduction}} = \sigma E_c \tag{4.7}$$

in which the conduction current density is equal to the product of the conductor conductivity σ and the electric field strength E_c in the conductor.

4.3. Displacement Currents. Current *always* flows in a closed path. If this is generally true, it is clear that in the process of charging a condenser we must recognize a new type of current. Charge movement is confined to the metallic portion of the circuit including the battery. Inside the condenser, at least with a vacuum medium, there is no free charge movement; but as the charge on the plates increases so also does the electric field, and we must add a time-changing electric field to our concept of current. The detailed discussion is given in Art. 16.3.

The current density is

$$J_{\text{displ}} = \frac{\partial D}{\partial t} = \epsilon \frac{\partial E}{\partial t} \tag{4.8}$$

and since

$$D = P + \epsilon_0 E$$

$$J_{\text{displ}} = \frac{\partial P}{\partial t} + \epsilon_0 \frac{\partial E}{\partial t} \tag{4.9}$$

The first term we recognize as a moving-charge current density related to the *bound* or polarized charge of an insulator, but the second term, sometimes called the *ethereal current density*, is present even in evacuated space.

Displacement currents play little role in metallic conduction. In metals they are dwarfed into insignificance by the magnitude of the ordinary Ohm's law type of conduction; but in insulators where the conduction currents are weak, the displacement current may well play a role even at low frequencies and most certainly when the frequencies are elevated.

4.4. Total Current Density. Steady (direct) current densities are either of the convection ($J = \rho v$) or the conduction ($J = \sigma E$) types. Under transient conditions, either one of these types may flow together with the displacement current density. Thus

$$J = \rho v + \frac{\partial D}{\partial t} \tag{4.10}$$

or

$$J = \sigma E + \frac{\partial D}{\partial t} \tag{4.11}$$

4.5. Classification by Time Relations. Electric currents may be grouped according to their variation in time as:

(1) *Steady-State Currents.* Steady-state currents are those which repeat themselves in succeeding equal periods of time. They may be:

(*a*) *Steady Direct Currents.* A steady direct current is a unidirectional current of unvarying magnitude.

(*b*) *Pulsating Direct Currents.* Pulsating currents are unidirectional but have short periodic variations. Rectified alternating currents are frequently of this type.

(*c*) *Alternating Currents.* The ordinary alternating current varies as a sine function of the time. Necessarily it is cyclic or periodic.

(2) *Transient Currents.* When a change of any kind is made in an electric circuit the current changes from one steady-state condition to a different steady-state value. During the intervening period the transient current flows.

Lightning is responsible for transient currents and potentials that have a very destructive effect upon electrical apparatus, unless controlled. Switching operations on power lines are also the sources of many troublesome transients.

4.6. Effects of Current Flow. An electric current cannot be seen, but evidence of its existence appears in many forms. Some of the effects important from an engineering standpoint are:

(1) *Heating.* Any material carrying a current becomes heated, removing energy from the electric field and dissipating it to surrounding bodies. Electric furnaces of certain designs operate on this principle.

(2) *Electrochemical Effect.* The passage of a current through an electrolyte (say silver nitrate) causes a deposit of one of the chemical elements (silver). Electroplating is based upon this fact. This effect may also be used as a means of defining the magnitude of the unit of current.

(3) *Magnetic Effect.* An electric current sets up a magnetic field in the region surrounding it, which exhibits itself by exerting forces upon other currents or magnets lying in the field. The action of a motor depends upon this fact.

Furthermore, if the current is changing in time or moving in space to a new position, through its magnetic field it creates electric intensities in nearby conductors and causes currents to flow in the latter. The action of transformers and generators rests upon this phenomenon. We shall defer further discussion of these points to Parts II and III, which are entirely devoted to study of the magnetic effect.

(4) *Rearrangement of Charge.* When a current flows in a conductor it alters the distribution of surface charges with a consequent change in the electric field in the surrounding dielectric. This matter is considered further in part D of the present chapter. Conversely, when electric charges are being redistributed, their movement must constitute a current.

(5) *Radiation of Energy.* Electric charge radiates energy in the form of an electromagnetic wave when it is accelerated or decelerated. With an alternating current there is acceleration and deceleration continually and continuous radiation of energy. Around a circuit carrying alternating current there is a regular redistribution of charge and, in the surrounding space, a continual time-changing electric field which we recognize as a displacement current. But the latter creates a time-changing magnetic field and this in turn, by transformer action, induces in the insulating region a time-changing voltage or electric field. So the electric and magnetic fields mutually create and support each other and in interacting propel radiant energy out into space. This phenomenon becomes of great importance at the higher (radio) frequencies. This subject is considered more extensively in Part IV.

B. SOURCES OF ELECTRIC FIELD ENERGY

4.7. Sources of Electric Energy. Any battery, generator, frictional apparatus, or other machine that has the ability to separate positive and negative electricity is a source of electric energy. Since energy is never lost or made, a source of electric energy merely transforms energy from some other form to that of storage in an electric field.

However, in order to utilize the energy thus stored, current must flow through the source and on around a circuit to the load or receiver of energy.

Intensities about a Source. Force is required to separate electricity. Then, in the neighborhood of a source of electric energy (both internal and external regions), there are electric intensities and nonelectric forces, as illustrated in Fig. 4.2. We are assuming that no current is flowing.

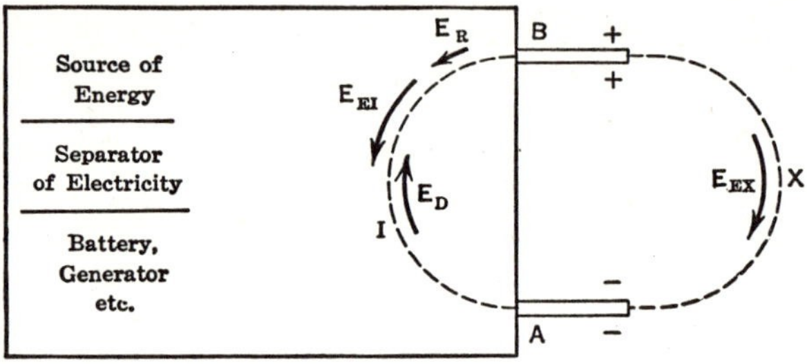

FIG. 4.2. Intensities about a source of energy.

Inside the source itself, along path AIB, from negative terminal A to positive terminal B, there are two forces per unit positive charge (intensities):

(1) E_D, the *driving intensity*, directed from A to I to B, which is *nonelectrical* in character, arising from the chemical energy of the battery, etc., which is responsible for having separated the charges.

(2) E_{EI}, the *electrostatic intensity*, directed from B to I to A, which is *electrical* in character, arising from the fact that the electricity is separated. Considered over the whole path AIB, this intensity must be equal and opposite to E_D with no current flowing. If they were not equal, E_D would continue to drive electrical particles to charge the terminals until the balance did obtain.

Outside the source, along some external path BXA, we find also the usual electric intensity, E_{EX}, which is directed from the positive to the negative charge, that is, from B to X to A.

Potential Differences about a Source. Potential differences, being scalar quantities, are usually easier to handle than the vector electric intensities. They have also the advantage of being directly associated with energy (work per unit charge). Let us compute the potential differences between terminals A and B by moving a unit positive charge from A to I to B along the internal path and from A to X to B along the external path (Fig. 4.2).

$\Delta V_{BA} = V_B - V_A = \frac{\text{joules}}{\text{coulomb}}$ (volts) *done* by the nonelectrical force E_D in moving a unit positive charge from A to I to B, assuming no current flow.

$= \frac{\text{joules}}{\text{coulomb}}$ *received* by the electrostatic field (moving the positive unit charge against E_{BI} from A to I to B).

$= \frac{\text{joules}}{\text{coulomb}}$ *received* by the electrostatic field (moving the unit positive charge against E_{BX} from A to X to B). This is true as the potential difference between two points is the same, irrespective of the path.

Another way of looking at the matter is possible. Let us move the unit positive charge over the path $AIBXA$. Then the work per unit charge given to the electric field over the internal source path AIB is given up by the field which moves the charge (by means of E_{BX}) along the path BXA external to the source. Hence the total change in potential over a complete path is zero.

$$\Sigma V_{\text{closed path}} = 0 \tag{4.12}$$

This is Kirchhoff's law of potentials for a closed path.

4.8. Energy Transfer, Source to Field to Load. The preceding discussion brings out two important points:

(1) A positive charge driven by external means *against* the electrostatic field intensity (going from lower to higher absolute potentials) indicates a transfer of energy from the external source to the electric field.

(2) A positive charge moved by the electric field intensity *in its own direction* (going from higher to lower absolute potentials) indicates a transfer of energy from the electric field to some external device or load.

4.9. Source Conditions with Current Flow. In case a current does flow through the source made of conducting material, positive charges are urged from A to I to B and negative charges from B to I to A. Both charge movements are caused by the driving intensity E_D and constitute a current from A to I to B, resulting in resistive intensity E_R (from B to I to A) caused by the flow of charges through the conducting material. (See Art. 4.13.) With current flow, the driving intensity must equal both opposite intensities.

$$E_D = E_{EI} + E_R \tag{4.13}$$

Consequently the intensity E_{EI} is reduced during current flow and a lower charge is maintained on the terminals of the source.

Put in more familiar language, the internal EMF or potential of the source (proportional to E_D) is reduced by the iR drop (proportional to E_R) and gives a lower terminal potential difference (proportional to E_{EI}).

$$\text{EMF} = V_T + iR \tag{4.14}$$

Under these conditions the source supplies power, part of which appears in heat form and the remainder is delivered externally.

$$\underset{\substack{\text{generated}\\\text{power}}}{(\text{EMF})i} = \underset{\substack{\text{externally}\\\text{delivered}\\\text{power}}}{V_T i} + \underset{\substack{\text{heat}\\\text{power}}}{i^2R} \tag{4.15}$$

C. STEADY CURRENTS OF CONDUCTION

4.10. The Transient Period Preceding Steady Conduction. Let us consider a battery or generator of constant potential, which means also a constant driving intensity E_D. Bring a conducting body C to a fixed position in the electric field of the battery. The electric field is approximately that shown in Fig. 4.3. Introducing conductor C increases the capacitance of the electric system between A and B. If Q were fixed in number, the increased capacitance would tend to decrease the electric potential between A and B ($Q = CV$), decreasing the electric intensity E_{EI} in the battery. At once the driving intensity E_D would propel extra charges to the terminals until the potential V was again established, with a consequent increase in Q.

Now let conductor C which possesses induced surface charges $+q$ and $-q$ approach the battery terminals. Charges Q and q both increase in magnitude. Consequently there are currents both in the battery and in the conductor C so long as it is moving. This is the situation during the transient period before C touches the battery.

When C touches both terminals of the battery, a continuous conducting circuit is formed through C and the battery and the charges $+Q$ and $-q$ tend to annul each other. So also do $-Q$ and $+q$, producing a tendency to decrease the potential difference between B and A, as well as the internal electrostatic intensity E_{EI}. Hence we find the driving intensity E_D moving charges to the terminals and beyond, through the conductor C. As C and the battery form a closed circuit, the charges continue to move round and round the circuit, and a continuous current is built up.

Thus the current has grown from an initial steady-state value of zero to a final steady-state value depending upon the resisting forces of the conducting circuit. We shall now investigate the reaction of a conductor to a steady flow of current.

4.11. The Steady-State Period. In order to study the steady-state condition we must first consider the resistive or reaction forces in a conductor which arise to impede the current. Conductors may be

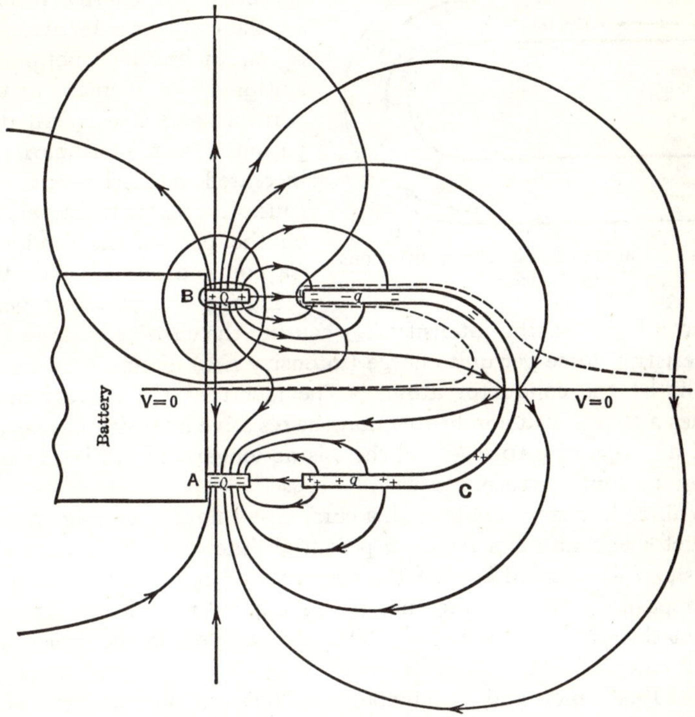

FIG. 4.3. Field about a source of energy.

gases, liquids, or solids, but we shall limit the discussion to solids, as their phenomena are most frequently encountered in engineering practice.

4.12. Steady Conduction Currents in Solid Conductors. We have already seen (Art. 4.2) that the current in solids is really due to the drift motion of a cloud of free electrons moving against the intensity. The positively charged atoms are rigidly fixed and cannot enter into drift motion and thus do not enter the current. In Fig. 4.4, let *BXA* be a conductor connected to the terminals of a battery (*B* to the positive and A to the negative terminal). Then there is an intensity E_{EX} directed from B to X to A, which would send positive charges the same way if they were free to move, but actually succeeds in driving the free electronic cloud from A to X to B. This is described as a positive current from B to X to A. The current flows from higher to lower potentials. The electric field gives up, to the conductor, energy which it received from the source. The energy received by the conductor increases the atomic vibratory motion, resulting in an increase in temperature and consequent dissipation of heat. Ordinarily the increased atomic activity causes a greater resistive effect on the part of the conductor. Carbon is a notable exception.

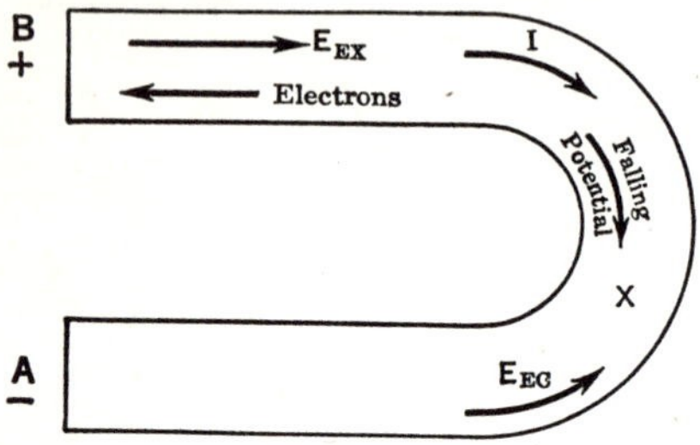

FIG. 4.4. Current and intensity directions in the load.

During the transient period of current growth the intensity E_{EX} causing the current is larger than the resisting force per unit charge (intensity E_{EC}) directed against the current by the conductor atoms. The fact that the current finally reaches a steady value indicates that the resistive intensity has become equal and opposite to E_{EX} and there is no longer any net intensity to cause a current increase.

Speaking in terms of potential, a current $+I$ flowing through a given conductor encounters a resisting potential difference $-V_C$ due to the intensity E_{EC} directed against the current, and requires that an equal and opposite potential difference V_R be applied in order to force the current through the conductor. The current flows in the direction of falling potential V_R.

4.13. Resistance and Conductance. Now we turn to experiment to determine the exact relation between the potential difference applied to the ends of a conductor and the resulting current. Ohm's law states the result for *solid* conductors.

Ohm's Law (Experiment). A given solid conductor, maintained at a given temperature, requires an electric potential difference across its ends that is proportional to the steady current flowing.

$$\text{(Applied)} \quad V_R = RI \quad \text{volts} \tag{4.16}$$

$$\text{(Resistive)} \quad V_C = -V_R = -RI \tag{4.17}$$

Two very useful definitions follow from this experimental relation.

Resistance (Definition). The resistance of a given conductor is the ratio of the potential difference across its ends to the steady current resulting.

$$R = \frac{V}{I} \quad \frac{\text{volts}}{\text{ampere}} = \text{ohms} \qquad \text{(Scalar)} \tag{4.18}$$

The dimensional formula is

$$\text{Resistance } R = \frac{\text{volts}}{\text{ampere}} = \frac{L^2M^1T^{-2}Q^{-1}}{T^{-1}Q} = L^2M^1T^{-1}Q^{-2} \tag{4.19}$$

Conductance (Definition). The conductance of a given conductor is the ratio of the steady current flowing to the potential difference across its ends. The conductance is thus the reciprocal of the resistance.

$$G = \frac{I}{V} = \frac{1}{R} \quad \frac{1}{\text{ohms}} = \text{mhos} \tag{4.20}$$

$$I = GV \tag{4.21}$$

Further experimental study shows that the resistance of a conductor depends upon its temperature. The relationship is rather complicated, so we resort to the empirical expression:

$$R_\theta = R_1[1 + \alpha_1(\theta - \theta_1)] \tag{4.22}$$

in which R_1 is the resistance at temperature θ_1.
R_θ is the resistance at temperature θ.
α_1 is an experimental value depending on the material of the conductor and is applicable over only a small range of temperature about θ_1. α_1 itself is not constant, except within this range.

Table 4.1 gives a few values of α. Manganin is noteworthy for its low value, making it a desirable material for the construction of resistance standards.

4.14. Heat Loss in a Conductor. If a steady current of I amperes flows from B to X to A (Fig. 4.4) over a time T seconds, we may say that $Q = IT$ coulombs move over this path through a falling potential V_R. Consequently the electric field gives up energy to the conductor which appears as heat, and the conductor rises in temperature and dissipates the heat. The heat energy thus produced may be expressed as follows:

$$\text{Heat energy } H = V_RQ = V_RIT \quad \text{joules or watt-seconds} \tag{4.23}$$

$$= I^2RT$$

The time rate of heat production in the conductor is

$$\text{Power } P = V_R \frac{Q}{T} = V_R I \quad \text{watts} \tag{4.24}$$

$$= I^2R = V^2G$$

4.15. Cylindrical Conductors. Not too near the ends of a cylindrical conductor, symmetry shows that the current flows parallel to the axis. Thus the intensity E_{EX} responsible for the current must be parallel to the axis. Then the equipotential surfaces are parallel disks normal to the intensity and are equally spaced for equal increments of potential, giving a uniform electric field. The current density is equal over all parts of the conductor cross section. These statements are true only for steady currents. Currents changing in time introduce other forces that produce a redistribution—notably the skin and proximity effects in alternating currents.

Experiments with cylindrical conductors carrying steady currents show that the resistance of any given conductor is

$$R = \rho \frac{s}{A} = \frac{1}{\sigma} \frac{s}{A} \quad \text{ohms} \tag{4.25}$$

in which ρ is the resistivity, a quantity whose magnitude depends on the material of the conductor and its temperature.
$\sigma = 1/\rho$ is the conductivity.
s is the length of the conductor along the current.
A is the cross section normal to the current.

From this relation we define:

Resistivity (Definition). The resistivity ρ of a given material at a given temperature is the resistance of a 1-meter cube of the material to a current flowing normally to two opposite faces. Its unit is the ohm-meter.

Conductivity (Definition). The conductivity σ of a given material at a given temperature is the conductance of a 1-meter cube, for a current flowing normally to two opposite faces. The conductivity is the reciprocal of the resistivity.

$$\sigma = \frac{1}{\rho} = \frac{1}{\text{ohm-meter}} = \frac{\text{mho}}{\text{meter}} \tag{4.26}$$

Table 4.1 gives the conductivities of various materials.

TABLE 4.1

PROPERTIES OF CONDUCTORS

CONDUCTIVITY, TEMPERATURE COEFFICIENT

Material	Conductivity (σ) Mhos/Meter at 20° C	Temperature Coefficient (α) at 20° C
Aluminum	3.54×10^7	0.0039
Brass (Cu 66%, Zn 34%)	1.39×10^7	.002
Carbon	2.86×10^4	−.0005
Copper, Annealed	5.8×10^7	.00393
Hard Drawn	5.65×10^7	.00382
Iron, 99.98% pure	1.0×10^7	.005
Lead	0.45×10^7	.004
Manganin (Cu 84, Mn 12, Ni 4)	0.23×10^7	.000002
Mercury	0.104×10^7	.00089
Nichrome	0.1×10^7	.0004
Silver	6.14×10^7	.0038
Steel (4% Si)	0.161×10^7	.0008
Tungsten	1.81×10^7	.0045

4.16. Conductor Volume Conditions for Steady Current. Let us consider a small right cylinder completely inside the conductor with its axis parallel to the steady current. The volume is electrically neutral even though a current is flowing, and as many electrons flow in one end as leave by the other during any given time interval. Applying Ohm's law to the cylindrical element of length s and cross section A, the relation between the current density J and electric intensity E_{EX} is obtained. See Fig. 4.5.

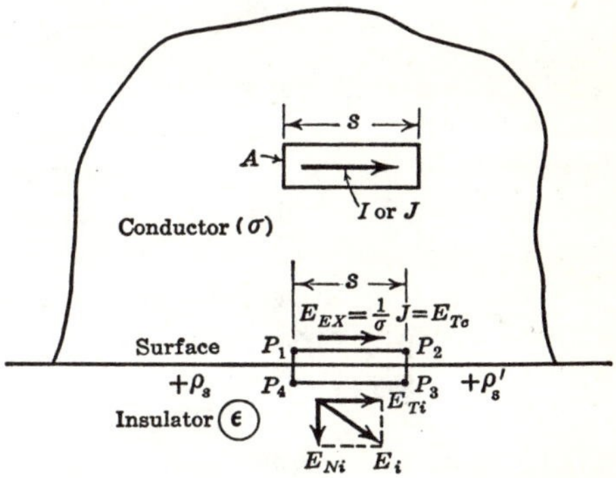

FIG. 4.5. Conductor-insulator boundary conditions with current flowing.

$$V_R = RI \tag{4.27}$$

$$E_{EX} \cdot s = \left(\rho \frac{s}{A}\right)(JA)$$

$$E_{EX} = \rho J$$

$$= \frac{1}{\sigma} J \tag{4.28}$$

The intensity E_{BX} is equal to the current density J times the resistivity, or the current density divided by the conductivity.

4.17. Steady Current in Insulators. There is no perfect insulator, because every material conducts to some extent through its volume. Table 1.5 gives the volume conductivities of a number of insulators. Insulators also permit a leakage of charge over their surfaces. In making apparatus for accurate measurements or in designing high-voltage insulators this must be considered.

4.18. Kirchhoff's Laws for Circuits. Kirchhoff's two laws together with Ohm's law provide a sufficient basis for obtaining a solution of any direct-current or steady-current circuit problem. That is to say, if the potential differences and the resistances of a circuit are given, these laws properly applied give the current solution.

$$\text{Ohm's law:} \qquad V_R = RI \quad \text{(See Art. 4.13)} \tag{4.29}$$

$$\text{Kirchhoff's law I:} \qquad \sum_{\substack{\text{closed}\\ \text{path}}} V = 0 \quad \text{(See Art. 4.7)} \tag{4.30}$$

$$\text{Kirchhoff's law II:} \qquad \sum_{\substack{\text{branch}\\ \text{point}}} I = 0 \tag{4.31}$$

Kirchhoff's second law applies to a branch point where one conductor divides into two or more parts. With a steady current flow, as much current flows up to the point in certain branches as leaves in other branches. As Kirchhoff put the matter, the sum of all currents flowing into the point (or away from the point) is zero.

In translating these laws into algebraic equations, algebraic signs must be carefully observed. In the first law, if we proceed around the circuit in a given direction and if rising potentials are arbitrarily classed as positive, then falling potentials must be considered negative. Similarly in the second law, if currents toward a point are called positive, those away must be called negative.

D. EFFECTS OF CURRENT ON THE ELECTRIC FIELD

4.19. Boundary Conditions. Special study is required of the boundary surfaces separating any two materials having different values of resistivity. These surfaces may be divided into three classes.

(1) *Conductor–Insulator Surface.* (*a*) *Current Flowing in Conductor, Parallel to Surface.* Let us consider a small closed rectangular path $P_1P_2P_3P_4$ as shown in Fig. 4.5. P_1 and P_2, distant s, lie just inside the conductor, while P_3 and P_4 lie just inside the insulator. P_2P_3 and

P_4P_1 are chosen as infinitesimally short distances so that there can be no finite change in potential along them. Now let E_{Tc} and E_{Ti} equal the tangential components of the intensities in the conductor and in the insulator, respectively. Around a closed path we have seen that the potential drop totals zero. Since the difference in potential from P_1 to P_2 is equal to $E_{Tc}s$, and from P_4 to P_3 is $E_{Ti}s$, then $E_{Tc} = E_{Ti}$. The tangential components E_T of the intensities at a conductor–insulator surface are equal and are equal to the current density times its resistivity. (See equation 4.28.)

$$E_{Tc} = E_{Ti} = \rho_c J_c = \frac{1}{\sigma_c} J_c \qquad (4.32)$$

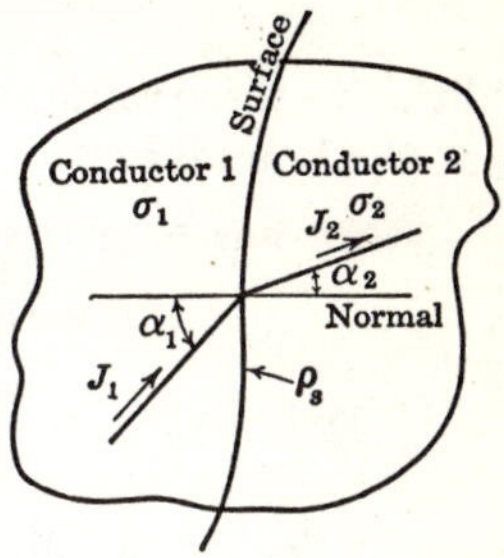

FIG. 4.6. Current density refraction.

On the insulator side of the surface there is, of course, a normal component of intensity E_{Ni} that depends both for its magnitude and direction upon the difference in potential between this particular conductor section and some other conductor. If this conductor is at a higher potential than the other, it possesses a surface charge density ρ_s that is positive, and the intensity E_{Ni} just off the surface is outward.

$$E_{Ni} = \frac{\rho_s}{\epsilon} \quad \text{(Outward for } +\rho_s\text{, occurring on the conductor of higher potential)} \qquad (4.33)$$

The total intensity in the insulator E_i is the vector sum of E_{Ti} and E_{Ni}. The electric flux lines thus leave a conductor carrying a current at an angle, and the conductor is no longer an equipotential.

(*b*) *No Current in Conductor.* In this case, the conductor is an equipotential and the flux lines leave the surface normally.

(2) *Conductor–Conductor Surface.* With different conductivities the vector current density may be refracted at the surface, as shown in Fig. 4.6.

$$\rho_1 J_{1T} = E_{1T} = E_{2T} = \rho_2 J_{2T}$$

or

$$\frac{J_{1T}}{J_{2T}} = \frac{\rho_2}{\rho_1} = \frac{\sigma_1}{\sigma_2} \qquad (4.34)$$

and

$$J_{1N} = J_{2N} \qquad (4.35)$$

or

$$\frac{E_{1N}}{\rho_1} = \frac{E_{2N}}{\rho_2}$$

$$\frac{\tan \alpha_1}{\tan \alpha_2} = \frac{\left(\dfrac{E_{1T}}{E_{1N}}\right)}{\left(\dfrac{E_{2T}}{E_{2N}}\right)} = \frac{E_{2N}}{E_{1N}} = \frac{\rho_2}{\rho_1} = \frac{\sigma_1}{\sigma_2} \tag{4.36}$$

In addition to these refraction conditions, we may note a surface charge layer at the boundary.

$$\begin{aligned}\rho_{s12} &= \epsilon_0(E_{2N} - E_{1N}) = \epsilon_0(\rho_2 J_{2N} - \rho_1 J_{1N}) \\ &= \epsilon_0 J_N(\rho_2 - \rho_1) \\ &= \epsilon_0 J_N\left(\frac{1}{\sigma_2} - \frac{1}{\sigma_1}\right) \quad \frac{\text{coulombs}}{\text{square meter}}\end{aligned} \tag{4.37}$$

We here use ϵ_0 for the dielectric constant since experiment indicates that $\epsilon \cong \epsilon_0$ for the metallic conductors.

(3) *Insulator–Insulator Surface.* The above conductor–conductor surface values apply to an insulator–insulator surface when the insu-

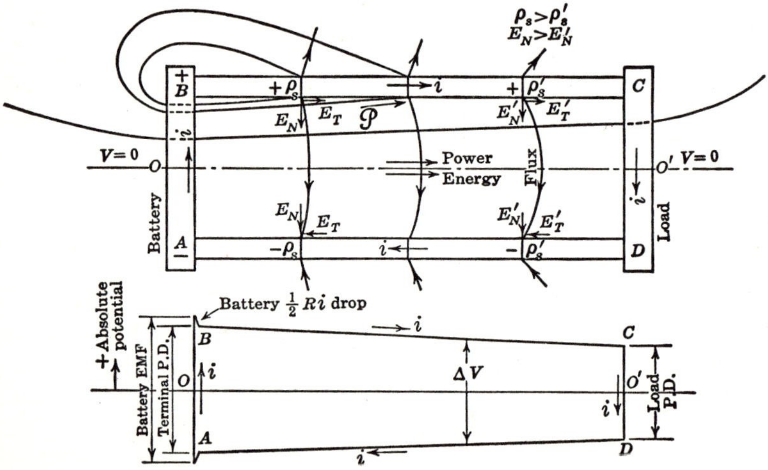

FIG. 4.7. Electric field of a line carrying current.

lators carry a current under steady conditions. Under rapidly alternating field conditions, the refraction is governed almost entirely by the dielectric constants rather than by the resistivities.

4.20. Electric Field of Battery, Line, and Load. For simplicity, consider a two-parallel-wire line carrying current, connected to a battery and a load, as shown in Fig. 4.7. Because of the Ri drop in the circuit the wire is no longer an equipotential. As a consequence, the closed equipotential surfaces, which all cross through the battery, meet either the wire or the load at an angle with the surface, changing direction sharply to close on themselves across the wire or load conductor. The surface equipotential situation is determined by the conductor–insulator boundary conditions.

Near the battery the potential difference is larger than near the load, giving a larger E_N and ρ_s at the wire surface near the battery ($\rho_s > \rho_s'$ and $E_N > E_N'$. E_T is the same at the two ends for equal current densities. Consequently the flux lines are bowed to a larger degree at the load end of the line. Figure 4.7 shows the potential distribution along the line. (See also Art. 16.4.)

E. THE TRANSIENT STATE

4.21. The Transient State. The condition between two differing steady states is called the transient state. Altering the applied voltage or the circuit constants (resistance, inductance, and capacitance) produces temporary currents and voltages that usually die out rather rapidly. The discussion of transient effects caused by an alteration of the magnetic quality of the circuit (inductance) is reserved for Part II.

4.22. Charging a Condenser at Constant Voltage—Time Constant. Let us connect a condenser of capacitance C farads through a wire resistance R to a battery of constant voltage V, as shown in Fig. 4.8*a*. We are assuming that we may separate the resistance and the capacitance of a circuit and "lump" them in separate places. Practice has shown that in many cases this is permissible. As soon as the switch is closed, changing current i flows through the resistance requiring a portion (Ri) of the applied voltage. The remaining voltage is applied to the condenser, or $V_c = q/C$. The condenser charge q grows until its voltage V_c rises to equal the battery voltage, at which time the current becomes zero.

We use Kirchhoff's law $\Sigma V = 0$ to set up the voltage relation. Starting at point A and proceeding clockwise around the circuit with the current, we obtain an equation for voltage that is true for all instants of the charging period.

$$\Sigma V = 0 = +V - Ri - V_c = 0$$

$$+V - R\frac{dq}{dt} - \frac{q}{C} = 0 \tag{4.38}$$

The algebraic signs are:

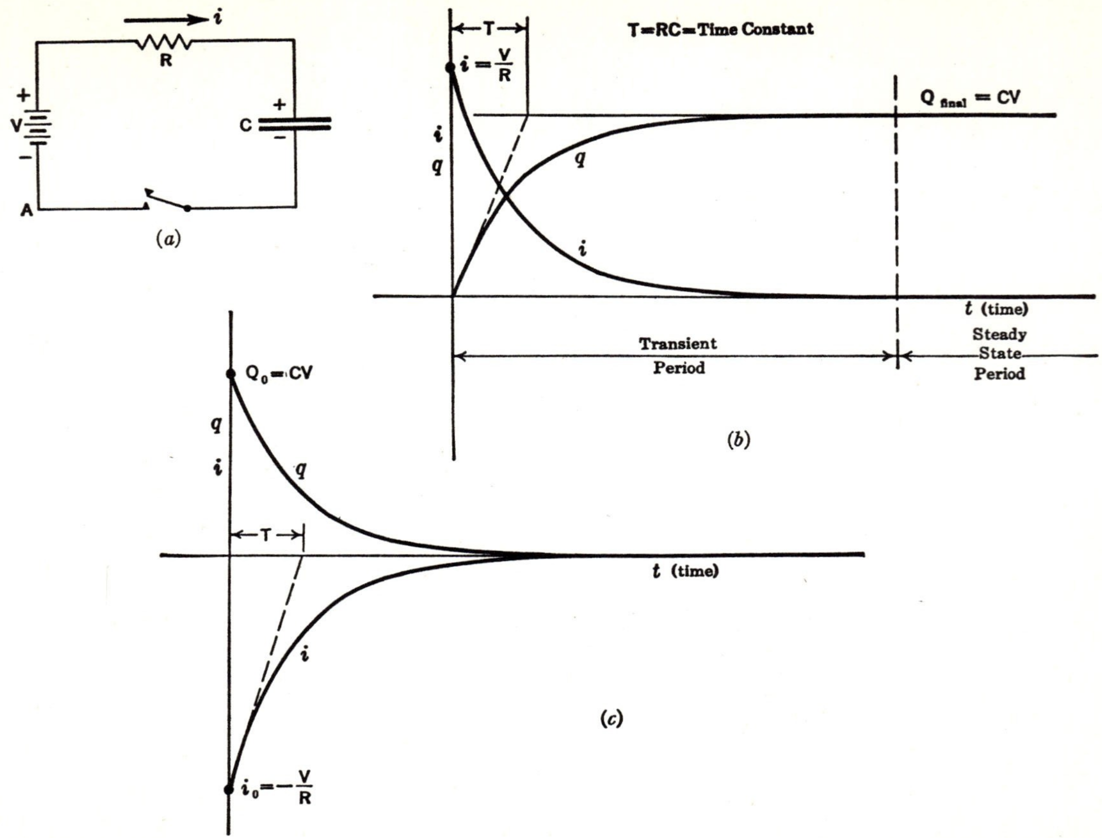

FIG. 4.8. Battery-resistance-capacitance circuit. (*a*) Circuit. (*b*) Capacitance charging characteristic. (*c*) Capacitance discharging characteristic.

$+V$, as a positive charge carried through the battery with the driving intensity, but against the electrostatic intensity, rises to a higher voltage. We shall assume a rise in voltage as positive.

$-Ri$, as a positive charge moving with the current, reaches lower voltages.

$-V_c$, as a positive charge assumed to cross the condenser, goes with the field to lower voltages.

Equation 4.38 may be solved for q as a function of time by separation of the variables.

$$\frac{dq}{q - CV} = -\frac{1}{RC}\,dt$$

$$\ln (q - CV) = -\frac{1}{RC}t + K$$

The integration constant K can be evaluated only if we know the charge at some given instant. Assuming that we count time from the instant of closing the switch, we have this initial condition: At $t = 0$, $q = 0$ (for an uncharged condenser). Then

$$\ln (0 - CV) = 0 + K$$

or

$$K = \ln (-CV)$$

Then the final equation in q is

$$\ln \left(\frac{q - CV}{-CV}\right) = -\frac{1}{RC}t$$

$$q = CV(1 - e^{-(1/RC)t}) \tag{4.39}$$

From this expression the equation for current is obtained by differentiation:

$$i = \frac{dq}{dt} = \frac{V}{R}e^{-(1/RC)t} \tag{4.40}$$

Theoretically, the current never reaches its final value zero, nor the charge its ultimate magnitude $Q = CV$. Practically, the transient period is usually over in a very short time. The current starts with its greatest value, $i = V/R$, as the condenser back voltage at this time is zero. Again practically, the current can never start with a value other than zero. It does so here, apparently, only because the inductive effect of the current which is associated with the inertia of the charges has been neglected.

The transient voltage on the condenser is

$$V_c = \frac{q}{C} = V(1 - e^{-(1/RC)t}) \tag{4.41}$$

In the final state, the battery has given up VQ joules and the condenser has received $\frac{1}{2}VQ$ joules. Hence, under the conditions of this problem only, with constant voltage applied to an initially uncharged condenser, the resistance has received $\frac{1}{2}VQ$ joules which were transformed into heat and dissipated. Under other conditions of charging the heat value will not equal the energy stored in the condenser.

The time constant of a circuit (like this) is defined as the amount of time that would be required to reach the final condition (charge) if the initial rate of change (current) had been continuously maintained.

$$T = \frac{Q_{\text{final}}}{(dq/dt)_{t=0}} = \frac{CV}{(V/R)} = RC \quad \text{seconds} \tag{4.42}$$

4.23. Discharging a Condenser. The discharge equation is obtained by substituting zero for V in equation 4.38.

$$-R\frac{dq}{dt} - \frac{q}{C} = 0$$

$$\frac{dq}{q} = -\frac{1}{RC}\,dt$$

$$\ln q = -\frac{1}{RC}t + K$$

Counting time from the instant of opening the switch,

$$t = 0 \quad \text{when} \quad q = Q = CV$$

$$\ln CV = 0 + K$$

Then

$$\ln\frac{q}{CV} = -\frac{1}{RC}t$$

$$q = CVe^{-(1/RC)t} \tag{4.43}$$

The current is

$$i = \frac{dq}{dt} = -\frac{V}{R}e^{-(1/RC)t} \tag{4.44}$$

During the discharge, the energy stored in the condenser is completely dissipated in heat by the resistance.

4.24. Absorption Current, Capacitance, Voltage Division in a Two-Layer Condenser. This is a continuation of the discussion given at the end of Art. 3.4 where a *constant voltage* is suddenly applied to a condenser. Here let us consider a parallel plate condenser with two layers of insulation, each s meters thick. The plate area of the condenser is A. The two dielectrics have conductivities and dielectric constants σ_1, σ_2, ϵ_1, ϵ_2. First there is the initial charging current, i_N in Fig. 3.3, which has a short duration, corresponding to the geometric capacitance, determined by the two dielectric constants only, and given by two condensers in series. The geometric capacitance is

$$\frac{1}{C_g} = \frac{1}{A}\left(\frac{s}{\epsilon_1} + \frac{s}{\epsilon_2}\right)$$

$$C_g = \frac{A}{s}\left(\frac{\epsilon_1\epsilon_2}{\epsilon_1 + \epsilon_2}\right) \tag{4.45}$$

At this initial instant there is no interface charge density ρ_{s12}, and

$$D = D_1 = D_2 = \epsilon_1 E_1 = \epsilon_2 E_2 \tag{4.46}$$

$$\Delta V = s(E_1 + E_2) \tag{4.47}$$

$$E_1 = \frac{\Delta V}{s}\left(\frac{\epsilon_2}{\epsilon_1 + \epsilon_2}\right) \quad \text{and} \quad E_2 = \frac{\Delta V}{s}\left(\frac{\epsilon_1}{\epsilon_1 + \epsilon_2}\right) \tag{4.48}$$

but the two conduction current densities are not equal,

$$J_1 = \sigma_1 E_1 \neq J_2 = \sigma_2 E_2 \tag{4.49}$$

The inequality of these two currents will later on build up interface charge ρ_{s12}. (We are ignoring the displacement currents; with them included the two currents are equal at all times.)

i_N falls rapidly to the final conduction value i_c, determined entirely by the conductivities. Now the current densities are equal.

$$i_c = J_c A = A\sigma_1 E_1 = A\sigma_2 E_2 \tag{4.50}$$

This is the condition at infinite time. Here also

$$\Delta V = s(E_1 + E_2) \tag{4.51}$$

$$E_1 = \frac{\Delta V}{s}\left(\frac{\sigma_2}{\sigma_1 + \sigma_2}\right)$$

and

$$E_2 = \frac{\Delta V}{s}\left(\frac{\sigma_1}{\sigma_1 + \sigma_2}\right) \tag{4.52}$$

and the interface charge density is obtained from

$$D_2 - D_1 = \rho_{s12} = \epsilon_2 E_2 - \epsilon_1 E_1 \tag{4.53}$$

The capacitance C_0 is greater than C_g because of the interface charge which has been absorbed over an infinite time, or

$$C_0 = C_g + \frac{A\rho_{s12}}{\Delta V} \tag{4.54}$$

Figure 4.9 illustrates the field at $t = \infty$.

In between $t = 0$ and $t = \infty$, the absorption current falls slowly and the capacitance grows slowly from C_g to C_0. To compute the absorption current over its entire range, the displacement current must be included, but this depends on $\partial E/\partial t$, and we shall not carry this problem through to obtain this quantity.

4.25. Nonlinear Resistance. A linear resistance is one in which the value of the resistance remains constant irrespective of the magnitude of the current carried by it. To state it another way, in this type of resistance the potential drop over the resistance is linearly related to the current magnitude. A wire wound resistor fulfills this condition if we ignore the relatively small influence of temperature. See Figs. 4.10*b* and 4.10*e*.

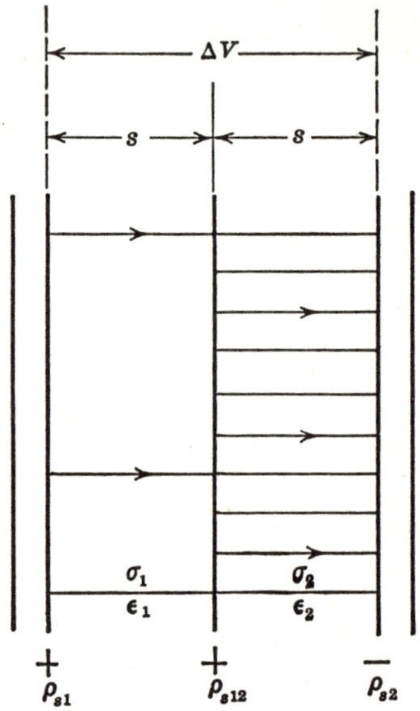

FIG. 4.9. Field in a two-layer condenser with conductivity, at $t = \infty$.

Two characteristics of nonlinear resistances are illustrated in Fig. 4.10. In (*c*) and (*f*), the potential–current and resistance–current relations for Thyrite[1] are shown. The potential increases with current but less than linearly, while the resistance decreases with the growth of current. Thyrite, fabricated in the form of disks 6 inches in diameter by ¾ inch thick, may be used in lightning arresters, shunted across contacts to prevent sparking, placed across field windings to prevent overvoltages and for other similar applications. A 6- by ¾-inch disk conforms to the following laws, expressed in volts, amperes, and ohms:

$$R_T I^{0.72} = 580$$

$$V_T = R_T I = 580\, I^{0.28} \tag{4.55}$$

$$I = 1.365 \times 10^{-10} V_T^{\,3.57}$$

[1] T. Brownlee, "The Calculation of Circuits Containing Thyrite," *Gen. Elec. Rev.*, **37**, 175 (April 1934), and 218 (May 1934).

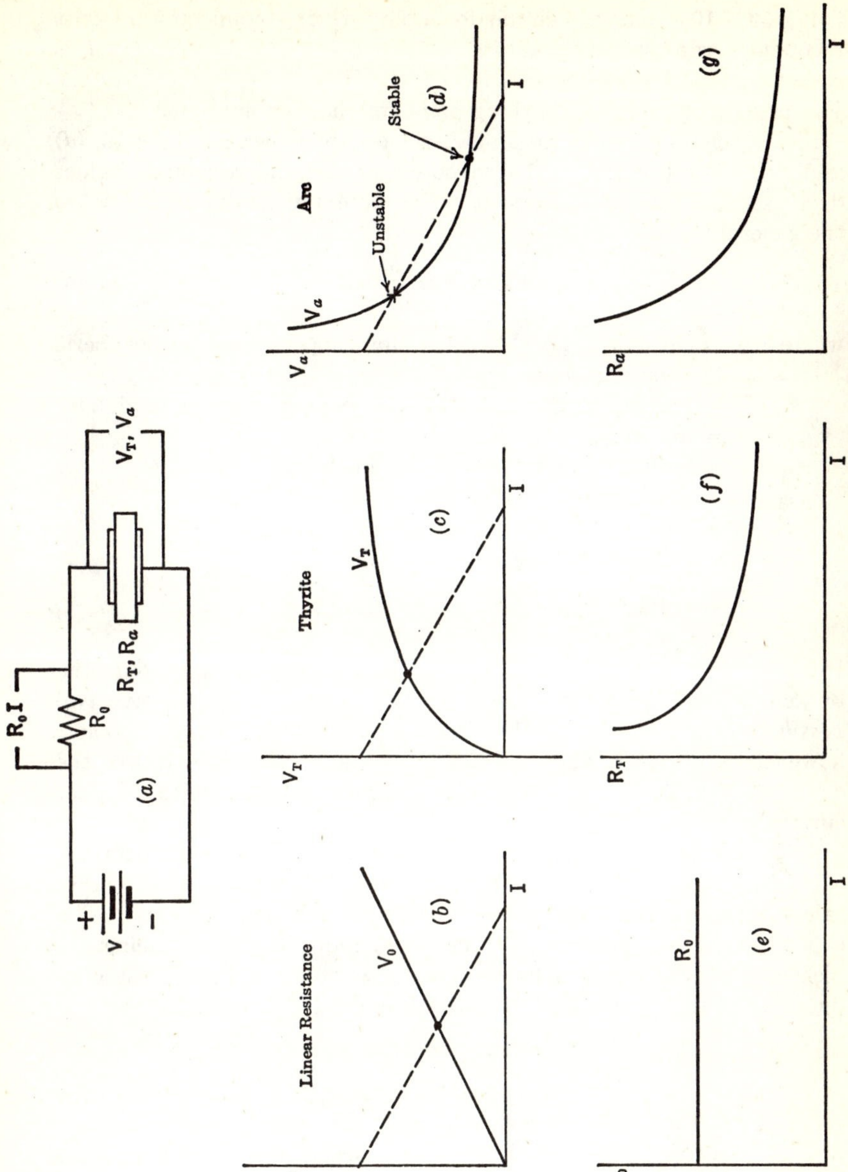

FIG. 4.10. Characteristics of linear and nonlinear resistances.

Figure 4.10*a* gives the circuit for a Thyrite disk connected in series through a wire wound resistance R_0 to a battery V. $V - R_0I$ is shown in the broken line in Fig. 4.10*c*. Where this line crosses the V_T versus I curve is the working point for the combination.

Similar characteristics for an electric arc are shown in Fig. 4.10, (*d*) and (*g*). For the arc, both the potential and the resistance values decrease with increasing current. The potential across an arc takes the general form[1]

$$V_a = A + \frac{Ms}{I^n} \tag{4.56}$$

where A = cathode drop (14.7 volts for copper in air, atmospheric pressure).

M = arc vapor constant (257.6 for copper contacts in air).

s = arc length in inches.

T = boiling temperature of the anode, absolute (2560° K for copper).

$n = 2.62 \times 10^{-4}\,T$ ($n = 0.67$ for copper).

I = current in amperes.

$\dfrac{Ms}{I^n}$ = voltage drop over the arc column.

In series with a wire wound resistor and a battery, $V - R_0I$, the broken line in Fig. 4.10*d*, crosses the V_a versus I curve to give two *possible* working points for the circuit as a whole. The upper point, however, is unstable, and the least disturbance promptly causes the circuit to shift to the lower or stable point of operation with a higher current value.

4.26. Geometric Equivalence of Capacitance and Conductance. The capacitance of two conductors and dielectric is not readily calculable except for simple conductor shapes and arrangements. The capacitance may, however, be measured indirectly by replacing the dielectric with a liquid of known conductivity σ, measuring the conductance of the system G, and calculating C from the relation

$$C = \frac{\epsilon}{\sigma} G \tag{4.57}$$

Equation 4.57 is easily proved. Consider the same voltage difference ΔV between the conductors, first with the dielectric and then with the conducting liquid. The electric flux lines in the dielectric coincide with the flow lines of current in the liquid, and the electric equi-

[1] W. B. Kindy, "Control Contacts and Arc Lengths," *Electric Journal*, **27**, 727 (December 1930).

potential surfaces are identical in both cases. Throughout the media $D = \epsilon E$ and $J = \sigma E$, so that everywhere

$$D = \frac{\epsilon}{\sigma} J \tag{4.58}$$

Integrating D and J over any equipotential surface gives

$$\iint_A D\,dA = \psi = Q = C\,\Delta V \tag{4.59}$$

and

$$\frac{\epsilon}{\sigma} \iint_A J\,dA = \frac{\epsilon}{\sigma} I = \frac{\epsilon}{\sigma} G\,\Delta V \tag{4.60}$$

By virtue of equation 4.58, the two integrals are equal and equation 4.57 follows at once.

The location of the equipotential surfaces in the liquid are readily obtained by using a vacuum tube voltmeter and tracing out points of common voltage. A voltmeter requiring no current for operation is obviously necessary for this task.

Graphical methods for obtaining the dielectric and current flow fields, particularly useful in two-dimensional problems, are discussed in Chapter 7.

PROBLEMS

1. If a current density of 2000 amperes per square centimeter flows through copper, and if it is assumed that each atom contributes one electron to the drifting electronic cloud, compute the drift velocity of the electrons.

2. Explain how positive charges (or negative) flowing around a complete circuit can remove energy from the source and deliver it to the load. What role does resistance play in this process?

3. Suppose that a battery of higher potential is used to force a current backward through the source. What changes take place in the values of the intensities and potentials in the source? How now is energy transferred? In what application is this condition found?

4. Find the ratio of conductivity at 60° C to that at 20° C for annealed copper; for manganin.

5. What is the surface charge density between annealed copper and brass for a normal current density of 200 amperes per square centimeter?

6. A 0.1-microfarad condenser is charged (at constant potential) to 60,000 volts through a resistance of 1.0 megohm. How much energy is lost in heat? How long does it take the condenser to receive one-half its final charge? What would be the effect of using 10 megohms? Compute the time constants.

7. Consider a condenser of 1 microfarad already charged to 2000 volts. Now connect it to a battery with a potential of 3000 volts through a resistance of 1 megohm. Compute the energy lost in the resistance. Repeat for the case in which the polarity of the 3000 volt battery is reversed with respect to the polarity of the condenser.

8. Three condensers, of 1, 2, and 3 microfarads, are placed in parallel. Compute the capacitance. Repeat for a series connection.

9. Write (but do not solve) the differential equations governing the following: A condenser C_1 carrying a charge Q_1 is connected through resistance R to condenser C_2. What is the charge q_2 as a function of time? Current? Sketch the curves.

10. Consider the parallel plate condenser described in Art. 4.24. For A = 1000 sq cm, $s = 0.1$ cm, $\epsilon_{r1} = 2$, $\epsilon_{r2} = 8$, $\sigma_1 = 10^{-7}$, $\sigma_2 = 10^{-6}$, $\Delta V = 1000$ volts, compute:

(a) Vacuum capacitance C_v; geometric capacitance C_g.
(b) At $t = 0$, the
  (1) Conduction current in each layer.
  (2) Division of potential across the two layers.
(c) At $t = \infty$, the
  (1) Final conduction current i_c.
  (2) Interface charge ρ_{s12}.
  (3) Division of potential across the two layers.
  (4) Final capacitance C_0.

11. A 4000-volt battery is connected through constant resistance R_0 to a 6- by ¾-inch Thyrite disk. Compute R_0 if $I = 10$ amperes; if $I = 100$ amperes.

12. A 100-volt battery is connected through constant resistance R_0 to copper electrodes in air, with a spacing of $s = 0.1$ inch. Plot the V_a versus I and R_a versus I characteristics. If the stable current is 1.0 ampere, compute the required value of R_0, the R_0I drop, the arc drop, the cathode drop. Find the values of I and V_a for the unstable point with the same value of R_0. What value of R_0 would permit coincidence of the points of stability and instability?

CHAPTER 5

SOLUTION BY METHOD OF IMAGES

In Chapter 2 only those problems are solved for which the distribution of the charges is assumed to be known, except for the case of two parallel wires carrying equal and opposite types of electricity. It frequently happens in electrical engineering practice that, when a difference of potential is established between two (or more) conductors, the charge distribution on each conductor is altered by the proximity of the other conductor, as a result of induction. The general problem of electrostatic induction is explained in part B of Chapter 3. In this chapter it is proposed to examine certain problems in which the charge distribution is originally unknown, but which may be determined by use of the method of images. Relatively few problems can be solved by this method, but when applicable it reduces a very complex problem to one of simple proportions.

Solution by the method of images consists essentially in replacing the complex charge distribution, conductors, and insulators by a substitute arrangement of charges, conductors, and insulators which has at least one equipotential surface of the same geometric shape as the actual distribution.

5.1. Wire and Ground. Consider a round cylindrical wire (A in Fig. 5.1) of radius R, carrying $+\rho_l$ coulombs per meter, placed center height H above ground. The wire possesses a positive absolute potential V_A while the ground (earth) has zero potential. The electric flux lines, though appearing to emanate from the center of influence at height h above ground, actually run from the positive charges on the wire surface to an equal number of negative charges on the ground surface. The surface charge density $-\rho_s$ on the ground has a maximum directly under the wire at O and declines rapidly as the distance from the wire increases.

The problem as it stands is rather complex, inasmuch as the exact variation of the ground charge density $-\rho_s$ is not known. However, the upper half of Fig. 5.1 is identical with the right half of Fig. 2.14 in Art. 2.13 (for two parallel equal wires) inasmuch as the conductor boundaries of the former figure (wire A and ground) are identical with the wire surface and zero potential of the latter. Consequently, if the conducting volume ground of Fig. 5.1 is removed and is replaced by an insulating volume of the same dielectric constant as that of the upper half of the figure, and if a conductor A' (of the same size as A but carrying charge $-\rho_l$) is placed symmetrically with respect to the zero potential surface, the two wires A and A' give the same field shape as actually exists between wire A and ground. Wire A' is then

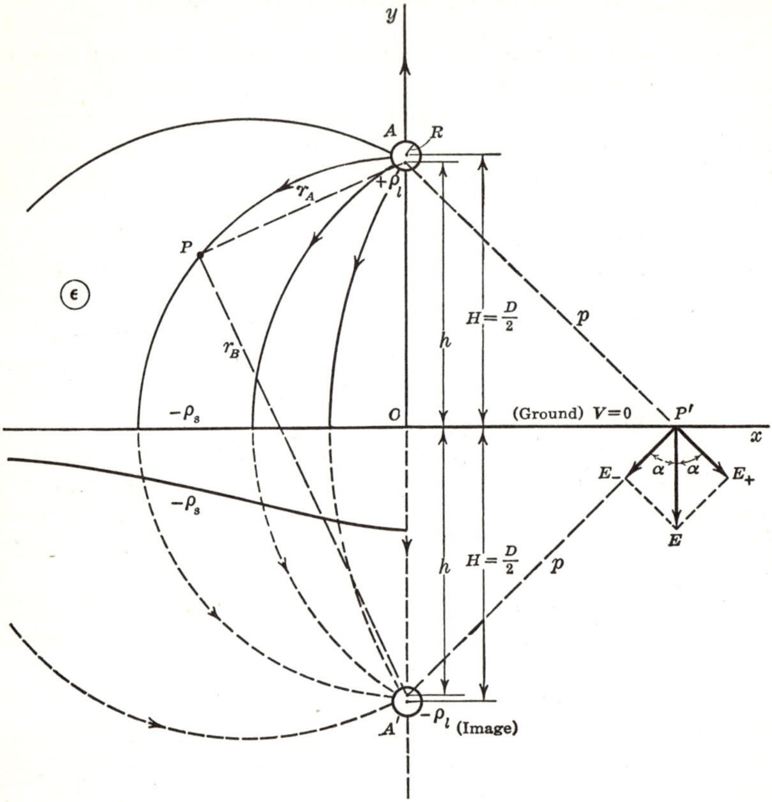

FIG. 5.1. Electric flux lines of wire and ground.

called the *image* of wire A. The method of images may be used only when the unknown situation can be replaced by another having a known solution.

Using values taken from Art. 2.12: The height of the center of influence is

$$h = \sqrt{\left(\frac{D}{2}\right)^2 - R^2} = \sqrt{H^2 - R^2} \tag{5.1}$$

The absolute potential of wire A is

$$V_A = \frac{\rho_l}{2\pi\epsilon} \ln \left[\frac{D}{2R} + \sqrt{\left(\frac{D}{2R}\right)^2 - 1} \right] \tag{5.2}$$

which, for wires small compared with the height $H = D/2$ and $h \cong H$, becomes

$$V_A = \frac{\rho_l}{2\pi\epsilon} \ln \frac{D}{R} \tag{5.3}$$

The absolute potential of any field point P is

$$V_p = \frac{\rho_l}{2\pi\epsilon} \ln \frac{r_B}{r_A} \tag{5.4}$$

The capacitance of wire A to ground, per meter, is

$$C_{AG} = \frac{\rho_l}{V_A} = \frac{2\pi\epsilon}{\ln \left[\frac{D}{2R} + \sqrt{\left(\frac{D}{2R}\right)^2 - 1} \right]} \quad \frac{\text{farads}}{\text{meter}} \tag{5.5}$$

or

$$C_{AG} = \frac{2\pi\epsilon}{\ln \frac{D}{R}} \quad \text{for } D \gg R \tag{5.6}$$

This capacitance is double that between the two wires A and A'.

The charge density on the ground surface is easily computed. For a given charge density $-\rho_s$ on the ground, the electric intensity just off the ground is

$$E = \frac{\rho_s}{\epsilon} \tag{5.7}$$

But E can be computed from the knowledge of $\pm\rho_l$ and their locations. Using point O as origin, the ground surface as the x direction, and OA as the y direction, the intensity at point P' (Fig. 5.1) is

$$E_+ = \frac{\rho_l}{2\pi\epsilon p} = \frac{\rho_l}{2\pi\epsilon(x^2 + h^2)^{1/2}}$$

and its vertical component is

$$E_{+V} = \frac{\rho_l}{2\pi\epsilon p} \cos \alpha = \frac{\rho_l}{2\pi\epsilon} \frac{h}{p^2} = \frac{\rho_l}{2\pi\epsilon} \frac{h}{(x^2 + h^2)}$$

E_- has the same vertical component.

The total intensity E, just off the ground, at P' is

$$E = 2E_{+V} = \frac{\rho_l}{\pi\epsilon} \frac{h}{(x^2 + h^2)} \tag{5.8}$$

and the charge density is

$$\rho_s = -\epsilon E = -\frac{\rho_l}{\pi}\left(\frac{h}{x^2 + h^2}\right) \quad \frac{\text{coulombs}}{\text{square meter}} \tag{5.9}$$

which is given by the curve $-\rho_s$ in Fig. 5.1. The area under the ρ_s curve gives the total charge on a strip of ground which is 1 meter parallel to the wire direction and of infinite width. This charge must therefore equal the line charge per meter ρ_l.

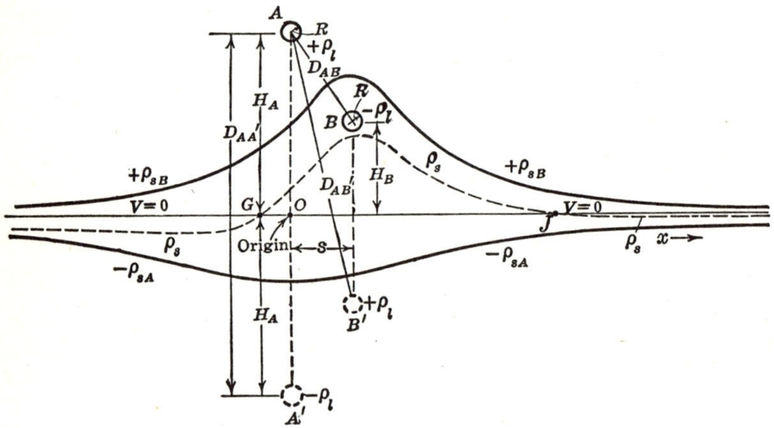

FIG. 5.2. Two wires and ground.

It is easy to see that any distribution of electricity above ground may be imaged by a symmetrical distribution of electricity of the opposite type placed below ground.

5.2. Interference. An *isolated* conducting wire, parallel to A, placed at any field point as P in Fig. 5.1, must assume the absolute potential of that point, given by equation 5.4. If, now, the absolute potential of wire A is changing in time (as well as ρ_l), owing to the connection of the wire and ground to an alternating-current generator, the absolute potential of the wire P must also change. The creation and disturbance of the potential of one conductor by the electric field of another conductor is known as *electric interference.*

In addition there is another form of troublesome interaction between parallel lines, known as *inductive interference*. A changing current in one line creates a changing *magnetic* field which induces a potential in the second line. Frequent transposition of the wires in each circuit materially reduces the undesirable effects of inductive interference.

5.3. Two Wires and Ground. Consider two nongrounded circular wires A and B, of radii R_A and R_B, placed heights H_A and H_B above ground and separated by a horizontal distance s, as shown in Fig. 5.2. For simplicity, let us assume that the wires are small compared with the heights above ground ($R \ll H_A$ or H_B). Now suppose that a direct-current generator of constant potential is connected to wires A and B (making $A+$), establishing between them a potential difference of ΔV_{AB}. The value of the charge per meter of length ($+\rho_l$ on A and $-\rho_l$ on B) is different, in the presence of ground, from that for the two wires acting alone. Obviously, in the light of the method of images, we must study the field not only of the two wires A and B but also that of their two images A' and B'.

The absolute potential of wire A is determined in part by the charges on A and A' and in part by charges on B and B' and is

$$V_A = +\frac{\rho_l}{2\pi\epsilon}\ln\frac{D_{AA'}}{R_A} - \frac{\rho_l}{2\pi\epsilon}\ln\frac{D_{AB'}}{D_{AB}} = \frac{\rho_l}{2\pi\epsilon}\ln\frac{D_{AA'}D_{AB}}{R_A D_{AB'}} \tag{5.10}$$

Similarly, the absolute potential of wire B is

$$V_B = -\frac{\rho_l}{2\pi\epsilon}\ln\frac{D_{BB'}}{R_B} + \frac{\rho_l}{2\pi\epsilon}\ln\frac{D_{BA'}}{D_{AB}} = -\frac{\rho_l}{2\pi\epsilon}\ln\frac{D_{BB'}D_{AB}}{R_B D_{BA'}} \tag{5.11}$$

The difference in potential between the wires A and B is

$$\Delta V_{AB} = V_A - V_B = \frac{\rho_l}{2\pi\epsilon}\ln\frac{D_{AA'}D_{BB'}D_{AB}^2}{R_A R_B D_{AB'}^2} \tag{5.12}$$

The capacitance per meter of wire A to ground (in the presence of B) is

$$C_{AgB} = \frac{\rho_l}{V_A} = \frac{2\pi\epsilon}{\ln\dfrac{D_{AA'}D_{AB}}{R_A D_{AB'}}} \quad \frac{\text{farads}}{\text{meter}} \tag{5.13}$$

that of wire B to ground (in the presence of A) is

$$C_{BgA} = \frac{-\rho_l}{V_B} = \frac{2\pi\epsilon}{\ln\dfrac{D_{BB'}D_{AB}}{R_B D_{BA'}}} \quad \frac{\text{farads}}{\text{meter}} \tag{5.14}$$

and that of the two wires (in the presence of ground) is

$$C_{ABg} = \frac{\rho_l}{\Delta V_{AB}} = \frac{2\pi\epsilon}{\ln \frac{D_{AA'}D_{BB'}D_{AB}^2}{R_A R_B D_{AB'}^2}} \quad \frac{\text{farads}}{\text{meter}} \tag{5.15}$$

The total charge on the ground must net zero. However, on a certain area directly under wire B, the charge density is positive, but on both sides of this area the charge density is negative. Using the

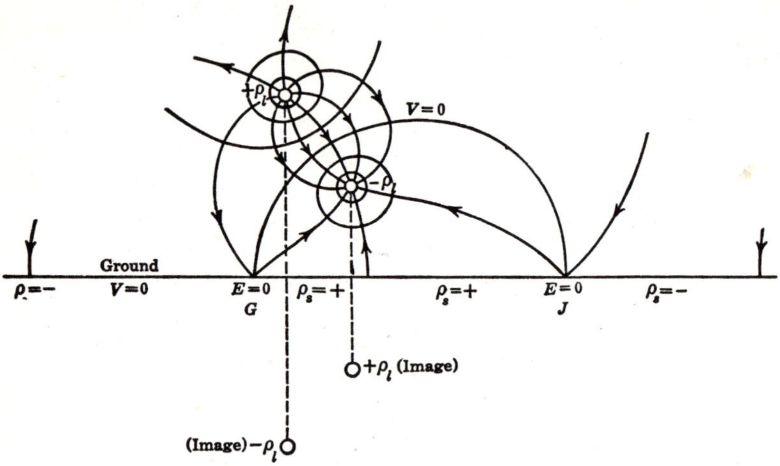

FIG. 5.3. Two-wire transmission line and ground.

point O in Fig. 5.2 as origin, the charge density is computed readily as the sum of the separate effects of wires A and B.

$$\rho_s = \rho_{lA} + \rho_{lB}$$

$$\rho_s = -\frac{|\rho_{lA}|}{\pi}\frac{H_A}{x^2 + H_A^2} + \frac{|\rho_{lB}|}{\pi}\frac{H_B}{(x-s)^2 + H_B^2} \quad \frac{\text{coulombs}}{\text{square meter}} \tag{5.16}$$

The flux lines and equipotential surfaces for this case are shown in Fig. 5.3. At lines G and J, the charge density, being neither positive nor negative, is zero, and these lines are singular lines.

5.4. Two Wires of Same Height and Ground. This is evidently a special case of the preceding situation. It is worth noting, however, that in this case, as shown in Fig. 5.4, the zero potential surface is formed by two perpendicular planes bisecting the space between the lines and their images. The charge density follows the curve ρ_{sg}.

5.5. Wire Parallel to 90° Air Corner. Referring again to Fig. 5.4, it is readily seen that, if wire B is placed near a conductor whose immediately adjacent surface is composed of two planes intersecting at 90°, the preceding case provides a solution, if A, A', and B' are chosen

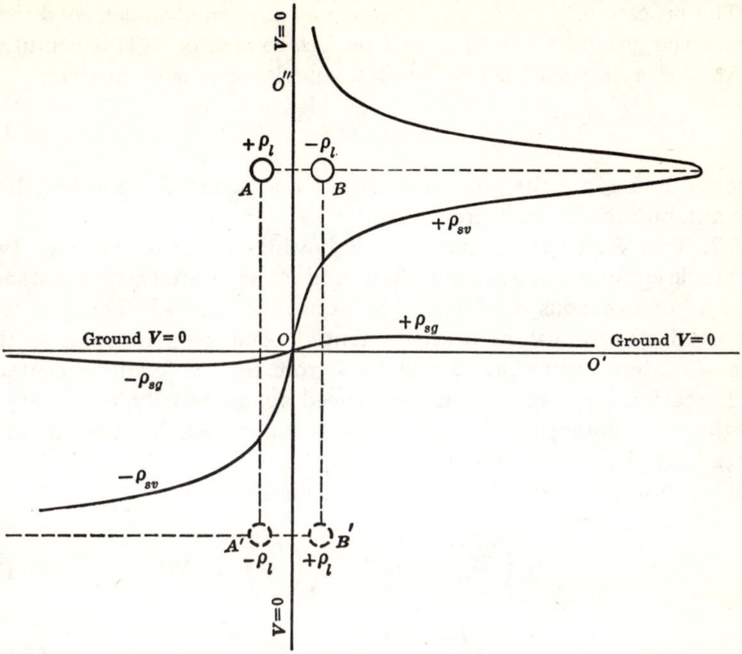

FIG. 5.4. Two wires at the same height above ground. (Also wire in a 90° corner.)

as images. In addition to the charge density ρ_{sg} along the plane $O'O$, there is also a charge density ρ_{sv} along the plane OO''.

5.6. Imaging Horizontal and Vertical Currents. In the foregoing articles we have seen that a positive charge above ground is imaged by an equal negative charge equidistant below ground. Extending this principle to moving charges or currents it may be noted that:

(1) Horizontal currents above ground must be imaged by *reverse* currents symmetrically disposed below ground.

(2) Vertical currents, conversely, must be imaged by currents in the *same* direction.

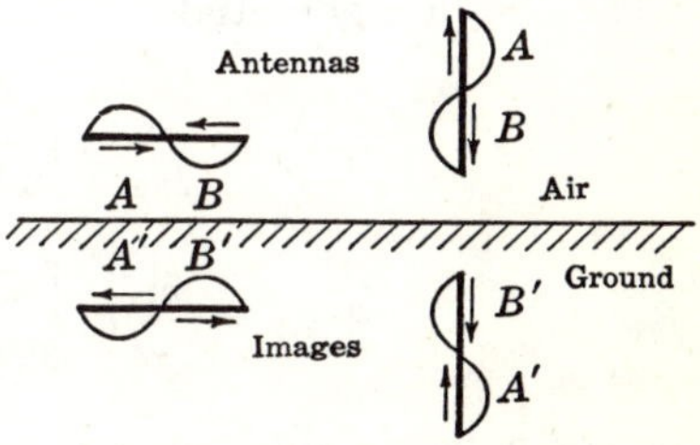

FIG. 5.5. Imaging horizontal and vertical currents. (Full wave length antenna illustrated.)

As an illustration, Fig. 5.5 shows the correct method of imaging a full wave length antenna with a sinusoidally distributed current, when placed both horizontally and vertically above ground.

The process of imaging as outlined above is valid under conditions where the ground approximates a perfect conductor. This condition is fulfilled with static or low-frequency fields, or in general when

$$60\sigma\lambda_0 \gg \epsilon_r \tag{5.17}$$

where σ and ϵ_r are the ground constants and λ_0 is the wave length of the antenna radiation in free space.

5.7. Two Eccentric Cylinders. Following Fig. 5.6, consider two eccentric circular cylinders of radii R_1 and R_2, separated center distance s by a homogeneous insulator of dielectric constant ϵ. This problem is solved by establishing a zero potential plane perpendicular to the line of centers, distant $D_1/2$ and $D_2/2$ from the two cylinder centers, and creating two image cylinders placed symmetrically with respect to the zero potential plane. Then the equations for two parallel wires (Art. 2.12) may be applied to this case.

First locate the position of the zero potential plane.

$$h = \sqrt{\left(\frac{D_2}{2}\right)^2 - R_2^2} = \sqrt{\left(\frac{D_1}{2}\right)^2 - R_1^2} \tag{5.18}$$

$$s = \frac{D_2}{2} - \frac{D_1}{2} \tag{5.19}$$

These two equations may be solved for D_2 and D_1, and from the first h may be found, which locates the positions of the centers of influence, where charge ρ_l is assumed.

The absolute potentials of the two wires are

$$+V_1 = \frac{\rho_l}{2\pi\epsilon} \ln \left[\frac{D_1}{2R_1} + \sqrt{\left(\frac{D_1}{2R_1}\right)^2 - 1}\right] \tag{5.20}$$

$$+V_2 = \frac{\rho_l}{2\pi\epsilon} \ln \left[\frac{D_2}{2R_2} + \sqrt{\left(\frac{D_2}{2R_2}\right)^2 - 1}\right] \tag{5.21}$$

and the potential difference is

$$\Delta V_{12} = V_1 - V_2 = \frac{\rho_l}{2\pi\epsilon} \ln \frac{\dfrac{D_1}{2R_1} + \sqrt{\left(\dfrac{D_1}{2R_1}\right)^2 - 1}}{\dfrac{D_2}{2R_2} + \sqrt{\left(\dfrac{D_2}{2R_2}\right)^2 - 1}} \quad \text{volts} \tag{5.22}$$

The capacitance between the two cylinders is

$$C_{12} = \frac{\rho_l}{\Delta V_{12}} = \frac{2\pi\epsilon}{\ln \dfrac{\dfrac{D_1}{2R_1} + \sqrt{\left(\dfrac{D_1}{2R_1}\right)^2 - 1}}{\dfrac{D_2}{2R_2} + \sqrt{\left(\dfrac{D^2}{2R_2}\right)^2 - 1}}} \quad \frac{\text{farads}}{\text{meter}} \tag{5.23}$$

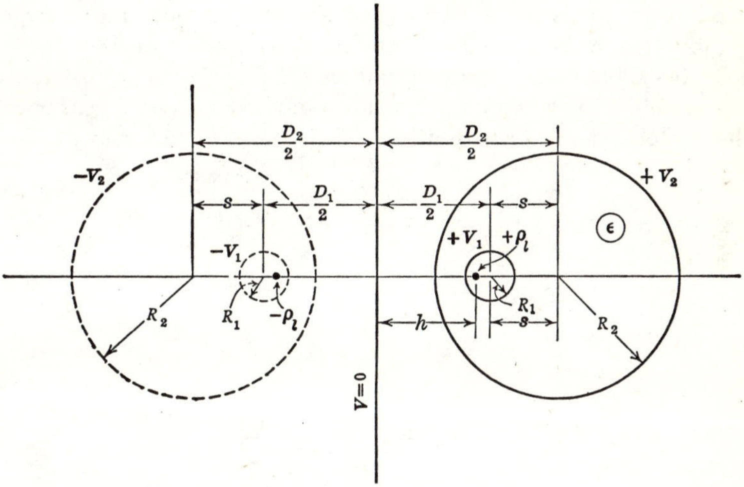

FIG. 5.6. Two eccentric cylinders.

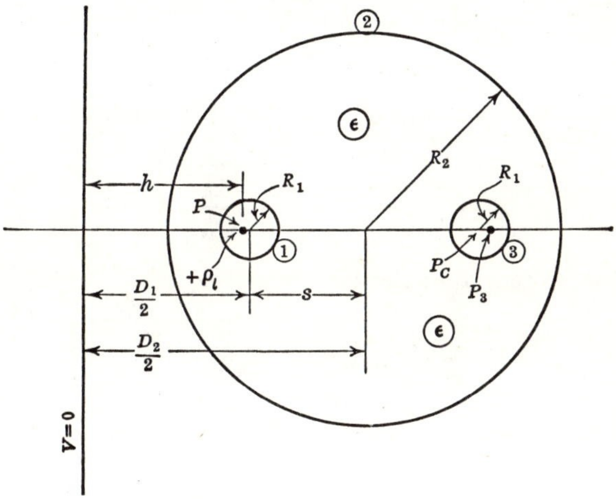

FIG. 5.7. Two-wire cable.

5.8. Two-Conductor Cable. The preceding case may be extended to cover the case of two equal circular wires placed symmetrically inside a metallic sheath, provided that the wire radius R_1 is small compared with that of the sheath R_2, and that the insulation is strictly homogeneous. See Fig. 5.7.

Assume a line charge ρ_l placed inside wire 1 at the center of influence P, as determined by the geometry of wire 1 and the sheath. This charge creates absolute potentials $+V_1$ and $+V_2$ on wire 1 and the sheath respectively, if the presence of wire 3 is ignored. Line charge ρ_l, by induction, creates two equal and opposite charges on wire 3, whose inductive effect back on wire 1 and on the sheath charge distribution is negligible if the wire radius is small compared with the sheath radius. The line charge creates potential $+V_3$ on wire 3. Assuming that

$$R_1 \ll R_2$$

$$h = \sqrt{\left(\frac{D_1}{2}\right)^2 - R_1^2} \cong \frac{D_1}{2}$$

$$= \sqrt{\left(\frac{D_2}{2}\right)^2 - R_2^2}$$

$$h + s \cong \frac{D_1}{2} + s = \frac{D_2}{2}$$

and

$$\frac{D_1}{2R_1} + \sqrt{\left(\frac{D_1}{2R_1}\right)^2 - 1} \cong \frac{2h}{R_1}$$

Assume that the line charge at P is $+\rho_l$. Then the absolute potential of wire 1 is

$$+V_1 = \frac{\rho_l}{2\pi\epsilon} \ln \frac{2h}{R_1} \tag{5.24}$$

The absolute potential of the sheath is

$$+V_2 = \frac{\rho_l}{2\pi\epsilon} \ln \left[\frac{D_2}{2R_2} + \sqrt{\left(\frac{D_2}{2R_2}\right)^2 - 1}\right] \tag{5.25}$$

The absolute potential at point P_C, the center of wire 3, is

$$+V_3 = \frac{\rho_l}{2\pi\epsilon} \ln \frac{2(h+s)}{2s} = \frac{\rho_l}{2\pi\epsilon} \ln \frac{h+s}{s} \tag{5.26}$$

The difference of potential between wires 1 and 3 is

$$\Delta V_{13} = V_1 - V_3 = \frac{\rho_l}{2\pi\epsilon}\left(\ln\frac{2h}{R_1} - \ln\frac{h+s}{s}\right)$$

$$= \frac{\rho_l}{2\pi\epsilon}\ln\left(\frac{2s}{R_1}\cdot\frac{h}{h+s}\right)$$

$$= \frac{\rho_l}{2\pi\epsilon}\ln\left(\frac{2s}{R_1}\cdot\frac{R_2^2 - s^2}{R_2^2 + s^2}\right) \tag{5.27}$$

The last step is readily shown, as

$$R_2^2 = \left(\frac{D_2}{2}\right)^2 - \left(\frac{D_1}{2}\right)^2$$

$$s^2 = \left(\frac{D_2}{2} - \frac{D_1}{2}\right)^2$$

and a few steps will show that

$$\frac{R_2^2 - s^2}{R_2^2 + s^2} = \frac{h}{h+s}$$

Now reverse the situation by putting $-\rho_l$ at the center of influence P_3 on wire 3, which is symmetrically placed with respect to P on wire 1, removing $+\rho_l$ from P. This gives the same potentials (but with reversed sign) on wires 3, 2, and 1 that were previously found on wires 1, 2, and 3, respectively.

Superposing the two fields gives the field that exists when a potential difference is established between the two wires under conditions of normal cable operation. The absolute potential of wire 1 becomes

$$V_1' = (+V_1) - (+V_3) = \frac{\rho_l}{2\pi\epsilon}\ln\left(\frac{2s}{R_1}\cdot\frac{R_2^2 - s^2}{R_2^2 + s^2}\right) \tag{5.28}$$

$$V_1' = -V_3'$$

and equals, with negative sign attached, the absolute potential of wire 3. The sheath, subjected equally to the two separate field effects, is zero potential

$$V_{\text{sheath}} = 0 \tag{5.29}$$

The difference in potential between wires 1 and 3, for normal cable use, is

$$\Delta V_{13}' = 2\Delta V_{13} = \frac{\rho_l}{\pi\epsilon}\ln\left(\frac{2s}{R_1}\cdot\frac{R_2^2 - s^2}{R_2^2 + s^2}\right) \quad \text{volts} \tag{5.30}$$

and the capacitance between the two wires (in the presence of the sheath) is

$$C_{13s} = \frac{\rho_l}{\Delta V_{13}'} = \frac{\pi\epsilon}{\ln\left(\dfrac{2s}{R_1}\cdot\dfrac{R_2^2 - s^2}{R_2^2 + s^2}\right)} \quad \frac{\text{farads}}{\text{meter}} \tag{5.31}$$

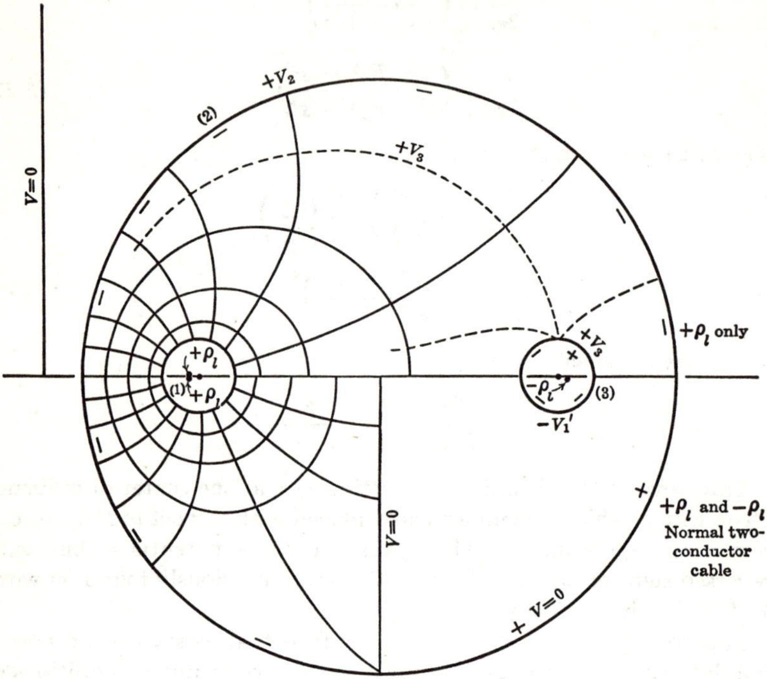

Fig. 5.8. Field of eccentric cylinders (upper part). Field of a two-wire cable (lower part).

The upper part of Fig. 5.8 shows the field that exists in the presence of $+\rho_l$ alone, and the lower part shows the field of both $+\rho_l$ and $-\rho_l$.

For the purpose of calculating the charging current in each wire according to

$$i = C_{10s}\frac{dV_{10}}{dt} \tag{5.32}$$

we define the capacitance to neutral as

$$C_{10s} = 2C_{13s} = \frac{2\pi\epsilon}{\ln\left(\dfrac{2s}{R_1}\cdot\dfrac{R_2^2 - s^2}{R_2^2 + s^2}\right)} \quad \frac{\text{farads}}{\text{meter}} \tag{5.33}$$

with V_{10} equal to the potential difference between one wire and the neutral (center) point. V_{10}, for this geometry, is $\frac{1}{2}V_{13}$.

5.9. Capacitance of Three-Conductor Cable. The three-conductor cable field can be solved in a similar manner to that used in the preceding paragraph except that three images at 120° angular spacing are required. The capacitance to neutral then proves to be

$$C_{10\,s} = \frac{4\pi\epsilon}{\ln\left(\dfrac{3s^2}{R_1^2} \cdot \dfrac{(R_2^2 - s^2)^3}{(R_2^6 - s^6)}\right)} \quad \frac{\text{farads}}{\text{meter}} \tag{5.34}$$

subject to the restriction that $R_1 \ll R_2$.

5.10. The Point and Sphere. Consider a point charge $+Q$ brought up to a distance f from the center of a grounded metallic sphere of radius R, as shown in Fig. 5.9. This problem may be solved by replacing the grounded sphere by an image charge $-q$ placed at the center of influence P_0. If the image charge has the right value and is correctly placed, the two charges establish a zero potential surface which is the sphere. The absolute potential of any point P in space is

$$V_p = \frac{+Q}{4\pi\epsilon r_1} + \frac{-q}{4\pi\epsilon r_2} \tag{5.35}$$

If the zero potential surface in particular is desired,

$$V_p = 0 = \frac{+Q}{4\pi\epsilon r_1} + \frac{-q}{4\pi\epsilon r_2} \tag{5.36}$$

or

$$\frac{r_1}{r_2} = \frac{Q}{q} = p \qquad \text{(a constant)} \tag{5.37}$$

which is the equation of a sphere. p is found as follows:

At point P' on the zero potential surface,

$$\frac{r_1}{r_2} = \frac{Q}{q} = p = \frac{f - R}{R - s}$$

At point P'', also on the zero potential surface,

$$\frac{r_1}{r_2} = \frac{Q}{q} = p = \frac{f + R}{R + s}$$

Therefore

$$p = \frac{f - R}{R - s} = \frac{f + R}{R + s}$$

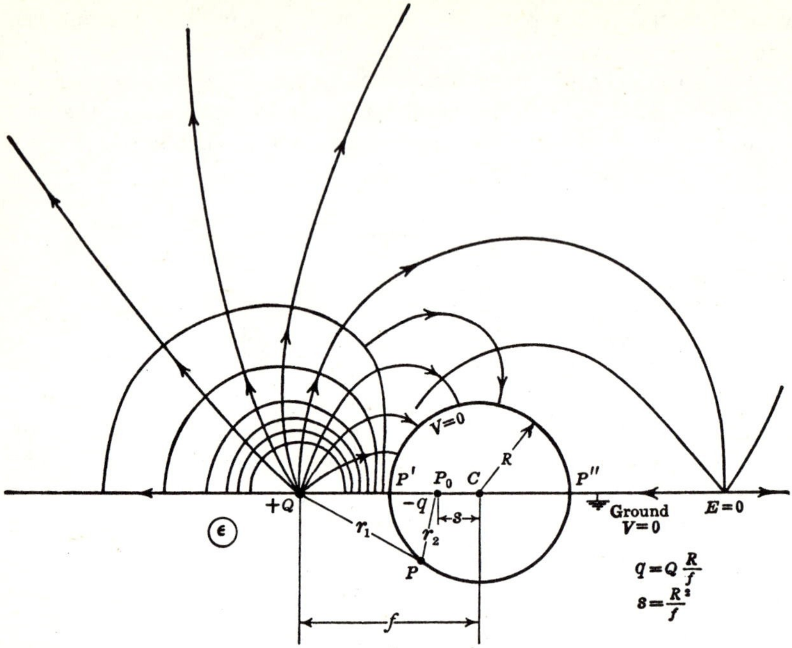

Fig. 5.9. Field of a point charge and grounded sphere. (See also Fig. 1.27.)

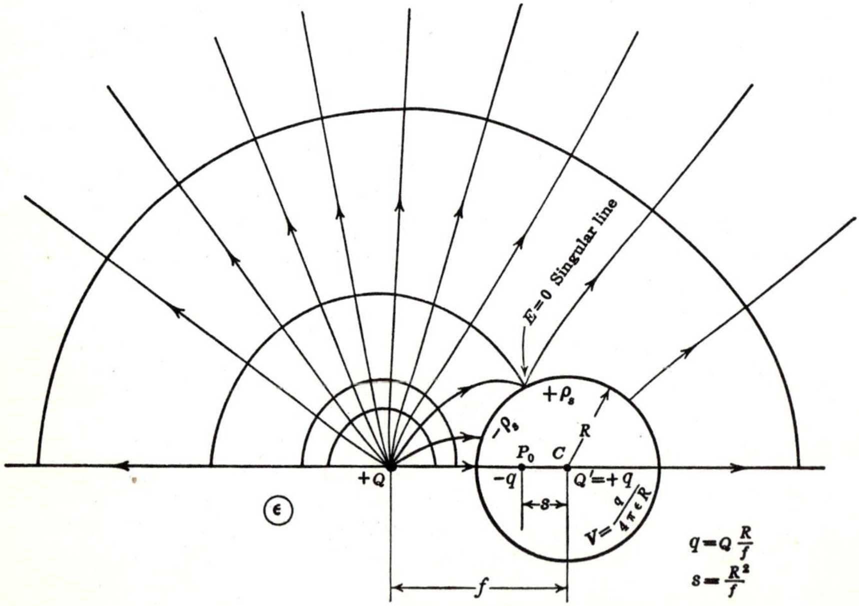

Fig. 5.10. Field of a point charge and an isolated sphere.

from which

$$s = \frac{R^2}{f} \tag{5.38}$$

Also

$$p^2 = \frac{f - R}{R - s} \cdot \frac{f + R}{R + s} = \frac{f^2 - R^2}{R^2 - s^2} = \frac{f^2 - R^2}{R^2 - \left(\dfrac{R^2}{f}\right)^2}$$

$$= \frac{f^2}{R^2}$$

and

$$p = \frac{f}{R} = \frac{Q}{q} \tag{5.39}$$

Therefore, in order to obtain the field when a definite charge $+Q$ is placed center distance f from a zero potential sphere of radius R, we must put

$$q = -Q\frac{R}{f} \tag{5.40}$$

at point P_0, which is distant s from the sphere center, where

$$s = \frac{R^2}{f} \tag{5.41}$$

The charge density on the sphere is obtained from a study of the triangles QPP_0 and PP_0C in Fig. 5.9. The components of intensity along QP and PP_0 are

$$E_1 = \frac{Q}{4\pi\epsilon r_1^2} \tag{5.42}$$

$$E_2 = \frac{q}{4\pi\epsilon r_2^2} = \frac{Q\dfrac{R}{f}}{4\pi\epsilon r_1^2\left(\dfrac{R}{f}\right)^2} = E_1\frac{f}{R} \tag{5.43}$$

Thus the components of intensity stand to each other in the ratio

$$\frac{E_2}{E_1} = \frac{f}{R} = \frac{r_1}{r_2} = p \tag{5.44}$$

and, by similar triangles, the whole intensity E stands to E_1 as QP_0 stands to PP_0. Substituting for s and r_2, we get

$$E = E_1\frac{f - s}{r_2} = \frac{Q}{4\pi\epsilon r_1^3 R}(f^2 - R^2) \tag{5.45}$$

which is a vector along the inward normal to the sphere surface. The charge density is

$$\rho_{sP} = D = \epsilon E = -\frac{Q}{4\pi r_1{}^3 R}(f^2 - R^2) \tag{5.46}$$

which varies from a maximum at P' to a minimum at P''.

The force of attraction between Q and the grounded sphere is

$$\text{Force} = \frac{Qq}{4\pi\epsilon(f - s)^2} \tag{5.47}$$

The sphere may now be changed to any potential except zero by putting a charge Q' at C, the sphere center, and removing the ground.

$$\pm V_{\text{sphere}} = \frac{\pm Q'}{4\pi\epsilon R} \tag{5.48}$$

In this event, the total sphere charge becomes

$$Q_{\text{sphere}} = -q \pm Q' \tag{5.49}$$

For the isolated sphere, shown in Fig. 5.10, $Q' = +q$.

5.11. Two Equal Spheres. High voltages in air are usually measured by changing the spacing of a sphere gap, composed of equal-sized

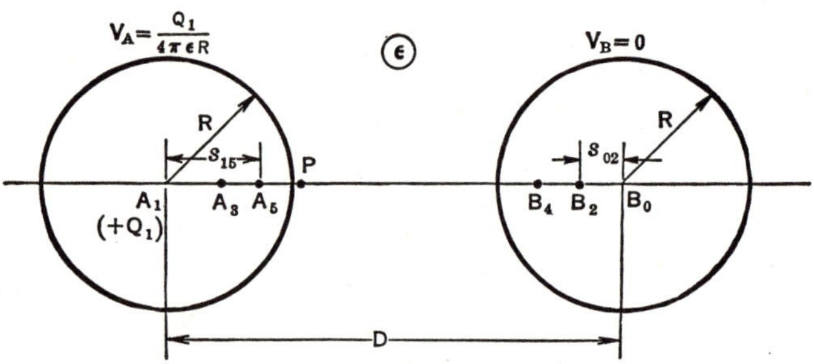

FIG. 5.11. Images for two equal spheres.

spheres, until spark-over occurs. The problem of two spheres may be solved with the help of the preceding case of the point and sphere and by using the process of *successive imaging*. This method is explained with the aid of Fig. 5.11. Consider two spheres, each of radius R, separated center distance D. To solve this problem, first find a series of image charges which gives sphere A a definite positive potential and makes sphere B zero potential, and then, by reversing the image system and changing the signs of the charges, makes sphere B acquire an equal (to A) negative potential while making sphere A

zero. Superpose these two image sets to establish twice the potential difference between A and B, and now A has as much positive potential as B has negative value. Finally, fill in the spheres with conducting material.

To make A positive potential and B zero, using the simplification $p = R/D$: At center point A_1, put charge $+Q_1$. This gives A potential

$$V_A = +\frac{Q_1}{4\pi\epsilon R}$$

and gives B potential

$$V_B = \frac{Q_1}{4\pi\epsilon D}$$

At point B_2, distance

$$s_{02} = \frac{R^2}{D} = B_0B_2 = R \cdot p$$

put charge

$$Q_2 = -Q_1\frac{R}{D} = -Q_1p$$

This makes

$$V_B = 0$$

and makes

$$V_A = +\frac{Q_1}{4\pi\epsilon R} + \frac{Q_2}{4\pi\epsilon\left(D - \dfrac{R^2}{D}\right)} = +\frac{Q_1}{4\pi\epsilon R} + \frac{Q_2}{4\pi\epsilon D(1 - p^2)}$$

At point A_3, distant

$$s_{13} = \frac{R^2}{\left(D - \dfrac{R^2}{D}\right)} = A_1A_3 = R \cdot \frac{p}{1 - p^2}$$

put charge

$$Q_3 = -Q_2 \cdot \frac{R}{\left(D - \dfrac{R^2}{D}\right)} = -Q_2 \cdot \frac{p}{1 - p^2}$$

$$= +Q_1\frac{R}{D} \cdot \frac{R}{\left(D - \dfrac{R^2}{D}\right)} = +Q_1\frac{p^2}{1 - p^2}$$

This makes V_A its original value $V_A = +Q_1/(4\pi\epsilon R)$ and makes

$$V_B = 0 + \frac{Q_3}{4\pi\epsilon\left[D - \dfrac{R^2}{\left(D - \dfrac{R^2}{D}\right)}\right]} = 0 + \frac{Q_3}{4\pi\epsilon D\left(1 - \dfrac{p^2}{1 - p^2}\right)}$$

At point B_4, distant

$$s_{04} = \frac{R^2}{\left[D - \dfrac{R^2}{\left(D - \dfrac{R^2}{D}\right)}\right]} = R \cdot \frac{p}{1 - \dfrac{p^2}{1 - p^2}}$$

put charge

$$Q_4 = -Q_3 \cdot \frac{R}{D - \left(\dfrac{R^2}{D - \dfrac{R^2}{D}}\right)} = -Q_3 \cdot \frac{p}{1 - \dfrac{p^2}{1 - p^2}}$$

This makes

$$V_B = 0$$

and makes

$$V_A = +\frac{Q_1}{4\pi\epsilon R} + \frac{Q_4}{4\pi\epsilon\left\{D - \left[D - \dfrac{R^2}{\left(D - \dfrac{R^2}{D}\right)}\right]\right\}}$$

$$= \frac{Q_1}{4\pi\epsilon R} + \frac{Q_4}{4\pi\epsilon D\left[1 - \left(1 - \dfrac{p^2}{1 - p^2}\right)\right]}$$

and so on. Theoretically, an infinite number of images is necessary to make

$$V_A = \frac{Q_1}{4\pi\epsilon R}$$

and

$$V_B = 0$$

For large spacings a small number may suffice. Each succeeding image is smaller than its predecessor, so that all except the first few may be neglected.

Reversing the images for A and B and changing the charge signs makes

$$V_A = 0$$

and

$$V_B = -\frac{Q_1}{4\pi\epsilon R}$$

Superposing the two sets of images makes

$$V_A = +\frac{Q_1}{4\pi\epsilon R} \tag{5.50}$$

$$V_B = -\frac{Q_1}{4\pi\epsilon R} \tag{5.51}$$

and

$$\Delta V = V_A - V_B = 2\frac{Q_1}{4\pi\epsilon R} \tag{5.52}$$

Table 5.1 shows the locations and values of the images necessary to give sphere A potential $V_A = Q_1/(4\pi\epsilon R)$ and sphere B potential zero.

TABLE 5.1

SPHERE GAP IMAGES

LOCATION AND VALUES TO MAKE $V_A = \frac{Q_1}{4\pi\epsilon R}$ AND $V_B = 0$

Image Location—At Point	Image Charge Value	Distance
A_1	$+Q_1$	$s_{11} = 0$
B_2	$Q_2 = -Q_1 p$	$s_{02} = Rp$
A_3	$Q_3 = +Q_1 \frac{p^2}{1-p^2}$	$s_{13} = R\frac{p}{1-p^2}$
B_4	$Q_4 = -Q_1 \frac{p^3}{(1-p^2)\left(1-\frac{p^2}{1-p^2}\right)}$	$s_{04} = R\frac{p}{1-\frac{p^2}{1-p^2}}$
A_5	$Q_5 = +Q_1 \frac{p^4}{(1-p^2)\left(1-\frac{p^2}{1-p^2}\right)\left(1-\frac{p^2}{1-\frac{p^2}{1-p^2}}\right)}$	$s_{15} = R\frac{p}{1-\frac{p^2}{1-\frac{p^2}{1-p^2}}}$
etc.		

$$p = \frac{R}{D}$$

The total charge on sphere A is obtained by adding the image charges $Q_1 + Q_3 + Q_5 ++$ and the reversed set of images $-Q_2 - Q_4 - Q_6 --$. An equal charge of opposite sign is found on sphere B. The procedure is somewhat laborious, so only the result is given.

$$Q_A = -Q_B = Q_1(1 + p + p^2 + p^3 + 2p^4 + 3p^5 ++) \tag{5.53}$$

The capacitance of the two spheres is

$$C = \frac{Q_A}{\Delta V} = 2\pi\epsilon R\,(1 + p + p^2 + p^3 + 2p^4 + 3p^5 + +) \qquad (5.54)$$

The highest gradient in the field E_P occurs at point P and may be computed by summing the separate intensity effects of the various images. This requires considerable mathematical rearrangement to get the result in usable form, if the spheres are close together. We use therefore the forms given by Russell[1] and Dean[2] and Peek,[3] two formulas for E_P which are numerically practically equal,

$$E_P = \frac{\Delta V}{X} f_1$$

$$= \frac{\Delta V}{X}\left[1 + \frac{1}{3}\frac{X}{R} + \frac{1}{45}\left(\frac{X}{R}\right)^2 + \frac{2}{945}\left(\frac{X}{R}\right)^3 + \frac{17}{14{,}175}\left(\frac{X}{R}\right)^4 + \cdots\right] \qquad (5.55)$$

and

$$E_P = \frac{\Delta V}{X} f_2 = \frac{\Delta V}{X}\left[\frac{\frac{X}{R} + 1 + \sqrt{\left(\frac{X}{R} + 1\right)^2 + 8}}{4}\right] \qquad (5.56)$$

in which X = spacing between adjacent sphere surfaces.
R = sphere radius.

The function f_1 has been plotted in Fig. 5.12, for values of X/R up to 2. f_1 is the ratio of the maximum gradient at P to the average value of the gradient.

5.12. Sphere Gap Used as a Voltmeter—Spark-Over. Theoretically, when the spacing is greater than the sphere diameter ($X > 2R$), the sphere gap, with rising potential ΔV, first exhibits a visual corona around each sphere. Only when the potential is raised considerably above the corona value does spark-over occur. The spark-over voltage exceeds the corona voltage more and more as X exceeds $2R$. For $X < 2R$ spark-over takes place without formation of corona and in this range of spacings may be used as a voltmeter.

Both theory and experiment show that the field intensity or gradient E_s at the point P in Fig. 5.11 required for spark-over is greater than

[1] Russell, *Alternating Currents*, Columbia University Press, 1914, Vol. I, p. 257.

[2] Dean, *Gen. Elec. Rev.*, **16**, 148 (1913).

[3] Peek, *Dielectric Phenomena in High Voltage Engineering*, McGraw-Hill, 1929, p. 69.

the value demanded in a uniform field, or to put it another way the uniform field breakdown gradient must be exceeded over a certain distance out from the sphere surface. The *experimental* value of

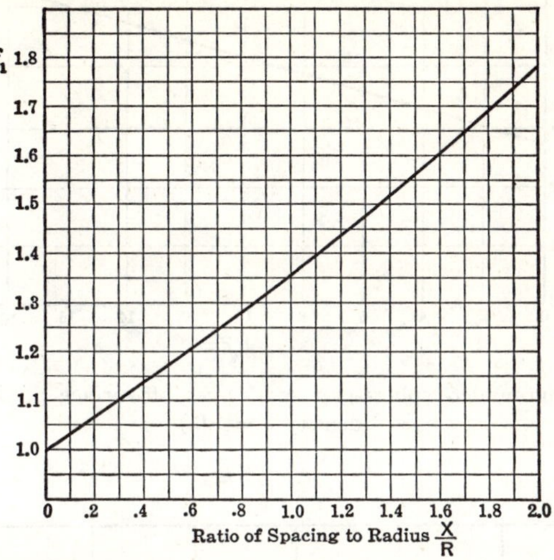

FIG. 5.12. Ungrounded sphere gap. Ratio of maximum to average potential gradient, f_1 versus X/R.

spark-over gradient E_s for power frequencies, corresponding to spark-over potential ΔV_s, has been determined by Peek and is

$$E_s = 2{,}720{,}000\ \delta \left(1 + \frac{0.054}{\sqrt{\delta R}}\right) \frac{\text{volts}}{\text{meter}} \tag{5.57}$$

in which

$$\delta = \frac{3.92b}{273 + T} \tag{5.58}$$

is the relative air density at barometer reading b in centimeters of mercury and at temperature $T°$ C. At $T = 25°$ C and $b = 76$ centimeters of mercury, $\delta = 1.0$. E_s is plotted against R in Fig. 5.13 for $\delta = 1.0$.

With E_P set equal to E_s, $\Delta V = \Delta V_s$, the spark-over value ΔV_s is plotted against X for spheres of diameter $D = 6.25$ or $R = 3.125$ centimeters, with $\delta = 1.0$. See Fig. 5.14. This is the *ungrounded* sphere-gap curve.

The foregoing discussion has assumed that the spheres are placed symmetrically with respect to all nearby objects including ground so that the field is not distorted. However, in many applications the

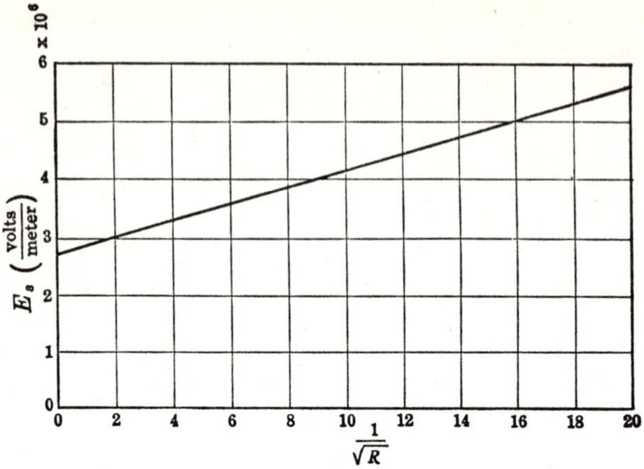

Fig. 5.13. Ungrounded sphere-gap spark-over off-surface gradient. (For air, density $\delta = 1.0$.)

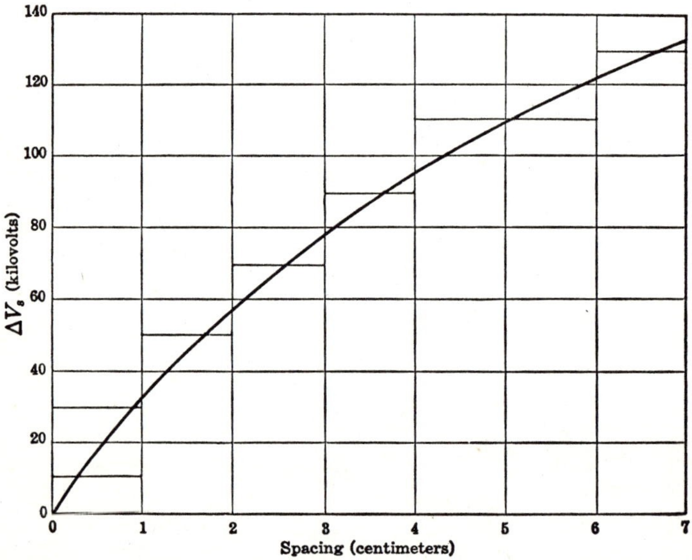

Fig. 5.14. Ungrounded sphere-gap spark-over voltage versus spacing X. Sphere radius $R = 3.125$ centimeters. (For air, density $\delta = 1.0$.)

sphere gap is placed with its axis vertical and with the lower sphere grounded and placed no more than four or five diameters above ground. In this case the field is distorted, some of the flux lines from the upper sphere passing to ground rather than entirely to the lower sphere, and the spark-over potential depends upon the *polarity* of the *upper* sphere. At larger values of X/R the positive potential spark-over values are somewhat higher than for negative polarity.

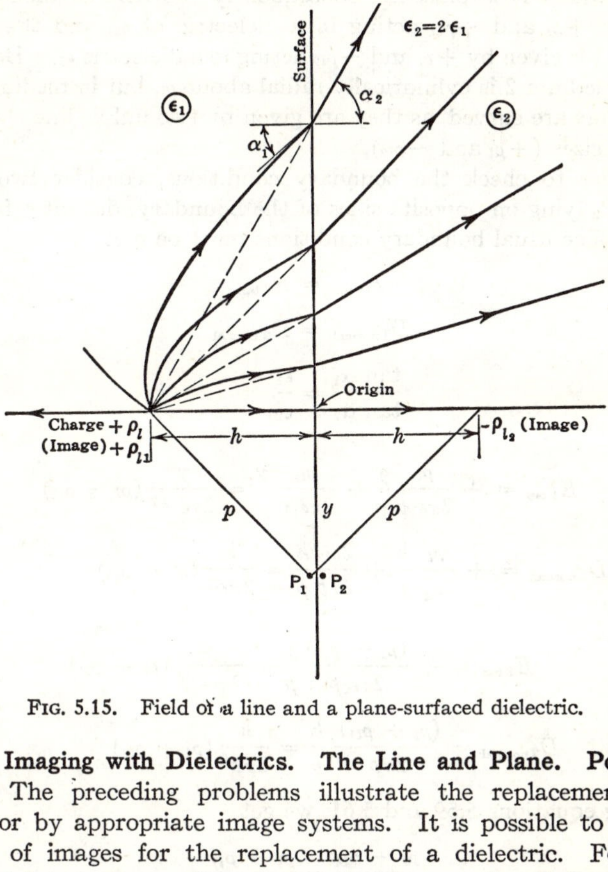

FIG. 5.15. Field of a line and a plane-surfaced dielectric.

5.13. Imaging with Dielectrics. The Line and Plane. Point and Plane. The preceding problems illustrate the replacement of a conductor by appropriate image systems. It is possible to use the method of images for the replacement of a dielectric. Following Fig. 5.15, let us place a line charge $+\rho_l$ on a small cylinder lying in medium 1 and distant h from the plane intersection between two dielectrics of differing dielectric constants. In order to find the solution it is necessary to guess the sizes and locations of the images, and to prove that the guess is correct it is necessary to show that the surface conditions at the boundary between the two dielectrics are fulfilled.

The proper way to guess the location of the images is to place line charge $-\rho_{l2}$ in medium 2 distant h from the boundary, and to place line charge $+\rho_{l1}$ in medium 1 at the same location as $+\rho_l$. The absolute values of images $+\rho_{l1}$ and ρ_{l2} are equal.

$$|\rho_{l1}| = |\rho_{l2}|$$

An image must not be used to give the field in the same region as that in which it is placed. Consequently the field in medium 2 is given by $+\rho_l$ and $+\rho_{l1}$ acting in a dielectric of ϵ_2, and the field in medium 1 is given by $+\rho_l$ and $-\rho_{l2}$ acting in a dielectric ϵ_1. Hence the field in medium 2 is cylindrically radial about ρ_l, but in medium 1 the lines of flux are curved, as they are given by two unlike line charges of differing sizes ($+\rho_l$ and $-\rho_{l2}$).

In order to check the boundary conditions, consider two points P_1 and P_2 lying on opposite sides of the boundary, distant y from the origin. The usual boundary conditions must be met.

$$E_{1\,\text{tan}} = E_{2\,\text{tan}}$$

$$D_{1\,\text{normal}} = D_{2\,\text{normal}}$$

$$\frac{\tan \alpha_1}{\tan \alpha_2} = \frac{\epsilon_1}{\epsilon_2}$$

At P_1:

$$E_{1\,\text{tan}} = +\frac{\rho_l}{2\pi\epsilon_1 p}\frac{y}{p} - \frac{\rho_{l2}}{2\pi\epsilon_1 p}\frac{y}{p} = \frac{y}{2\pi\epsilon_1 p^2}(\rho_l - \rho_{l2}) \tag{5.59}$$

$$D_{1\,\text{normal}} = +\frac{\rho_l}{2\pi p}\frac{h}{p} + \frac{2\rho_{l2}}{2\pi p}\frac{h}{p} = \frac{h}{2\pi p^2}(\rho_l + \rho_{l2}). \tag{5.60}$$

At P_2:

$$E_{2\,\text{tan}} = +\frac{(\rho_l + \rho_{l1})}{2\pi\epsilon_2 p}\frac{y}{p} = \frac{y}{2\pi\epsilon_2 p^2}(\rho_l + \rho_{l1}) \tag{5.61}$$

$$D_{2\,\text{normal}} = \frac{(\rho_l + \rho_{l1})}{2\pi p}\frac{h}{p} = \frac{h}{2\pi p^2}(\rho_l + \rho_{l1}) \tag{5.62}$$

Equating equations 5.59 and 5.61, we get

$$\frac{\rho_l - \rho_{l2}}{4\pi\epsilon_1} = \frac{\rho_l + \rho_{l1}}{4\pi\epsilon_2}$$

and from equations 5.60 and 5.62

$$\rho_l + \rho_{l2} = \rho_l + \rho_{l1}$$

and

$$\rho_{l2} = \rho_{l1} \tag{5.63}$$

Hence

$$\frac{\rho_l - \rho_{l2}}{4\pi\epsilon_1} = \frac{\rho_l + \rho_{l2}}{4\pi\epsilon_2}$$

and

$$\rho_{l2} = \rho_l \frac{\epsilon_2 - \epsilon_1}{\epsilon_2 + \epsilon_1} \tag{5.64}$$

Equation 5.64 specifies the size of the image ρ_{l2} in terms of the original charge ρ_l, and the boundary conditions are satisfied. The upper part of Fig. 5.15 shows the field that results when $\epsilon_2 = 2\epsilon_1$.

It can be shown that equation 5.64 gives also the correct image size for a small *sphere* brought near a plane dielectric surface.

PROBLEMS

1. A cylindrical wire (A in Fig. 5.1) of 1-cm radius is placed in the air·100 cm above ground. The potential difference between the wire and ground is 60,000 volts. Compute:

(*a*) The charge on the wire, and locate the center of influence.
(*b*) The capacitance of the wire to ground.
(*c*) The ground charge density directly below the wire, and also at 100 cm along the ground.
(*d*) The maximum potential gradient in the air.
(*e*) The absolute potential of a point B, 60 cm above ground and 40 cm to the right of the wire A.

2. In addition to wire A of Prob. 1, put the same-size wire at B. Assuming the same charge on wire A as found in Prob. 1 and an equal and opposite charge on wire B, compute:

(*a*) The absolute potential of each wire, and their difference in potential.
(*b*) The capacitance of each wire to ground, and their capacitance to each other.
(*c*) The location of the singular lines on the ground.
(*d*) The location and value of the maximum ground charge density.

3. Now move wire B of Prob. 2 to the same height above ground as that of wire A, and answer the same questions.

4. A single-phase power line, radius $R_1 = 1$ cm and horizontal spacing $D_1 = 100$ cm, is placed 1000 cm above ground. Directly beneath it is placed a second line, wire radius $R_2 = 0.5$ cm and horizontal spacing $D_2 = 50$ cm, at a height of 500 cm above ground. Compute the potential difference induced between the wires of the second line, if a potential difference of 12,000 volts is established between the wires of the first line.

5. Following Figs. 5.6, 5.7, and 5.8, assume a two-conductor cable ($\epsilon_r = 3$) with wire radii of 0.2 cm, sheath radius of 4 cm, and $s = 2.4$ cm. The potential difference between the wires is 15,000 volts. Compute:

(*a*) The charge ρ_l, and its location.
(*b*) The capacitance between the two wires.
(*c*) The capacitance of each wire to sheath.

(*d*) The capacitance to neutral.
(*e*) The 60-cycle charging current.

6. Repeat Prob. 5 for a three-conductor cable.

7. Following Fig. 5.9, bring a point charge ($Q = +10^{-8}$ coulomb) up to 10 cm from the center of a grounded conducting sphere of 4 cm radius. Locate the size and position of the image. Find the charge density on the sphere. Compute the force of attraction.

If the sphere ground is removed and its potential is raised to +6,000 volts, find the extra charge which must be added, the total charge density, and the force of attraction.

8. Consider two spheres of 10-cm radius placed in air, separated center distance of 40 cm, with a potential difference of 10,000 volts. Locate the images and compute the maximum potential gradient in the air.

9. Compute the capacitance of two spheres, each of radius R, spaced center distance D, under the condition that $R \ll D$.

10. Compute and plot the spark-over potential ΔV_s against spacing X for an ungrounded sphere gap in air, with $D = 12.5$ cm and $\delta = 1.0$. Compute the capacitances for $X = 0.5R$ and for $X = 10R$.

11. Place a line charge ($+\rho_l = 10^{-9}$ coulomb per meter) in air 8 cm from a plane surfaced rubber volume ($\epsilon_r = 3$). Locate the images and give their magnitudes. Compute the normal and tangential components of E and D on the air–rubber surface 6 cm above the foot of the perpendicular from the charge to the plane, and give the flux refraction angles at this point.

CHAPTER 6

SPACE CHARGE

LAPLACE AND POISSON EQUATIONS

So far in the text we have discussed fields in which the free charge has been distributed mainly over the surfaces of conductors. The field between conductors has been studied throughout regions in which any concentration of free electric charge has been zero— that is, so called "space-charge-free" regions. In this chapter we shall study some of the effects of the presence of space charge. In the first place, the very presence of the space charge modifies the spatial distribution of the potential and the intensity, and since the charges are free they move under the action of the field and constitute a convection current.

6.1. Space Charge. Negative space charge (comprised of electrons, $e = 1.602 \times 10^{-19}$ coulomb) is to be found in the region between the filament and plate of a high-vacuum tube. In tubes of this kind the space charge is supplied by the heated filament. The effect of the heating is to increase the energies of a fair number of the filament electrons sufficiently to enable them to leap over the surface barrier or work function which, at low temperatures, effectively restrains the electrons from emerging from the filament into outer space. This process is known as *thermionic emission.*

Negative space charge may originate also by *photoelectric emission.* If a beam of light with sufficiently high frequency, usually in the ultraviolet range, is directed upon the surface of a conductor, electrons are emitted into the outer space. Or electrons may be pulled out of a cold metal by direct field action, but this process, known as *field emission*, occurs to an appreciable extent only with fields of the order of 10^8 volts per meter.

In a gas at low field strengths electrons may be "knocked" out of a cold cathode by *positive ion bombardment*, since any gas volume contains a few positive ions which can be accelerated by the field. Negative space charge may be produced also by *secondary emission* of electrons. Primary electrons moving up to the anode may deliver sufficient energy to the anode to release new or secondary electrons into the adjoining space.

Space charges of both types, positive and negative, are to be found in gas-filled tubes and in electric discharges such as glow discharges and sparks. Positive space charge is composed of positive ions created by ionizing collisions between electrons and neutral atoms in which additional electrons are freed from the atoms. The positive charge of an ion most frequently is equal numerically to the charge of the electron. Both positive and negative space charge density may occur simultaneously in the same portion of an electric discharge. The *net* space charge density is the excess of the positive over the negative value. Thus

$$\rho_{\text{net}} = \rho_+ - \rho_- = n_+e_+ - n_-e_- \quad \text{coulombs per cubic meter} \tag{6.1}$$

where n_+ and n_- represent the number of positive ions and electrons per cubic meter.

The *net* space charge density, in an arc for illustration, is positive near the cathode, negative near the anode and approximately zero throughout the great portion of the center region. Though the net value may be zero in the center portion, it must not be forgotten that there are, in fact, tremendously large concentrations of both signs, but $n_+ = n_-$.

The electric field effect is determined solely by the *net* space charge density. Zero *net* space charge, for instance, creates no electric field of its own in its neighborhood. However, there is the current effect of space charge because the charges of opposite sign move in opposite directions, the positive space charge with whatever field is present and the negative space charge against, both movements, however, constituting a convection current density in the field direction. In general the positive charge is likely to move with a very much lower velocity than the negative. The current density is

$$J = \rho_+v_+ + \rho_-v_- = n_+e_+v_+ + n_-e_-v_- \tag{6.2}$$

Unfortunately it is not possible to state any general rules which will prescribe values for ρ_+, ρ_-, v_+, and v_-. Values for these quantities vary with the type of electric discharge, type of gas present, gas pressure, magnitude of the current, geometry of electrodes, etc.

6.2. Equations of Laplace and Poisson. Consider a portion of space in an electric field included in a small cube with sides dx, dy, and dz. On the left-hand face, $dy\,dz$, in Fig. 6.1, let the x component of the electric intensity be E_x. Since it is normal to the face, the flux crossing the face *into* the cube is the normal flux density times the area, $(\epsilon E_x)(dy\,dz)$. On the right-hand face, $dy\,dz$, the flux *emerging* from the cube is $\epsilon\left(E_x + \frac{\partial E_x}{\partial x}\,dx\right)(dy\,dz)$. In the last expression

$\epsilon \frac{\partial E_x}{\partial x}$ represents the *space rate* at which the x component of the flux density is changing in the x direction (partial derivations are required since E_x is a function of x, y, and z and we want only the x variation), and this quantity multiplied by dx gives the total change in flux crossing the two faces. Hence the *net outward* flux from the cube over the two faces, $dy\, dz$, is

$$\left(\epsilon \frac{\partial E_x}{\partial x} dx\right)(dy\, dz) \tag{6.3}$$

Similarly over the other two pairs of faces, the net outward flux is

$$\left(\epsilon \frac{\partial E_y}{\partial y} dy\right)(dz\, dx)$$

and

$$\left(\epsilon \frac{\partial E_z}{\partial z} dz\right)(dx\, dy)$$

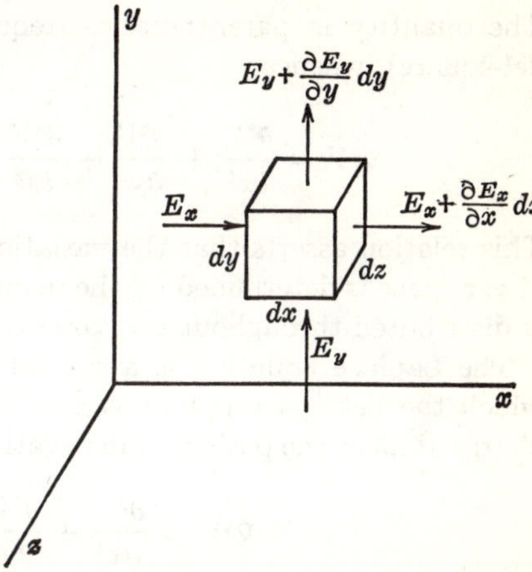

FIG. 6.1. Cubic volume used for calculating net outward flux.

and from the whole cube the net outward flux is

$$d\psi_{\substack{\text{net}\\\text{out}}} = \left[\epsilon\left(\frac{\partial E_x}{\partial x} + \frac{\partial E_y}{\partial y} + \frac{\partial E_z}{\partial z}\right)\right](dx\, dy\, dz) \tag{6.4}$$

The quantity in square brackets is then the *net outward flux per unit volume* and is called in vector analysis the divergence of the vector $D = \epsilon E$. This quantity must, of course, be equal to the net space charge density ρ_{net}.

$$\frac{d\psi_{\substack{\text{net}\\\text{out}}}}{dv} = \text{div } D = \epsilon \text{ div } E$$
$$= \epsilon\left(\frac{\partial E_x}{\partial x} + \frac{\partial E_y}{\partial y} + \frac{\partial E_z}{\partial z}\right) = \rho_{\text{net}} \tag{6.5}$$

Equation 6.5 is the equivalent of the Gauss theorem given in Art. 1.26.

E_x, the x component of the intensity, is negatively equal to the space derivative of the electric potential taken in the x direction. Similarly for the other components,

$$E_x = -\frac{\partial V}{\partial x}, \quad E_y = -\frac{\partial V}{\partial y}, \quad E_z = -\frac{\partial V}{\partial z} \tag{6.6}$$

Upon introducing these quantities into equation 6.5 we obtain the Poisson equation,

$$-\frac{d\psi_{\substack{\text{net}\\ \text{out}}}}{dv} = \epsilon\left(\frac{\partial^2 V}{\partial x^2} + \frac{\partial^2 V}{\partial y^2} + \frac{\partial^2 V}{\partial z^2}\right) = -\rho_{\text{net}} \tag{6.7}$$

The quantity in parentheses is frequently written $\nabla^2 V$ (pronounced del-square), whence

$$\nabla^2 V = \frac{\partial^2 V}{\partial x^2} + \frac{\partial^2 V}{\partial y^2} + \frac{\partial^2 V}{\partial z^2} = -\frac{\rho_{\text{net}}}{\epsilon} = -\frac{1}{\epsilon}\frac{d\psi_{\substack{\text{net}\\ \text{out}}}}{dv} \tag{6.8}$$

This relation asserts that the variation of the potential V throughout x, y, z space is determined by the manner in which the net space charge is distributed throughout the same region.

The Laplace equation is a special case and applies to a region in which the net space charge is zero, because either there is no space charge at all or the positive and negative values balance exactly. Then

$$\nabla^2 V = \frac{\partial^2 V}{\partial x^2} + \frac{\partial^2 V}{\partial y^2} + \frac{\partial^2 V}{\partial z^2} = 0 \tag{6.9}$$

prescribes the possible variation in potential throughout this space in which ρ_{net} is zero, when taken in conjunction with the boundary conditions at the electrodes.

6.3. Space Charge and Current with Parallel Plane Electrodes. Space-Charge-Limited Current. Two parallel plane electrodes are supposed to be placed inside an evacuated space, with area A which is large compared with their separation s. Let one of the electrodes, the cathode, be heated to a high temperature T (degrees absolute) and a constant potential difference V_s be established between the electrodes. For convenience we shall call the hot-cathode potential $V_c = 0$ and the potential of the cold electrode, the anode, V_s. V is then the potential, and E the intensity, at any intermediate point distant x from the cathode.

We shall assume that the cathode temperature is high enough for the cathode to be able to supply all the electrons which can be carried over to the anode. Since the space between the electrodes is filled with electrons or negative space charge, there is a limit to the current and to the number of electrons per second which can be passed across to the anode. Hence the current is said to be "space-charge-limited." This condition is also known as one in which there is "complete space charge." If the cathode emits more electrons per second than the space can accept, the remainder are reabsorbed by the cathode. The electrons are emitted with a "distribution" of velocities, that is, some faster and some slower. Actually the situation is quite complicated and in order to find a solution that is not too burdensome for the magnitudes of the current, potential, and field strength throughout the electron-filled space, we must resort to simplifying assumptions.

We assume that the electrons are emitted from the cathode with zero velocity and energy. With this assumption the electrons could not advance against an adverse or retarding field intensity. On the other hand, as we shall see later, if there were an accelerating field the electrons would be supplied faster than the space, with its potential distribution, could accept them. Therefore, with a free supply of electrons, we assume the field intensity at the cathode to be zero. Away from the cathode, E and V acquire values which are functions of the distance x and the density of the space charge.

An electron of charge e loses potential energy eV in moving to a position where the potential is V, and acquires an equal amount of kinetic energy $\frac{1}{2}mv^2$. Then the energy balance and velocity are

$$eV = \tfrac{1}{2}mv^2 \quad \text{joules} \tag{6.10}$$

and

$$v = \sqrt{\frac{2e}{m}}\,V^{\frac{1}{2}} \quad \frac{\text{meters}}{\text{second}} \tag{6.11}$$

Since the field is uniform or unidirectional, and the potential is a function only of the x distance, Poisson's equation 6.8 reduces to

$$\nabla^2 V = \frac{d^2V}{dx^2} = -\frac{\rho}{\epsilon} \tag{6.12}$$

The current density of electrons is $J = \rho v$, which is constant over all cross sections and independent of x. Then

$$-\rho = -\frac{J}{v} = -J\sqrt{\frac{m}{2e}}\,V^{-\frac{1}{2}} \tag{6.13}$$

and

$$\frac{d^2V}{dx^2} = -\frac{J}{\epsilon}\sqrt{\frac{m}{2e}}\,V^{-\frac{1}{2}} \tag{6.14}$$

Multiply the left side by $2\,(dV/dx)\,dx$ and the right side by $2\,dV$, and integrate, remembering that the current density J is a *negative* constant value for a given set of conditions, independent of the value of x. (Hint: Let $z = dV/dx$.) The result is

$$\left(\frac{dV}{dx}\right)^2 = \frac{-4J}{\epsilon}\left(\frac{m}{2e}\right)^{\frac{1}{2}} V^{\frac{1}{2}} + \text{constant} \tag{6.15}$$

Since $E = -dV/dx = 0$, and $V = 0$, at the cathode, the constant is zero. Taking the square root

$$\frac{dV}{dx} = \left(\frac{-4J}{\epsilon}\right)^{1/2} \left(\frac{m}{2e}\right)^{1/4} V^{1/4} \tag{6.16}$$

or

$$V^{-1/4}\, dV = \left(\frac{-4J}{\epsilon}\right)^{1/2} \left(\frac{m}{2e}\right)^{1/4} dx$$

and

$$\tfrac{4}{3} V^{3/4} = \left(\frac{-4J}{\epsilon}\right)^{1/2} \left(\frac{m}{2e}\right)^{1/4} x + \text{constant}$$

At $x = 0$, $V = 0$, and the constant is zero, and

$$\begin{aligned} J &= -\frac{4\epsilon}{9}\left(\frac{2e}{m}\right)^{1/2} \frac{V^{3/2}}{x^2} \\ &= -2.335 \times 10^{-6} \left(\frac{V^{3/2}}{x^2}\right) \end{aligned} \tag{6.17}$$

This is the equation for the "space-charge-limited" current density, which, under the assumptions, is the maximum value that the space charge will permit. This equation gives the distribution of potential as a function of x. At the anode, $x = s$, and $V = V_s$, and

$$\begin{aligned} J &= -\frac{4\epsilon}{9}\left(\frac{2e}{m}\right)^{1/2} \frac{V_s^{3/2}}{s^2} \\ &= -2.335 \times 10^{-6} \frac{V_s^{3/2}}{s^2} \quad \frac{\text{ampere}}{\text{square meter}} \end{aligned} \tag{6.18}$$

Then

$$V = 5681.5(-J)^{2/3} x^{4/3} \tag{6.19}$$

and

$$E = -\frac{dV}{dx} = -7575(-J)^{2/3} x^{1/3} \tag{6.20}$$

$$\rho = -\epsilon \frac{d^2V}{dx^2} = -2525\epsilon(-J)^{2/3} x^{-2/3} \tag{6.21}$$

$$v = \frac{J}{\rho} = \frac{(-J)^{1/3}}{2525\epsilon} x^{2/3} \tag{6.22}$$

The quantities V, $+E$, $-\rho$, v, and $-J$ are plotted in relative values, compared with the values at the anode, in Fig. 6.2. The values in parentheses in the figure indicate the powers of x followed by the corresponding functions. The straight broken line gives the potential variation for the corresponding space-charge-free condition. In a uniform field, the presence of negative space charge is revealed by the downward flexion $\left(\frac{d^2V}{dx^2}\right)$ of the potential. With positive space charge the flexion of the potential would be upward.

The curve for ρ indicates the weakness in the underlying assumptions; since the current density $J = \rho v$ is the same over all cross sections, it follows that the assumption of zero velocity at the cathode

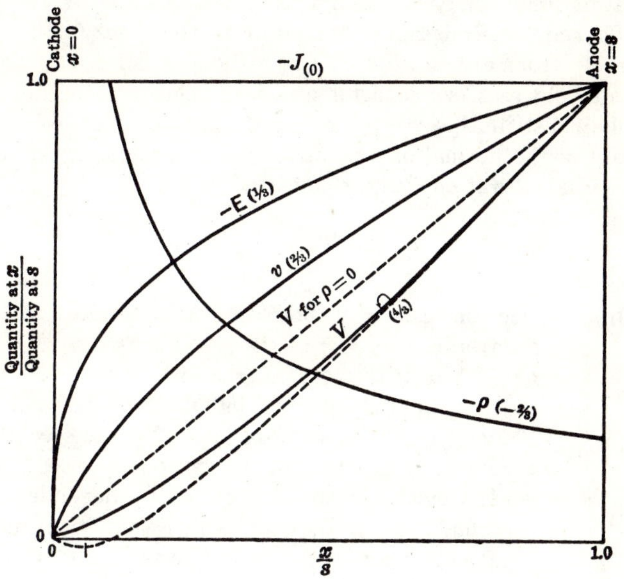

FIG. 6.2. Values of quantities at x relative to values at anode, for space-charge-limited current between parallel plane electrodes.

necessarily implies an infinite space charge concentration at the cathode, an absurdity of course. Nevertheless these results are not far wrong for the regions which are not too close to the cathode. As the anode is approached, the space charge density becomes smaller, the velocity and field intensity become larger.

The correct analysis of this problem, though somewhat involved, has been carried out and shows that there is a potential minimum at some distance off the cathode. The true potential variation is shown in the curved broken line in Fig. 6.2. A plane, parallel to the cathode, at the potential minimum is called a "virtual cathode." Between the minimum point and the cathode, the intensity is retarding and tends to drive electrons back toward the cathode. Electrons get out of the cathode then only because they *do* have initial velocities, but only those electrons with sufficient initial velocity and energy to overcome the retarding field get past the virtual cathode; the rest are turned back at various stages before reaching this point. If the potential of the anode is raised with respect to the cathode the magnitude of the potential minimum below the cathode potential becomes smaller and the position of the minimum point moves closer to the cathode. When the anode potential is more than about 10 volts above the cathode the potential minimum may well be ignored.

6.4. Thermionic Emission. According to the theory of thermionic emission electrons are emitted by the cathode in all directions with a variety of velocities, some faster and some slower, according to the Maxwellian distribution laws (see Art. 1.7). However, we are interested only in the normal or x components of these electron velocities which yield a saturation current density

$$J_{sat} = \frac{4\pi mek^2}{h^3} T^2 \epsilon^{-\frac{\phi}{kT}} \tag{6.23}$$

in which m = electron mass = 9.1066×10^{-31} kilogram.
e = electron charge = 1.602×10^{-19} coulomb.
ϵ = natural logarithmic base = 2.718
k = Boltzmann constant = 1.380×10^{-23} joule per degree.
h = Planck's constant = 6.624×10^{-34} joule-second.
T = absolute temperature of the cathode.
ϕ = work function or barrier in joules, dependent on the cathode material, over which the electrons in the cathode must be raised before they may be emitted. ϕ varies from about 1 electron-volt to 6 electron-volts, according to the cathode material and the character of the surface. (See Art. 1.29.)

$$\frac{4\pi mek^2}{h^3} = 120.1 \times 10^4 \ \frac{\text{amperes}}{\text{square meter-degree}^2} \tag{6.24}$$

This equation gives the theoretical maximum emission current density for clean metal surfaces. The emission from certain metal cathodes appears to be only half as large. Of this current only that portion given by equation 6.18 can flow across the space charge region, and the remainder flows back into the cathode.

6.5. Parallel Plane Electrodes. Temperature-Limited Current. For each value of cathode temperature there is a maximum current which can be obtained from the cathode; and for each value of positive anode potential, relative to the cathode, there is a maximum space-charge-limited current which the space can pass. Either of these factors can limit the current. Let us assume a definite cathode temperature T_1 and gradually raise the anode potential V_s. At first the current is space-charge-limited and follows the $\frac{3}{2}$ power law as shown in Fig. 6.3*a*. At low values of V_s, the potential distribution has the space-charge-limited potential minimum shown in Fig. 6.2, which gradually approaches the cathode and becomes smaller for increasing anode potentials. For still greater values of V_s the entire cathode current output can be, and is, passed through the intervening space,

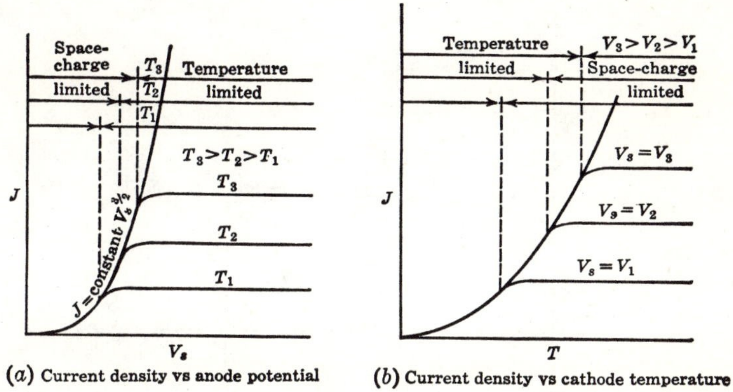

FIG. 6.3. Dependence of current density on cathode temperature and anode voltage.

the potential acquires a positive slope at the cathode, the electrons speed up, and the space charge is less dense. The *current is now limited by the temperature of the cathode* and even with increasing V_s no more current can flow. The potential distribution lies between the values of V corresponding to $\rho = 0$ and the $\frac{4}{3}$ power line. See equation 6.19 and Fig. 6.2. For higher values of temperature the cutoff value of current occurs at higher values of current. Figure 6.3*b* illustrates the matter in a different way.

6.6. Gas-Filled Tubes. If a small quantity of gas is admitted to the space previously occupied only by electrons and if the anode potential is high enough (say 20 to 30 volts or more) the situation is profoundly altered. Driven by the field the electrons collide with the gas atoms, exciting or ionizing some of them until the space is filled with a mixture of electrons, neutral atoms, and positive ions. The latter move toward the cathode but, on account of their far larger values of mass, with much smaller velocities than those acquired by the electrons. The positive ions reduce the *net* space charge concentration and since they spend much longer times between the electrodes than do the electrons, a few positive ions can negate the space charge effect of a large number of electrons.

Finally a balance is reached and a large concentration of positive ions in a *thin* positive-ion sheath or layer is built up near the cathode. This region has a positive flexion for the potential and almost all the potential drop between the plates is to be found here. Throughout the remainder of the discharge up to the anode the net space charge is very small, approaching zero, and the potential is almost constant through this region, as shown in Fig. 6.4.

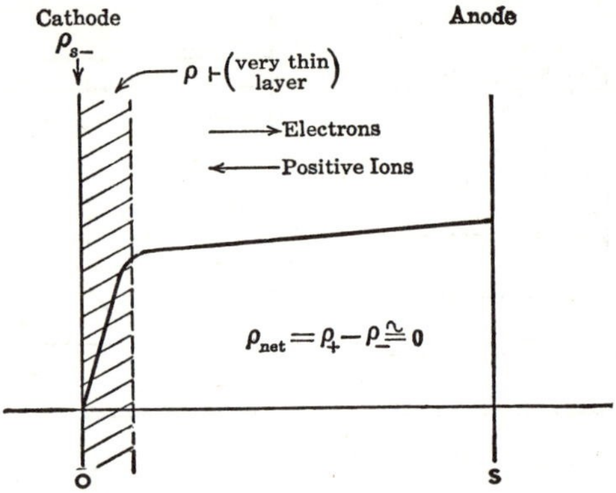

FIG. 6.4. Distribution of charge and voltage in a gas-filled tube.

Under these conditions the current is no longer space-charge-limited and may reach values of hundreds of amperes instead of the small fractions of an ampere permitted when negative space charge limits the flow.

PROBLEMS

1. Consider parallel plane electrodes separated 0.2 cm with a potential difference of 150 volts. The cathode is tungsten heated to 2500° K. The work function of tungsten is 4.5 electron-volts. Compute:

(*a*) The thermionic current density emitted by the cathode.

(*b*) The space-charge-limited current density. What happens to the difference between this value and the preceding value? Is there a potential minimum in this case?

(*c*) At the anode, values of E, ρ, v, n_-.

(*d*) At the midpoint between electrodes, values of E, ρ, v, V, n_-.

(*e*) At the midpoint, values of E and V, if ρ were zero.

(*f*) The anode potential at which the thermionic current becomes temperature limited.

2. Change the cathode temperature in Prob. 1 to 2000° K and calculate the thermionic current density. Sketch the curves of Fig. 6.2 for this condition.

3. The potential throughout a certain region which is space-charge-free, we assume, is a function only of the x coordinate. Using the LaPlace equation, show that the potential is a linear function of x. Since the equation is a second-order differential, there must be two arbitrary constants. How are these evaluated?

4. Solve the Poisson equation for a region between parallel plates in which the charge density is constant valued. How are the arbitrary constants evaluated?

5. Solve the LaPlace equation for a space-charge-free region between concentric cylinders, with V a function of r only. Under these conditions

$$\nabla^2 V = \frac{1}{r}\frac{\partial}{\partial r}\left(r\frac{\partial V}{\partial r}\right) = 0$$

6. Solve the Laplace equation for a space-charge-free region between concentric spheres, with V a function of r only. Under these conditions

$$\nabla^2 V = \frac{1}{r^2}\frac{\partial}{\partial r}\left(r^2\frac{\partial V}{\partial r}\right) = 0$$

CHAPTER 7

MAPPING ELECTRIC FIELDS

The solution of the electric field may always be found, at least theoretically, if the quantity and the location of all the charges are given, in addition to the character and location of the boundaries. This information is sufficient to determine the intensity and the potential at every point in the electrostatic field. Again, if the potential and shape of the boundaries and the dielectric constants are known, the problem is always theoretically soluble.

Practically, however, the solution may easily become mathematically burdensome if not impossible. A great many problems of importance in engineering are of this character. We are fortunate in having available the graphical method known as field mapping[1] for the solution of problems of complex geometrical character.

The process of field mapping divides into two parts—field sketching (see Art. 1.40) which gives a fairly good picture of the flux lines and the orthogonal equipotential surfaces, and the mapping proper that gives a flux and equipotential surface drawing of high accuracy.

7.1. Two-Dimensional Fields. Two-dimensional fields are those in which the spatial distribution of potential occurs only in two dimensions. No matter what its shape, any flux line lies in a plane including these two dimensions. Consequently, a sketch of the equipotentials and flux lines drawn in this plane gives a complete picture of the field. The discussion in this chapter concerns only this type of field.

7.2. Curvilinear Squares. Figure 7.1 shows two conductors which are assumed to have a potential difference, so that there is an electric field in the intervening region. The flux lines run from the positive charge on A's surface to the negative charge on B's surface. A unit positive charge, released from any point on the A surface, would, owing to the field intensity, trace a flux line path to conductor B. Obviously, an infinite number of these paths may be drawn. Any surface everywhere normal to the flux line paths is an equipotential; therefore an infinite number of equipotentials may be drawn.

Let us, however, draw only a few of the many possible flux line paths and a few equipotentials, but so chosen that in the left end

[1] A. D. Moore, *Fundamentals of Electrical Design*, McGraw-Hill, 1927, Chapters V to IX.

of Fig. 7.1 they divide the *uniform* field into "squares." Three equipotential surfaces are drawn dividing the potential difference between the conductors into four equal parts, though any other integral division might have been made. This division then determines the spacing between flux line paths, if "squares" are to be obtained.

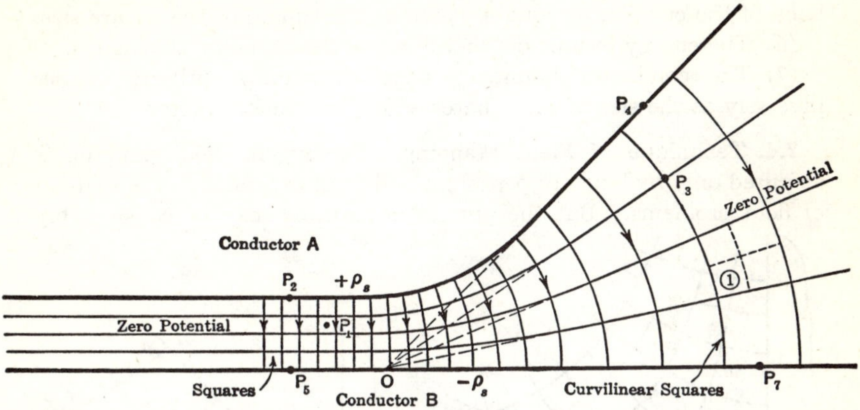

FIG. 7.1. Field-mapping technique.

Proceeding toward the right in Fig. 7.1, the flux lines, which leave and enter the conductors normally, are curved. Likewise the equipotentials spread out. Soon after the curved corner of conductor A is passed, the equipotentials become asymptotic to straight lines through point O, located at the intersection of the projection of A's sloping surface upon B. Away from the curved corner the flux lines tend to approximate circles about point O. Near the corner the flux lines are, of course, not circular.

If, throughout the entire field, the flux lines have been drawn with such a spacing that any of the field sections may be subdivided by 50 per cent equipotentials and 50 per cent flux lines and thereby the subsections tend to become more nearly "true squares," the field is correctly drawn, and the sections are called "curvilinear squares." Section 1 is shown subdivided in this manner.

7.3. Properties of Curvilinear Squares. Considering the method of construction of the curvilinear squares, the following properties stand out:

(1) The potential difference across any "curvilinear square" is equal to that across any other square. Consequently,

(2) The intensity and flux density vary throughout the field inversely as the size of the squares. Hence,

(3) Equal quantities of flux pass through all volumes formed by taking unit depth perpendicular to the squares. Hence,

(4) Equal quantities of free charge are found on the conductor unit-depth-surface at the end of each square.

(5) The energy density varies inversely as the square of the size of the curvilinear square ($\omega_e \sim E^2$). But as the volume of unit depth back of the curvilinear square varies as the square of the square size,

(6) The energy in unit-depth-volume is the same for all squares.

(7) The conductor boundary force over unit-depth-area varies inversely as the size of the squares ending on the conductor.

7.4. Technique of Field Mapping. Facility in field mapping is acquired only by hours of persistent endeavor expended in the solution of field problems. But the process of learning may be considerably

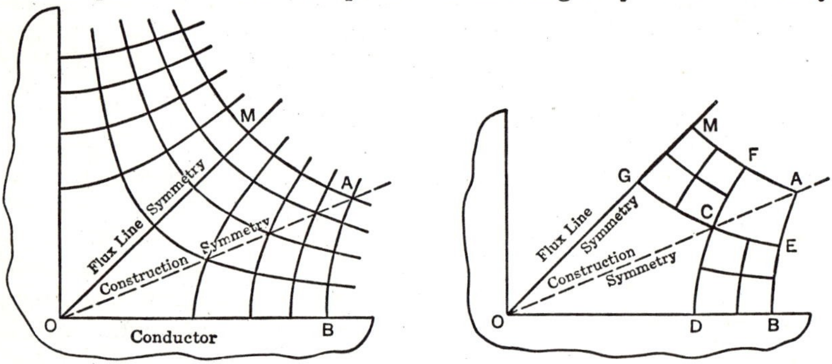

(a) (b)

FIG. 7.2. The 90°-corner field.

shortened by the solution of certain special cases. It frequently happens that a complex field can be reduced to a series of simple special cases properly correlated. We proceed then to the study of basic special cases.

7.5. The 90°-Corner Field. Consider a conductor with two infinite plane equipotential surfaces which meet at 90°, as shown in Fig. 7.2*a*, over which a positive charge has been distributed. In this figure the field has been correctly mapped and the final result is given. However, the first step in mapping is to make a field sketch (see Art. 1.40), only qualitatively correct but much like Fig. 7.2*a* in appearance, indicating the general field form. It may be noted that the charge density on the conductor surface and the field intensity become vanishingly small as the singular line at O is approached. The flux lines leave the conductor normally, then bend rather sharply, and finally approach parallelism at a considerable distance from the conductor. Obviously, the field is symmetrical about the straight flux line *OM*, which leaves the singular line (perpendicular to the paper) through *O* at 45° to the conductor. Line *OM* may be called a "flux line symmetry." On the basis of an ultimate "curvilinear square" map, the equipotential surfaces cross the flux line *OM* with identically the same shapes as those of the flux lines from the conductor. In other words, the two halves of the octant *MOB* are symmetrical about line *OA*, which may be called a "construction symmetry."

With these facts in mind and following Fig. 7.2*b*, proceed to construct a field map of the octant *MOB*. Choose some point *D* and draw in a line *DCF* that represents the best "guess" (based on knowledge gathered from the field sketch) as to the location of a flux line. Then a symmetrically disposed line *GCE* represents the best "guess" as to an equipotential. These two lines must cross each other normally and must cross the construction symmetry *OA* at 45°. They must also leave points *D* and *G* normally. Lines *BEA* and *MFA* may now be drawn, with the purpose of producing curvilinear squares. Starting point *D* should have been chosen so that as the sketcher proceeds outward toward the converging field the square size does not become too small for convenience. A square not much larger than the weight of pencil line used is likely to be highly inaccurate. The next step is to test the map so far drawn to see if the "squares" obtained are really "curvilinear squares." The test is this: *If, upon subdivision by half-equipotentials and by half-flux-lines, the parts of the square tend to become "true squares," the squares are curvilinear squares, and the field is mapped correctly.* It is an interesting fact that, if the squares are incorrectly drawn, subdivision produces "rectangles" and not "squares." So, in Fig. 7.2*b*, it is useless to draw more squares than are shown until the subdivision test has demonstrated their correctness. It is usually better to erase an incorrect start and try again than to try to repair a bad job.

At first glance the section *OGCDO* (which looks like a boot) may not appear to be a curvilinear square. However, upon subdivision, three squares will appear, leaving over a very much smaller boot. In the limit of repeated subdivision the boot becomes vanishingly small. Consider square *OMABO*, which is similarly subdivided. It should be noted that, in a converging field (such as Fig. 7.2), the half-equipotentials and half-flux-lines are never just halfway across the square, but should be placed slightly to the converging field side of the halfway mark.

7.6. The 270°-Corner Field. The 270°-corner field shown in Figs. 7.3*a* and 7.3*b* introduces no new point of importance. It should be noted, however, that the field is generally diverging, and that the surface charge density approaches an infinite value as O is reached. In this case O is not a singular line but is to be regarded as a circular corner of zero radius.

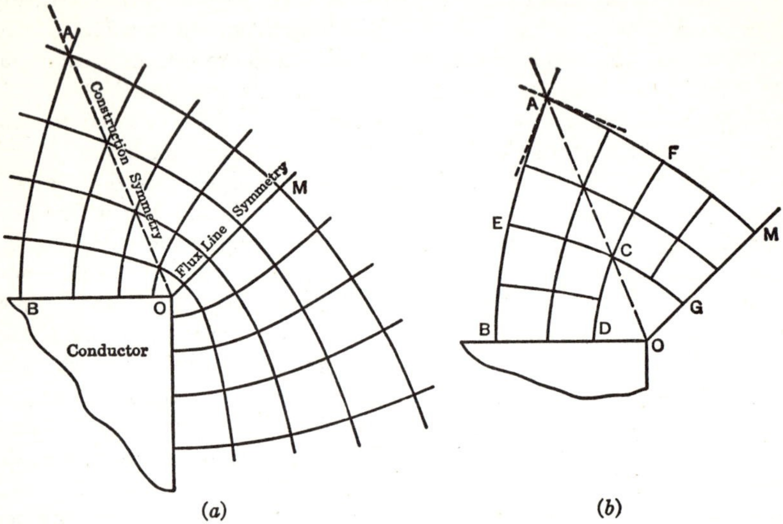

FIG. 7.3. The 270°-corner field.

7.7. The 90°-Slot Fields. Figure 7.4a shows the field between two conductors having a potential difference, in which the conductors are separated by two parallel faced slots of the same width which meet at

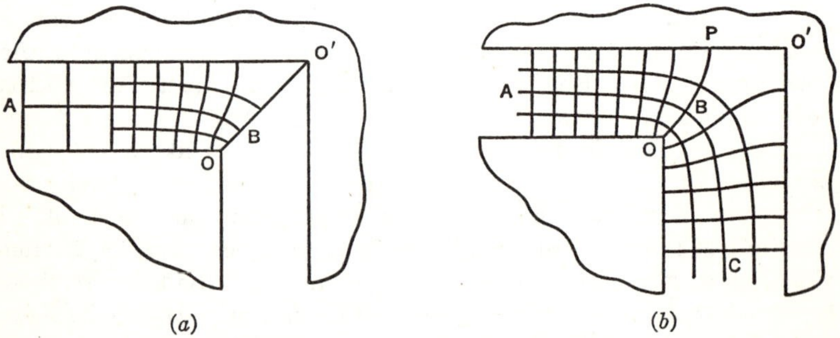

FIG. 7.4. Fields of 90° slots.

90°. The field is symmetrical about the straight flux line running from O to O'. This field may be regarded as a composition of two like uniform fields in the space far from OO', the 90°-corner field near O', and the 270°-corner field near O. It is merely a matter of ingenuity to bring these separate solutions together to form a harmonious whole. The midpotential AB, which is exactly halfway at A, crosses OO' closer to O, owing to the convergence of the field toward O.

Figure 7.4*b* illustrates a similar field in the case where one portion of the slot has a greater spacing than the other portion. The flux line leaving O at 135° extends to point P on the nearer conductor surface.

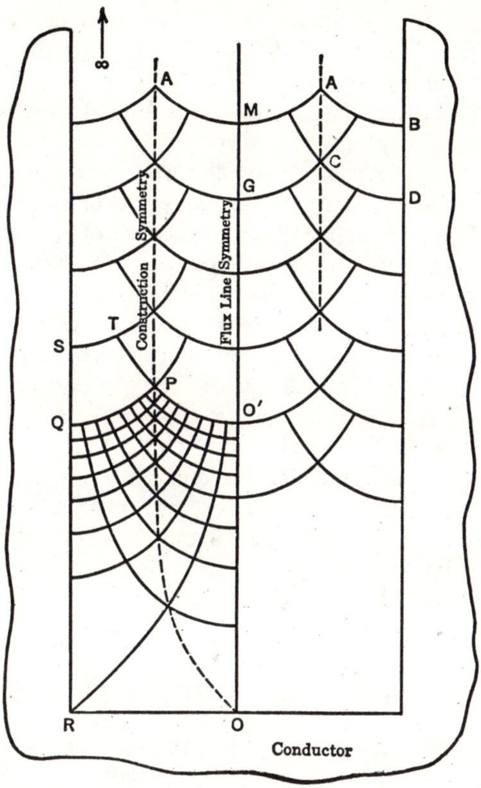

FIG. 7.5. Field of an infinite slot.

The midpotential ABC is not symmetrically disposed about the conductor containing O.

7.8. The Infinite Slot. Figure 7.5 shows the field at the square bottom of a parallel faced slot of infinite depth in one direction. OM is a flux line symmetry, and OA is a construction symmetry. Not too near the bottom of the slot the equipotentials which cross OM and the flux lines which leave the conductor are symmetrical about the construction symmetry. Near the bottom of the slot the construction symmetry OA must become curved in order to reach the junction between the conductor and the flux line symmetry at O.

In a rapidly converging field it is often desirable to change the size of the square. Curvilinear square $O'PQROO'$ has been subdivided in order to show local details, but at the upper end the squares have become too small for handling. QP is divided into eight parts. This means that square $PQST$ is the equivalent of 64 squares of the kind found in $O'PQROO'$. If further subdivision were carried out, the field about R would be like that of the 90° corner, while that about O would become a uniform field.

7.9. Improving Mapping Technique. A field sketch, qualitatively correct, should always precede any effort spent on mapping. The better the sketch the easier it is to produce a good map quickly. In general, it is easier to work from a region of small squares to one of larger squares. Straight lines and curves of known shape should be drawn in correctly when their locations are known. The process of subdivision as a check for squares should not be carried too far.

Maps should be made on plain white paper with sharp pencils of medium hardness. An eraser is one of the most important tools—it is much easier to start afresh than to try to correct a map that is sadly astray. The one rule that is absolutely unbreakable, if good maps are to be produced, requires that equipotentials and flux lines be drawn as nearly 90° as possible by use of the eye. Many angle errors are revealed by frequently turning the map and looking at it sideways or upside down. The field mapper should make a practice of looking squarely at the map—a view from an angle is likely to be deceptive.

TABLE 7.1

RELATED QUANTITIES IN STREAMLINE FIELDS

Type of Field	*Flux or Flow Lines*	*Vector Quantity*	*Scalar Quantity*
Electric	Electric flux	Electric flux density	Electric potential
Magnetic	Magnetic flux	Magnetic flux density	Magnetic potential
Electric current	Electric current	Electric current density	Electric potential
Heat conduction	Heat units per unit time	Temperature gradient	Temperature
Hydro and aero*	Water or air per unit time	Water or air velocity	Velocity potential

* Hydrodynamic and aerodynamic fields cannot be treated by field mapping methods if the velocity of the fluid particles is high enough to create turbulence.

7.10. Applications to Other Fields. Mapping technique may be carried over and applied to any field that has streamline qualities—

current flow through a conductor, heat flow through a material, water flow and air flow where turbulence is absent, and magnetic fields. Table 7.1 gives the quantities which are comparable in a number of streamline fields.

PROBLEMS

1. At point P_1 in Fig. 7.1 the flux density is $1000\epsilon_0$ lines per square meter and $\epsilon_r = 1$. Find the charge density and pressure at points P_2 and P_4, the intensity at P_3. Find the energy density and total energy per square at points P_1 and P_3. How many flux lines cross the field between points P_2 and P_4? What is the total force on conductor B between points P_5 and P_7?

2. Plot the field between the conductors shown in Fig. P2. What is the charge density at point C?

3. Plot the field of Fig. P3, starting from the upper end.

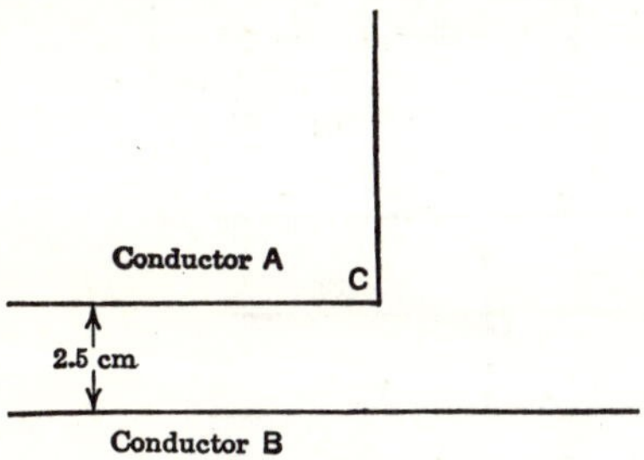

FIG. P2

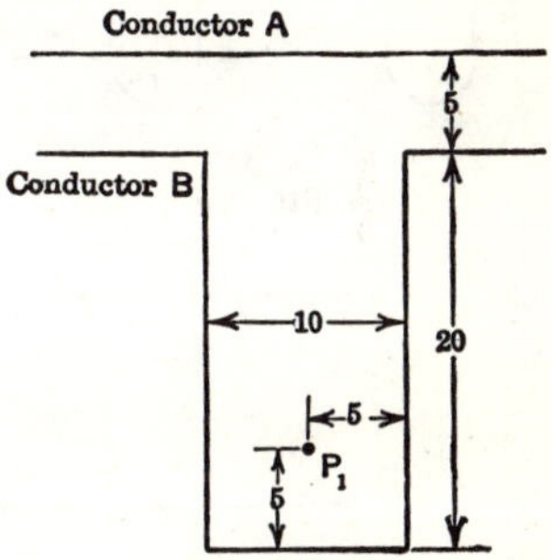

FIG. P3

4. Place a conductor at point P_1 in Fig. P3. Assume conductors A and B at ground potential and the conductor at P_1 at some positive potential. Plot the field.

5. Plot the current flow field and equipotentials in Fig. P5. The current density near A is 1000 amperes per square centimeter. Conductivity $\sigma = 5.8 \times$

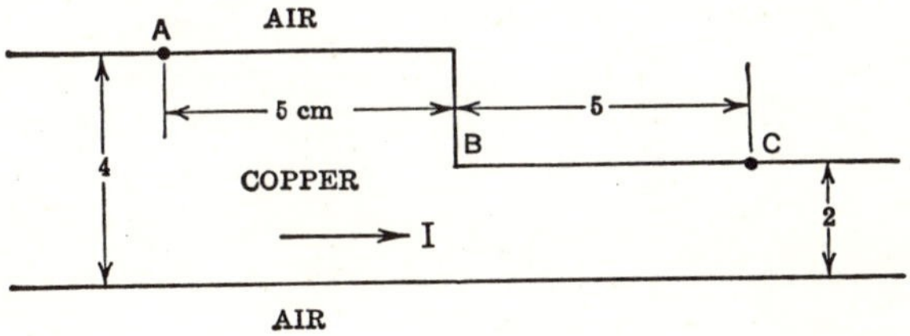

FIG. P5

10^7 mhos per meter. Find the resistance between points A and C, and the voltage drop from A to B, and B to C.

6. Plot the field of Fig. P6.

7. With equal positive and negative charges on conductors A and B, and with C grounded, plot the field of Fig. P7.

8. Plot the field of Fig. P8 for a half infinite plane at potential $V = 1000$ volts placed halfway between two full infinite planes each at zero potential.

9. Plot the field of Fig. P8 if the uppermost plane is removed.

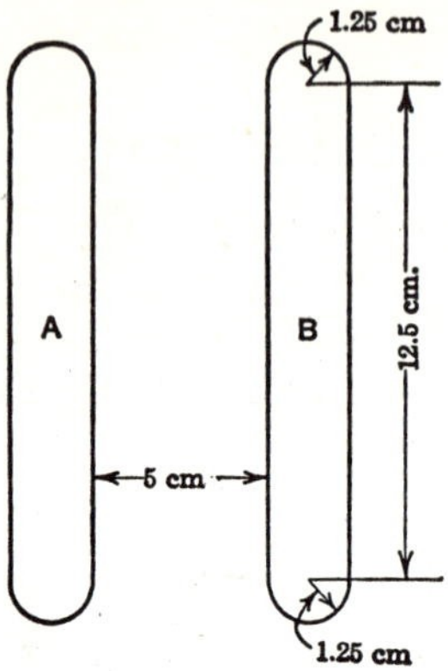

FIG. P6

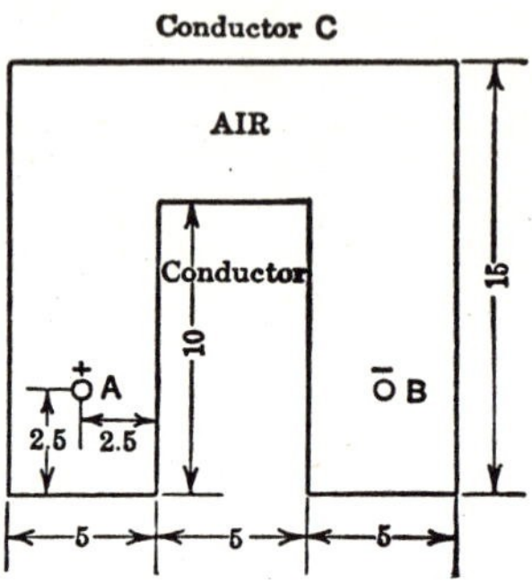

FIG. P7

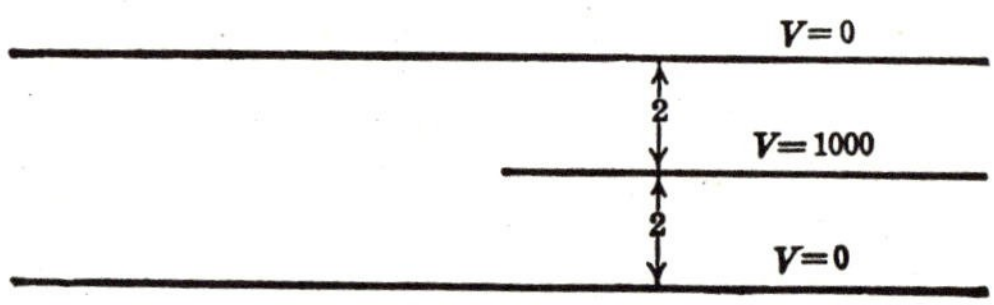

FIG. P8

CHAPTER 8

ENERGY AND FORCES IN CONDENSERS

The fact that forces are required to separate positive and negative electricity indicates that potential energy is stored in the system. When a condenser is connected to a battery or other source of electric potential, the condenser plates serve as an extension of the battery terminals and become charged with equal amounts of electricity of opposite types, according to the relation

$$Q = CV \tag{8.1}$$

where V is the condenser potential. It is explained in Art. 1.53 that the energy stored in the condenser may be computed from the equation

$$W_e = \frac{1}{2}QV = \frac{1}{2}CV^2 = \frac{1}{2}\frac{Q^2}{C} \quad \text{joules} \tag{8.2}$$

and that this energy may be regarded as distributed throughout the volume of the dielectric with a density equal to

$$\omega_e = \frac{\epsilon E^2}{2} = \frac{D^2}{2\epsilon} = \frac{DE}{2} \quad \frac{\text{joules}}{\text{cubic meter}} \tag{8.3}$$

8.1. Condenser Energy Storage Is a Minimum. Like charges repel each other and unlike charges mutually attract. But this is equivalent to saying that positive charges seek to reach lower absolute potentials, while negative charges are urged toward higher absolute potentials. Still another way of stating the matter is that charges of electricity exert forces upon each other in such a way that they attempt to *reduce* the difference of potential between themselves.

Applying these observations to a charged condenser, we may say that on each conductor of the condenser the charges by mutual repulsion seek the boundary, and then, influenced by the attraction of the charges on the opposite conductor, still further alter their distribution until there is the least possible difference of potential between the conductors.

Since the energy stored in the electric field of a condenser is $W_e = \frac{1}{2}QV$, the distribution of charges Q over the boundaries of the condenser is the particular one that gives the lowest possible difference of potential and an energy storage that is lower than that of any other distribution of the charges.

A. FORCES IN CONDENSERS

8.2. Forces in Condensers with Constant Charge Q. Let us consider a condenser composed of two conductors which are not connected to a battery so that the charges $\pm Q$ are fixed in number and are separated by a region containing two insulators, as shown in Fig. 8.1. As a consequence of the energy stored in this system a force is exerted

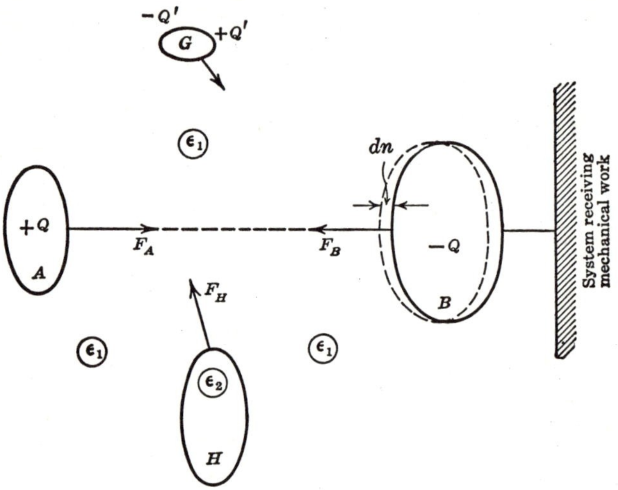

FIG. 8.1. Forces on a condenser with constant charge.

upon each part of the condenser, which causes a motion of that part if it is free to move. However, the force exists whether motion results or not. The electric field by its forces tries to do mechanical work upon the boundaries.

To evaluate the force on any part (say body B in Fig. 8.1), let us assume that the force on B causes a small displacement dn. Then the mechanical work dMW is done at the expense of the energy stored in the field, if no heat losses are presumed.

$$0 = dMW + dW_e$$

or

$$dMW = -dW_e \tag{8.4}$$

The force F_B equals the mechanical work divided by the displacement.

$$F_B = \frac{dMW}{dn} = -\frac{dW_e}{dn} \quad (Q \text{ assumed constant}) \tag{8.5}$$

$$= -\frac{Q^2}{2}\frac{d(1/C)}{dn} = +\frac{Q^2}{2}\frac{1}{C^2}\frac{dC}{dn} \tag{8.6}$$

The forces of the electric field, acting on the charges, try to do mechanical work on the boundaries and reduce the energy storage. If the energy storage is

$$\underset{\downarrow}{W_e} = \frac{1}{2}\frac{Q^2}{\underset{\uparrow}{C}} \tag{8.7}$$

and Q is held constant, the force is such as to reduce the energy storage or increase the capacitance C of the condenser, and it has the direction that gives the greatest increase in mechanical work or capacitance for a given small movement.

8.3. Forces in Condensers with Constant Voltage Applied. The only difference between this case and the preceding is that the charges do not remain constant as the condenser is given a small motion. As the condenser moves a small distance and changes from a smaller to a larger capacitance, it does external mechanical work at the expense of the energy storage, and the condenser potential difference *tends* to decline, but the least tendency for the condenser voltage to decline is the signal for the constant voltage battery to send electric charges over to the condenser to keep it at a constant potential. The net result, for a small motion, is that for every unit of energy that goes to mechanical work there is an energy storage decline of one unit, which is promptly followed by the delivery of two units of energy to the condenser by the battery. As a consequence the energy storage *increases* one unit while the mechanical work is performed. This is readily shown:

The energy from the battery

$$= dW_B = V_b\,dQ = V_b^2\,dC$$

The energy storage change is

$$dW_e = d(\tfrac{1}{2}CV_b^2) = \tfrac{1}{2}V_b^2\,dC = \tfrac{1}{2}\,dW_B$$

The energy balance during the motion is

$$dW_b = 2dW_e = dW_e + dMW \tag{8.8}$$

or

$$dW_e = +dMW$$

and the force is

$$F_n = \frac{dMW}{dn} = +\frac{dW_e}{dn} \quad \text{(constant potential applied)} \tag{8.9}$$

$$= +\frac{V^2}{2}\frac{dC}{dn} \tag{8.10}$$

8.4. Constant Charge or Constant Voltage Condensers. The preceding cases show that *any* condenser tends to:

(1) Distribute its charge over its conducting boundaries so as to: (*a*) establish the least possible potential difference between boundaries; (*b*) store a minimum amount of energy.

(2) Do mechanical work on any external mechanical system connected to one or another of the boundaries.

(3) Permit its boundaries to move in a direction which increases the capacitance.

With *constant charge* the condenser tends to:

(4) Move in a direction that *decreases* its energy storage.

With *constant voltage* the condenser tends to:

(5) Move in a direction that *increases* its energy storage.

8.5. Changing the Capacitance of a Condenser. It is logical to inquire about the various geometrical changes in a condenser which alter its capacitance. The following list indicates methods that may be used to *increase* the capacitance (following Fig. 8.1):

(1) Moving the conductors A and B closer together.

(2) Introducing a dielectric H (ϵ_2) into the dielectric (ϵ_1) already present, where $\epsilon_2 > \epsilon_1$ (since the capacitance is always directly proportional to the dielectric constant).

(3) Introducing a conductor G into the dielectric (ϵ_1) already present. (For this purpose, a conductor may be regarded as a dielectric with an infinite dielectric constant.)

(4) Moving the dielectric H ($\epsilon_2 > \epsilon_1$) or the conductor G from a weaker to a denser part of the field.

(5) Turning a dielectric or a conductor, of ellipsoidal shape so that its longest axis points along the direction of the flux lines.

(6) Moving a dielectric ϵ_3 (where $\epsilon_3 < \epsilon_1$) to a weaker part of the field, or removing it entirely.

All these methods *increase* the capacitance and, with constant charge Q, *reduce* the stored energy and potential difference between A and B.

$$\underline{Q} = \underset{\uparrow\downarrow}{CV} \tag{8.11}$$

$$\underset{\downarrow}{W_e} = \tfrac{1}{2}\underline{Q}\underset{\downarrow}{V} \tag{8.12}$$

With constant voltage, they *increase* the capacitance, the charge, and the energy stored.

$$\underset{\uparrow}{Q} = \underset{\uparrow}{C}\underline{V} \tag{8.13}$$

$$\underset{\uparrow}{W_e} = \tfrac{1}{2}\underset{\uparrow}{Q}\underline{V} \tag{8.14}$$

B. MOVABLE CONDENSERS

8.6. Movement with Constant Charge. Under this condition, the relations of Art. 8.2 may be extended to cover a condenser movement of *finite* size. Integrating equation 8.4 for the total energy changes that accompany the entire motion, we get

$$0 = \int_{W\,\text{initial}}^{W\,\text{final}} dW_e + \int_0^{\Delta MW} dMW \tag{8.15}$$

$$0 = (W_f - W_i) + \Delta MW \tag{8.16}$$

and, if the total motion is called Δn, the average force is

$$F_{\text{average}} = \frac{\Delta MW}{\Delta n} = -\frac{(W_f - W_i)}{\Delta n} \tag{8.17}$$

$$= -\frac{Q^2}{2\Delta n}\left(\frac{1}{C_f} - \frac{1}{C_i}\right) \tag{8.18}$$

8.7. Movement with Constant Voltage Applied. $R = 0$. With no resistance between the battery and the condenser, a finite movement of the latter causes energy changes obtained by integrating equation 8.8 over the entire motion. See Fig. 8.2.

$$\int dW_{\text{battery}} = \int_{\text{initial}}^{\text{final}} dW_e + \int_0^{\Delta MW} dMW \tag{8.19}$$

$$W_{\text{battery}} = 2(W_f - W_i) = (W_f - W_i) + \Delta MW \tag{8.20}$$

If the total motion is Δn, the average force is

$$F_{\text{average}} = \frac{\Delta MW}{\Delta n} = + \frac{(W_f - W_i)}{\Delta n} \tag{8.21}$$

$$= \frac{V^2}{2\Delta n}(C_f - C_i) \tag{8.22}$$

8.8. Movement with Constant Voltage Applied. *R* = *R*. With resistance in the circuit between the battery and the condenser (Fig. 8.2), the condenser potential difference and that of the battery differ by the value of the iR drop, which occurs simultaneously with the charge flow that accompanies the condenser movement, and leads

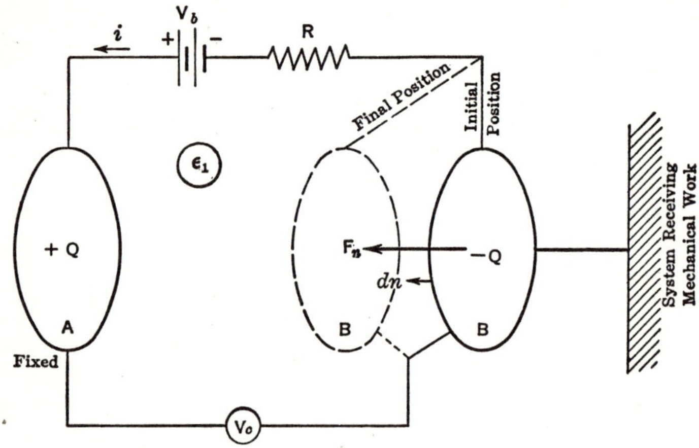

FIG. 8.2. Forces on a condenser with constant voltage applied.

to a loss in heat in the resistance, thereby decreasing the amount of mechanical work that can be performed during the motion. Let us consider the energy changes that occur during a "smooth" motion of the condenser. A "smooth" motion is one that starts at zero time with zero velocity and after time T is brought again to zero velocity by whatever restraining forces are necessary.

Let V_b = constant battery voltage.
V_c = condenser back voltage (variable).
C = condenser capacitance (variable).
R = resistance.

$$i = \text{current (variable)} = \frac{V_b - V_c}{R}.$$

t = elapsed time during movement.
T = total time of completed movement.

q = charge on condenser at time t (variable) $= CV_c$.
ΔQ = change in condenser charge $= (Q_2 - Q_1)$.
W_b = total energy supplied by battery $= V_b(Q_2 - Q_1)$.
ΔW_e = total change of energy storage in condenser $= W_2 - W_1 = \frac{1}{2}V_b(Q_2 - Q_1)$.

H = total heat energy lost in resistance $= \int_0^T i^2R\,dt$.

ΔMW = total mechanical work performed.

The energy supplied by the battery less that transformed into heat increases the energy storage in the condenser and performs mechanical work.

$$W_b - H = (W_2 - W_1) + \Delta MW \tag{8.23}$$

which equals

$$V_b(Q_2 - Q_1) - H = \tfrac{1}{2}V_b(Q_2 - Q_1) + \Delta MW$$

Hence

$$\Delta MW + H = \tfrac{1}{2}V_b(Q_2 - Q_1) = \Delta W_e$$

But instantaneously,

$$\frac{dH}{dW_b} = \frac{i^2R\,dt}{V_bi\,dt} = \frac{Ri\,dt}{V_b\,dt} = \frac{R\,dq}{V_b\,dt} \tag{8.24}$$

For the entire movement

$$\frac{H}{W_b} = \frac{\int R\,dq}{\int V_b\,dt} = \frac{R(Q_2 - Q_1)}{V_bT} = \frac{R\,\Delta Q}{(\Delta Q/\Delta C)T} = \frac{R\,\Delta C}{T}$$

or

$$H = W_b\frac{R(Q_2 - Q_1)}{V_bT} = \frac{R(Q_2 - Q_1)^2}{T} \tag{8.25}$$

Finally, the mechanical work is

$$\Delta MW = \Delta W_e - H = \tfrac{1}{2}V_b(Q_2 - Q_1) - \frac{R(Q_2 - Q_1)^2}{T}$$

$$= \tfrac{1}{2}V_b\,\Delta Q - \frac{R(\Delta Q)^2}{T} \tag{8.26}$$

From this it is seen that, if the motion takes place very slowly ($T = \infty$), the mechanical work reduces to the value of the change in energy storage, as given in the preceding article.

Figure 8.3 shows qualitatively the manner in which the charge, capacitance, potential, and current change throughout the movement, as well as the power from the battery which supplies power to the condenser energy storage, heat, and mechanical work.

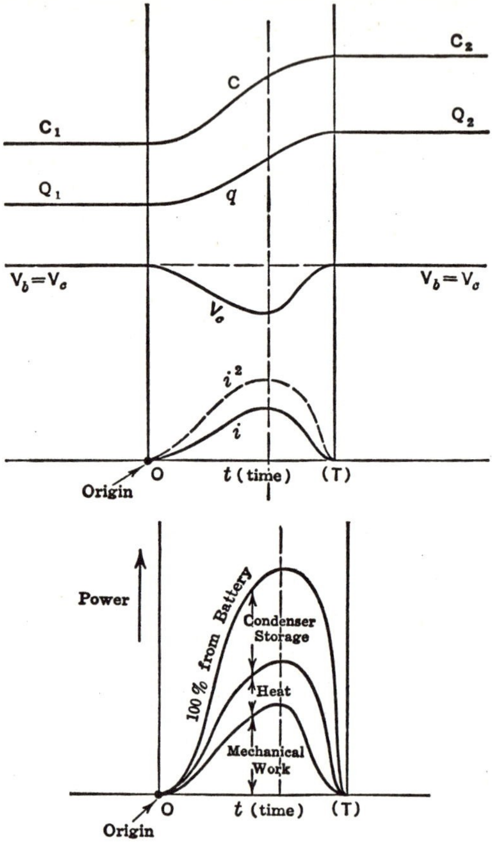

FIG. 8.3. Changes in a condenser during a "smooth" motion.

C. FORCES ON DIELECTRICS

8.9. Mechanical Pressure on Boundaries. At the boundary between two media in the presence of an electric field, there exists a normal mechanical pressure which tends to cause one of them to move into the space occupied by the other. It is the object in this section to develop formulas for this pressure covering the two types of boundary—the conductor–insulator and the insulator–insulator types.

8.10. Pressure inside an Insulator. First let us choose a small volume inside an insulator remote from the boundary, throughout which we may consider the electric intensity E to be sensibly constant. In this region we may picture a very large number of atoms uniformly distributed which have been polarized or stretched by the action of the electric field. Since each atom is composed of positive and negative electricity, we may also think of a cloud of uniformly distributed negative electricity which has been slightly shifted with respect to an equal and uniformly distributed cloud of positive electricity. The negative electricity by the mutual repulsion of the many particles sets up an outward pressure equally in all directions. It tends to swell the small volume into surrounding territory, acting normally upon any boundary. The positive charges act likewise.

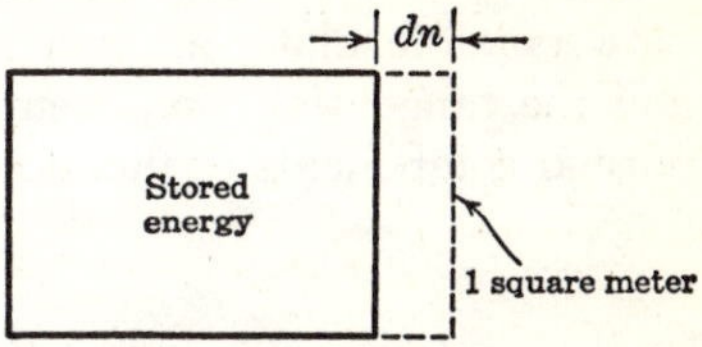

FIG. 8.4. Insulator volume subject to boundary pressures.

The magnitude of this pressure is

$$\frac{(\epsilon - \epsilon_0)E^2}{2} \frac{\text{newtons}}{\text{square meter}} \qquad \text{(Normal to boundary–outward)} \tag{8.27}$$

This is easily seen. Assuming that we hold the equipotential surfaces of the field constant whether we have the insulator present or not, we note that $\epsilon E^2/2$ joules per cubic meter is the energy density in the insulator and that $\epsilon_0 E^2/2$ joules per cubic meter is the energy density in vacuum without the insulator. Consequently the difference, $(\epsilon - \epsilon_0)E^2/2$ joules per cubic meter, is the increase in energy density upon introducing the insulator which becomes polarized.

If we now permit unit area of the small volume to expand a small distance dn as shown in Fig. 8.4, the increase in energy (at the same density assumed) is $\dfrac{(\epsilon - \epsilon_0)E^2}{2}\, dv$ joules, but dv is $1 \times dn$ cubic meters. Under the assumed condition of constant voltage, the increase in energy is equal to the mechanical work done by the volume where no heat is lost. The mechanical work divided by the movement is the force per unit area, or in this case the

$$\text{Pressure } p_h = \frac{(\epsilon - \epsilon_0)E^2}{2} \quad \frac{\text{newtons}}{\text{square meter}}$$

$$\text{(Normal to boundary–outward. Insulator } \epsilon \text{ to vacuum } \epsilon_0) \tag{8.28}$$

When two insulators meet at a boundary, each tries to push the boundary out normally. Then the difference in pressure is the final result.

$$p_h = \frac{(\epsilon_1 - \epsilon_0)E_1^2}{2} - \frac{(\epsilon_2 - \epsilon_0)E_2^2}{2} \quad \frac{\text{newtons}}{\text{square meter}}$$

(Pushing boundary from medium 1 to 2) (8.29)

8.11. Film Pressure at Insulator Boundary. At the boundary between an insulator and vacuum the electric flux is usually refracted, and the tangential components of intensity are equal, as well as the normal components of flux density.

$$E_{1T} = E_{2T}$$

$$D_{1N} = D_{2N}$$

In the insulator just back of the film the polarization or shift of the bound charge per square meter along the line of the flux is

$$P_1 = D_1 - \epsilon_0 E_1 \quad \frac{\text{coulombs}}{\text{square meter}} \tag{8.30}$$

The component of polarization normal to the boundary surface is

$$P_{1N} = D_{1N} - \epsilon_0 E_{1N} \quad \frac{\text{coulombs}}{\text{square meter}} \qquad \text{(Normal)} \tag{8.31}$$

as shown in Fig. 8.5. The polarization in the vacuum is zero. Now let us investigate the film itself closely. Since only a normal motion of the film will expand the boundary, we need consider only the normal component of the polarization. In the film itself the normal flux density remains constant, but the normal intensity and polarization vary gradually and assume the values known for them just outside the film. Figure 8.6 illustrates these values. Throughout the film the dielectric constant varies gradually from ϵ_1 to ϵ_0.

The normal polarization is

$$P_N = D_N - \epsilon_0 E_N \tag{8.32}$$

whence

$$dP_N = -\epsilon_0 \, dE_N \tag{8.33}$$

as D_N is a constant throughout the film. Now consider a thin layer portion of the film, s meters from the origin and ds thick, where ds is considered to be the distance between the negative and positive charges in the atoms in layer ds. At distance s, the negative ends of the atoms lie in field strength E_N. At distance $(s + ds)$ the positive ends of the atoms lie in field strength $E_N + dE_N$. Hence each double charge is subjected to a net force per unit charge of dE_N toward the right in Fig. 8.6. But there are P_N charges per square meter. Thus the force per square meter acting to the right on the layer ds is

$$dF = P_N\,dE_N = -\frac{1}{\epsilon_0} P_N\,dP_N \tag{8.34}$$

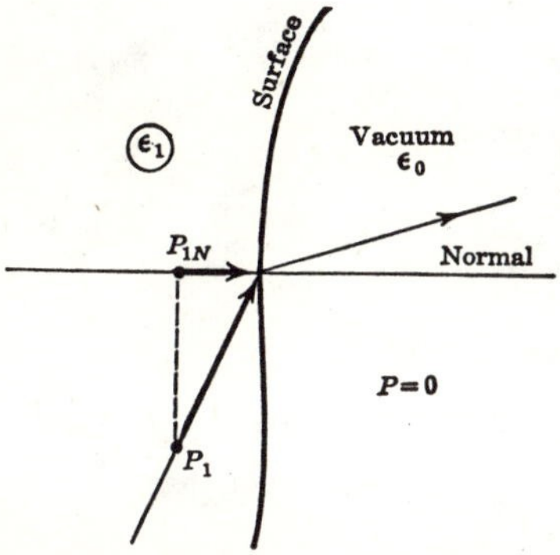

FIG. 8.5. Polarization at an insulator boundary.

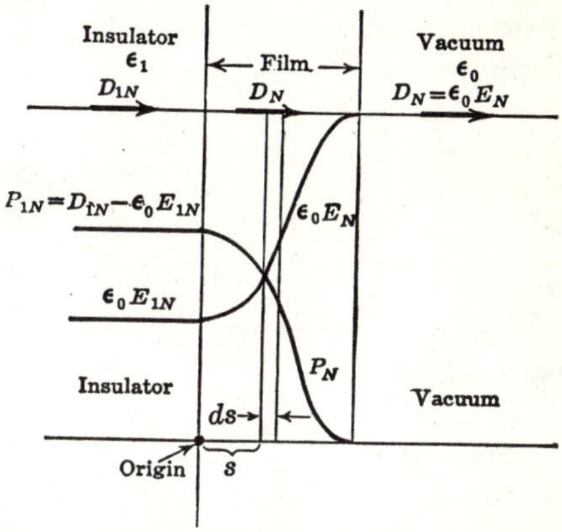

FIG. 8.6. Pressures on an insulator boundary film.

The whole pressure upon the film is then

$$p_{\text{film}} = \int_{P_N=P_{1N}}^{P_N=0} -\frac{1}{\epsilon_0} P_N\,dP_N = \frac{1}{2\epsilon_0} P_{1N}^2 \quad \frac{\text{newtons}}{\text{square meter}} \qquad \text{(Normal)} \tag{8.35}$$

or since

$$P_{1N} = D_{1N} - \epsilon_0 E_{1N} = (\epsilon_1 - \epsilon_0)E_{1N}$$

$$p_{\text{film}} = \frac{(\epsilon_1 - \epsilon_0)^2 E_{1N}^2}{2\epsilon_0} \quad \frac{\text{newtons}}{\text{square meter}} \qquad \text{(Normal to boundary—pushing film into vacuum)} \tag{8.36}$$

When two insulators meet at a boundary, the normal pressure on the film urging it from medium 1 to medium 2 is

$$p_{\text{film}} = \frac{(\epsilon_1 - \epsilon_0)^2 E_{1N}^2}{2\epsilon_0} - \frac{(\epsilon_2 - \epsilon_0)^2 E_{2N}^2}{2\epsilon_0} \quad \frac{\text{newtons}}{\text{square meter}}$$

$$\text{(Pushing boundary from medium 1 to 2)} \tag{8.37}$$

8.12. Total Pressure at Insulator–Insulator Boundary. The total boundary pressure between two insulators, urging the boundary from medium 1 to medium 2, is the sum of the two pressures:

$$p_{\text{boundary}} = \frac{1}{2}\left[(\epsilon_1 - \epsilon_0)E_1^2 - (\epsilon_2 - \epsilon_0)E_2^2 + \frac{(\epsilon_1 - \epsilon_0)^2 E_{1N}^2}{\epsilon_0} - \frac{(\epsilon_2 - \epsilon_0)^2 E_{2N}^2}{\epsilon_0}\right] \frac{\text{newtons}}{\text{square meter}} \qquad (8.38)$$

This can be simplified by rewriting it in terms of flux density, remembering that

$$D_1^2 = D_{1N}^2 + D_{1T}^2$$

$$E_1^2 = E_{1N}^2 + E_{1T}^2$$

$$D_{1N} = D_{2N}$$

$$E_{1T} = E_{2T}$$

reducing to

$$p_{\text{boundary}} = \frac{1}{2}\left(\frac{D_{1T}^2}{\epsilon_1} - \frac{D_{2T}^2}{\epsilon_2} + \frac{D_{2N}^2}{\epsilon_2} - \frac{D_{1N}^2}{\epsilon_1}\right) \frac{\text{newtons}}{\text{square meter}} \qquad (8.39)$$

urging medium 1 toward 2 if the result is positive.

Practically, the result is always to urge the boundary toward the medium having the lower dielectric constant. Rewrite equation 8.39 as

$$p_{\text{boundary}} = \frac{1}{2}\left[E_{1T}^2(\epsilon_1 - \epsilon_2) + D_{1N}^2\left(\frac{1}{\epsilon_2} - \frac{1}{\epsilon_1}\right)\right] \frac{\text{newtons}}{\text{square meter}} \qquad \text{(Normal)} \quad (8.40)$$

Now if $\epsilon_1 > \epsilon_2$ the result is positive, urging 1 toward 2.

8.13. Attraction at a Conductor–Insulator Boundary. In Art. 1.48 we saw that, when an insulator meets a conducting surface upon which there are ρ_s "free" charges per square meter, there is a normal attraction

$$p_c = \frac{1}{2}\frac{\rho_s^2}{\epsilon} \frac{\text{newtons}}{\text{square meter}} \qquad \text{(Toward insulator normally)} \quad (8.41)$$

tending to pull the conducting surface into the insulator, whether ρ_s is positive or negative.

Since the surface charge density is equal to the normal flux density

$$p_c = \frac{D_N^2}{2\epsilon} \frac{\text{newtons}}{\text{square meter}} \qquad \text{(Toward insulator normally)} \quad (8.42)$$

PROBLEMS

1. Two metallic circular plates, 40 cm in diameter, are parallel to each other on a common axis in air.

(*a*) Constant charge, $+Q = 10^{-6}$ coulomb.

(1) Prove that the force of attraction is independent of the spacing, and compute its value.

(2) Find the mechanical work performed in changing the spacing from 1.0 cm. to 0.5 cm.

(3) Find the change of energy stored as a consequence of the motion.

(*b*) Constant potential, V = value above for 1.0-cm spacing.

(1) Prove that the force of attraction varies inversely as the square of the separation. Compute the force for $s = 1.0$, and $s = 0.5$ cm.

(2) Same as part 2 above.

(3) Same as part 3 above.

2. If the difference of potential between two conductors is doubled, how much is the force on each increased?

3. Given two thin concentric brass cylinders of radii 5 and 6 cm and length 30 cm each. The inner cylinder projects 10 cm into the outer one. Their potential difference is 1200 volts, maintained constant.

(*a*) Find the pull on the inner cylinder.

(*b*) If the inner cylinder moves in 10 cm more in 5 seconds, find the average current.

4. Given two circular parallel plates, 40 cm in diameter and 1 cm apart. The plates are parallel to the earth. If the support for the under plate is taken away, would the plates come together or would the under plate fall to earth? Each plate is charged with 10^{-6} coulomb, opposite signs. The under plate weighs 500 grams.

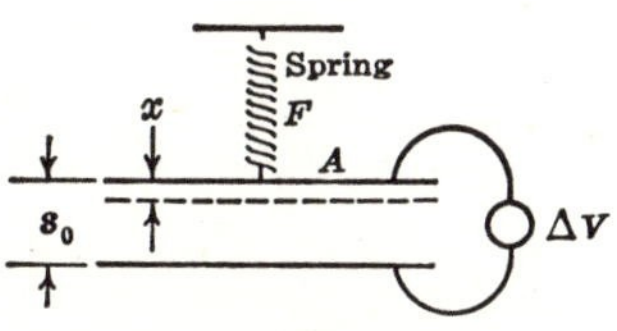

FIG. P6

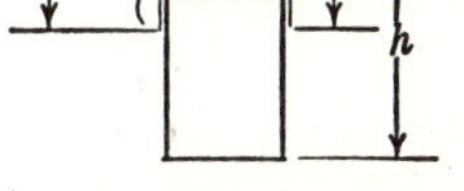

FIG P7

5. Three square parallel plates 5 cm apart in air and 20 cm on a side form an electrostatic voltmeter. The two outer plates are stationary and are connected by a metallic wire. The voltage to be measured is placed across the middle and one of the outer plates. When the middle plate is a quarter way in, the applied voltage is 15,000; when halfway in, the voltage is 40,000. If we neglect any fringing flux and if the deflection of the spring attached to the middle plate is directly proportional to the force upon it, what voltage will pull the middle plate in completely?

6. Consider the electrostatic voltmeter shown in Fig. P6. s_0 is the initial separation, with no voltage applied. x is the deflection when under voltage. $h = F/x$ newtons per meter is the spring constant. A equals the plate area (large compared with separation). Compute ΔV for any deflection. Is the movable plate stable in all positions?

7. A parallel plate air condenser with rectangular plates ($m \cdot h = 50 \cdot 100$ cm) as shown in Fig. P7, with separation $s = 2$ cm, pulls a rubber dielectric ($\epsilon_{r2} = 3$) into itself distance n. Compute the pull P on the rubber for any insertion n, for:

(*a*) Constant potential $\Delta V = 10{,}000$ volts.
(*b*) Constant charge Q, corresponding to 10,000 volts with $n = 0$.

8. For the condenser of Prob. 1*b*, let the motion take place in 0.1 second, and let $R = 100$ megohms. Compute the heat loss and the mechanical work performed.

9. A circular rubber cylinder ($\epsilon_r = 3$) of radius $R = 5$ cm is placed with its axis normal to an originally uniform air field $E_0 = 500{,}000$ volts per meter. The field inside the rubber is uniform, parallel to E_0, and equal to $E_r = E_0 \dfrac{2}{\epsilon_r + 1}$. Compute the pressure around the cylindrical boundary. Compute the force over half the surface.

10. Why does an electric charge attract a neutral conductor? An insulator?

PART II
THE MAGNETIC FIELD

CHAPTER 9

THE MAGNETOSTATIC FIELD

A steady or unvarying current, which requires a continuous flow of charge around a closed circuit, is accompanied by a magneto*static* field which pervades all space. The purpose of this chapter is to indicate the methods and techniques commonly used to describe and calculate the magnitude of the magnetostatic field both inside and outside current-bearing regions.

9.1. Qualitative Description of the Field. Simple experiments indicate the "circulatory" character of the magnetic field. One

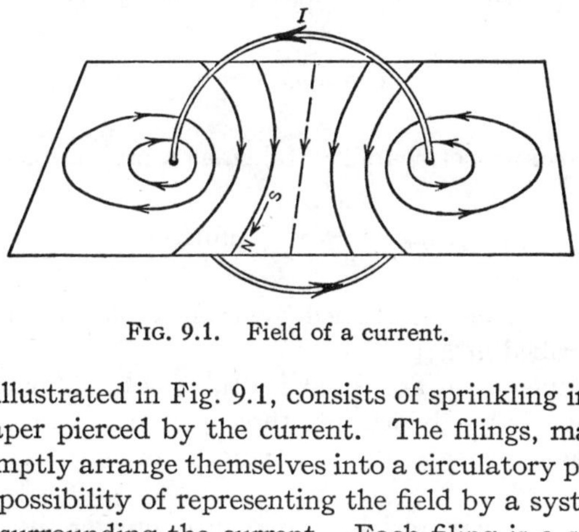

FIG. 9.1. Field of a current.

experiment, illustrated in Fig. 9.1, consists of sprinkling iron filings on a sheet of paper pierced by the current. The filings, magnetized by the field, promptly arrange themselves into a circulatory pattern which suggests the possibility of representing the field by a system of closed "flux" lines surrounding the current. Each filing is a small magnet with north- and south-seeking poles oriented as indicated in Fig. 9.1. The directional "sense" of the field, as a matter of convention, is indicated by placing arrows on the flux lines so that they pass *through* the magnetized particles from south to north, and then out of the north poles, around the current, and back to the south poles. The flux lines thus form a right-handed whorl about the current—that is, with the right thumb placed along the current, the fingers point in the conventional direction of the flux lines.

The total amount of the magnetization, as represented by the number of flux lines, depends on the magnitude of the current (or current-turns), on the shape of the circuit carrying the current, and, to a marked degree, on the magnetic characteristics of the medium as represented by the permeability μ.

9.2. Permeability of the Medium. The total quantity of magnetic flux circulating around a given amount of current varies with the medium. Each medium is therefore assigned a value μ, called the permeability, which is a measure of this property. The permeability is comparable in many respects with the dielectric constant ϵ.

The simplest medium is free space (a vacuum) to which is assigned the value

$$\mu = \mu_0 = \frac{4\pi}{10^7} \ \frac{\text{henrys}}{\text{meter}} \tag{9.1}$$

The theoretical reasons underlying the choice of this value and the units must be deferred until later. (See Art. 9.34 and Appendix I.)

All other media are composed of material substances—gases, liquids, or solids. For these media the permeability is written

$$\mu = \mu_0\mu_r \ \frac{\text{henrys}}{\text{meter}} \tag{9.2}$$

with μ_r, the *relative* permeability, taken as a pure numeric.

$$\mu_r = \frac{\mu}{\mu_0} \qquad \text{(Numeric)} \tag{9.3}$$

Values of μ_r must be determined by appropriate experiments, one of which is described in Art. 13.4.

Typical values of μ_r are listed in Table 9.1. Magnetic materials may conveniently be divided into three groups:

(1) Diamagnetic materials; for these, μ_r is very slightly less than unity.

(2) Paramagnetic materials: μ_r is slightly greater than unity.

(3) Ferromagnetic materials: μ_r is enormously greater than unity. The physical bases for these marked differences are discussed briefly in Arts. 9.44 and 13.20.

The ferromagnetic materials obviously occupy an important place in engineering techniques; the other types are scarcely used at all. μ_r is not a true constant for each material of this class but is a function of the field strength and thus may vary from point to point. The special methods required to handle this problem are discussed in Chapter 13.

9.3. Definition of Field Quantities. The magnetic field is described quantitatively by a number of quantities, some vector and some scalar,

TABLE 9.1

RELATIVE PERMEABILITIES OF MAGNETIC MATERIALS

	μ_r
Free Space (Vacuum)	1.0
Diamagnetic Materials	
Antimony	$1 - 7 \times 10^{-5}$
Bismuth	$1 - 17 \times 10^{-5}$
Copper	$1 - 0.94 \times 10^{-5}$
Lead	$1 - 1.7 \times 10^{-5}$
Mercury	$1 - 3.2 \times 10^{-5}$
Quartz	$1 - 1.5 \times 10^{-5}$
Silver	$1 - 2.6 \times 10^{-5}$
Water	$1 - 0.88 \times 10^{-5}$
Argon (NTP)	$1 - 0.95 \times 10^{-8}$
Hydrogen (NTP)	$1 - 0.21 \times 10^{-8}$
Paramagnetic Materials	
Aluminum	$1 + 0.21 \times 10^{-4}$
Neodymium	$1 + 30 \times 10^{-4}$
Oxygen (liquid at −182° C)	$1 + 40 \times 10^{-4}$
Palladium	$1 + 8.2 \times 10^{-4}$
Platinum	$1 + 2.9 \times 10^{-4}$
Air (NTP)	$1 + 3.6 \times 10^{-7}$
Oxygen (NTP)	$1 + 17.9 \times 10^{-7}$
Ferromagnetic Materials	
Cobalt	270
Nickel	500
Iron (soft—low H)	200
Iron (very pure—laboratory specimen)	275,000 (max)
Steel (commercial)	6,000 (max)
Steel (special silicon)	66,000 (max)
Heusler's alloy (15% Al, 26% Mn, 59% Cu_2)	500 (max)
Permalloy (79% Ni, 21% Fe)	105,000 (max)
Perminvar (45% Ni, 30% Fe, 25% Co) (under 0.08 weber/m²)	300
Hipernik (special) (50% Ni, 50% Fe)	167,000 (max)
Hipernik (standard)	60,000 (max)
Supermalloy (79% Ni, 5% Mo, 16% Fe)	1,050,000 (max)

which are defined below. Additional quantities will be defined as needed in later paragraphs.

9.4. Intensity (Field Strength) *H*. The magnetic intensity H is a vector point function whose negative line integral over any closed line path is equal to the current enclosed by the path, irrespective of the magnetic quality of the medium. Thus

$$\oint_{\text{closed path}} (-H)\, ds \cos \alpha = nI \quad \text{amperes or ampere-turns} \tag{9.4}$$

where n is the number of turns, each carrying current I, and α is the angle between $(-H)$ and an element ds of the path. The units of H then are

$$H \quad \frac{\text{amperes}}{\text{meter}} \tag{9.5}$$

The dimensional formula for current is $T^{-1}Q$ and for H is $L^{-1}T^{-1}Q$ These forms are listed in Appendix I. Equation 9.4 is valid both inside

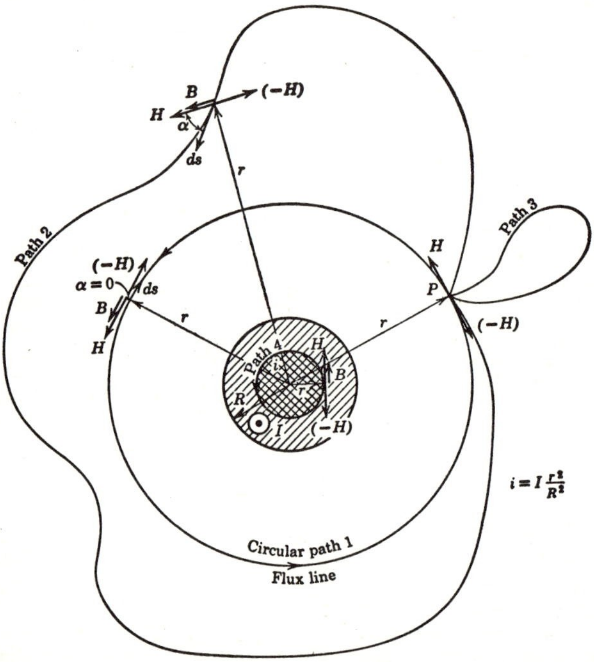

FIG. 9.2. Paths of integration of $\oint H\, ds \cos \alpha$ about current I.

and outside current-bearing regions. Magnetic intensity H is comparable, in many ways, with electric intensity E.

To illustrate, let us consider the field of an infinitely long straight wire of circular cross section carrying a steady current I (Fig. 9.2). The portion of the circuit carrying the return current is assumed to be at an infinite distance, or alternatively to be a concentric conducting tube.

By virtue of the circular geometrical symmetry the flux lines lie in concentric circles, and H is a right-handed circumferential vector whose magnitude is proportional to I and is a function only of the radial distance r from the wire center. Evaluated over the closed circular path 1, *outside* the current, equation 9.4 takes the simple form

$$H \cdot 2\pi r = I \quad \text{amperes} \tag{9.6}$$

and

$$H = \frac{I}{2\pi r} \quad \frac{\text{amperes}}{\text{meter}} \qquad \text{(Outside current)} \tag{9.7}$$

since each path element ds is point by point parallel to $(-H)$ so that $\alpha = 0$ and $\cos \alpha = 1$. Equation 9.7 shows that H, for the region outside the current, declines hyperbolically with r.

Integration of $(-H)$ over closed path 2, which lies outside but links the current, yields the same result as equation 9.6, but integration over any path such as 3, enclosing no current, yields zero.

Integration over concentric circular path 4, *inside* the current-bearing region, gives

$$H = \frac{i}{2\pi r} \quad \frac{\text{amperes}}{\text{meter}} \qquad \text{(Inside current)} \tag{9.8}$$

where i is the magnitude of the current enclosed. For a steady current, the current density $J = I/\pi R^2 = i/\pi r^2$ is constant over the cross section, and

$$H = \frac{i}{2\pi r} = \frac{I}{2\pi R^2} r = \frac{1}{2} Jr \quad \frac{\text{amperes}}{\text{meter}} \tag{9.9}$$

Inside the current-bearing region H increases linearly with r. At the surface $r = R$, the inner and outer values of H are equal, or $H_R = I/(2\pi R) = \frac{1}{2}JR$. The magnitude of H is illustrated in Fig. 9.3.

9.5. Flux Density *B*. The magnetic flux density B is defined as a vector point function related to the intensity through the equation

$$B = \mu H = \frac{d\phi}{dA_n} \quad \frac{\text{webers}}{\text{square meter}} \tag{9.10}$$

In isotropic media (same properties in all directions) B is parallel to H, as indicated in Fig. 9.2. The numerical relationship between B and H, in general, is not constant but depends on the variable value of μ_r in $\mu = \mu_0\mu_r$. This point arises particularly with the ferromagnetic materials and is considered in Chapter 13. (In anisotropic materials B and H are not parallel and μ is no longer a scalar quantity.)

B may be represented in field drawings, as illustrated in Fig. 9.2, by vectors placed at points in the field. The method commonly used, however, is to draw magnetic flux lines parallel to B at every point so that the flux line density in lines per square meter (or webers per square meter) crossing a normal area element is numerically equal to the magnitude of B. Thus $B = d\phi/dA_n$ webers per square meter.

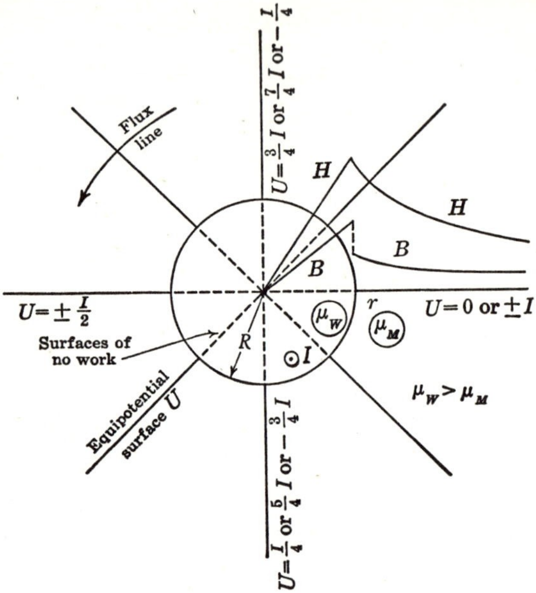

Fig. 9.3. ***B, H,*** **and** ***U*** **near an infinite wire current. (The equipotential surfaces *U* end at the wire surface.)**

The magnitude of B is proportional to μ_r and consequently is markedly dependent on the character of the magnetic medium. The physical meaning is that the total magnetization at any point, as represented by B, is the combined magnetizing effect of the currents carried in wire circuits and of the magnetizing effect of the medium itself. The latter is proportional to $(\mu_r - 1)$. For ferromagnetic materials the magnetization of the medium is thousands or even hundreds of thousands of times greater than that of the currents carried in wires. In a sense, the wire currents merely "trigger" or "release" the magnetization always present in the medium. The physical basis of magnetization of various materials is presented in Arts. 9.44 and 13.20.

The dimensional formula for B is obtained from that for ϕ and the latter is developed from $\phi = V_gT$ or W/i (flux is equal to the volt-second or the joule per ampere.) These forms are listed in Appendix I.

Alternatively, the dimensions of μ are determined in Art. 9.35, whence the dimensions of B are those of H multiplied by those of μ.

To illustrate, we shall calculate the flux density B both inside and outside the infinite wire discussed in Art. 9.4 and plot both B and H in Fig. 9.3.

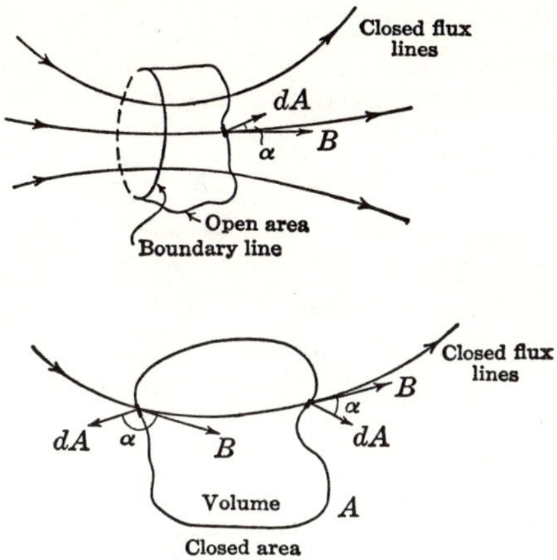

FIG. 9.4. Integration of B over open and closed areas.

Inside the wire

$$B_W = \mu_W H = \frac{\mu_W i}{2\pi r} = \frac{\mu_W I}{2\pi R^2} r = \frac{\mu_W}{2} Jr \quad \frac{\text{webers}}{\text{square meter}} \tag{9.11}$$

for constant current density J and wire permeability μ_W.

Outside the wire, with permeability μ_M,

$$B_M = \mu_M H = \frac{\mu_M I}{2\pi r} \quad \frac{\text{webers}}{\text{square meter}} \tag{9.12}$$

H, independent of the medium by definition, is continuous across the wire surface boundary, whereas B depends upon the permeability of the medium and therefore is discontinuous if the permeability is different on the two sides of the boundary.

9.6. Magnetic Flux ϕ. The magnetic flux ϕ is defined as a scalar quantity obtained by integrating the normal component of the flux density B over any area.

$$\phi = \int\!\!\int_A B_n \, dA \quad \text{webers} \tag{9.13}$$

This is illustrated in Fig. 9.4, where the closed flux lines are shown crossing an open area ending in a boundary line.

For a closed area, without a boundary line, and hence enclosing a volume, the *net* flux crossing the area *outward* is zero, or the outward flux minus the inward flux totals zero. This is a necessary consequence of the fact that the magnetic flux lines are closed.

9.7. Basic Equation for *B*. The fact that magnetic flux lines are *always* closed enables us to write an equation for B that must be satisfied everywhere. The argument parallels that given in Art. 6.2, where a similar expression is given for electric flux density D.

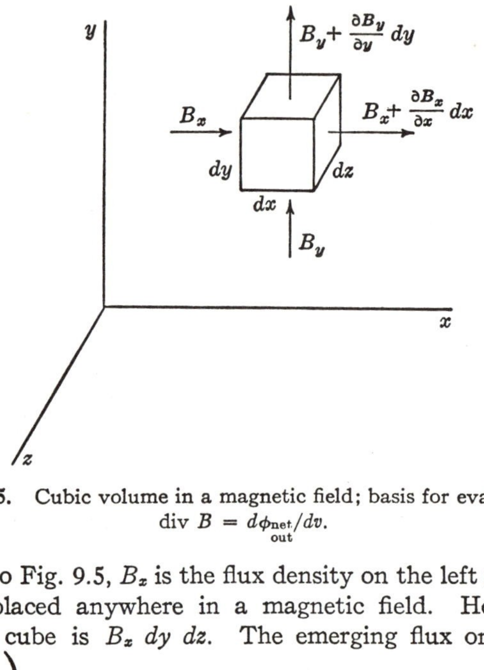

FIG. 9.5. Cubic volume in a magnetic field; basis for evaluating $\text{div } B = d\phi_{\substack{\text{net}\\ \text{out}}}/dv$.

Referring to Fig. 9.5, B_x is the flux density on the left face $dy\,dz$ of a small cube placed anywhere in a magnetic field. Hence the flux entering the cube is $B_x\,dy\,dz$. The emerging flux on the right is $\left(B_x + \frac{\partial B_x}{\partial x} dx\right)(dy\,dz)$. The net flux, excess outward over inward, on these two faces is the difference $\frac{\partial B_x}{\partial x} dx\,dy\,dz$. Taken over all the faces,

$$d\phi_{\substack{\text{net}\\ \text{out}}} = \left(\frac{\partial B_x}{\partial x} + \frac{\partial B_y}{\partial y} + \frac{\partial B_z}{\partial z}\right)(dx\,dy\,dz) = 0 \tag{9.14}$$

which is zero since the flux lines are closed and the same number must enter as emerge.

The quantity in parentheses is the *net outward flux per unit volume* and is called in vector analysis the divergence of the vector $B = \mu H$.

Then

$$\frac{d\phi_{\substack{\text{net}\\\text{out}}}}{dv} = \text{div } B = \text{div } (\mu H) = 0$$

$$= \frac{\partial B_x}{\partial x} + \frac{\partial B_y}{\partial y} + \frac{\partial B_z}{\partial z} = 0 \qquad (9.15)$$

If μ is a constant, independent of field strength, this takes the form

$$\frac{d\phi_{\substack{\text{net}\\\text{out}}}}{dv} = \mu\left(\frac{\partial H_x}{\partial x} + \frac{\partial H_y}{\partial y} + \frac{\partial H_z}{\partial z}\right) = 0 \qquad (9.16)$$

9.8. Magnetomotive Force (MMF) $\mathfrak{F}$ and Magnetostatic Potential *U*. These quantities are bracketed together because they have the same *LMTQ* dimensions and have certain common properties. There are, however, differences in definition and in the use of these quantities which warrant assigning them different names and symbols.

(1) *Magnetomotive Force* (*MMF*) $\mathfrak{F}$. The magnetomotive force is a scalar quantity (and hence not a force) obtained by evaluating the line integral of $(-H)$ over any path between two points in the field, irrespective of the medium, and is valid both inside and outside current-bearing regions. (See Fig. 9.2.)

$$\mathfrak{F} = \mathfrak{F}_2 - \mathfrak{F}_1 = \int_1^2 H\,ds \cos\alpha \quad \text{amperes} \qquad (9.17)$$

Evaluated over a *closed* path equation 9.17 takes the form

$$\mathfrak{F} = \oint_{\text{closed path}} H\,ds \cos\alpha = nI \quad \text{amperes} \qquad (9.18)$$

and thus is precisely equal to equation 9.4. The magnetomotive force, taken over a closed path, is equal to the quantity of current enclosed or linked, irrespective of the character of the medium.

The name, magnetomotive force, is a misnomer since the quantity is scalar and is closely related to work or energy, not force. This is true also of electromotive force. These names, however, are so extensively used in the literature that it is difficult to find satisfactory substitutes.

(2) *Magnetostatic Potential* U. The absolute magnetostatic potential at a field point is a scalar quantity, *valid only in non-current-bearing regions*, obtained by evaluating the line integral of $(-H)$ over a path extending from infinite distance (or outside the field) to the field point. The unit of this quantity is the ampere.

$$U_P = \int_\infty^P (-H)\,ds \cos\alpha \quad \text{amperes} \qquad (9.19)$$

Taken along a flux line, $ds \cos \alpha = dn$, and

$$U_P = \int_{\infty}^{P} (-H)\, dn \quad \text{amperes} \tag{9.20}$$

It is convenient to consider also the difference of potential between two points, but this obviously is equal to the MMF between these points if a path connecting the points lies entirely in a non-current-bearing region.

$$U_2 - U_1 = \mathfrak{F}_2 - \mathfrak{F}_1 \quad \text{amperes} \quad \text{(Non-current-bearing region)} \tag{9.21}$$

As a consequence of these definitions the potential at a point proves to be a multivalued function. For instance, let U_P be calculated at a point by following a path from infinity to the point, then calculate, using equation 9.18, the additional potential for an additional closed path linking current I. The closed path may be followed any integral number of times k. The potential at the point then is

$$U = U_P \pm kI \quad \text{amperes} \tag{9.22}$$

the + sign being used if the closed path journey links the current in the same sense as $(-H)$. The dimensional formula for U or $\mathfrak{F}$ is the same as that for current. (See Appendix I.)

Magnetic equipotential surfaces (U = constant) are always perpendicular, point by point, to the vector H. They are however restricted to regions outside current-bearing volumes and thus are open surfaces terminating at the surfaces of current-bearing regions.

For the infinite straight wire case the equipotential surfaces, illustrated in Fig. 9.3, are radial coaxial planes running outward from the wire surface. From symmetry any of these planes may arbitrarily be assigned the value zero potential and the values of potential on the other planes calculated relatively. The several values on each plane differ from each other always by an integral multiple of I. That the equipotential surfaces (here planes) may *not* be extended into the current-bearing region follows from the fact that $\int (-H)\, ds \cos \alpha$, carried out over a concentric circular arc of angle θ inside the current-bearing region, yields a different value than if carried out over a concentric circular arc of the same angle lying outside the current.

(3) *Surfaces of No Work. Kernel.* The surfaces which are equipotentials outside the current-bearing regions may be extended into the current-bearing regions where they are called "surfaces of no work." Here they continue perpendicular to H and meet in a line called the "kernel," which is the center of the magnetic flux system

where $H = B = 0$. The kernel in the current-bearing region is the equivalent of the singular line sometimes found outside this region. The condition for either singular lines or points is $H = B = 0$.

9.9. Magnetic Field Gradient. This quantity is defined as a vector quantity equal to the maximum change in $-$MMF $(-d\mathfrak{F})$ or magnetic potential (dU) per unit distance (dn) taken parallel to a flux line or perpendicular to a surface of constant $\mathfrak{F}$ or U. The process here defined is the reverse of the integration of equation 9.4 or equation 9.20. Consequently the gradient is equal to $(-H)$.

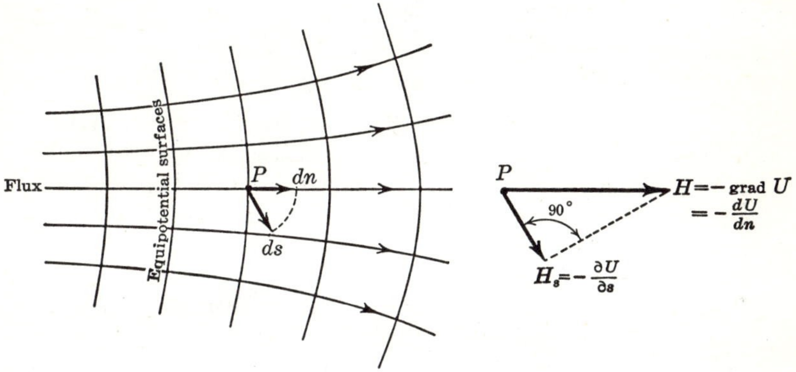

FIG. 9.6. Magnetic field gradient.

Inside or outside a current-bearing area

$$\text{P.G.} = -H = -\frac{d\mathfrak{F}}{dn} \quad \frac{\text{amperes}}{\text{meter}} \tag{9.23}$$

Outside the current-bearing area it is

$$\text{P.G.} = -H = -\frac{d\mathfrak{F}}{dn} = \frac{dU}{dn} \quad \frac{\text{amperes}}{\text{meter}} \tag{9.24}$$

If $\mathfrak{F}$ or U is partially differentiated in some other direction s, the value obtained is a vector having the s direction and a magnitude equal to the component of the gradient in that direction.

$$\begin{aligned} \text{P.G.}_{,s} = -H_s &= -\frac{\partial \mathfrak{F}}{\partial s} = \frac{\partial U}{\partial s} \\ &= -\frac{d\mathfrak{F}}{dn}\cos\alpha \\ &= \frac{dU}{dn}\cos\alpha \quad \frac{\text{amperes}}{\text{meter}} \end{aligned} \tag{9.25}$$

These operations are illustrated in Fig. 9.6.

9.10. Laplace Equation for Magnetostatics. The Laplace equation is valid only outside a current-bearing region where there is a potential function U. From equation 9.25,

$$-H_x = \frac{\partial U}{\partial x}, \quad -H_y = \frac{\partial U}{\partial y}, \quad -H_z = \frac{\partial U}{\partial z} \tag{9.26}$$

Equation 9.16 then transforms to

$$-\frac{1}{\mu}\frac{d\phi_{\substack{\text{net}\\ \text{out}}}}{dv} = \frac{\partial^2 U}{\partial x^2} + \frac{\partial^2 U}{\partial y^2} + \frac{\partial^2 U}{\partial z^2} = 0 \tag{9.27}$$

$$= \nabla^2 U = 0 \tag{9.28}$$

Equation 9.27, frequently written in the form of equation 9.28, is called the Laplace equation for magnetostatics and may be used only in non-current-bearing regions. The solutions of the Laplace equation must be consistent with the values of U occurring over the surfaces of conductors carrying current and over interface boundaries separating media of differing permeabilities.

9.11. Field of a Current Filament. A steady current I, it is assumed, is flowing around a conducting closed line filament of any shape and of negligible cross section. The medium is assumed to extend through all space. The flux lines circulate in closed lines about the current, as illustrated in Fig. 9.7. The equipotential surfaces, which are necessarily perpendicular to the flux lines, thus are open surfaces which have the filament of current as a common boundary line.

The immediate problem is to find an expression for potential U and for field strength H consistent with the geometry of a filament and with the magnetic qualities of a field. These requirements call to mind the properties of a solid angle. At any field point the solid angle is determined by the cone drawn from the point to the filament and by the shape of the filament. Moving the field point in a closed path linking the filament, the change in solid angle is 4π steradians, irrespective of the path followed. Also, integrating $(-H)$ around a path linking the current filament gives the value I, irrespective of the path followed. These requirements are jointly satisfied by assuming that the magnetic potential of the field is proportional to the solid angle Ω subtended at any field point by the filament.

This problem is solved by making the potential at any point proportional to Ω and the closed path (linking a current) change in potential equal to I. Then

$$U = \frac{I}{4\pi}\Omega \quad \text{amperes} \tag{9.29}$$

and

$$\sum_{\substack{\text{closed}\\\text{path}}} -U = \mathfrak{F} = \frac{I}{4\pi} 4\pi = I \quad \text{amperes} \tag{9.30}$$

since the change of solid angle over a closed path linking a current is $\pm 4\pi$.

The magnetic field intensity is

$$H = -\frac{dU}{dn} = +\frac{I}{4\pi}\frac{d\Omega}{dn} \quad \frac{\text{amperes}}{\text{meter}} \tag{9.31}$$

The flux density is then

$$B = \mu H \quad \frac{\text{webers}}{\text{square meter}} \tag{9.32}$$

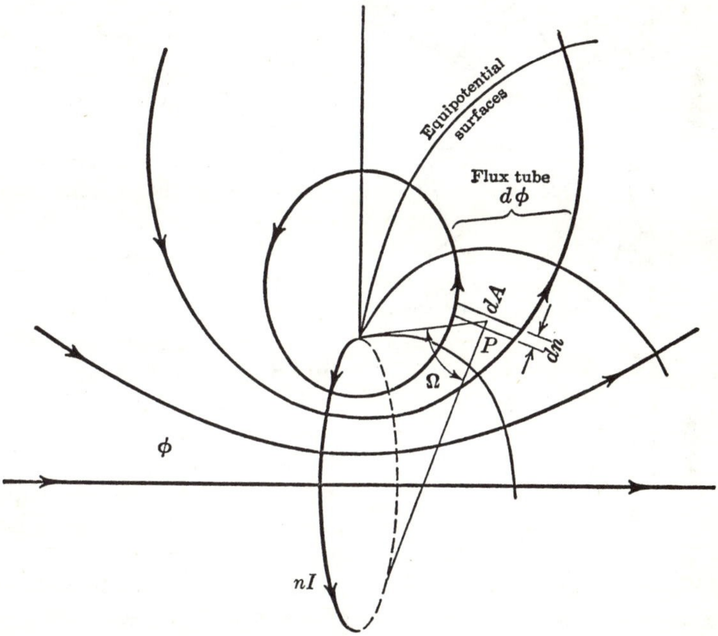

FIG. 9.7. Field of a current filament.

An alternative approach to this basic problem is given in Art. 9.18. The general shape of the field is illustrated in Fig. 9.7.

9.12. Field of a Circular Current. The flux lines and equipotential surfaces outside a circular filamentary current are illustrated in Fig. 9.8. The equipotential surfaces, representing equal increments of potential, leave the filamentary current at equal angles and then are located according to the condition Ω = constant given in equation

9.29. The perpendicular flux lines are everywhere parallel to H, which is given by equation 9.31.

9.13. Field of a Magnet. A piece of material inserted into a magnetic field at once becomes a magnet and sets up its own magnetic field. The final field is a composite of the original field and the magnet field. The piece of material is weakly magnetized if it is one of the diamagnetic or paramagnetic materials, or it may be strongly magnetized if it is made of iron, cobalt, nickel, or any of the ferromagnetic group of materials. Upon removal of the original field, the diamagnetic and paramagnetic materials lose their induced state of magnetization, but ferromagnetic materials remain *permanently* magnetized—for instance, a compass needle or any permanent magnet.

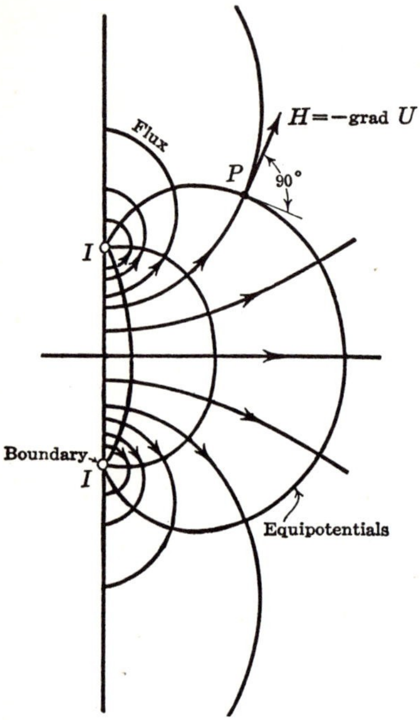

FIG. 9.8. Field of a circular current filament.

The magnetic fields about poles and those about currents are experimentally indistinguishable. This requires that field formulas based on one or the other point of view must be correlated. The precise relations are brought out in Art. 9.40 on Rowland's ring.

A permanent magnet has certain properties easily discovered in simple experiments. When the magnet is immersed in iron filings, the filings tend to cluster near the ends of the magnet. The effective centers of these regions are called poles. Suspended by a thread, a magnet orients itself along the line of the earth's magnetic field. The pole at the north end is called a north or positive pole, the other a south or negative pole. The fact that there is no translation of the magnet indicates that the poles are always of equal strength. Cutting a magnet fails to separate the poles since each part now appears as a complete magnet with equal and opposite poles.

The pole (unlike the electric charge) is not a physical entity. It is rather an external manifestation of the state of magnetization existing throughout the interior of the magnet. Nevertheless, poles exert forces on each other according to the inverse square distance law enunciated by Coulomb in 1785.

It is both convenient and natural to state the Coulomb law of force between poles in the same form used earlier to express the law of force between charges.

9.14. Coulomb's Law. Unit Pole. Dipole. With the aid of a torsion balance Coulomb in 1785 discovered the inverse square law of force between magnetic poles. In mks rationalized units the law is stated as follows:

(1) *Coulomb's Law.* Two magnetic point poles in free space exert forces upon each other directed along their common line, equal to the product of the pole strengths divided by the square of the separation distance and by $4\pi\mu_0$, repulsive for poles of the same sign and attractive for poles of unlike sign.

$$F_0 = \frac{p_1 p_2}{4\pi\mu_0 r^2} \quad \text{newtons} \tag{9.33}$$

Since $\mu_0 = 4\pi/10^7$ henrys per meter, unit pole is defined as follows.

(2) *Unit Pole.* Unit pole repels a like pole at a distance of one meter in free space with a force of $10^7/(16\pi^2)$ newtons (equal to 63,325 newtons). This is an enormous force, much too large to be encountered in practice. We shall then expect to encounter only fractional pole strengths on actual magnets. The pole has the same unit, the weber, as a line of magnetic flux. See Art. 9.16.

(3) *Dipole and Moment.* In Art. 9.13 it has been pointed out that it is impossible to break a magnet and effectively separate the two equal poles of opposite sign. The simplest physical magnetic element is a dipole, composed of equal poles of opposite sign, separated a small distance h. The moment of the dipole is defined as a vector quantity directed along the line of the poles from negative (south) to positive (north) pole, equal to the product

$$m = ph \quad \text{weber-meter} \qquad \text{(Vector)} \tag{9.34}$$

According to the modern theory of magnetism, the magnetic effect, whether created by measurable currents in wires or by magnets, is ultimately traceable to some form of current. In magnets and magnetizable media the underlying origin of the magnetic field is seen to be in the circulation of electrons in their atomic and molecular orbits or in the spinning of electrons about their own axes. Orbital electron motions and electron spins create dia- and paramagnetism; the electron spin is primarily responsible for the ferromagnetic effect. This is discussed further in Arts. 9.45 and 13.20.

Consequently, the dipole, which appears externally as a magnet, is assumed in this respect to be the precise equivalent of a small current circulating around the dipole axis. The value of the current then is *proportional* to the value of the dipole moment. The full relation is given in Art. 9.18, equation 9.54.

$$i \sim m = ph \tag{9.35}$$

The axis of the current coincides with the dipole axis.

9.15. Alternative Definitions of H and U_P. The field strength H may be defined, alternatively, as the force exerted at a point on a unit positive pole placed at the point.

$$H = \frac{dF}{dp_+} = \frac{F}{+1p} \quad \frac{\text{newtons}}{\text{weber}} \text{ or } \frac{\text{amperes}}{\text{meter}} \qquad \text{(Vector)} \tag{9.36}$$

To use this definition we must envisage the (theoretical) possibility of cutting a magnet and using the unit positive pole in a manner similar to that of the unit positive charge, simply for the purpose of specifying the field strength. The units of H are equally well newtons per weber or amperes per meter.

The magnetic potential U_P (outside a current-bearing region), alternatively, is

$$U_P = \int_\infty^P (-H)\, ds \cos\alpha \quad \frac{\text{joules}}{\text{weber}} \text{ or amperes} \tag{9.37}$$

9.16. Field Strength of p Point Poles. Consider the theoretical possibility of gathering p positive unit poles together at a point in free space. The field strength, using equation 9.33, is

$$H = \frac{dF_0}{dp_+} = \frac{p}{4\pi\mu_0 r^2} \quad \frac{\text{newtons}}{\text{weber}} \text{ or } \frac{\text{amperes}}{\text{meter}} \qquad \text{(Vector)} \tag{9.38}$$

which is a vector directed radially outward from p. The units for H, newtons per weber where poles are concerned, or amperes per meter for currents, must ultimately prove to be the same.

The flux density in free space is

$$B_0 = \mu_0 H = \frac{p}{4\pi r^2} \quad \frac{\text{magnetic lines}}{\text{square meter}} \text{ or } \frac{\text{webers}}{\text{square meter}} \qquad \text{(Vector)} \tag{9.39}$$

and the flux crossing a sphere of radius r about p is

$$\begin{aligned} \phi = B_0 A &= \frac{p}{4\pi r^2} \cdot 4\pi r^2 \\ &= p \quad \text{poles or lines or webers} \qquad \text{(Scalar)} \end{aligned} \tag{9.40}$$

One line of magnetic flux or one weber is associated with each unit pole.

The magnetic absolute potential is

$$U_P = \int_{\infty}^{r} (-H)\, ds \cos\alpha \quad \frac{\text{joules}}{\text{weber}} \text{ or amperes} \qquad \text{(Scalar)} \quad (9.41)$$

Evaluated along a radial path in free space, ds parallel to $(-H)$ and $\cos\alpha = 1$,

$$U_P = \frac{p}{4\pi\mu_0 r} \quad \frac{\text{joules}}{\text{weber}} \text{ or amperes} \qquad \text{(Scalar)} \quad (9.42)$$

9.17. Volumetric Distribution of Poles. If the p poles are distributed throughout a volume of space with a volume density dp/dv, the vector intensity at any field point is given by the vector integration

$$H = \frac{1}{4\pi\mu_0} \iiint_v \frac{\pm dp}{r^2} \quad \frac{\text{amperes}}{\text{meter}} \qquad \text{(Vector summation)} \quad (9.43)$$

The absolute potential is given by the scalar integration

$$U_P = \frac{1}{4\pi\mu_0} \iiint_v \frac{\pm dp}{r} \quad \text{amperes} \qquad \text{(Scalar summation)} \quad (9.44)$$

Equations 9.43 and 9.44 replace equations 9.38 and 9.42. Due regard must be had for the signs of the poles in making the summations indicated. r is the distance from element dp to the field point where H or U_P is evaluated.

9.18. Fields of a Dipole and Magnetic Shell. Let us first consider the field created in free space by a dipole and then extend the results to a magnetic shell or layer of dipoles.

(1) *Dipole.* The dipole is a small magnetized element with $+p$ and $-p$ poles at the ends, separated a small distance h. The magnetic potential at point P in Fig. 9.9 is

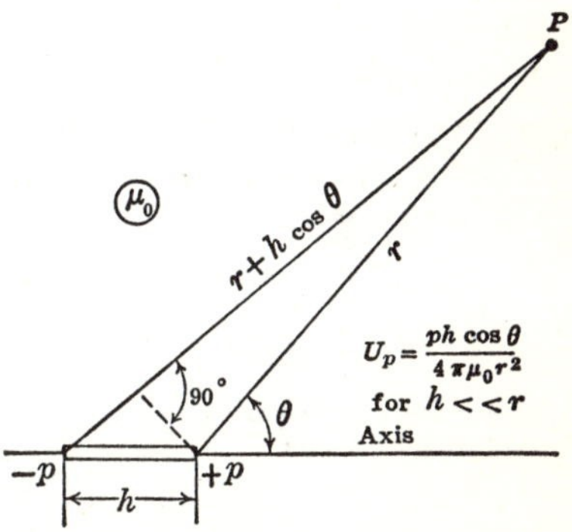

FIG. 9.9. Magnetic potential of a dipole in free space.

$$U_P = \frac{p}{4\pi\mu_0}\left(\frac{1}{r} - \frac{1}{r + h\cos\theta}\right) \quad \frac{\text{joules}}{\text{weber}} \qquad \text{(Scalar)} \quad (9.45)$$

$$= \frac{p}{4\pi\mu_0}\left(\frac{h\cos\theta}{r^2 + rh\cos\theta}\right) \quad \text{amperes} \qquad (9.46)$$

If we now make the restriction $h \ll r$,

$$U_P = \frac{ph}{4\pi\mu_0} \frac{\cos\theta}{r^2} = \frac{m}{4\pi\mu_0} \frac{\cos\theta}{r^2} \quad \text{amperes} \tag{9.47}$$

The restriction is more academic than real since U_P is proportional to the moment $m = ph$ and it is always possible to obtain the same value of moment by decreasing h and increasing p so that the product $m = ph$ is maintained constant.

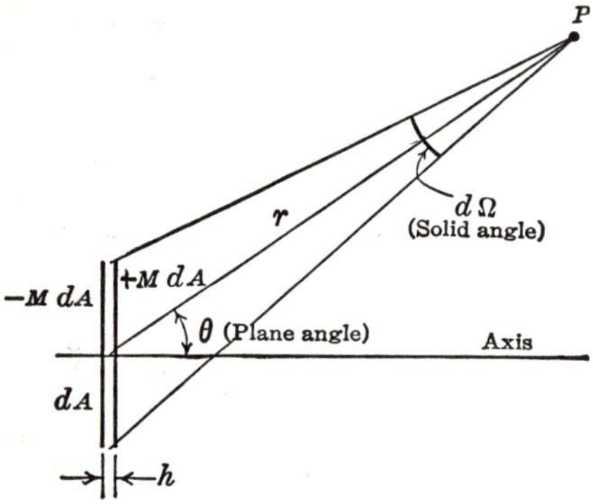

Fig. 9.10. Magnetic potential of a small plane shell.

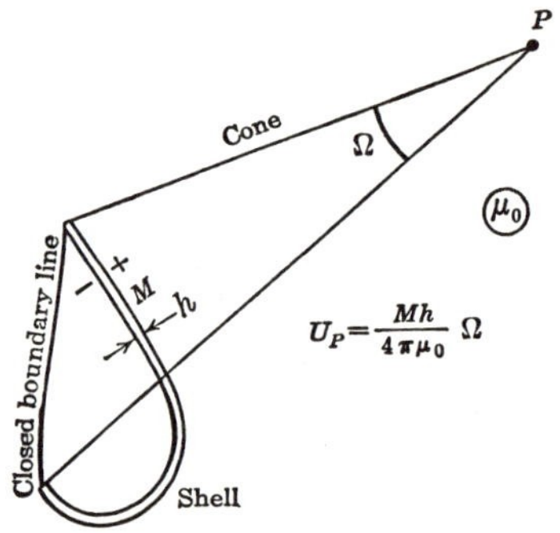

Fig. 9.11. Magnetic potential of a shell of any shape.

(2) *Small Plane Shell.* Over a small area dA let us arrange a uniformly dense layer of dipoles, with axes perpendicular to the area (like pins stuck normally into a pin cushion). The density of poles (or dipoles of length h) per unit area is

$$M = \frac{dp}{dA} = \frac{d(ph)}{h\,dA} = \frac{dm}{h\,dA} \quad \frac{\text{poles}}{\text{square meter}} \text{ or } \frac{\text{webers}}{\text{square meter}} \tag{9.48}$$

The free space potential at P in Fig. 9.10 then is

$$dU_P = \frac{Mh}{4\pi\mu_0}\left(\frac{dA\cos\theta}{r^2}\right) \tag{9.49}$$

The term in parentheses is the solid angle $d\Omega$ subtended at P by the boundary line of the small plane shell, giving

$$dU_P = \frac{Mh}{4\pi\mu_0}\,d\Omega \quad \frac{\text{joules}}{\text{weber}} \text{ or amperes} \tag{9.50}$$

The potential is a function, geometrically, only of solid angle, provided that the shell is *uniformly and normally magnetized.*

(3) *Magnetized Shell of Any Shape.* Equation 9.50 may be integrated to give the potential in free space for any size or shape of magnetized shell, uniformly and normally magnetized. Following Fig. 9.11,

$$U_P = \frac{Mh}{4\pi\mu_0}\,\Omega \quad \frac{\text{joules}}{\text{weber}} \text{ or amperes} \tag{9.51}$$

The potential is a function of the boundary line of the shell, but otherwise it is independent of the shell shape or size. Any and all shells having the same boundary line and the same normal and uniform magnetization $Mh/(4\pi\mu_0)$ have the same potential at P.

Equipotential surfaces are open, ending in the boundary line. Throughout space the equation of shape is given by

$$\left(\frac{Mh}{4\pi\mu_0}\right)\Omega = \text{constant} \tag{9.52}$$

The flux lines, perpendicular to the equipotential surfaces, therefore are closed lines surrounding the boundary line.

At any field point, the intensity is

$$H = -\frac{dU_P}{dn} = -\frac{Mh}{4\pi\mu_0}\frac{d\Omega}{dn} \quad \frac{\text{newtons}}{\text{weber}} \text{ or } \frac{\text{amperes}}{\text{meter}} \tag{9.53}$$

Referring to Art. 9.11 and comparing Figs. 9.7 and 9.11, we see that the free space field of a current nI is identical with the free space field of a magnetic shell having the current as a boundary line, provided that the current and the shell magnetization are related through

$$nI = \frac{Mh}{\mu_0} \tag{9.54}$$

Implied also is the fact that H, defined as force per unit pole in equation 9.36 and based on Coulomb's law, is identical with the definition given in equation 9.4 based on a current relation.

The identity of the two approaches to evaluation of the magnetostatic field is explored further in Art. 9.40 on Rowland's ring.

9.19. Equivalent Current Mesh and Dipole Layer. A filamentary current can be replaced by a current mesh, having the filament as a boundary line, for the purpose of calculating an external field. A current mesh, illustrated in Fig. 9.12, is somewhat like a section of wire screen carrying internal currents, all equal, around each closed wire path. The internal currents all cancel, irrespective of the shape of the screen area, leaving the boundary or filament current as the only current not cancelled by adjacent sections of the screen. From this it is evident that the screen area may be deformed at will, subject only to the condition that the shape of the boundary line be retained unchanged.

Recalling the equivalence of current and dipole moment expressed in equation 9.54

$$i = nI = \frac{Mh}{\mu_0} \qquad (9.55)$$

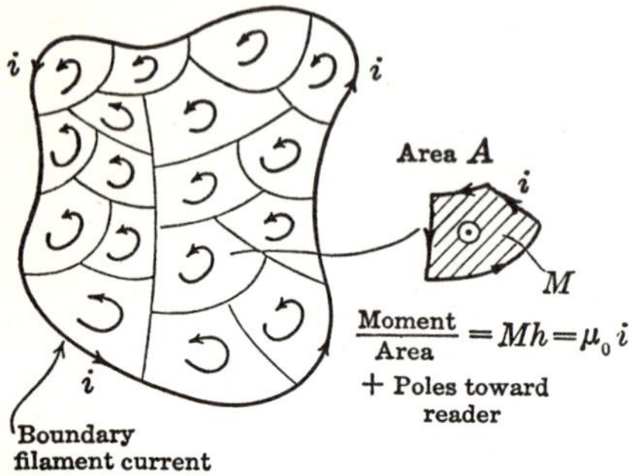

FIG. 9.12. Equivalence of current filament, current mesh, and dipole layer.

it is seen that the filamentary current and its mesh may be replaced by a layer of dipoles of uniform strength placed everywhere perpendicular to the mesh area. Thus a dipole layer, uniformly and normally magnetized, produces the same external field as a current flowing around the boundary line of the layer. This is subject to the restriction, developed in Art. 9.18, that the length h of each dipole is very small compared with the distance out to the field point where the field strength is to be calculated.

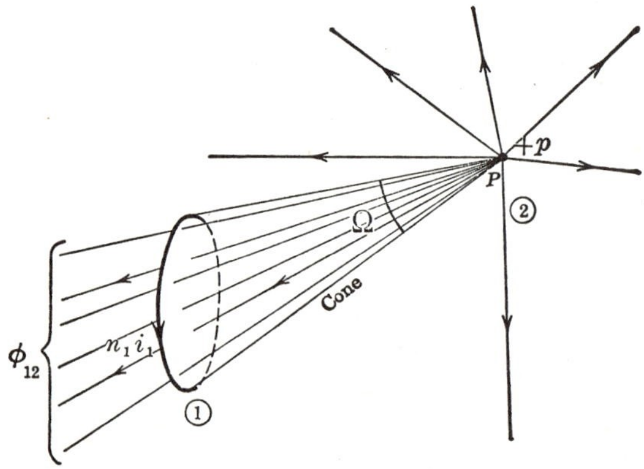

FIG. 9.13. Pole field linking a current.

9.20. Mutual Energy of Filamentary Current and Pole. Following Fig. 9.13 and equations 9.37 and 9.29, it is seen that to bring unit positive pole through free space from infinity to point P, forming solid angle Ω with coil current $n_1 i_1$, requires work per unit pole equal to

$$U_P = \frac{n_1 i_1}{4\pi}\,\Omega \quad \frac{\text{joules}}{\text{weber}} \text{ or amperes} \qquad (9.56)$$

If p unit poles are brought up to P the work performed is

$$W_m = pU_P = n_1 i_1 \left(\frac{\Omega}{4\pi}\right) p \quad \text{joules} \tag{9.57}$$

This is the mutual energy required to bring the two systems, $n_1 i_1$ and p poles, together. (This assumes that the two systems are brought together slowly so that there is no induced voltage in the coil that might cause a change in the current $n_1 i_1$ during the process.)

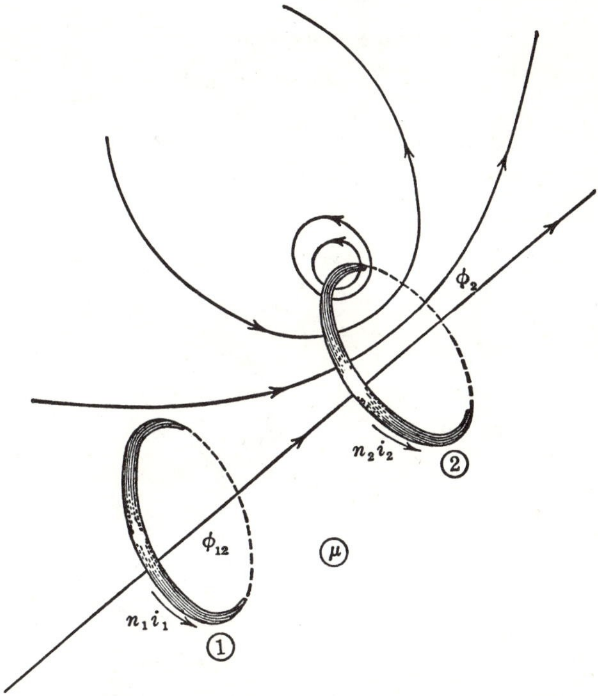

FIG. 9.14. Mutual flux linkages. (Flux shown is created by $n_2 i_2$ alone.)

From equation 9.40 we note that the flux ϕ from p poles is $\phi = p$ lines or webers. $\Omega/(4\pi)$ is the fraction of the flux, and $(\Omega/4\pi)p$ is the total amount of flux ϕ_{12}, emerging from p poles, which passes through the cone and links current $n_1 i_1$. Thus the mutual energy is

$$W_{m12} = n_1 i_1 \phi_{12} \quad \text{joules} \tag{9.58}$$

which is the product of the current $n_1 i_1$ and the flux ϕ_{12} which links system 1 coming from some other system 2.

Flux linkages λ are defined as the product of the number of turns n and the number of flux lines ϕ which surround or link the turns.

$$\lambda = n\phi \quad \text{webers} \tag{9.59}$$

Using this quantity, the mutual energy is

$$W_{m12} = i_1\lambda_{12} \quad \text{joules} \tag{9.60}$$

For incremental changes

$$d\lambda = n\, d\phi \quad \text{webers} \tag{9.61}$$

and

$$dW_{m12} = n_1 i_1\, d\phi_{12} \tag{9.62}$$

$$= i_1\, d\lambda_{12} \quad \text{joules} \tag{9.63}$$

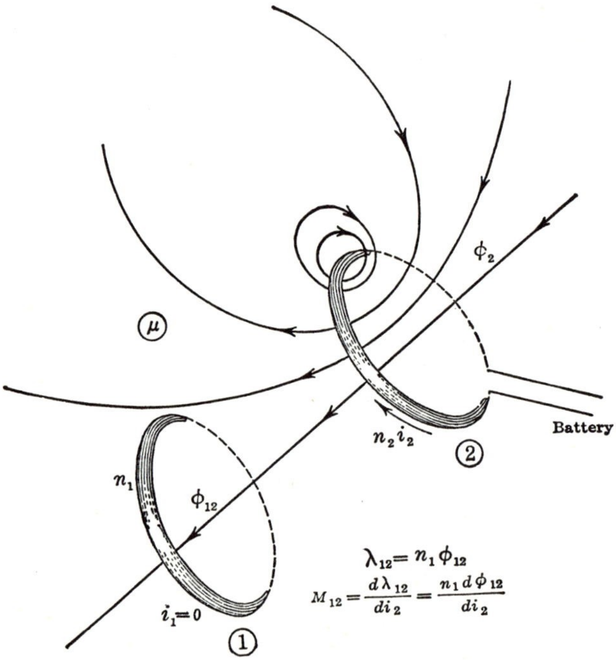

FIG. 9.15. Mutual relations of two coils. (Flux created by coil 2.)

9.21. Mutual Energy of Two Currents. The preceding relations may readily be extended to evaluate the mutual energy of two currents. First consider Fig. 9.14 or Fig. 9.15. Here the current i_2 is caused to flow in coil 2 of n_2 turns, producing flux ϕ_2. If the medium is other than free space, the flux, for a given current, is increased in proportion to the relative permeability μ_r. Of the total flux ϕ_2 created by current

n_2i_2, a fraction, ϕ_{12}, links coil 1. The mutual energy in joules then is

$$W_{m12} = i_1\lambda_{12} = n_1i_1\phi_{12} \qquad (\mu_r = \text{constant}) \tag{9.64}$$

This assumes that the permeability has remained constant as the flux has increased.

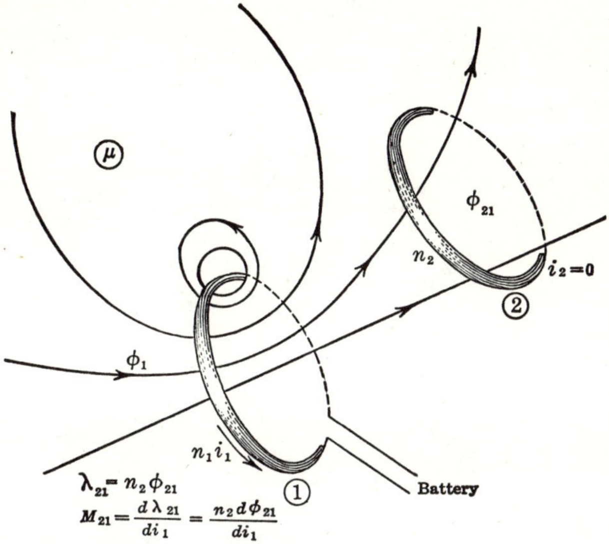

FIG. 9.16. Mutual relations of two coils. (Flux created by coil 1.)

A more general form, involving only incremental changes, is valid even with a ferromagnetic medium (for which the relative permeability μ_r depends on the field strength).

$$dW_{m12} = i_1\, d\lambda_{12} = n_1i_1\, d\phi_{12} \quad \text{joules} \tag{9.65}$$

or

$$W_{m12} = \int i_1\, d\lambda_{12} = \int n_1i_1\, d\phi_{12} \tag{9.66}$$

The *same* value of mutual energy is obtained by causing an appropriate amount of current n_1i_1 to flow in coil 1, creating total flux ϕ_1, of which a portion ϕ_{21} links coil 2. See Fig. 9.16. The comparable equations are

$$W_{m21} = i_2\lambda_{21} = n_2i_2\phi_{21} \qquad (\mu_r = \text{constant}) \tag{9.67}$$

for a medium of constant permeability. For variable μ_r

$$dW_{m21} = i_2\, d\lambda_{21} = n_2i_2\, d\phi_{21} \tag{9.68}$$

and

$$W_{m21} = \int i_2 \, d\lambda_{21} = \int n_2 i_2 \, d\phi_{21}$$

Figures 9.15 and 9.16 illustrate the separate fields created by each current independently of the other. The *total* field of both currents acting simultaneously is illustrated in Figs. 9.17 and 9.18. The total field is a superposition of the individual fields and depends on the relative directional sense of the two currents and, for ferromagnetic

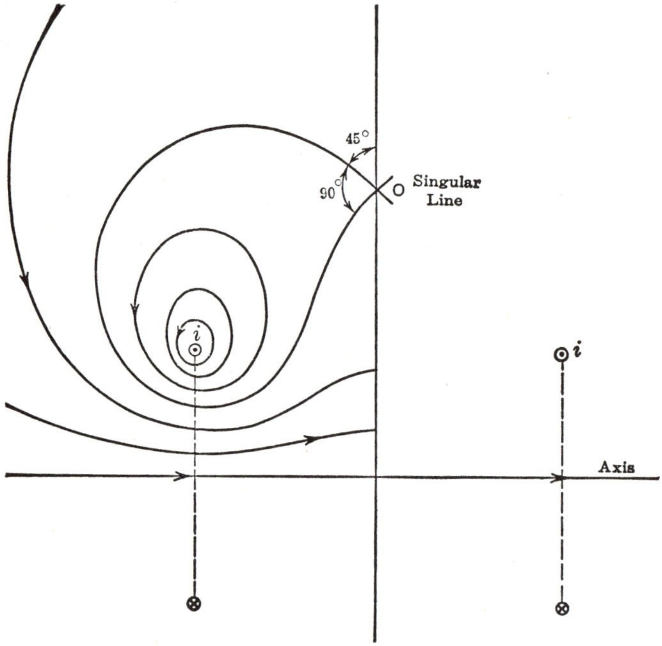

FIG. 9.17. Total field of two circular coils, boosting.

materials, involves the point by point dependence of μ_r on field strength.

Interesting features of the current arrangement of Fig. 9.17 are the existence of a circular singular line (where $H = 0$) and the fact that the field near the center is substantially uniform. The latter point is utilized in the arrangement of coils in wattmeters and a-c ammeters. With the coils split as shown, the moving coil is inserted at the center and operates in a nearly uniform field. In a wattmeter this leads to the advantages of a uniformly distributed scale. In an ammeter, the scale is approximately quadratic.

In the arrangement of Fig. 9.18, we find a singular point at the center of the field.

9.22. Mutual Inductance. Forces and Torques on Coils. The mutual inductance M of two coils is defined as the incremental flux linkages in one coil per incremental change of current in the other.

$$M_{12} = \frac{d\lambda_{12}}{di_2} = \frac{dW_{m12}}{i_1\,di_2} \quad \frac{\text{webers}}{\text{ampere}} \text{ or henrys} \tag{9.69}$$

and

$$M_{21} = \frac{d\lambda_{21}}{di_1} = \frac{dW_{m21}}{i_2\,di_1}$$

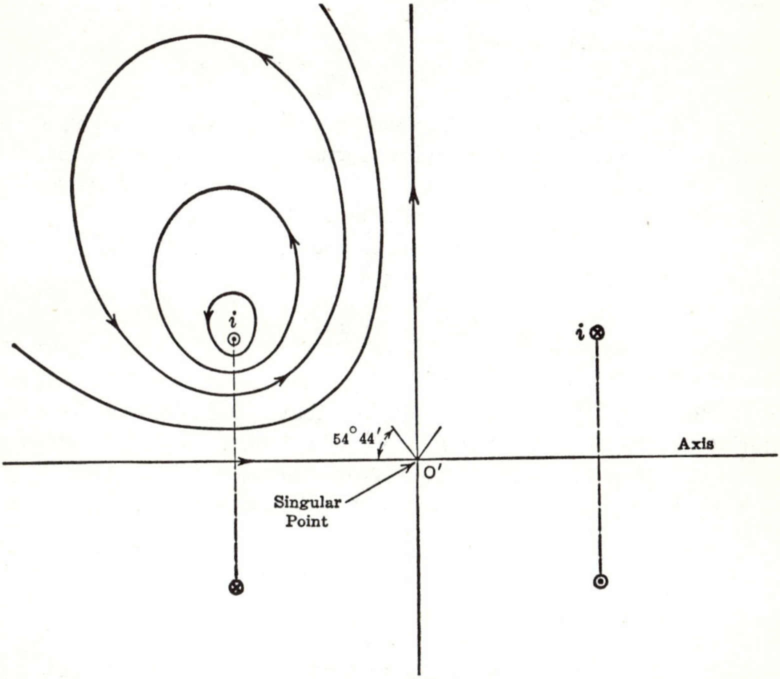

FIG. 9.18. Total field of two circular coils, bucking.

For equal incremental changes di_1 or di_2, the change $dW_{m12} = dW_{m21}$, and

$$M = M_{12} = M_{21} \quad \text{henrys} \tag{9.70}$$

For constant permeability

$$M = M_{12} = \frac{\lambda_{12}}{i_2} = \frac{W_{m12}}{i_1 i_2}$$

$$= M_{21} = \frac{\lambda_{21}}{i_1} = \frac{W_{m21}}{i_2 i_1} \quad \text{henrys} \tag{9.71}$$

The mutual inductance is a function of the geometry of the two circuits and of the permeability of the medium and has a maximum value if the coils are "closely wrapped" so that the turns are linked by all the flux possible. If the permeability depends on field strength, so also does the inductance. The inductance is a magnetic quantity comparable with the capacitance in the electric field.

Additional inductance relations, involving time changing fluxes and currents, are given in Chapter 13.

There is a force on each coil, directed oppositely to that on the other, given by

$$F = \frac{dW_m}{dn} = \pm i_1 i_2 \frac{dM}{dn} \quad \text{newtons} \tag{9.72}$$

where dW_m is the change in mutual energy and dM is the change in mutual inductance, accompanying a small change in coil spacing dn taken in the direction of the force. The direction of the force and of dn is that which produces the maximum *increase* in dW_m or dM.

For a movement of one coil in an *arbitrary* direction ds, the *component* of the force in that direction is equal to

$$F_s = \frac{dW_m}{ds} = \pm i_1 i_2 \frac{dM}{ds} \quad \text{newtons} \tag{9.73}$$

It is assumed in these force equations that i_1 and i_2 remain constant during the displacement dn or ds, which is true if the displacement is made very slowly so that no EMF's are generated in the coils.

There may also be a torque on each coil equal to

$$\mathfrak{T} = \frac{dW_m}{d\theta} = \pm i_1 i_2 \frac{dM}{d\theta} \quad \text{newton-meters} \tag{9.74}$$

where $d\theta$ is the angular change corresponding to the maximum *increase* in dW_m or dM. An angular displacement at any other angle θ_s gives a component torque

$$\mathfrak{T}_s = \frac{dW_m}{d\theta_s} = \pm i_1 i_2 \frac{dM}{d\theta_s} \quad \text{newton-meters} \tag{9.75}$$

The forces and torques discussed above are the vectorially totalized values of the forces and torques on each current-length element given in Art. 9.31. A reversal of current in either coil causes reversal of the forces and torques. These points are elaborated in Chapter 12.

9.23. Self-Energy of One Current. Integrating equation 9.66 to obtain the *mutual* energy of two coils leads to

$$W_m = i_1 \int d\lambda_{12} = i_2 \int d\lambda_{21} \quad \text{joules} \tag{9.76}$$

Here the current is removed from under the integral sign since the integration, based on the fundamental definition of these quantities, is of the flux from one coil made to link the current (or current-turns) of the other coil with the current assumed constant. To put it another way, the current comes from one source and the flux from another.

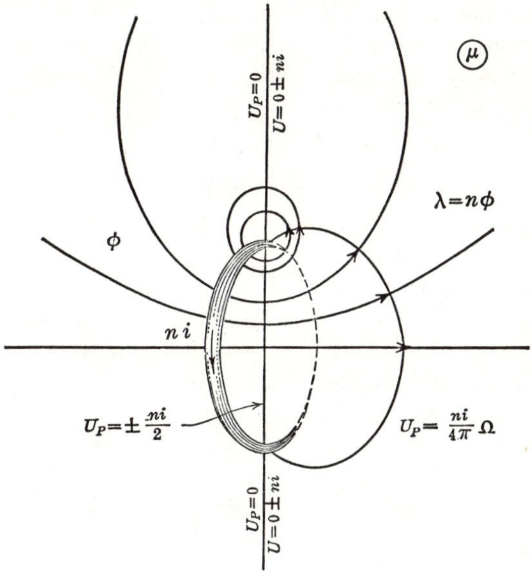

FIG. 9.19. Field of a closely wrapped coil.

To calculate the *self*-energy of one current, the current must be retained under the integral sign since the current and the flux which it creates are functionally related. Then

$$W_s = \int i_1\, d\lambda_1 = \int n_1 i_1\, d\phi_1 \quad \text{joules} \tag{9.77}$$

If the permeability of the medium is a constant, independent of field strength, the flux is linearly related to the current and

$$W_s = \tfrac{1}{2} i_1 \lambda_1 = \tfrac{1}{2} n_1 i_1 \phi_1 \quad \text{joules} \tag{9.78}$$

Figure 9.19 illustrates the field about one coil for the case of n turns "closely wrapped," so that all the flux links all the turns.

There is a force on each current-length element of the coil proportional to the current length and to the flux density at the element created by the other elements. (See Art. 9.32.)

9.24. Self-Inductance. The self-inductance L of a coil is defined as the incremental flux linkages per incremental change in current in the coil, giving

$$L = \frac{d\lambda}{di} = \frac{n\,d\phi}{di} = \frac{dW_s}{i\,di} \quad \frac{\text{webers}}{\text{ampere}} \text{ or henrys} \tag{9.79}$$

The self-inductance of a coil is a function of the geometry of the coil and of the permeability of the medium. It has a maximum value if the turns are closely wrapped so that all the flux links all the turns.

For constant permeability

$$L = \frac{\lambda}{i} = \frac{n\phi}{i} = \frac{2W_s}{i^2} \quad \frac{\text{webers}}{\text{ampere}} \text{ or henrys} \tag{9.80}$$

For further inductance relations, involving variable fluxes and currents, consult Chapters 11 and 13.

The forces on all the current-length elements of the coil are so directed as to *increase* the self-energy and the self-inductance. For a further discussion of these matters, consult Chapter 12.

9.25. Coefficient of Coupling of a Mutual Inductance. The coefficient of coupling K of a mutual inductance is the ratio of the mutual inductance to the square root of the product of the self-inductances.

$$K = \frac{M}{\sqrt{L_1 L_2}} \tag{9.81}$$

or

$$K^2 = \frac{M^2}{L_1 L_2}$$

Introducing

$$M = \frac{\lambda_{21}}{i_1} = \frac{\lambda_{12}}{i_2}$$

$$L_1 = \frac{\lambda_1}{i_1}$$

and

$$L_2 = \frac{\lambda_2}{i_2}$$

we get

$$K^2 = \frac{\left(\frac{\lambda_{21}}{i_1}\right)\cdot\left(\frac{\lambda_{12}}{i_2}\right)}{\left(\frac{\lambda_1}{i_1}\right)\cdot\left(\frac{\lambda_2}{i_2}\right)} = \left(\frac{\lambda_{21}}{\lambda_1}\right)\cdot\left(\frac{\lambda_{12}}{\lambda_2}\right) \tag{9.82}$$

The square of the coupling coefficient is the product of two flux linkage ratios—the flux linkages at coil 2 caused by coil 1 to the flux linkages at coil 1 due to its own current, (λ_{21}/λ_1), and the flux linkages at coil 1 caused by coil 2 to the flux linkages at coil 2 due to its own current, (λ_{12}/λ_2).

In the particular case when the coils are "closely wrapped" so that all the flux of one of the coils links all its own turns, equation 9.82 may be simplified. In this case

$$\lambda_{21} = n_2\phi_{21}$$

$$\lambda_1 = n_1\phi_1$$

$$\lambda_{12} = n_1\phi_{12}$$

$$\lambda_2 = n_2\phi_2$$

Hence

$$K^2 = \frac{n_2\phi_{21}}{n_1\phi_1}\cdot\frac{n_1\phi_{12}}{n_2\phi_2} = \left(\frac{\phi_{21}}{\phi_1}\right)\cdot\left(\frac{\phi_{12}}{\phi_2}\right) \leqq 1 \tag{9.83}$$

The square of the coefficient of coupling is the product of two ratios: the percentage of the flux of coil 1 that links coil 2, (ϕ_{21}/ϕ_1), and the percentage of the flux of coil 2 that links coil 1, (ϕ_{12}/ϕ_2). Neither of these ratios can ever be greater than 1. A "tightly coupled" pair of coils has a higher coefficient of coupling than a pair which is "loosely coupled." From this development it is seen that the mutual inductance is equal to or less than the square root of the product of the self-inductances.

9.26. Total Inductance of Coils in Series. Two coils may be connected in series in two ways: boosting, so that the magnetic field of each coil aids that of the other; or bucking, so that the two fields oppose, as shown in Fig. 9.20.

The total energy stored is:

Boosting:

$$W_m = \tfrac{1}{2}L_1i^2 + Mi^2 + \tfrac{1}{2}L_2i^2$$

$$= \tfrac{1}{2}i^2(L_1 + 2M + L_2)$$

$$L_{\text{total}} = L_1 + 2M + L_2 \tag{9.84}$$

Bucking:

$$W_m = \tfrac{1}{2}L_1i^2 - Mi^2 + \tfrac{1}{2}L_2i^2$$

$$= \tfrac{1}{2}i^2(L_1 - 2M + L_2)$$

$$L_{\text{total}} = L_1 - 2M + L_2 \tag{9.85}$$

If the two coils are connected in parallel the problem is considerably more difficult because the current division between the coils is determined partly by the resistances in the two branches as well as by the self- and mutual inductances. For two coils in parallel connected to a battery, see Art. 11.12.

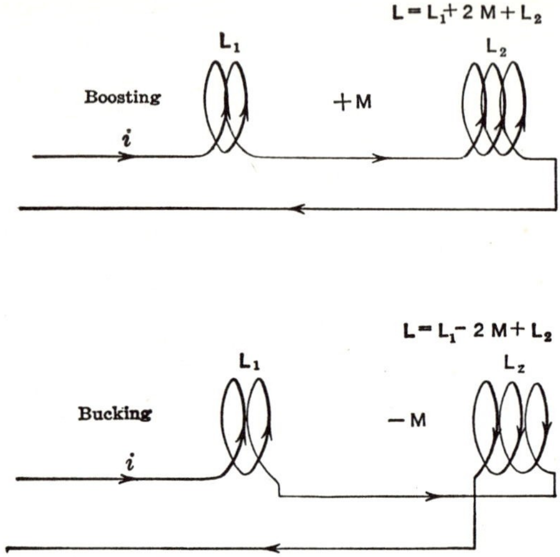

FIG. 9.20. Inductance of coils in series.

9.27. Total Reluctance. The total reluctance of a coil is the ratio of the closed-path magnetomotive force to the total magnetic flux produced.

$$\mathcal{R}_{\text{total}} = \frac{\Sigma \mathcal{F}_{\text{closed path}}}{\phi_{\text{total}}} = \frac{ni}{\phi_{\text{total}}}$$

$$= \frac{n^2}{(n\phi/i)} = \frac{n^2}{L} \quad \frac{\text{amperes}}{\text{weber}} \text{ or henrys}^{-1} \tag{9.86}$$

Under conditions of constant permeability, the reluctance is a concept which depends only upon the permeability of the medium and the geometry of the coil.

9.28. Reluctance of a Small Volume—Permeance. Much more useful than the concept of total reluctance is the reluctance of a small volume. Consider a small volume, as shown in Fig. 9.21, comprising a portion of a tube of flux, bounded on the sides by lines of flux (length Δn) and on the ends by portions of magnetic equipotential surfaces of area ΔA. Then the reluctance is the difference of magnetic potential divided by the flux.

$$\mathcal{R} = \frac{\Delta U}{\Delta \phi} = \frac{H \, \Delta n}{\mu H \, \Delta A} = \frac{\Delta n}{\mu \, \Delta A} \tag{9.87}$$

Several small volumes placed end to end and bounded by the same lines of flux have a reluctance

$$\mathcal{R} = \mathcal{R}_1 + \mathcal{R}_2 + \mathcal{R}_3 ++$$

$$= \frac{\Delta n_1}{\mu_1 \Delta A_1} + \frac{\Delta n_2}{\mu_2 \Delta A_2} + \frac{\Delta n_3}{\mu_3 \Delta A_3} ++ \tag{9.88}$$

and the flux through this tube is

$$\phi = \frac{\Delta U_1 + \Delta U_2 + \Delta U_3 ++++}{\mathcal{R}} \tag{9.89}$$

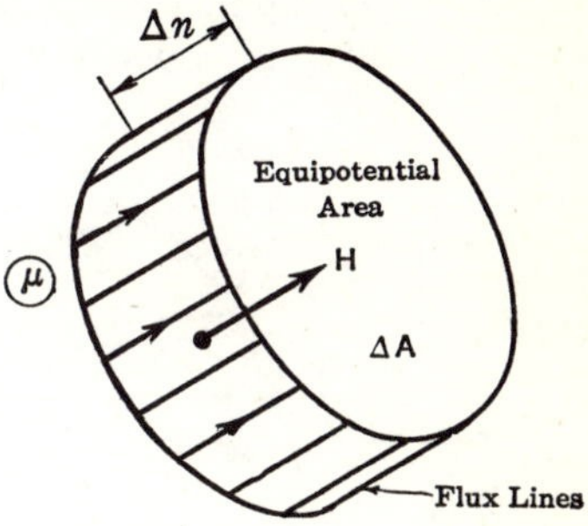

FIG. 9.21. Reluctance of a small volume.

The permeance $\mathcal{P}$ of a certain volume in a magnetic field is the reciprocal of the reluctance.

9.29. Reluctivity—Permittivity. The reluctance of a cube of unit volume (with a uniform field) is called the reluctivity ν, and it is equal (from equation 9.87) to the reciprocal of the permeability.

$$\nu = \frac{1}{\mu} = \text{reluctivity}$$

Permittivity is the reciprocal of reluctivity and consequently is equal to the permeability.

9.30. Energy Density in the Magnetic Field. The incremental storage of magnetic energy about a current ni (see equation 9.77) as the flux changes by $d\phi$ is $dW = ni\,d\phi$, where ni is the MMF about a closed path linking the current, or $dW = \mathcal{F}_{\text{closed path}}\,d\phi$. For a small cylindrical volume $\Delta n \cdot \Delta A$ of the type illustrated in Fig. 9.21, bounded by flux lines of length Δn and by equipotential surfaces of area ΔA, the stored energy is

$$\Delta W = \Delta\mathcal{F} \cdot d\phi$$

$$= H\,\Delta n \cdot dB\,\Delta A$$

The magnetic energy density then is

$$d\omega_m = \frac{dW}{dv} = \frac{\Delta W}{\Delta n \cdot \Delta A} = H\,dB$$

For a finite change in flux density,

$$\omega_m = \int H\,dB \quad \frac{\text{joules}}{\text{cubic meter}} \tag{9.90}$$

If the permeability is constant and B is proportional to H,

$$\omega_m = \int_0^H \mu H\, dH = \frac{\mu H^2}{2} = \frac{HB}{2} = \frac{B^2}{2\mu} \quad \frac{\text{joules}}{\text{cubic meter}} \tag{9.91}$$

9.31. Force on a Current Element. The mechanical work performed in moving $+p$ unit poles from infinity to a point P (Fig. 9.22) where they subtend solid angle Ω from a current i is M.W. $= W_{\text{mutual}} = U_P p =$ joules, provided that the motion is carried out slowly so that

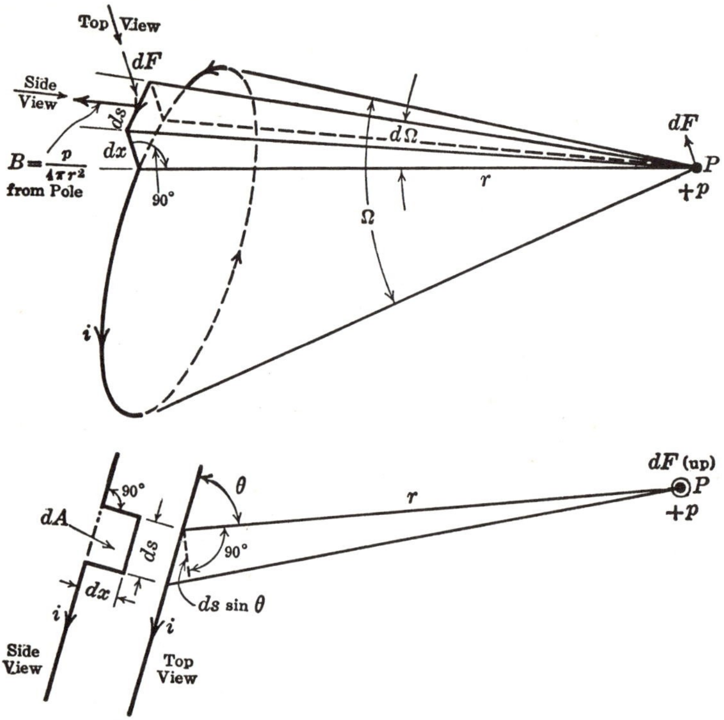

FIG. 9.22. Force on a current element.

the current is maintained constant. Under this condition the mechanical work is equal to the mutual energy between the current and the poles. The mutual energy is the product of the current by the flux from the poles that link it. From equation 9.57

$$\text{M.W.} = W_{\text{mutual}} = U_P p = i\phi = (i\Omega/4\pi)p$$

Now suppose we perform the step over again after having opened up angle Ω to $\Omega + d\Omega$, by moving a small element of current of length

ds outward a distance dx measured off 90° both to ds and to the radial distance r. Then the increment of mechanical work done upon the poles equals the increment of mutual energy and is equal to

$$\Delta\text{M.W.} = \Delta W_{\text{mutual}} = p\,\Delta U_P = i\,d\phi = ip\,d\Omega/4\pi \quad \text{joules}$$

But

$$dA = ds \cdot dx$$

$$\theta = \text{plane angle between the current and the radial line } r$$

$$dA \sin\theta = ds \sin\theta \cdot dx$$

$$d\Omega = \frac{dA \sin\theta}{r^2} = \frac{ds \sin\theta \cdot dx}{r^2}$$

Then

$$\Delta\text{M.W.} = \Delta W_{\text{mutual}} = ip\,\frac{ds \sin\theta \cdot dx}{4\pi r^2}$$

But

$$\frac{p}{4\pi r^2} = B = \text{flux density at the current, due to } p \text{ poles}$$

Hence

$$\Delta\text{M.W.} = \Delta W_{\text{mutual}} = Bi\,ds \sin\theta \cdot dx$$

The force in the dx direction upon the wire is

$$dF_x = \frac{\Delta\text{M.W.}}{dx} = Bi\,ds \sin\theta \quad \text{newtons} \tag{9.92}$$

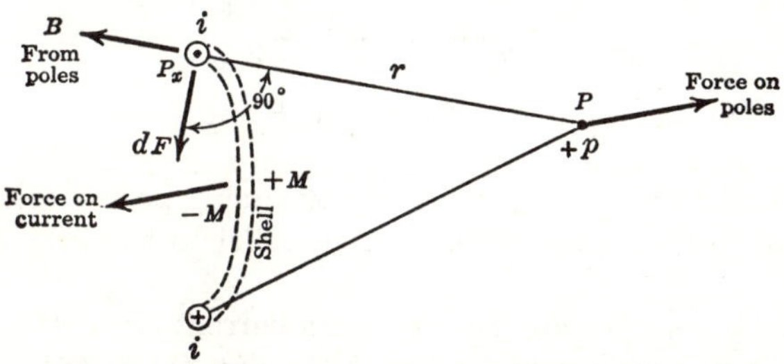

FIG. 9.23. Force on a current element.

Making the movement dx perpendicular to both the current and to r gives the largest possible increase in angle Ω; therefore the whole force on the current is obtained, not merely a component. The force is normal to i and to B. To obtain the direction of the force, consider Fig. 9.23. With the current and its equivalent magnetized shell as shown and $+p$ at point P, we note that the force of the current or shell on $+p$ is toward the right, while the force of $+p$ upon the whole current is toward the left. At P_x the force upon a small current element due to $+p$ poles is perpendicular to the current and to r. Obviously then the force upon the current element must be downward, so that it has a component toward the left which contributes to the whole force on the current.

By putting the palm of the right hand over the current element at P_x so that the thumb points along the current direction, and extending the fingers along the flux density, we note that the current tends to move *away* from the palm. Let us summarize these points:

The magnitude of the force in newtons upon a current i amperes is given by the product of the magnetic flux density B webers per square meter in which it lies, the magnitude of the current i amperes, the length of current involved ds meters, and the sine of the angle θ between the flux density B and the current i. The force is normal to the flux density B and the current i, and its direction is such that, if the palm of the right hand is placed over the current element so that the thumb points along the current direction and the fingers are extended along the flux density, the current element tends to move *away* from the palm. (See Fig. 9.24.)

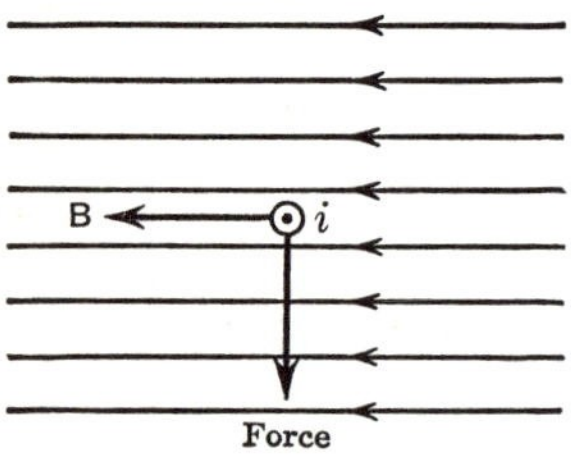

FIG. 9.24. Force on a current element.

$$dF = Bi\,ds \sin\theta \quad \text{newtons} \qquad \text{(Vector)} \tag{9.93}$$

$$dF \perp \text{ to } B \text{ and } i$$

$$\theta \text{ is the } \angle \text{ between } B \text{ and } i$$

A length of current experiences the maximum possible force when it lies normal to the field. When it is parallel to the field there is no force.

It should be noted that the force on a current *element* is not caused by the element's own field, and that the flux density at any current element created by the remaining portions of the current must be used in equation 9.93.

9.32. Field Intensity of a Current Element. Ampere's Rule. If p poles placed at a field point P (see Fig. 9.22) exert force on a current element, the current element exerts an equal and opposite force on the poles. The force per unit pole at P is the intensity or field strength at this point.

$$dF_{\text{on current}} = Bi\,ds \sin\theta = dF_{\text{on poles}} \qquad \text{(Oppositely directed)}$$

$$= \frac{p}{4\pi r^2}\, i\,ds \sin\theta$$

and the intensity is

$$dH_P = \frac{dF}{p} = \frac{i\,ds \sin\theta}{4\pi r^2} \quad \frac{\text{newtons}}{\text{weber}} \text{ or } \frac{\text{amperes}}{\text{meter}} \tag{9.94}$$

which is a vector at P normal to the plane including the current element and P.

Equation 9.94 is known as Ampère's rule. It may be extended to give the field strength of a finite current-length is where s is small compared with r.

$$H_P = \frac{is \sin\theta}{4\pi r^2} \quad \frac{\text{amperes}}{\text{meter}} \tag{9.95a}$$

(See Fig. 9.25.)

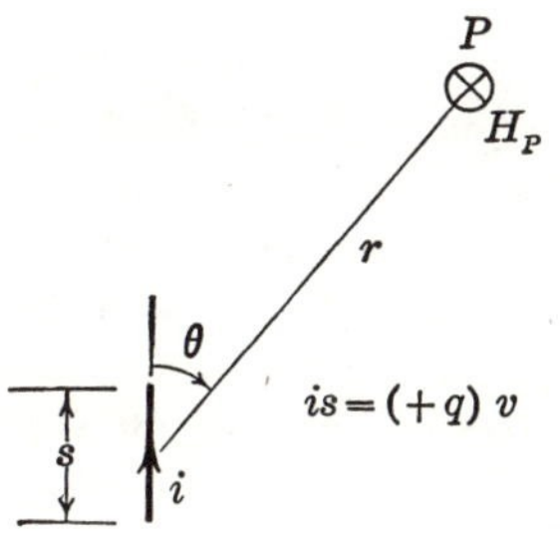

FIG. 9.25. Field strength of is and $(+q)v$. (Ampère's rule.)

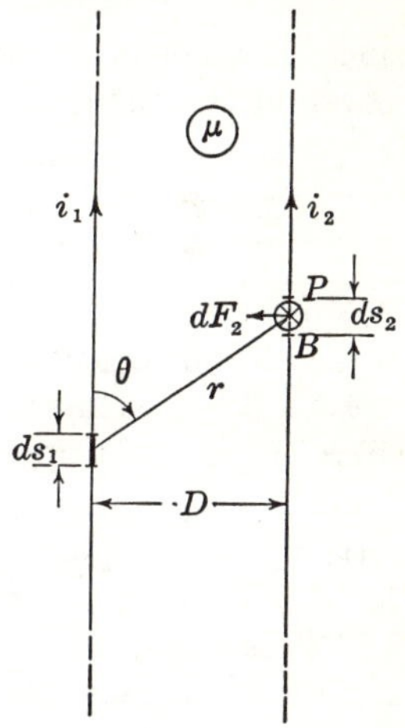

FIG. 9.26. Forces on parallel currents.

9.33. Equivalence of *is* and *qv*. Intensity of a Moving Charge. Current-length is and charge velocity $(+q)v$ are dimensionally equivalent. Consequently, the field strength of a charge q moving with velocity v is

$$H_P = \frac{qv \sin\theta}{4\pi r^2} \quad \frac{\text{amperes}}{\text{meter}} \tag{9.95b}$$

H is perpendicular to the plane of v and r with the vector sense given by the right-hand rule. For an electron of charge $(-e)$ moving with velocity v, the direction of H is, of course, reversed.

9.34. Forces on Infinite Parallel Wires. We first calculate H_2 and B_2 at point P on wire 2 in Fig. 9.26. Using equation 9.94,

$$B_2 = \mu H_2 = \frac{\mu i_1}{4\pi} \int_{s_1=-\infty}^{\infty} \frac{ds_1 \sin\theta}{r^2}$$

$$= \frac{\mu i_1}{4\pi D} \int_{\theta=0}^{\pi} \sin\theta \, d\theta = \frac{\mu i_1}{2\pi D} \quad \frac{\text{webers}}{\text{square meter}} \qquad (9.96)$$

since $ds_1 = (r\,d\theta/\sin\theta) = (D\,d\theta/\sin^2\theta)$ and $r = (D/\sin\theta)$. This is equal to the value given in equation 9.12 and assumes filamentary currents of small cross section compared with the center spacing D.

The elementary force dF_2 on length ds_2, caused by the field of the whole current i_1, by equation 9.93 is

$$dF_2 = B_2 i_2\, ds_2$$

since the angle θ, between B_2 and $i_2\, ds_2$, is 90°. Then the force per unit length of current i_2 is

$$\frac{F_2}{\text{meter}} = B_2 i_2 = \frac{\mu}{2\pi}\frac{i_1 i_2}{D} \quad \frac{\text{newtons}}{\text{meter}} \qquad (9.97)$$

By the right-hand rule, it is seen that parallel currents attract each other and antiparallel currents repel.

9.35. Evaluation of μ_0. The preceding case of infinite parallel wires offers an excellent opportunity to determine μ_0 experimentally. Conducted in free space with a center spacing of one meter and each wire carrying one ampere, the force per meter on each wire proves to be 2×10^{-7} newton per meter. On comparing this value with equation 9.97 it is seen that $\mu_0 = 4\pi/10^7$. Its dimensions are those of force divided by current squared, or LMQ^{-2} as given in Appendix I. Multiplying numerator and denominator by length, the dimensions are energy over I^2 (equal to henrys) divided by length. Hence

$$\mu_0 = \frac{4\pi}{10^7} \quad \frac{\text{henrys}}{\text{meter}} \qquad (9.98)$$

9.36. Force on a Moving Charge. Charge q moving with velocity v at an angle θ with the magnetic flux density is acted upon by a force

$$F = Bqv \sin\theta$$

$$= Bqv_n \quad \text{newtons} \qquad (9.99)$$

which is perpendicular to v and B by the right-hand rule as illustrated in Fig. 9.27. v_n is the component of v perpendicular to B. This is

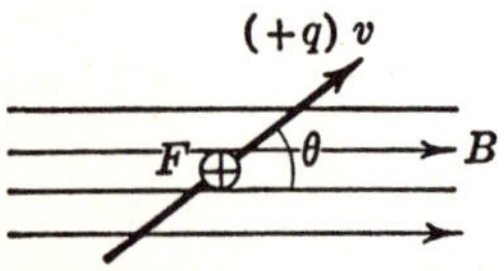

FIG. 9.27. Force on a charge moving across a magnetic field.

based on equation 9.93 and the equivalence of *is* and *qv* given in Art. 9.33.

A more general force formula includes the additional force effect of an electric field and is

$$F = qE \oplus qv_nB \quad \text{newtons} \tag{9.100}$$

in which the electric and magnetic force effects must be added vectorially.

9.37. Circular Motion of Charge in a Magnetic Field. Cyclotron. A charge "shot" normally into a uniform magnetic field describes a circular path since the force is constant and is always perpendicular to the motion. (If the charge enters the field at angle $\theta < 90°$, the component of velocity parallel to B remains unchanged and the path described is a helix.) The radially inward force of the field F_m is balanced by the radially outward centrifugal force F_c. (See Fig. 9.28.)

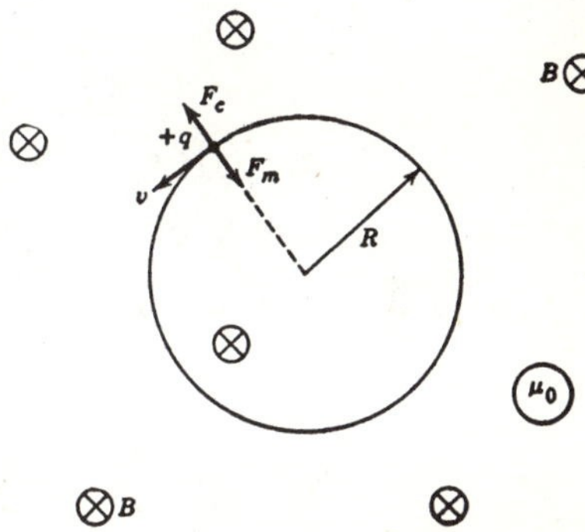

FIG. 9.28. Circular motion of a charge in a magnetic field.

$$\frac{mv^2}{R} = Bqv$$

The radius of the circle is

$$R = \frac{mv}{Bq} \tag{9.101}$$

and the time required for one revolution of the charge is

$$T = \frac{2\pi R}{v} = \frac{2\pi m}{Bq} \tag{9.102}$$

which is independent of the speed of the charge and of the radius of its path.

A cyclotron (Fig. 9.29) is a device with a constant and uniform magnetic field used for speeding up positive charged particles (usually the nucleus of heavy hydrogen, the deuteron, which is the combination of a proton and a neutron). Each half-revolution each particle follows a circular path of a certain radius as indicated above. Then the particle is speeded up over a short distance by an electric field and again follows a circular path of larger radius. Thus the particle is brought up to a high velocity in a succession of speed-ups and successively larger circular paths. The electric field must be reversed precisely in the time $T/2$ required for a half-revolution. The cyclotron cannot be used successfully with electrons since the speeds attained by these lighter particles alters the mass and the particles then get "out of phase" with the alternating electric field.

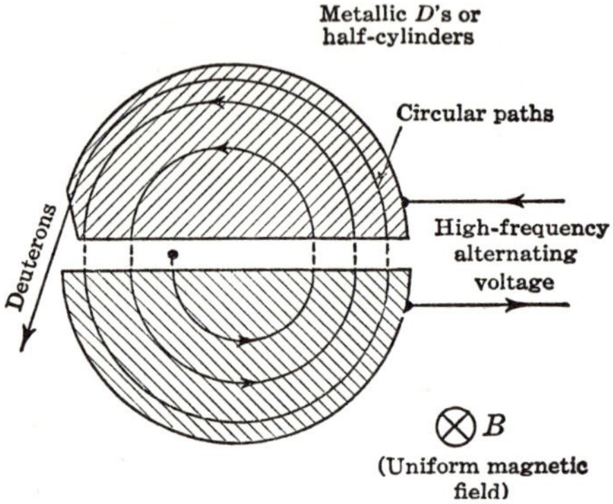

FIG. 9.29. Path of a charged particle in a cyclotron.

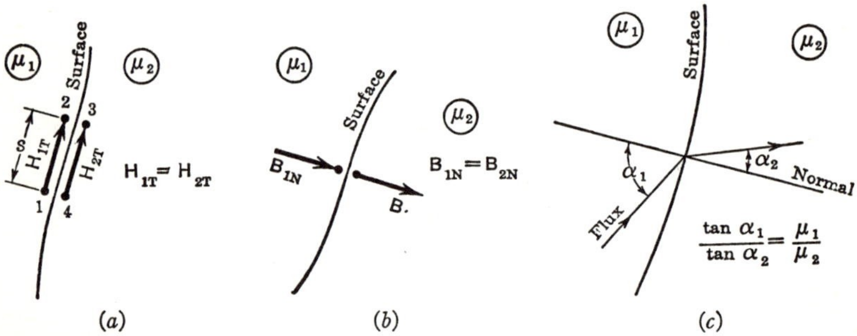

FIG. 9.30. Boundary conditions between two magnetic media. (*a*) Tangential intensities are equal. (*b*) Normal flux densities are equal. (*c*) Flux refraction.

9.38. Boundary Conditions. In much the same way that electric flux is refracted at a boundary between two dielectrics having different dielectric constants, magnetic flux is refracted at the interface between two media having different permeabilities. A unit positive pole carried over a small rectangular path 1–2–3–4–1 (Fig. 9.30*a*) requires as much work along one side in one medium as is returned along the other side in the other medium.

$$H_{1T}s = H_{2T}s$$

or

$$(1) \qquad H_{1T} = H_{2T} \tag{9.103}$$

This rule is based on the proposition that the work per unit pole taken over a closed path is zero. This condition is not violated even if current flows in both media (normal to the page) because the path width 1–4 or 2–3 is taken to be infinitesimal.

Since all the flux crosses the boundary surface, the normal component of the flux density is continuous (Fig. 9.30*b*).

$$(2) \qquad B_{1N} = B_{2N} \tag{9.104}$$

From these two rules, we get

$$(3) \qquad \frac{\tan \alpha_1}{\tan \alpha_2} = \frac{\left(\dfrac{B_{1T}}{B_{1N}}\right)}{\left(\dfrac{B_{2T}}{B_{2N}}\right)} = \frac{\mu_1 H_{1T}}{\mu_2 H_{2T}} = \frac{\mu_1}{\mu_2} \quad \text{(See Fig. 9.30}c\text{)} \tag{9.105}$$

The general shape of the magnetic field remains unchanged if we replace a part of one magnetic medium by another, provided only that the dividing surface lies either completely along flux lines or completely on surfaces normal to the flux. When the dividing surface lies along flux lines, the magnetic intensity H has the same value on both sides. If the division occurs along a surface normal to the flux, the flux density B has the same value on both sides.

9.39. Boundary Surfaces Parallel to Flux Lines or to Equipotential Surfaces. The general *shape* of the field (lines and equipotentials) remains unchanged if the boundary surface between two media lies everywhere: (1) along original flux lines (for one medium alone) or (2) along an original equipotential surface. This is illustrated with three cases.

(1) *Field about a Wire and Concentric Iron Pipe.* This is illustrated in Fig. 9.31. The intensity H rises linearly from the center to the wire surface (for a constant current) and falls hyperbolically with increasing r, irrespective of the medium. $B_0 = \mu_0 H$ follows the same form except in the iron where B_0 is multiplied by μ_r, the relative permeability of the iron. The flux lines are everywhere circular, but they are much more dense in the iron region. The boundary lies everywhere parallel to flux lines.

(2) *Iron Tube Bounded by Flux Lines.* This is similar in principle to the preceding case and differs only in the geometry. It is illustrated in Fig. 9.32.

(3) *Field about a Current in a Plane Iron-Free Space Boundary.* In this case, illustrated in Fig. 9.33, the boundary coincides with an original equipotential surface. The flux lines are still circles and the equipotentials are still radial planes. The total MMF around a flux line is I, but the work per unit pole in the iron portion is a very small

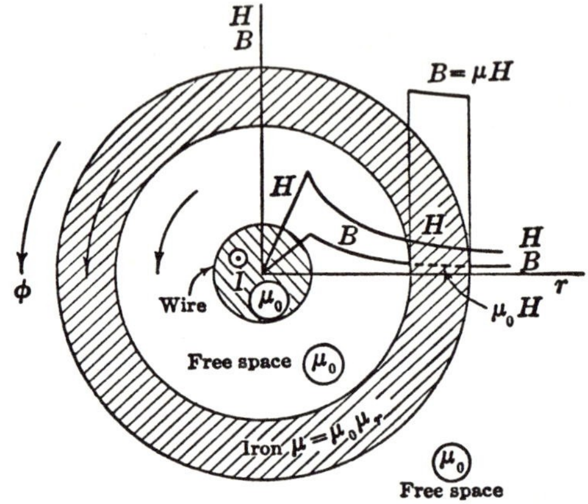

FIG. 9.31. Field of a wire and a concentric iron pipe.

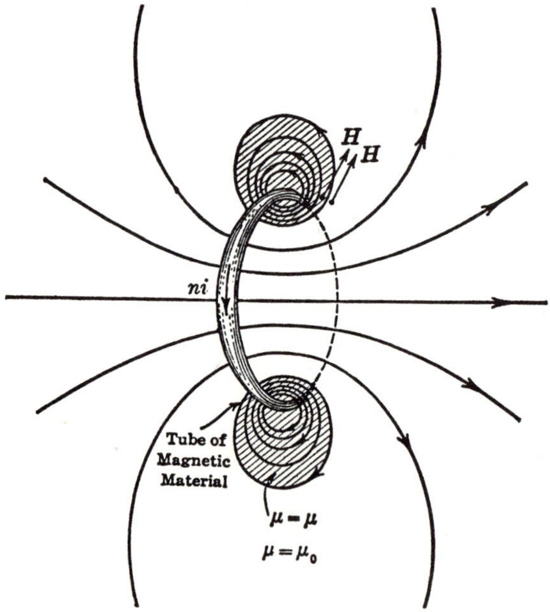

FIG. 9.32. Field of an iron tube bounded by flux lines.

fraction of the whole and most of the MMF is found in the free space region.

$$H_i \cdot \frac{2\pi r}{2} + H_0 \cdot \frac{2\pi r}{2} = I$$

and

$$\frac{H_i}{H_0} = \frac{1}{\mu_r} \tag{9.106}$$

The flux density is $B = \mu_0 H_0 = \mu H_i$.

9.40. Field of the Closed Solenoid (Rowland's Ring). Equivalent Current Sheet. We assume here a full closed solenoid, taken for convenience in circular form, with the n turns closely wrapped and the

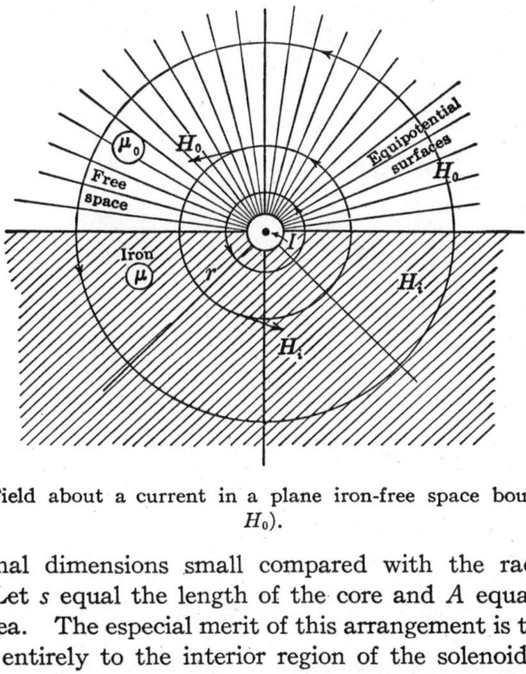

FIG. 9.33. Field about a current in a plane iron-free space boundary ($H_i \ll H_0$).

cross-sectional dimensions small compared with the radius of the solenoid. Let s equal the length of the core and A equal the cross-sectional area. The especial merit of this arrangement is that the flux is confined entirely to the interior region of the solenoid. We here idealize the problem slightly by assuming a uniform flux field and ignoring local field effects of the individual turns. (See Fig. 9.34.)

First assume a free space core and ni ampere-turns in the coil or ni/s ampere-turns per meter of core length. Then the MMF is $\mathfrak{F} = ni = Hs$, giving

$$H = \frac{ni}{s} \; \frac{\text{amperes}}{\text{meter}} \tag{9.107}$$

$$B_0 = \mu_0 H = \mu_0 \frac{ni}{s} \quad \frac{\text{webers}}{\text{square meter}} \tag{9.108}$$

$$\phi_0 = B_0 A = \mu_0 \frac{ni}{s} A \quad \text{webers} \tag{9.109}$$

and

$$L_0 = \frac{n\phi_0}{i} = \mu_0 \frac{n^2 A}{s} \quad \text{henrys} \tag{9.110}$$

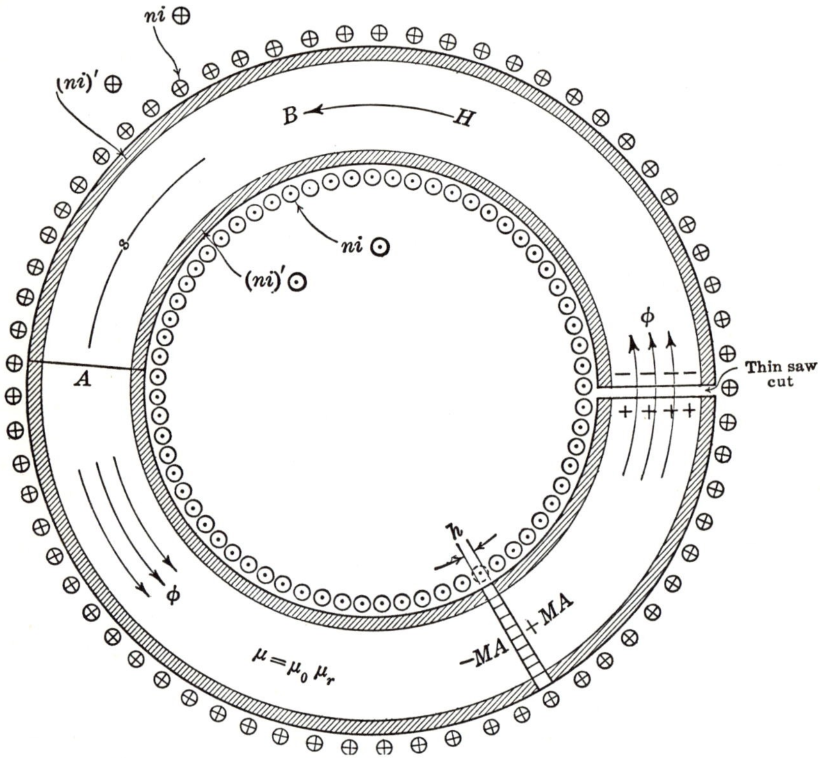

Fig. 9.34. Magnetization and equivalent current sheet in Rowland's ring.

We now fill the core with magnetizable material of relative permeability $\mu_r = \mu/\mu_0$. Then H remains the same, but B and ϕ are increased in the ratio μ_r (which may have values of slightly less than one for diamagnetic materials, values of slightly greater than one for paramagnetic materials, and values up in the thousands for ferromagnetic materials). For this core

$$H = \frac{ni}{s} \quad \frac{\text{amperes}}{\text{meter}} \tag{9.111}$$

$$B = \mu H = \mu \frac{ni}{s} \quad \frac{\text{webers}}{\text{square meter}} \tag{9.112}$$

$$\phi = BA = \mu \frac{ni}{s} A \quad \text{webers} \tag{9.113}$$

Since μ_r may be a function of field strength

$$L = \frac{d\lambda}{di} = n \frac{d\phi}{di} = n \frac{A\,dB}{(s/n)\,dH}$$

$$= \frac{n^2 A}{s} \frac{dB}{dH} \quad \text{henrys} \tag{9.114}$$

dB/dH must be obtained graphically from the hystero-eddy current loop discussed in Chapter 13. It varies with the value of the current and depends on whether the current is rising or falling.

For constant permeability, independent of field strength,

$$L = \frac{n\phi}{i} = \frac{n^2 A}{s} \mu \quad \text{henrys} \tag{9.115}$$

Next let us look more closely into the relatively enormous increase in magnetization that has resulted from introducing a core of permeable material. The increase in flux density (above that of free space for a given current i) is $B - B_0 = \beta$. The quantity β is called the ferric induction.

$$B - B_0 = \beta = B - \mu_0 H = (\mu - \mu_0) H \tag{9.116}$$

B_0 is the flux density created in free space by the current, β is the contribution of the magnetized medium, and B represents the combined effect of both wire current and medium. As pointed out earlier the magnetizing effect of the medium may be enormously greater (if μ_r is large) than that of the current. In this sense the current merely "triggers" the magnetization inherent in the permeable medium. $\mu_r = B/B_0$, in a sense, is the "amplification factor."

The physical basis for the magnetizing power of the permeable medium is given in Art. 9.45 for diamagnetic and paramagnetic materials and in Art. 13.20 for ferromagnetic materials.

Further insight into the magnetization of the permeable medium is obtained if the core is cut across normally with a *thin* saw cut. There will then appear on the interfaces, between the permeable medium and the free space cut, layers of poles, positive (north) on one face and an equal number of negative (south) poles on the other face. This is indicated in Fig. 9.34.

The number of poles per unit (normal) area $M = p/A$ (see equation 9.48) is equal to the flux per unit area created by the permeable medium, that is, $M = \beta = p/A$. Multiplying and dividing p/A by the core length s, it is seen that $M = \beta = p/A = ps/\text{volume}$, but ps is the magnetic moment of the core. Hence M is the magnetic moment of the permeable medium per unit volume. Combining these relations with equation 9.116 we obtain

$$B - B_0 = \beta = B - \mu_0 H = M = \frac{p}{A} = \frac{m}{\text{volume}} \tag{9.117}$$

The units are those of flux density (webers per square meter) or those of pole density (poles per square meter) or those of moment per unit volume (weber-meters per cubic meter). m is the total moment of the permeable medium.

Going a step further, we now imagine the permeable core to be divided into a series of thin magnetized disks, each of thickness h taken parallel to the flux lines. Each disk is a magnetized layer or shell of dipoles with $M = p/A$ poles per unit area and a moment per unit area equal to Mh. According to equation 9.54, the dipole layer disk of thickness h is equivalent to a current per unit length of core $(ni/s)'$ times the disk length h.

$$\left(\frac{ni}{s}\right)' h = \frac{Mh}{\mu_0} \quad \text{amperes} \tag{9.118}$$

The equivalent current circulates around the edge of the disk.

Per unit length of core we then have the free space magnetization of the wire current

$$\frac{ni}{s} = H = \frac{B_0}{\mu_0} \tag{9.119}$$

the contribution of the permeable medium (or equivalent surface current)

$$\left(\frac{ni}{s}\right)' = \frac{B - B_0}{\mu_0} = \frac{M}{\mu_0} \tag{9.120}$$

and the total magnetization

$$\frac{ni}{s} + \left(\frac{ni}{s}\right)' = \frac{B}{\mu_0} = H\mu_r \tag{9.121}$$

Figure 9.34 illustrates the equivalent surface current sheet $(ni)'$ as well as the wire current ni. It appears therefore that the magnetic effect of any magnetized permeable medium can be represented by an equivalent current sheet of appropriate strength (variable in magnitude if necessary) spread over the side area of the medium.

For paramagnetic materials $\mu_r > 1$, and for ferromagnetic materials $\mu_r \gg 1$, the magnetization of the permeable medium is directed as shown in Fig. 9.34. For diamagnetic materials $\mu_r < 1$, the wire current magnetization is in the direction shown, but that of the permeable medium is reversed, and the total effect is the difference.

A ring with an air gap of appreciable length requires a larger value of wire current ni to support the same flux density B, or, conversely, B is smaller for the same current ni. This case is considered in detail in Chapter 13.

To illustrate the magnitudes of the preceding case, let $s = 1$ meter, $A = 10^{-4}$ square meter, $B = 1$ weber per square meter, $\mu_r = 1000$ (typical of cast steel), and $\mu_0 = 4\pi/10^7$ henrys per meter. Then $\mu = 4000\pi/10^7$, $\phi = 10^{-4}$ weber, $H = B/\mu = ni/s = 10^7/(4000\pi) =$ 795.8 ampere-turns, and $B_0 = \mu_0 H = 0.001$ weber per square meter. $B - B_0 = \beta = M = 0.999$ weber per square meter = moment per unit volume. $p = MA = 0.999 \times 10^{-4}$ pole (or weber). The equivalent current sheet is $(ni)' = (B - B_0)s/\mu_0 = 0.999 \times 10^7/(4\pi) =$ 795,005 ampere-turns, which is 999 times the value of ni.

9.41. Intensity of Magnetization. From the preceding paragraph we may define the intensity of magnetization M as a vector given by

$$M = B - \mu_0 H = \frac{p}{A} \quad \frac{\text{poles}}{\text{square meter}} = \frac{\text{moment}}{\text{volume}} \tag{9.122}$$

M is comparable with polarization $P = D - \epsilon_0 E = \rho_s'$ moment per unit volume in the electric field.

9.42. Pressure on an Interface. The mechanical pressure on an interface between a permeable medium and free space, with flux lines either parallel or normal to the interface, is equal to the difference of energy densities on the two sides and is directed normally toward the free space side.

$$p_r = \omega_0 - \omega_m \quad \frac{\text{newtons}}{\text{square meter}} \tag{9.123}$$

This is comparable with the pressure on an interface between dielectrics, assuming either normal or parallel flux.

For the more general case of flux refracted at the interface, the expression for the pressure is considerably more complicated. See Arts. 8.12 and 13.14.

9.43. Field of a Coil and Iron of Any Shape. We have previously considered only the special cases of iron (or any magnetic material) so disposed that the iron boundary coincided either (1) with the flux lines, or (2) with equipotential surfaces. In either event the general shape of the field remains unchanged by the presence of the iron. We now consider the problem of the field set up by a current and iron of any shape and size and placed in any position relative to the current. The magnetic flux lines must be rearranged in such manner that they

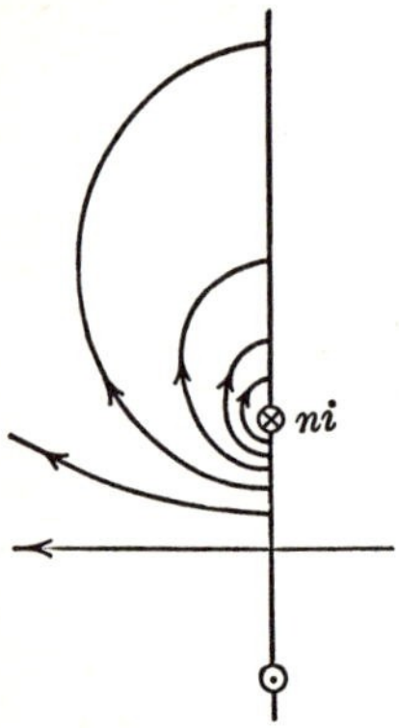

FIG. 9.35.

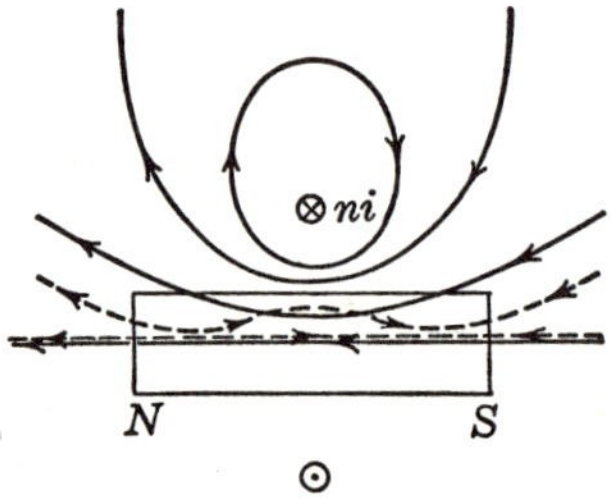

FIG. 9.36.

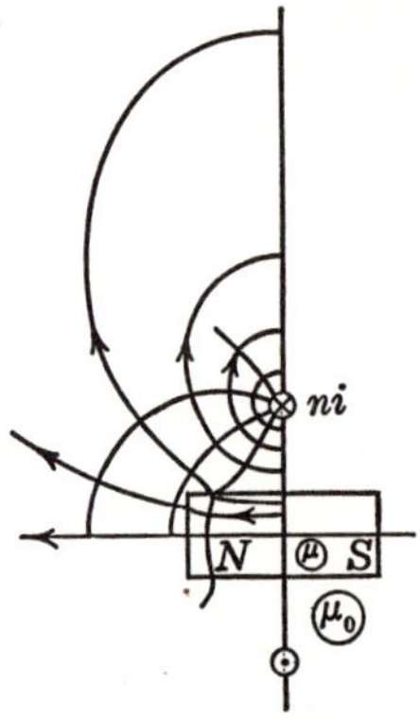

FIG. 9.37.

Field of a coil, and of a coil plus iron.

surround the current and yet satisfy the boundary conditions for refraction at the iron-free space interface as specified in Art. 9.38.

First consider the field of a coil in free space, shown in Fig. 9.35. Next place a cylinder of iron at the center of the coil. This, of course, becomes magnetized with north (positive) poles appearing at the left end and an equal number of south (negative) poles appearing at the right end, as shown in Fig. 9.36. The magnetizing effect of the coil (H) is precisely the same as before in Fig. 9.35 and is represented by the solid lines running from right to left. The magnetizing effect of the iron is shown in dotted lines. *Outside* the iron this effect is directed from right to left and adds to the effect of the coil, but *inside* the iron, the magnetizing effect of iron is directed from left to right and opposes the effect of the coil.

The net result is that the total number of flux lines ϕ is increased by virtue of the additional magnetization of the iron, and the flux density B is increased, particularly in the neighborhood of, and inside, the iron. Outside the iron H is increased as well as B, but inside the iron H is decreased since the magnetizing effect of the poles is directed oppositely to that of the coil. The final distribution of flux lines, though perhaps difficult to calculate, must satisfy the boundary conditions of Art. 9.38 at the iron–free space interface and at the coil conductor surface. Figure 9.37 illustrates the total field.

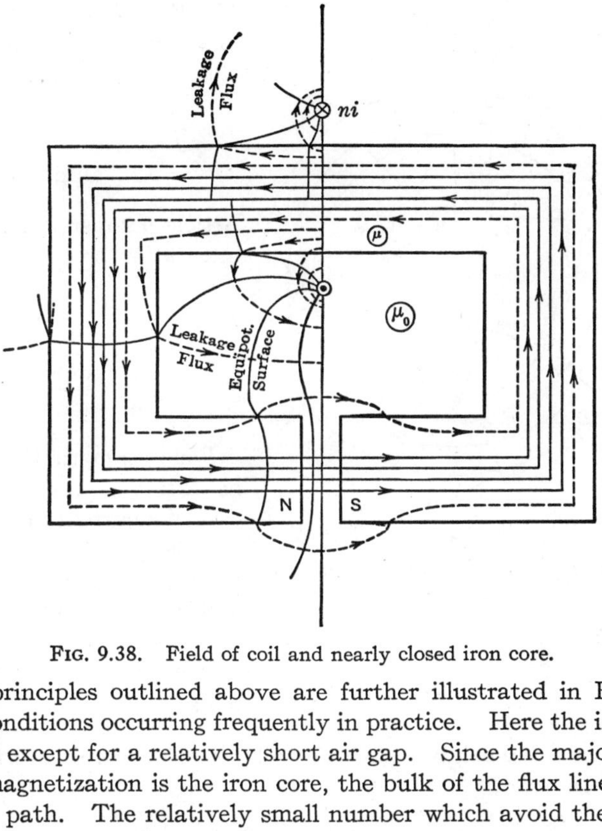

FIG. 9.38. Field of coil and nearly closed iron core.

The principles outlined above are further illustrated in Fig. 9.38 under conditions occurring frequently in practice. Here the iron path is closed except for a relatively short air gap. Since the major source of the magnetization is the iron core, the bulk of the flux lines follow the iron path. The relatively small number which avoid the air gap constitute the so-called leakage flux. The leakage portion may, of course, be reduced relatively by spreading the coil turns out over the iron path.

Figure 9.39 illustrates a variation of Fig. 9.38 in which a movable magnetic element is placed in the air gap. This effectively shortens the gap, increases the total flux, and concentrates the flux lines in and about the element.

9.44. General Definition of Intensity *H*. In this paragraph we shall give a general definition of field strength H, valid for all magnetostatic fields, and at all points whether inside or outside current-bearing regions, whether in free space or in permeable media, and valid for all combinations and shapes of magnetizable elements.

In Arts. 9.4 and 9.8 are given, not definitions of H, but rather conditions that H must fulfill under all conditions—namely, that the negative line integral of H over any closed line path is equal to the MMF and must equal the magnitude of the current enclosed by the path,

$$\mathfrak{F} = \oint_{\text{closed path}} (-H)\, ds \cos \alpha = nI \quad \text{amperes} \qquad (9.124)$$

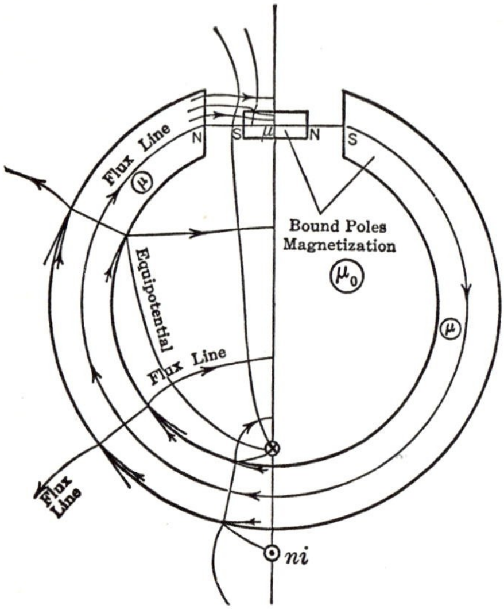

FIG. 9.39. Field of coil, core, air gap, and armature.

Article 9.9 merely inverts the integration process and gives

$$H = -\text{P.G.} = -\frac{d\mathfrak{F}}{dn} \quad \frac{\text{amperes}}{\text{meter}} \qquad (9.125)$$

Field strength H is defined in Art. 9.11 for the special case of a current filament surrounded by an all-extensive medium of constant permeability (independent of field strength).

$$H = -\frac{I}{4\pi}\frac{d\Omega}{dn} \quad \frac{\text{amperes}}{\text{meter}} \qquad (9.126)$$

It is again defined, in Art. 9.15, as the force per unit positive pole at any field point.

$$H = \frac{dF}{dp_+} \quad \frac{\text{newtons}}{\text{weber}} \qquad (9.127)$$

On this basis, the field strength in the neighborhood of p poles, by equation 9.38 of Art. 9.16, for free space, is

$$H = \frac{p}{4\pi\mu_0 r^2} \quad \frac{\text{newtons}}{\text{weber}} \qquad (9.128)$$

This form may be generalized to give the field of p poles distributed throughout a volume, by writing

$$H = \frac{1}{4\pi\mu_0} \iiint \frac{\pm dp}{r^2} \qquad \text{(Vector sum)} \tag{9.129}$$

in which the vector integration is taken throughout the volume containing poles of both signs. (See equation 9.43.)

For a magnetic shell, the form given in equation 9.53 is

$$H = -\frac{Mh}{4\pi\mu_0}\frac{d\Omega}{dn} \tag{9.130}$$

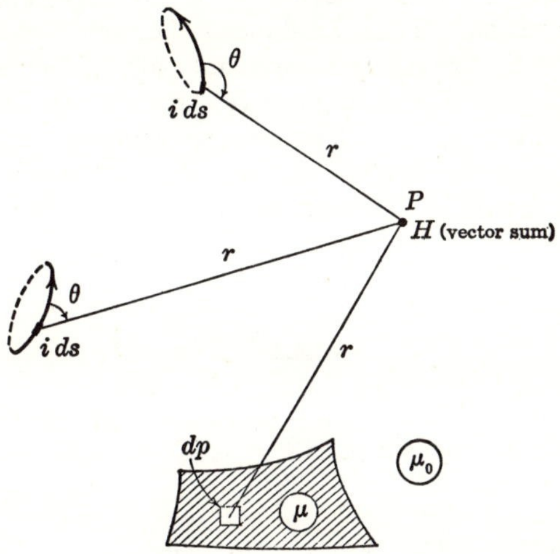

FIG. 9.40. Field strength H at any point. (See equation 9.132.)

The intensity for a filamentary current, obtained by integrating equation 9.94 over any desired portion of the filament, is

$$H = \frac{1}{4\pi} \int_s \frac{i\,ds \sin\theta}{r^2} \qquad \text{(Vector sum)} \tag{9.131}$$

Finally, the general definition of H at any point P, in Fig. 9.40, for any combination of currents and permeable media composed of positive and negative poles, is

$$H = \frac{1}{4\pi} \int_s \frac{i\,ds \sin\theta}{r^2} + \frac{1}{4\pi\mu_0} \iiint_v \frac{\pm dp}{r^2} \qquad \text{(Vector sum)} \tag{9.132}$$

where r is the distance from the point P to each element of current or pole. The summation is vector, each infinitesimal quantity dH being either perpendicular to the plane of $i\,ds$ and r, or in line with dp and r.

At all points, the flux density

$$B = \mu H \tag{9.133}$$

9.45. Physical Basis of Dia- and Paramagnetism. Both these types of magnetism are traceable ultimately to the rotation of electrons about the direction of a magnetic field H imposed externally upon the atoms and to the electron spin.

(1) *Diamagnetism.* To explain this type of magnetism we consider electron orbits to be "paired," with two electrons revolving in similar orbits (here assumed, for simplicity, to be circular) but with opposite directions of rotation. The orbital motion of each electron constitutes a current i flowing around the (circular) path of the orbit of radius R and area A in time T with angular velocity ω radians per second and circumferential velocity v. According to equation 9.54, the orbital current $i = e/T = ev/(2\pi R) = e\omega/(2\pi)$ equals the orbital magnetic moment per unit area of the orbit divided by μ_0. Then the magnetic moment of the orbital electron is

$$\text{Magnetic moment} = m_{\text{or}} = \mu_0 iA$$
$$= -\frac{\mu_0 eA}{T} = -\frac{\mu_0 e\omega A}{2\pi} = -\frac{\mu_0 evA}{2\pi R} \tag{9.134}$$

The orbital angular momentum is $p = mvR = m\omega^2 R$, in which m is the electron mass. The ratio

$$\frac{\text{Orbital magnetic moment}}{\text{Orbital angular momentum}} = -\frac{m_{\text{or}}}{p} = \frac{\mu_0 e}{2m} = 1.105 \times 10^5 \tag{9.135}$$

is a constant (in general, independent of the size or shape of the orbit). The magnetic moment is a vector taken normal to the plane of the orbit, with the sense given by the right-hand rule for current (opposite to the motion of the negative electron). The angular momentum is a vector in the opposite direction. See Fig. 9.41. The corresponding vectors, for the other electron of a pair, are reversed in direction.

The electron rotating in its orbit acts like a spinning top or gyroscope. The magnetic moment m_{or} is equivalent to a small magnet, S–N in Fig. 9.41. The field H exerts a torque $\mathfrak{I} = m_{\text{or}} H \sin(180° - \beta)$ on the two poles. The direction of the torque vector is perpendicular to m_{or} and H, in the right-hand screw order m_{or}, H, $\mathfrak{I}$. The torque, which is also the time rate of change of angular momentum dp/dt, is perpendicular to vector p. Since dp/dt is perpendicular to H and p, the torque can succeed in changing p only by a circular precession about H with a precessional angular velocity ω_p.

$$\mathfrak{J} = m_{or} H \sin (180° - \beta) = \frac{dp}{dt} = \omega_p \sin \beta \tag{9.136}$$

Hence the orbit and its vectors m_{or}, p, ω precess about H with angular velocity ω_p equal to

$$\omega_p = -\frac{m_{or}}{p} H = \frac{\mu_0 e}{2m} H \tag{9.137}$$

which is independent of angle β. The orbit and vectors of the paired electron precess in the reverse direction. These extra precessional motions constitute extra currents circulating about the H axis and produce extra magnetic moments Δm_{or} parallel and antiparallel to H.

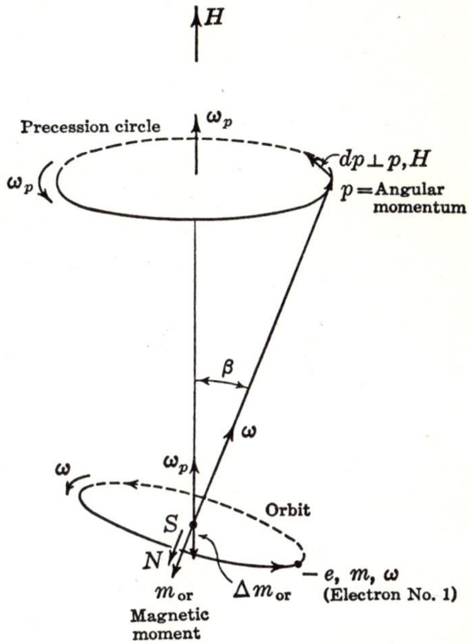

FIG. 9.41. Precession of an electron orbit about a magnetic field.

This can be simplified by noting that, of N electron orbits per unit volume, $\frac{2}{3}N$ precess about H at angle $\beta = 90°$ and produce no extra current about H, nor any magnetic moment parallel to H. The remaining $\frac{1}{3}N$ orbits have moments at $\beta = 0$, half parallel and half antiparallel to H, and do not precess about H. Let us consider the latter group in greater detail with the aid of Fig. 9.42.

Electrons 1 and 2 circle H as shown. With $H = 0$, the orbital angular velocity of both electrons is ω_0 and the attractive force of the nucleus is balanced by the centrifugal force $mv_0^2/R = m\omega_0^2 R$. With field H applied, electron 1 experiences an additional radial inward force $Bev = Be\omega R$ which must be balanced by an additional centrifugal force, which means that this electron speeds up, since the orbital radius according to quantum mechanical rules must remain constant. The additional force is

$$m\omega^2 R - m\omega_0^2 R = Be\omega R$$

or

$$m(\omega + \omega_0)(\omega - \omega_0) = Be\omega$$

The change in orbital angular velocity $\Delta\omega = \omega - \omega_0$ is small and $\omega + \omega_0 \cong 2\omega$, giving $2m\Delta\omega = Be$, or

$$\Delta\omega = \frac{e}{2m} B = \frac{\mu_0 e}{2m} H = 1.105 \times 10^5 H \quad \frac{\text{radians}}{\text{second}} \tag{9.138}$$

Since precession is denied to these groups of electrons the extra effects resulting from H must lead to a change in orbital angular velocity. Electrons 1 and 2 thus acquire angular velocities

$$\begin{aligned} \omega_1 &= \omega_0 + \Delta\omega \qquad \text{(Increase)} \\ \omega_2 &= \omega_0 - \Delta\omega \qquad \text{(Decrease)} \end{aligned} \tag{9.139}$$

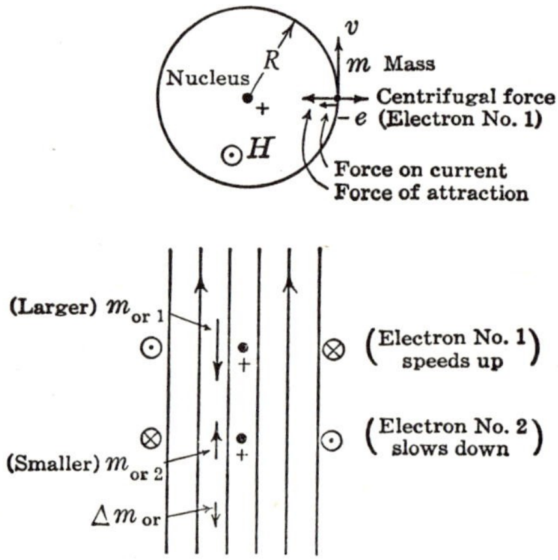

FIG. 9.42. Magnetic effects of paired electrons circulating around H.

The fractional change in velocity $\dfrac{\Delta\omega}{\omega_0} = \dfrac{\mu_0 e}{2m\omega_0} H$ is very small since ω_0 is of the order of 10^{16}. For $H = 100$ amperes per meter, $\dfrac{\Delta\omega}{\omega_0} \cong 10^{-9}$.

The magnetic moments of the two electrons, from equation 9.134, are

$$\begin{aligned} m_{\text{or}1} &= \frac{\mu_0 e A}{2\pi} \omega_1 \qquad \text{(Larger)} \\ m_{\text{or}2} &= \frac{\mu_0 e A}{2\pi} \omega_2 \qquad \text{(Smaller)} \end{aligned} \tag{9.140}$$

These are illustrated in Fig. 9.42. The differential moment of $N/3$ orbits per unit volume is

$$\frac{N}{3}\,\Delta m_{\text{or}} = -\frac{N}{3}\frac{\mu_0 e A}{2\pi}\,2\,\Delta\omega$$

$$= -\frac{\mu_0{}^2 e^2}{6\pi m} NAH = M \tag{9.141}$$

The differential moment is equal to M, the moment per unit volume, and is directed oppositely to H.

The net effect on a medium composed of paired electrons is to reduce the relative permeability below unity.

$$\mu_r = \frac{B}{\mu_0 H} = \frac{\mu_0 H + M}{\mu_0 H} = 1 + \frac{M}{\mu_0 H}$$

$$= 1 - \frac{\mu_0 e^2}{6\pi m} NA \tag{9.142}$$

$$= 1 - 1.768 \times 10^{-14} NA$$

To illustrate, consider solid copper, which is diamagnetic. It has 28 paired electrons (the 29th is free). There are 8.48×10^{28} atoms per cubic meter. The orbital area approximates 10^{-20} square meter, giving $NA \cong 2.37 \times 10^{10}$. Hence $\mu_r \cong 1 - 1.3 \times 10^{-5}$, compared with the approximate measured value of $1 - 0.94 \times 10^{-5}$. For gases at NTP, N is about $\frac{1}{1000}$ of the above value, giving $\mu_r \cong 1 - 1 \times 10^{-8}$. Consult Table 9.1 for values of μ_r for other diamagnetic substances.

Diamagnetism is a property that is practically independent of temperature.

(2) *Paramagnetism.* From this discussion it is clear that every substance having two or more electrons is inherently diamagnetic, as a result of the influence of H on paired electrons. But many atoms, however, have nonpaired electrons and electron spins which give the atom (or molecule) as a whole a *permanent* magnetic moment which is not dependent on H. For these atoms the paramagnetic effect proves to be stronger than the diamagnetic influence and is directed in the same direction as H, giving $\mu_r > 1$.

In a paramagnetic material we picture the permanent magnetic moments m_a of the molecules as randomly distributed, as many pointing in one direction as another, in the absence of a magnetic field. The total moment of a molecule m_a is a vectorial combination of the nonpaired orbital and spin momenta. With a field H applied all the moments precess about H with precessional angular velocities so directed that they yield components of magnetic moment in the *same* direction as H, thus giving a relative permeability greater than unity.

Paramagnetism is a property which is markedly dependent on temperature. The effect of thermal agitation is to keep the molecular moments randomly distributed, in opposition to the effect of the field H which produces a net moment in the field direction.

The theory of paramagnetism is too complicated and lengthy to be given here, but we may give the net result. For n molecules per unit volume, each with a magnetic moment m_a, the magnetic moment per unit volume M is equal to

$$M = \frac{nm_a^2H}{3kT} \tag{9.143}$$

with $k = 1.38 \times 10^{-23}$ joule per degree absolute (Boltzmann's constant), and T is the absolute temperature. Consequently the relative permeability

$$\mu_r = \frac{B}{\mu_0 H} = 1 + \frac{M}{\mu_0 H}$$

$$= 1 + \frac{nm_a^2}{3\mu_0 kT} \tag{9.144}$$

and is greater than unity.

As a step toward calculating the magnetic moment m_a, let us first obtain the moment of the Bohr magneton m_B. From equation 9.135, the orbital magnetic moment, ignoring the sign, is

$$m_{\text{or}} = p\frac{\mu_0 e}{2m} = 1.105 \times 10^5 p \tag{9.145}$$

but p, the orbital angular momentum, is restricted by quantum laws to integral values of $\frac{h}{2\pi}$, in which Planck's constant h has the value 6.624×10^{-34} joule-second. The least value of moment, corresponding to $1\left(\frac{h}{2\pi}\right)$, is called the Bohr magneton m_B.

$$m_B = \frac{\mu_0 e}{2m}\cdot\frac{h}{2\pi} = \frac{\mu_0 eh}{4\pi m} = 1.164 \times 10^{-29} \quad \text{weber-meter} \tag{9.146}$$

This is discussed further in Art. 1.6 (equation 1.15) and Appendix II.

For the electron spin, the unit of angular momentum is $\pm\frac{1}{2}\left(\frac{h}{2\pi}\right)$, half that for an orbital electron, giving an $\frac{m_s}{p}$ ratio twice as large and a spin magnetic moment equal to that of the orbital electron

$$m_s = \pm \frac{\mu_0 eh}{4\pi m} = \pm 1.164 \times 10^{-29} \quad \text{weber-meter} \tag{9.147}$$

The sign depends on whether the spin is parallel or antiparallel to H.

To calculate the order of magnitude, let us assume $n = 10^{29}$ atoms per cubic meter for a solid material, $m_a = 1m_B$, $T = 300°$ absolute. Then

$$\frac{nm_a^2}{3\mu_0 kT} = 0.87 \times 10^{-4} \tag{9.148}$$

and the relative permeability is

$$\mu_r = 1 + 0.87 \times 10^{-4} \tag{9.149}$$

For a paramagnetic gas at normal temperature and pressure, the excess above unity should be of the order of 10^{-7}.

For any actual atom or molecule, the value given in equation 9.148 will have to be altered in accordance with the number of orbits and spins that collaborate to give the magnetic moment m_a of the atom as a whole. The measured values of μ_r for various paramagnetic substances are given in Table 9.1.

From this we see that ferromagnetism, with relative permeabilities running up into the hundreds of thousands, cannot find its source in paramagnetism. Ferromagnetism must be based on phenomena so far not considered. The physical basis of ferromagnetism is given in Art. 13.20.

PROBLEMS

1. Refer to Fig. 9.3. Calculate the number of flux lines inside the wire, and outside the wire from $r = R$ to $r = 2R$, per meter of wire length. Calculate the MMF for $r = R/2$, $r = R$, and $r = 2R$. Outside the wire the MMF is equal to the potential U. Show how the latter can be differentiated to obtain H.

2. Refer to Figs. 9.7 and 9.8. For a circular filament of current I, of cross section A, show that the solid angle Ω at distance r large compared with the circle radius is given by $\Omega \cong A \cos\theta/r^2$, where θ is the angle between r and a line perpendicular to the circle area. Then $U \cong I\Omega/(4\pi)$ amperes.

3. Calculate the magnetic absolute potential at 1 meter from $p = 10^{-2}$ pole (or weber) in free space. Calculate H at the same point.

4. Refer to Art. 9.18. For a dipole with $p = 10^{-2}$, $h = 10^{-4}$, $r = 1$, calculate the potential in free space for $\theta = 0$ and $\theta = 60°$. Compute H at both points. Derive the expressions for U and H in rectangular coordinates. Find the locus of points where the intensity is perpendicular to the dipole axis. Show how any number of coils placed along the locus may be arranged so that no two will have any mutual coupling.

5. Refer to Prob. 4. If the dipoles are spread uniformly and normally over an area $A = 10^{-2}$ square meter, calculate the value of M. For one turn around the circumference of the magnetic shell calculate the equivalent current from equation 9.54.

6. Refer to Prob. 4. Place a hemispherical surface of radius 1 meter concentric with the dipole. The axis of revolution of the surface and that of the dipole coincide. Prove that the number of flux lines crossing the hemisphere is $\phi = ph/(2r)$.

7. Show how *three* identical coils may be placed so that there shall be no mutual inductance between any two.

8. A 5-turn closely wrapped circular coil carries a current of 10 amperes and sets up a flux of 0.0025 weber. Compute the number of flux linkages, the self-inductance, the energy, and the total reluctance. The medium is free space.

9. Place a coil, similar to the coil of Prob. 8 except that it has 10 turns, coaxially with the coil of Prob. 8, so that the coefficient of coupling is 0.5. The current is 10 amperes. Compute the inductance, flux linkages, self-energy, and reluctance of the second coil, and calculate the mutual inductance and energy. Compute the two values of total energy of the two coils.

10. Connect the coils of Probs. 8 and 9 in series. Compute the maximum and minimum values of total inductance.

11. Connect the coils of Probs. 8 and 9 in parallel. What factors control the division of current? To compute the total inductance for the parallel arrangement, consult Art. 11.12.

12. Suggest an arrangement of circular coils that will give a practically uniform field over a considerable volume.

13. Calculate the force on 1 meter of wire, carrying 10 amperes, lying at 30° to a magnetic field of $B = 1$ weber per square meter.

14. Calculate the maximum field of is at distance $r = 20$, with $s \ll r$.

15. Calculate the maximum field at $r = 1$ meter for an electron with velocity $v = 0.1c$.

16. Calculate the force per meter on parallel wires in air, carrying currents $i_1 = 1000$ and $i_2 = 2000$ amperes, at distance $D = 0.1$ meter.

17. A proton, with velocity $v = 0.1c$, is "shot" normally in a field of $B = 1$. Calculate the orbital radius and the period of a revolution.

18. At an iron–air interface the flux lines approach the surface from the iron side at an angle of 80° with the normal. If the iron relative permeability is 1000, find the angle with the normal in air. Compute the density of poles on the surface if $B = 1$ in the air. Recompute on the assumption that the iron $\mu_r = \infty$.

19. Refer to Fig. 9.31. The wire, of 1-cm radius, carries 100 amperes. The iron pipe has internal and external radii of 3 and 4 cm. Calculate, per meter of length, the flux in the wire, in the pipe, and in the enclosed air space.

20. Refer to Fig. 9.33. For $I = 10$ amperes and iron relative permeability of 1000, calculate the flux density at radius $r = 10$ cm. Calculate the MMF in the air and in the iron.

21. Using the illustrative figures given near the end of Art. 9.40, calculate the force on each face of the small air gap illustrated in Fig. 9.34.

22. Compute the magnetic potential for points inside and outside a closed spherical shell, uniformly and normally magnetized. Hint: Put a hole and a boundary line in the shell, compute the potential as the hole is shrunk to zero size.

23. Calculate the mutual energy of two dipoles with moments m_1 and m_2 separated distance r.

CHAPTER 10

MAGNETIC FIELDS OF SIMPLE GEOMETRIES

This chapter discusses magnetic fields produced by steady currents flowing in wires arranged in simple geometrical patterns, on the assumption that the permeability of each magnetic material is constant.

10.1. Field of a Circular Coil of *n* Turns, Closely Wrapped. The field of a closely wrapped coil of n turns, each carrying current i, is shown in Fig. 10.1. The lines of flux are closed and surround the current, and the magnetic equipotential surfaces are open surfaces ending on the wire carrying the current. In this case, the field is symmetrical about the axis of the coil, as well as about the plane containing the turns. The portion of the plane of the coil lying outside the coil is at zero magnetic potential because each point in it forms zero solid angle with the wire boundary. This surface is also at potential $\pm ni$. The remainder of the plane lying inside the coil is at a potential of $\pm ni/2$ because any point in this region forms a solid angle of 2π steradians with the wire boundary.

The magnetic field intensity can be computed by simple means only along the axis, and for this purpose two methods are employed:

(1) *Intensity as the Gradient of the Potential.* Equation 9.56 gives the magnetic potential at any point as

$$U_P = \frac{ni}{4\pi}\,\Omega \quad \text{amperes} \tag{10.1}$$

The axial potential (see Fig. 10.2) is readily computed, as the solid angle Ω formed by a right circular cone is $2\pi(1 - \cos\alpha)$, where α is the plane semi-angle of the cone. (See equation 1.52.)

$$\begin{aligned} U_P &= \tfrac{1}{2}(1 - \cos\alpha)ni \\ &= \frac{1}{2}\left(1 - \frac{r}{\sqrt{R^2 + r^2}}\right) ni \quad \text{amperes} \end{aligned} \tag{10.2}$$

The axis is everywhere normal to the equipotential surfaces. The space derivative along the axis (variable r) gives the maximum space rate of change of potential, or the gradient, which is the negative of the magnetic field intensity H.

$$H_{\text{axis}} = -\frac{dU_P}{dr}$$

$$= \frac{1}{2}\frac{R^2 ni}{(R^2 + r^2)^{3/2}} \quad \frac{\text{newtons}}{\text{weber}} \text{ or } \frac{\text{amperes}}{\text{meter}} \tag{10.3}$$

Curve H in Fig. 10.1 shows the variation of the axial intensity. This curve should be compared with the electric intensity of the ring of charge shown in Fig. 2.3.

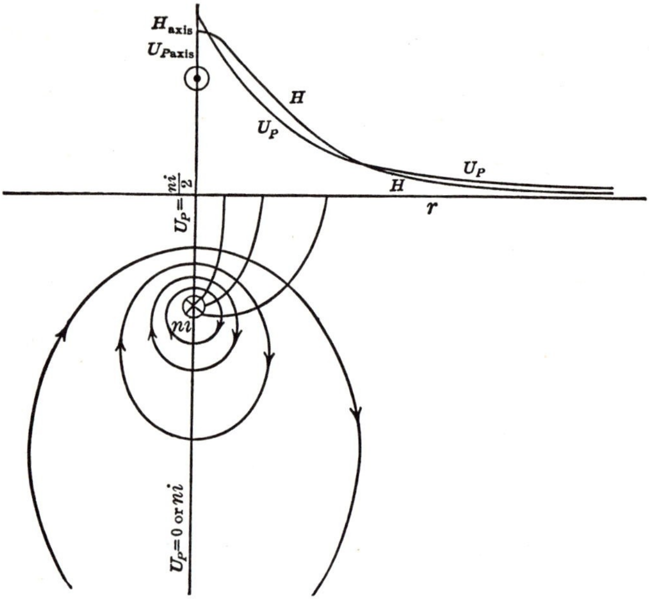

FIG. 10.1. Circular coil—axial intensity and potential.

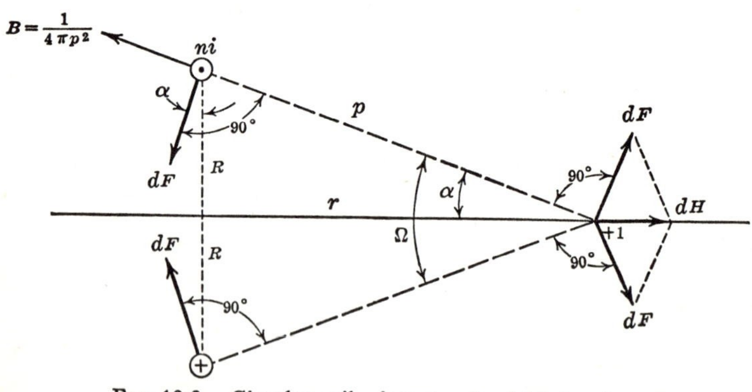

FIG. 10.2. Circular coil—forces and axial intensity.

(2) *Intensity by Use of the Force Law.* The axial intensity is readily computed by use of the equation for the force on a conductor in a magnetic field (equation 9.93). If a unit north pole is placed at the point P its radial field has a flux density at the wire of $B = \frac{1}{4\pi}\frac{1}{p^2} = \frac{1}{4\pi}\frac{1}{R^2 + r^2}$, which is normal to the wire.

The force on a small element ds of wire, from equation 9.93, is

$$dF = Bni\,ds$$

$$= \frac{1}{4\pi}\frac{ni\,ds}{R^2 + r^2}$$

normal to the wire and to the flux density. In summing up the differential forces upon the wire, it is obvious that we must consider only the components parallel to the axis, as the normal components cancel out. Hence

$$F = \frac{1}{4\pi}\int \frac{ni\,ds}{R^2 + r^2}\sin\alpha$$

$$= \frac{1}{2}\frac{R^2 ni}{(R^2 + r^2)^{3/2}} \quad \text{newtons}$$

as

$$\int ds = 2\pi R$$

This then is the force acting upon the wire directed to the left, and it is equal and opposite to the force upon the unit north pole at P. Hence the intensity at P is

$$H_{\text{axis}} = \frac{1}{2}\frac{R^2 ni}{(R^2 + r^2)^{3/2}} \tag{10.4}$$

At the center of the coil $r = 0$, and we get

$$H_{\text{axis center}} = \frac{1}{2}\frac{ni}{R} \quad \frac{\text{amperes}}{\text{meter}} \tag{10.5}$$

The difference of potential between any two axial points is

$$U_1 - U_2 = \int_{r_2}^{r_1} (-H)\,dr$$

$$= \int_{r_2}^{r_1} -\frac{1}{2}\frac{R^2 ni}{(R^2 + r^2)^{3/2}}\,dr = \frac{1}{2}\frac{-nir}{\sqrt{R^2 + r^2}}\Bigg]_{r_2}^{r_1}$$

$$= \frac{ni}{2}\left(\frac{r_2}{\sqrt{R^2 + r_2^2}} - \frac{r_1}{\sqrt{R^2 + r_1^2}}\right) \quad \frac{\text{joules}}{\text{weber}} \tag{10.6}$$

In particular, the difference of magnetic potential from ∞ to $-\infty$ is

$$\frac{ni}{2}(1 + 1) = ni \quad \frac{\text{joules}}{\text{weber}}$$

This is identically the value obtained in equation 9.30 as the work done in carrying unit pole once around the current ni.

With ordinary calculus methods we are not able to determine the intensity of field except on the axis. We have seen that lines of flux are closed loops surrounding the current, and that the magnetic equipotential surfaces start at the current and continue out into the field but are everywhere normal to the flux lines. We may, however, sketch the field qualitatively, as shown in Fig. 10.1.

(3) *Self-Inductance.* The self-inductance of a circular coil is calculable from the coil geometry only if the field strength is calculable. Since this is not possible by elementary methods, we give only the result without proof.

The inductance due to the flux inside the wire, for n closely wrapped turns, by equation 10.29 is

$$L_i \cong \frac{n\mu_W}{8\pi} 2\pi R = \frac{n\mu_W R}{4} \quad \text{henrys} \tag{10.7}$$

and that due to the external flux proves to be

$$L_o \cong n^2\mu_M R\left(\ln \frac{8R}{a} - 2\right) \tag{10.8}$$

giving a total value of

$$L \cong \frac{n\mu_W R}{4} + n^2\mu_M R\left(\ln \frac{8R}{a} - 2\right) \quad \text{henrys} \tag{10.9}$$

Here μ_W is the permeability of the wire, μ_M is the permeability of the external medium, R is the coil radius, and a is the equivalent radius of cross section of the group of n turns.

10.2. Field of a Cylindrical Coil of n Turns. Consider a cylindrical coil of length s and radius R, with n turns closely wrapped, and with the conductor cross-sectional area small compared with the area of the whole coil. Place a unit north pole at the point P where the intensity is sought, as shown in Fig. 10.3a. In order to add the effects of all the turns, we replace, for mathematical purposes, the actual coil by an idealized condition, namely, a cylindrical sheet of current ni in magnitude, and integrate the differential intensities of small elements of the cylinder.

Let us take a small cylinder of length dr and radius R distant $(b + r)$ from the origin taken at the coil center. Let b be the distance

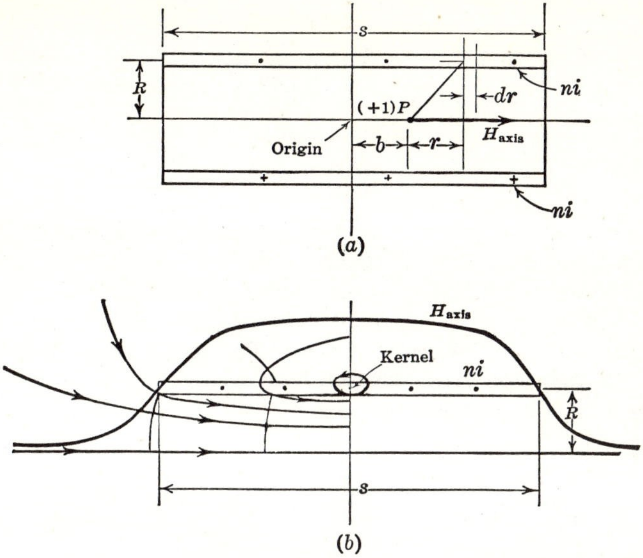

FIG. 10.3. Field of a cylindrical sheet of current.

from the origin to the point P, where the intensity is sought. From equation 10.3 the intensity is

$$dH_p = \frac{1}{2}\frac{R^2}{(R^2 + r^2)^{3/2}}\left(\frac{ni}{s}\,dr\right)$$

as $(ni\,dr/s)$ is the current in the elemental circular ring section.

$$H_p = \frac{ni}{2s}\int_{r=-\left(\frac{s}{2}+b\right)}^{r=\left(\frac{s}{2}-b\right)} \frac{R^2\,dr}{(R^2 + r^2)^{3/2}}$$

$$= \frac{ni}{2s}\frac{r}{\sqrt{R^2 + r^2}}\Bigg]_{-\left(\frac{s}{2}+b\right)}^{+\left(\frac{s}{2}-b\right)}$$

$$H_{axis} = \frac{ni}{2s}\left[\frac{\left(\frac{s}{2} - b\right)}{\sqrt{R^2 + \left(\frac{s}{2} - b\right)^2}} + \frac{\left(\frac{s}{2} + b\right)}{\sqrt{R^2 + \left(\frac{s}{2} + b\right)^2}}\right] \tag{10.10}$$

Note that, if we consider the intensity at the center ($b = 0$), equation 10.10 reduces to

$$H_{\text{axis center}} = \frac{ni}{\sqrt{4R^2 + s^2}} \quad \frac{\text{amperes}}{\text{meter}} \tag{10.11}$$

and further, if $s = 0$, it reduces to

$$H = \frac{ni}{2R} \tag{10.12}$$

which is the same expression as equation 10.5 for the intensity at the center on the axis of a closely wrapped coil.

The field shape is quite complicated in the immediate neighborhood of the wires of an actual solenoid, but for the equivalent cylindrical current sheet it is relatively simple. The axial value is shown in Fig. 10.3*b*. Except in the region very close to the wires the two fields are practically identical.

The potential on the axis is

$$U_P = \frac{ni}{2s}\left[\sqrt{R^2 + \left(\frac{s}{2} - b\right)^2} - \sqrt{R^2 + \left(\frac{s}{2} + b\right)^2} + s\right] \tag{10.13}$$

10.3. Long Solenoid. For the case of a very long solenoid ($s \gg R$), equation 10.10 reduces to

$$H = \frac{ni}{s} \quad \frac{\text{amperes}}{\text{meter}} \tag{10.14}$$

The flux lines are parallel to the axis, except near the ends of the solenoid, so that the result is a uniform field. Consequently equation 10.14 is very nearly true throughout the internal volume of the solenoid. Outside the long solenoid the reluctance is very low compared with the internal reluctance, and the flux density is low. If a unit north pole were carried around a closed flux line path, most of the magnetomotive force ni would be found in the internal part of the field.

The inductance of the long solenoid is readily computed by considering only the internal flux, on the assumption that it forms a uniform field. If the coil cross-sectional area is A, the flux through the coil is

$$\phi = BA = \mu HA = \frac{ni\mu A}{s} \tag{10.15}$$

and the inductance is

$$L = \frac{\lambda}{i} = \frac{n\phi}{i} = \frac{n^2\mu A}{s} \quad \text{henrys} \tag{10.16}$$

10.4. Field of the Closed Coil. The closed coil may be thought of as a long solenoid with its ends bent around until they close. In this case the magnetic field is completely confined inside the coil, as shown in Figs. 10.4 and 9.34. The intensity of the field is given by equation 10.14, where s is the circumferential coil length measured along the flux lines. The inner lines with shorter lengths lie in a region of higher intensity than the outer lines. However, if $s \gg R$, the field possesses practically uniform density throughout and s is taken as the average flux line length. The inductance of the closed coil is given by equations 10.16, 9.114, and 9.115.

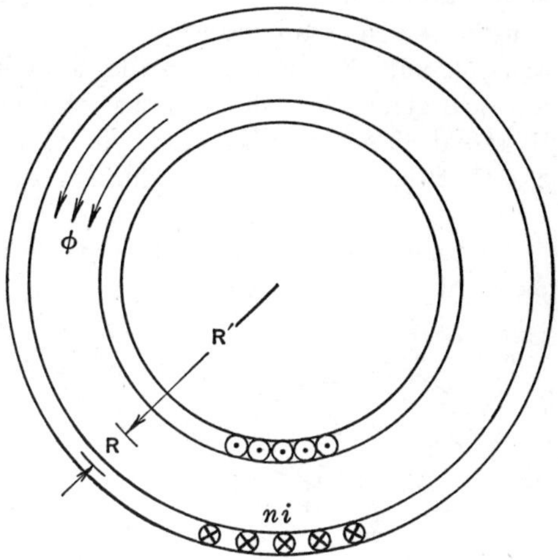

FIG. 10.4. Field of a closed solenoid. (Rowland's ring)

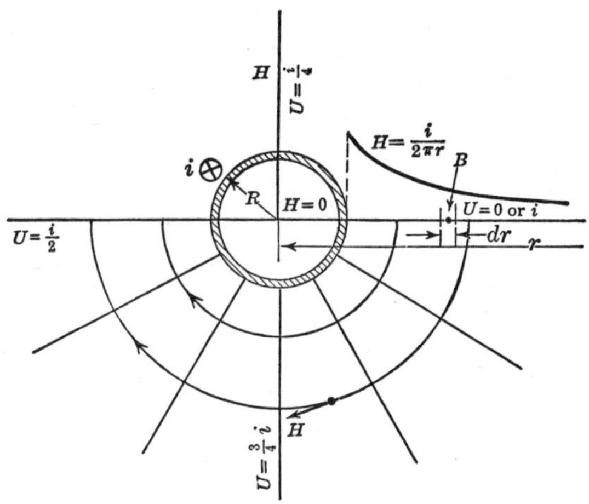

FIG. 10.5. Cylindrical tube of current.

10.5. Field of an Infinitely Long Circular Tube. Consider an infinitely long straight tube of current of circular cross section of radius R, as shown in Fig. 10.5, with the current flowing parallel to the axis and uniformly distributed. It is assumed that the current is steady and is flowing in a tube of negligible thickness. There must, of course, be a return conductor, which is, however, assumed to be at a distance sufficiently large so that its effect may be neglected. Geometrical symmetry demands that the magnetic equipotential surfaces be radial planes running outward from the outer surface of the tube, and that the flux lines lie in cylinders concentric about the center of the tube.

The magnetic field intensity outside the tube is readily computed. Remembering that the work per unit pole in moving over any closed path enclosing the current is i, and, choosing for simplicity a concentric circular path, we have

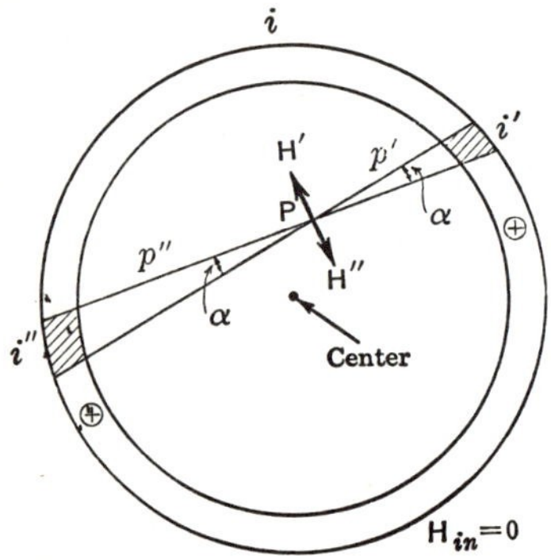

FIG. 10.6. Cylindrical tube of current. Internal intensity is zero.

$$\mathfrak{F}_{\text{closed path}} = i = H \cdot 2\pi r \quad \text{amperes}$$

whence the intensity, which is a circumferential vector, is

$$H_{\text{outside}} = \frac{i}{2\pi r} \quad \frac{\text{amperes}}{\text{meter}} \tag{10.17}$$

The intensity inside the tube is zero. As shown in Fig. 10.6, pass two planes parallel to the axis through any internal point P, separated by a small angle α. The magnitudes of the currents in the two filaments are to each other as the ratio of the distances from P.

$$\frac{i''}{i'} = \frac{p''}{p'} \quad \text{or} \quad \frac{i''}{p''} = \frac{i'}{p'}$$

Furthermore, the magnetic intensity effects of the filaments at P have the ratio unity.

$$\frac{H''}{H'} = \frac{\left(\dfrac{i''}{2\pi p''}\right)}{\left(\dfrac{i'}{2\pi p'}\right)} = 1 \tag{10.18}$$

These two filaments then create no field at any internal point, nor does the tube as a whole, as it is composed of filament pairs.

10.6. Field of a Cylindrical Wire Carrying a Current of Uniform Density—Field of a Tube in the Current Region. An infinitely long straight wire of circular cross section carrying a current i of uniform density has the same external field as a cylindrical tube of current. See Figs. 10.7 and 9.3.

$$H_{\text{outside}} = \frac{i}{2\pi r} \quad \frac{\text{amperes}}{\text{meter}} \tag{10.19}$$

This relationship is called the law of Biot and Savart.

Inside the wire there is a magnetic field that is readily computed. The flux lines lie in concentric cylinders. At any internal point P, a unit north pole is influenced by only those tubes of current that lie inside the radius r of the point, namely, by the current

$$i' = i\frac{r^2}{R^2}$$

The intensity at the point P is then

$$H_{\text{inside}} = \frac{i'}{2\pi r} = \frac{\left(i\dfrac{r^2}{R^2}\right)}{2\pi r} = \frac{i}{2\pi R^2}r \quad \frac{\text{amperes}}{\text{meter}} \tag{10.20}$$

and is thus seen to be linearly related to the radius. This is the magnetic situation for the case of a steady (direct but not pulsating) current.

For a tube of current, such as the one shown in Fig. 10.6, but with a wall thickness of appreciable value, the field intensity H varies from zero at the inner tube radius r_1 to the value $i/(2\pi r_2)$ at the outer radius r_2. In between we may divide the tube into an infinite number of infinitesimally thin cylindrical shells. At any point in this region, the outer shells contribute nothing to the field. Therefore we use the $i'/(2\pi r)$ law on the inner shells, where i' is that portion of the current lying between r_1 and r. Thus

$$H = \frac{i'}{2\pi r} = \frac{i}{2\pi r}\frac{(r^2 - r_1^2)}{(r_2^2 - r_1^2)} \tag{10.21}$$

for a current i of uniform density.

10.7. Skin Effect—Proximity Effect. The current density is not uniform but is larger in the outer tubes when the wire is carrying an alternating current. This phenomenon is called the skin effect and arises from the fact that the inner tubes of current have more flux lines surrounding them and hence more inductance than the outer tubes. With an alternating (or varying) current flowing, larger voltages are induced in the inner tubes, giving smaller net voltages and smaller current densities in the inner regions. In this case the magnetic intensity follows some such curve as shown by the broken line in Fig. 10.7.

There are two important consequences of skin effect. In the first place the alternating-current resistance[1] of the conductor is higher than the resistance to direct current. With the current crowded toward the outer layers, the inner core of the conductor is used to a smaller extent and the i^2R_{ac} loss is higher than the i^2R_{dc} loss for a given current value. The ratio of the a-c to the d-c resistance is a function of the wire radius, conductivity, and frequency, according

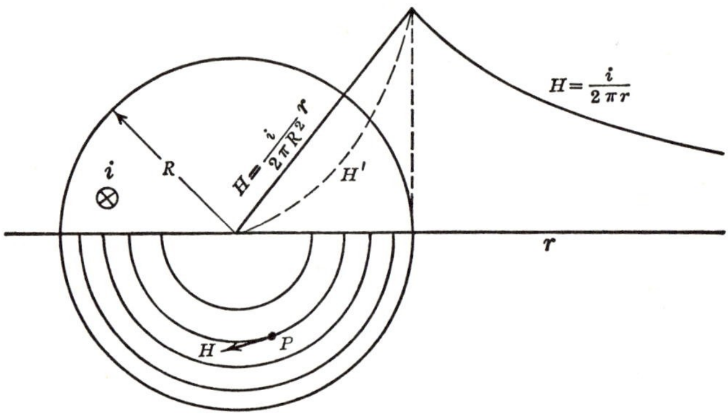

FIG. 10.7. Magnetic intensity about a solid wire.

to a formula developed by Lord Kelvin and evaluated in the National Bureau of Standards Scientific Paper No. 169. This quantity is plotted in Fig. 10.8 against the parameter kR, where R is the wire radius, $k = \frac{1}{2}\sqrt{\pi\mu\sigma f}$, and the conductivity is σ. Stranding the wire reduces the skin effect.

In the second place, the *internal* inductance of the wire, due to the internal flux only, is necessarily decreased as the current is crowded toward the outer layers of the conductor. Figure 10.8 shows also the ratio of the a-c to the d-c internal inductance, according to Kelvin's formula. The d-c value is given in equation 10.29.

Skin effect is the term applied to the redistribution of alternating current in a single conductor. With two or more conductors placed near together the alternating flux of each conductor's current influences the current distribution in the others. This influence is called the proximity effect, and is really the skin effect of several conductors. The analysis is decidedly complicated, but in general we can say that

[1] Woodruff, *Principles of Electric Power Transmission*, Second Edition, John Wiley & Sons, 1938, p. 53.

in a transmission wire pair, carrying oppositely directed currents, the current in each wire is crowded toward the adjacent wire. With sufficient crowding of current a concentration of heating of considerable value can be obtained.[1]

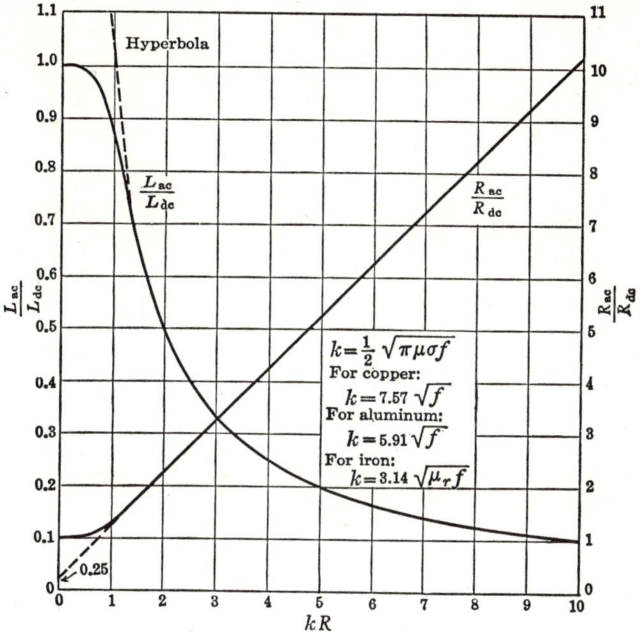

FIG. 10.8. Skin effect. Alternating resistance and internal inductance ratios for solid round wire.

10.8. Field of Two Infinitely Long Parallel Circular Wires Carrying Currents in Opposite Directions. The field of two infinitely long, straight, parallel wires of circular cross section carrying equal currents in opposite directions is shown in Fig. 10.9*a*. The circuit may be regarded as a very long rectangular coil. The magnetic equipotential surfaces are portions of circular cylinders running from wire to wire surface, which would, if extended, pass through the centers of the wires. The magnetic flux lines lie in eccentric cylindrical surfaces. The field is two dimensional.

Following Fig. 10.9*b* we may readily compute the magnetic potential U_P for points outside the wires along a line P, distant r_1 and r_2

[1] Bennett, "Concentration of Heating Currents," *Electrical Engineering*, **51,** 561 (August 1932).

respectively from the two wires, forming a solid angle β (or plane angle δ) with the wire centers.

$$\underset{\text{plane angle}}{2\delta \text{ radians}} = \underset{\text{solid angle}}{\beta \text{ steradians}} = \text{projection of 1 meter of line upon a cylinder of unit radius about } P$$

$$U_P = \frac{i\beta}{4\pi} = \frac{i\delta}{2\pi} \quad \text{amperes} \tag{10.22}$$

The plane through the wire centers is at zero potential outside the wire loop and is $\pm i/2$ inside the loop, as the maximum two-dimen-

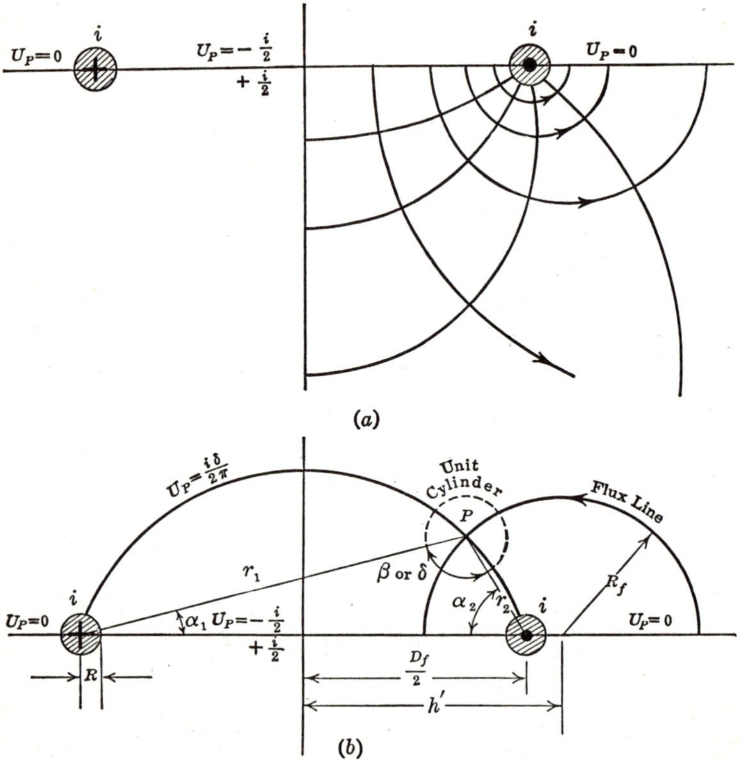

FIG. 10.9. Magnetic field of two parallel wires.

sional solid angle is 2π steradians. Intermediate values of potential lie on circular cylinders, which would pass through the centers of the wires if extended inside the wires. For any equipotential surface, β or δ is a constant, which makes $\alpha_1 + \alpha_2$ also a constant. Under this condition the locus of P, having an equipotential value, is a circle.

The set of surfaces orthogonal to a set of circular cylinders passing through two parallel straight lines is an eccentric system of circular cylinders having their centers in the plane of the two lines. This was shown in Art. 2.12 in developing the electric field of two parallel lines of charge of opposite sign. In order, then, to locate the eccentric cylinders containing the magnetic flux lines, we need only to replace the symbols given in equation 2.45 by their equivalent symbols in Fig. 10.9*b*.

Replace

h in equation 2.45 by $\frac{D_f}{2}$, one-half the distance between wire centers

$\frac{D}{2}$ in equation 2.45 by h', the distance from the center of the figure to the flux line center

R in equation 2.45 by R_f, the radius of the flux line

Then

$$\frac{D_f}{2} = \sqrt{(h')^2 - R_f^2} \tag{10.23}$$

10.9. Two Parallel Wires. Comparison of Electric and Magnetic Fields. The upper part of Fig. 10.10 shows the electric field about two circular wires of large radii compared with their separation, which are charged equally with electricity of opposite signs. The electric field is confined to the region outside the wires, but it is the same as though it did in fact center about the centers of influence which are closer together than the wire centers.

$$h = \sqrt{\left(\frac{D}{2}\right)^2 - R^2} \tag{10.24}$$

The lower part of the figure shows the magnetic field of two equal and opposite currents drawn on the assumption that the current density is uniform. Outside the wires, the field is composed of circular flux lines and equipotentials which have the same distribution in space as that of two small wires located at the large wire centers. The field inside the wires is obtained with the aid of field mapping methods given in Chapter 15. These methods require, first, the location of the kernel or line of zero magnetic intensity; second, the division of the current-carrying area into equal area subdivisions by lines which meet the equipotential surfaces outside the wire without refraction. The

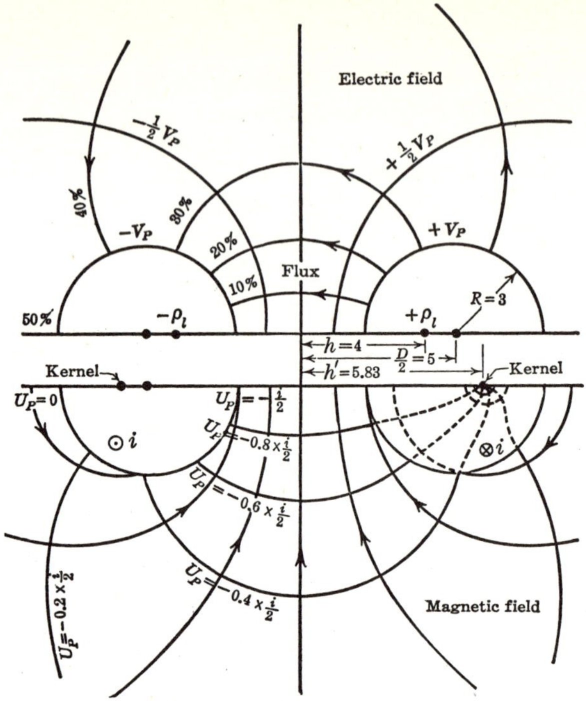

FIG. 10.10. Electric and magnetic fields of two large parallel wires. (Compare with Fig. 15.13.)

kernel lies in the plane of wire centers, distant r_1 and r_2 from the two centers, and its position is found as follows:

$$H = 0 = \frac{i}{2\pi r_1} - \frac{i}{2\pi R^2} r_2$$

$$r_1 = r_2 + D$$

which gives

$$r_2 = -\frac{D}{2} + \sqrt{\left(\frac{D}{2}\right)^2 + R^2} \tag{10.25}$$

With oppositely directed currents, the kernels lie outside the wire centers a distance r_2.

10.10. Force Action between Two Parallel Wires. Figure 10.11 shows the two separate fields of two parallel wires carrying equal currents in opposite directions. Let us obtain the force upon wire C due to the current in C and the field of A, where the wire centers are D

meters apart. If we assume that the wire radius is small compared with the distance between them, the flux density at C due to A is $B_A = i\mu_M/(2\pi D)$, directed upward as shown. $Bis \sin \theta$ is the force action

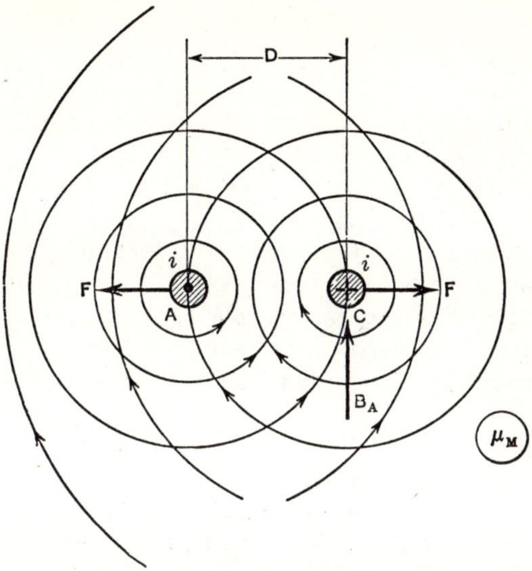

FIG. 10.11. Forces on parallel wires.

on C, with $\sin \theta = 1$, since the flux density is normal to the current in C. Then the force per unit length of wire is

$$F = \frac{i^2\mu_M}{2\pi D} \; \frac{\text{newtons}}{\text{meter of length}} \qquad \text{for } R \ll D \tag{10.26}$$

where F = force per unit length in newtons.
i = current per wire in amperes.
D = center distance between wires in meters.
μ_M = permeability of the *medium* around the wires.

The direction of the force, being normal to the wire and to the flux density, is in the plane of the wire centers, and normal to the wire. Hence on wire C it is toward the right, and on wire A the force is equal but toward the left. Oppositely directed currents then tend to force the wires apart. Similarly, two currents in the same direction tend to draw the wires together.

10.11. Inductance of Two Parallel Infinitely Long Straight Cylindrical Wires, Carrying Opposite and Uniformly Distributed Currents. It is convenient to divide this problem into three parts, finding separately the partial self-inductances due to the flux outside and inside the wires, and the mutual inductance between the wires, and adding the results. Let us consider only s meters of the infinite wires. See Fig. 10.12a. Let

L_{Ao} = partial self-inductance of A due to its own flux outside the wire.
L_{Ai} = partial self-inductance of A due to its own flux inside the wire.
M_{CA} = mutual inductance of wire C with respect to A.
μ_W = permeability of the wire.
μ_M = permeability of the surrounding medium.
R = radius of the wire in meters.
s = length of wire in meters (portion of the ∞ length).
D = distance between wire centers in meters.
i = current in the wire in amperes.
R is assumed to be small compared with D.

The definition of inductance is $L = n\phi/i$. Let us apply this relation to the problem of the external flux. The intensity H_A, distant r from the A wire center, due to the current in A alone is $H_A = i/(2\pi r)$, or $B_A = i\mu_M/(2\pi r)$. Then the elemental flux in the shaded area in Fig. 10.12a is $d\phi = B_A s\, dr$.

The total external flux in a length s due to the current in A is

$$\phi_A = \int B_A s\, dr = \int_{r=R}^{r=\infty} \frac{i}{2\pi r}\, \mu_M s\, dr = \frac{i}{2\pi}\, \mu_M s \ln r \Big]_{r=R}^{r=\infty}$$

Since there is but one turn, n equals 1. Then the partial inductance of one wire due to the external flux of that wire alone for a length s is

$$L_{AO} = \frac{n\phi}{i} = \frac{1}{2\pi}\, \mu_M s \ln r \Big]_{r=R}^{r=\infty} \tag{10.27}$$

In the foregoing development it was easy to use the definition of inductance as the flux linkages per unit of current because all the external flux links all the current. When the flux inside the wire is considered it is difficult to determine the number of flux linkages, so we resort to the energy definition $L_{Ai} = 2W/i^2$, where W equals the energy stored inside one wire. From equation 10.20, the intensity of field inside a wire due to a *uniform current distribution* is $H = ir/(2\pi R^2)$, and the flux density is $B = i\mu_W r/(2\pi R^2)$. But from equation 9.91, the energy density is $B^2/(2\mu_W)$. Then the energy stored in a hollow cylinder s meters long, of r radius, and dr thick is (Fig. 10.12b)

$$dW = \frac{B^2}{2\mu_W} \cdot 2\pi rs\,dr = \frac{\left(\dfrac{i\mu_W}{2\pi R^2}\right)^2}{2\mu_W} r^2 \cdot 2\pi rs\,dr$$

$$dW = \frac{i^2\mu_W}{4\pi R^4} sr^3\,dr$$

$$W = \frac{i^2\mu_W s}{4\pi R^4}\int_{r=0}^{r=R} r^3\,dr = \frac{i^2\mu_W s}{16\pi} \quad \text{joules} \tag{10.28}$$

and

$$L_{Ai} = \frac{2W}{i^2} = \frac{\mu_W s}{8\pi} \quad \text{henrys} \tag{10.29}$$

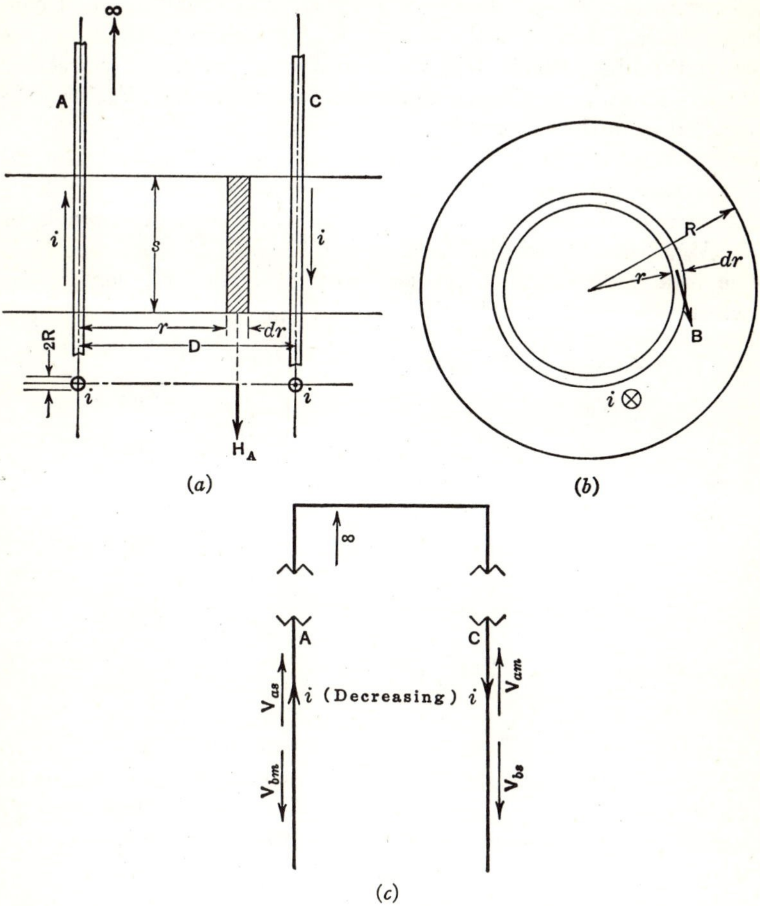

FIG. 10.12. Inductance of two parallel wires.

The mutual inductance of wire C with respect to A is the number of flux linkages around C due to unit current in A (see equation 9.71), but this yields the same value as equation 10.27 with its limits changed.

$$M_{CA} = \frac{1}{2\pi} \mu_M s \ln r \Big]_{r=D}^{r=\infty} = M_{AC} \tag{10.30}$$

To combine all these partial inductances properly, let us consider the induced voltages due to a decreasing current as shown in Fig. 10.12c. (See Chapter 11.) A voltage is induced in A due both to its external and to its internal flux and tends to maintain the current. Consequently these two effects are additive in wire A. Call this sum V_{as}, upward in wire A. The same decreasing external flux of A contributing to V_{as} induces in C a voltage V_{am} upward in wire C by virtue of the mutual inductance M_{CA}, but opposite in direction with respect to the circuit AC as a whole. Hence this mutual voltage V_{am} must be subtracted from the voltage V_{as}.

But because the current is decreasing in both A and C simultaneously, there is a second voltage $V_{bs} - V_{bm}$, which has the same direction and magnitude with respect to the circuit as a whole as $V_{as} - V_{am}$. Hence the total voltage is $2(V_{as} - V_{am})$.

The total inductance of the two parallel wires for a length s is therefore

$$L = 2(L_{Ai} + L_{AO} - M_{CA})$$

$$L = 2\left\{\frac{\mu_W s}{8\pi} + \left[\frac{1}{2\pi} \mu_M s \ln r\right]_{r=R}^{r=\infty} - \left[\frac{1}{2\pi} \mu_M s \ln r\right]_{r=D}^{r=\infty}\right\}$$

$$L = \frac{\mu_W s}{4\pi} + \frac{1}{\pi} \mu_M s \ln r \Big]_{r=R}^{r=\infty} + \frac{1}{\pi} \mu_M s \ln r \Big]_{r=\infty}^{r=D}$$

$$L = \frac{\mu_W s}{4\pi} + \frac{1}{\pi} \mu_M s \ln r \Big]_{r=R}^{r=D}$$

$$L = \frac{\mu_W s}{4\pi} + \frac{1}{\pi} \mu_M s \ln \left(\frac{D}{R}\right) \text{ henrys} \tag{10.31}$$

for $R \ll D$

10.12. Field of a Straight Wire of Finite Length. We have seen that a steady current flows only in a closed conductor. The intensity of field at any point P must therefore be caused by the entire current. It is, however, interesting and important to discuss the intensity caused by a portion of the current.

Consider a conducting path *AECD*, as shown in Fig. 10.13, carrying a current of i amperes. Let us study the force upon section *A* only, due to the unit north pole at *P*. The force on wire *A* is equal and opposite to the force on the unit north pole caused by the current in *A*. But the latter force is merely the intensity at *P* due to the current in *A*, which is the result we are seeking.

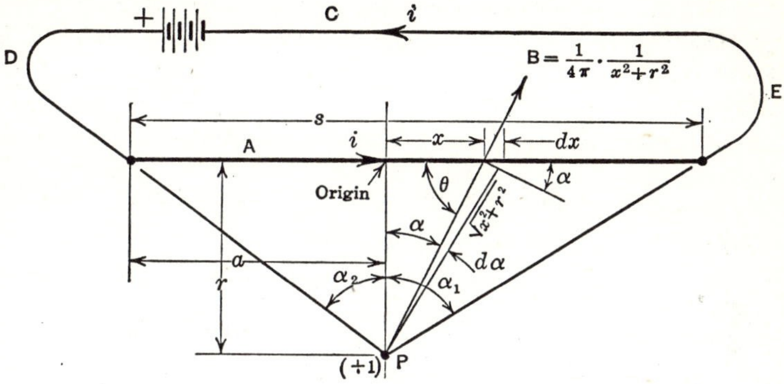

FIG. 10.13. Intensity of a straight current section.

Let the origin be taken on the wire at the foot of the perpendicular from P, a be the distance from one end of the wire A to the origin, r the perpendicular distance from P to the wire, s the length of wire A, x the distance from the origin to a small length of wire dx on which we will compute the force dF, and θ the angle between the pole flux density and current.

$$dF = Bi\,dx \sin\theta$$

but

$$B = \frac{1}{4\pi}\cdot\frac{1}{x^2+r^2} = \frac{1}{4\pi\left(\dfrac{r^2}{\cos^2\alpha}\right)} \qquad \text{from unit pole}$$

$$\sin\theta = \frac{r}{\sqrt{x^2+r^2}} = \cos\alpha$$

$$dx = \frac{\sqrt{(x^2+r^2)}\,d\alpha}{\cos\alpha} = \frac{r\,d\alpha}{\cos^2\alpha}$$

$$dF = \frac{1}{4\pi}\frac{\cos^2\alpha}{r^2}\,i\,\frac{r\,d\alpha}{\cos^2\alpha}\cos\alpha = \frac{1}{4\pi}\frac{i}{r}\cos\alpha\,d\alpha$$

and

$$F = \frac{i}{4\pi r}\int_{\alpha=\alpha_2}^{\alpha=\alpha_1}\cos\alpha\,d\alpha = \frac{i}{4\pi r}(\sin\alpha_1 \pm \sin\alpha_2)$$

which is the force on wire A upward perpendicular to the paper and is equal but opposite to the force on the unit pole, which is the intensity at P, downward. Then

$$H_p = \frac{i}{4\pi r}(\sin \alpha_1 \pm \sin \alpha_2) \quad \frac{\text{amperes}}{\text{meter}} \tag{10.32}$$

The + sign is chosen when the perpendicular from the point P to the wire meets the wire itself, the − sign when it meets the wire produced.

Since

$$\sin \alpha_1 = \frac{s - a}{\sqrt{(s - a)^2 + r^2}}$$

and

$$\sin \alpha_2 = \frac{a}{\sqrt{a^2 + r^2}}$$

$$H_p = \frac{i}{4\pi r}\left[\frac{s - a}{\sqrt{(s - a)^2 + r^2}} + \frac{a}{\sqrt{a^2 + r^2}}\right] \quad \frac{\text{amperes}}{\text{meter}} \tag{10.33}$$

If the perpendicular from P to the wire meets the wire produced, the value of a must be taken negative. These values hold only for points outside the wire radius.

Since the intensity at any point is perpendicular to a plane containing the wire and the point, we can see that a unit pole placed at a point acting under the force upon it would describe a circular path whose plane would be normal to the wire and including the point. Hence the lines of flux are circles with their centers at the wire center. The magnetic equipotential surfaces, everywhere normal to the flux, are coaxial planes centered at the wire.

As s is made larger until it approaches infinity, equation 10.33 reduces to

$$H_p = \frac{i}{2\pi r} \tag{10.34}$$

which is the equation for the intensity about an infinite straight line source. (See equation 9.7.)

10.13. Inductance of a Straight Finite Cylindrical Conductor. It is convenient to divide this problem into two parts, finding separately the partial self-inductances due to the flux inside and outside the wire, and adding the results. See Fig. 10.14. Let

L_i = partial self-inductance of the wire due to the flux inside the wire.

L_0 = partial self-inductance of the wire due to the flux outside the wire.
μ_W = permeability of the wire.
R = radius of the wire in meters.
s = length of the wire in meters.
i = current in the wire in amperes.

When considering the inductance due to flux inside the wire, we shall make the assumption that the radius of the wire is small com-

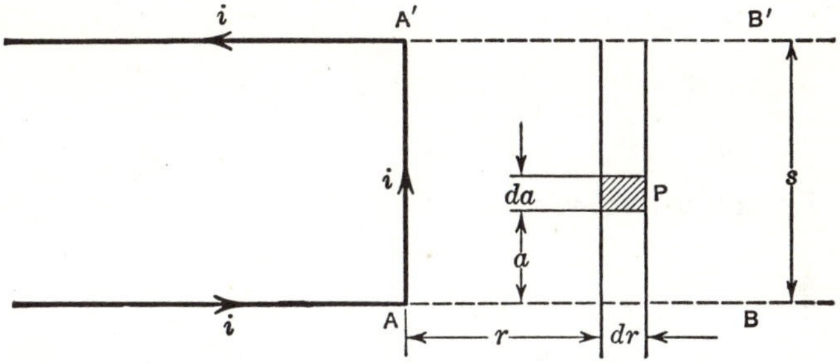

FIG. 10.14. Inductance of wire section AA'.

pared with the length of the wire. From equation 10.29, the partial inductance due to the internal flux is

$$L_i = \frac{2W}{i^2} = \frac{\mu_W s}{8\pi} \quad \text{henrys} \tag{10.35}$$

The partial self-inductance due to the flux external to the wire may be obtained readily by using the definition based on flux linkages per unit of current, equation 9.80. This equation and equation 10.33 give the magnetic flux density at point P due to a finite straight current section as

$$B_P = \frac{i}{4\pi r}\mu_M\left(\frac{s-a}{\sqrt{(s-a)^2+r^2}} + \frac{a}{\sqrt{a^2+r^2}}\right) \tag{10.36}$$

which is a vector at point P, downward normal to the plane including the wire section AA' and point P. Integrating the flux density over area $AA'B'B$, which is everywhere normal to the flux density, gives the total flux through that area. It should be noted that we are calculating, not all the flux, but only that part lying between the lines $A'B'$ and AB, since this is the flux which circles AA' and passes through the current loop.

Then $\phi_0 = \phi_{AA'B'B} = \int_{r=R}^{r=\infty} \int_{a=0}^{a=s} B_p \, dr \, da$

$$\phi_0 = \frac{i\mu_M}{4\pi} \int_{r=R}^{r=\infty} \frac{1}{r} dr \int_{a=0}^{a=s} \left(\frac{s-a}{\sqrt{(s-a)^2 + r^2}} + \frac{a}{\sqrt{a^2 + r^2}} \right) da$$

$$= \frac{i\mu_M}{4\pi} \int_{r=R}^{r=\infty} \frac{2}{r} \left(\sqrt{s^2 + r^2} - r \right) dr$$

$$= \frac{i\mu_M}{2\pi} \int_{r=R}^{r=\infty} \left(\frac{\sqrt{s^2 + r^2}}{r} - 1 \right) dr$$

$$\phi_0 = \frac{i\mu_M}{2\pi} \left[\sqrt{s^2 + r^2} - r - s \ln \frac{s + \sqrt{s^2 + r^2}}{r} \right]_{r=R}^{r=\infty}$$

This integral may be rewritten to advantage for the insertion of the limits:

$$\phi_0 = \frac{\mu_M i}{2\pi} \left[\frac{(\sqrt{s^2 + r^2} - r)(\sqrt{s^2 + r^2} + r)}{(\sqrt{s^2 + r^2} + r)} - s \ln \left(\frac{s}{r} + \sqrt{1 + \frac{s^2}{r^2}} \right) \left(\frac{\dfrac{s}{r} - \sqrt{\dfrac{s^2}{r^2} + 1}}{\dfrac{s}{r} - \sqrt{\dfrac{s^2}{r^2} + 1}} \right) \right]_{r=R}^{r=\infty}$$

$$= \frac{\mu_M i}{2\pi} \left[\frac{s^2}{\sqrt{s^2 + r^2} + r} - s \ln \left(\frac{-1}{\dfrac{s}{r} - \sqrt{\dfrac{s^2}{r^2} + 1}} \right) \right]_{r=R}^{r=\infty}$$

$$\phi_0 = \frac{\mu_M i}{2\pi} \left(s \ln \frac{s + \sqrt{s^2 + R^2}}{R} - \sqrt{s^2 + R^2} + R \right) \qquad \textbf{(10.37)}$$

and the partial self-inductance due to the external flux $L_0 = \phi_0/i$ becomes

$$L_0 = \frac{\mu_M}{2\pi} \left(s \ln \frac{s + \sqrt{s^2 + R^2}}{R} - \sqrt{s^2 + R^2} + R \right) \qquad (10.38)$$

Then the total self-inductance of the finite straight cylindrical wire is $L_s = L_i + L_0$.

$$L_s = \frac{\mu_M}{2\pi} \left(s \ln \frac{s + \sqrt{s^2 + R^2}}{R} - \sqrt{s^2 + R^2} + R \right) + \frac{\mu_W s}{8\pi} \qquad (10.39)$$

If R is very small compared with the wire length s, equation 10.39 reduces to

$$L_s = \frac{\mu_M s}{2\pi}\left(\ln\frac{2s}{R} - 1\right) + \frac{\mu_W s}{8\pi} \tag{10.40}$$

and further, if the wire is nonmagnetic, $\mu_W = \mu_0$, and

$$L_s = \frac{\mu_M s}{2\pi}\left(\ln\frac{2s}{R} - 1\right) + \frac{\mu_0 s}{8\pi} \tag{10.41}$$

10.14. Inductance of Two Parallel Finite Straight Cylindrical Conductors Carrying Equal Currents in Opposite Directions. Equation 10.37 gives the external flux about one wire out to infinity. Replacing the limits with $r = R$ to $r = D$, the distance between centers of the wires, gives the flux which is external to one wire and lies between the two wires. Both currents have fluxes which add together in the region between the wires. Therefore the total external flux which passes across the plane bounded by $AA'B'B$, in Fig. 10.15, is

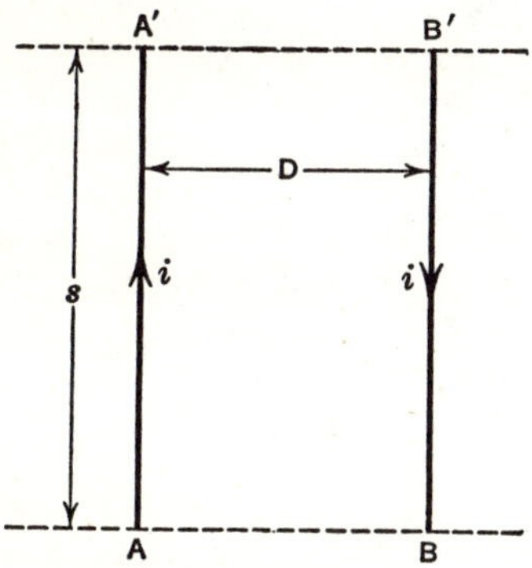

FIG. 10.15. Inductance of a short parallel wire pair.

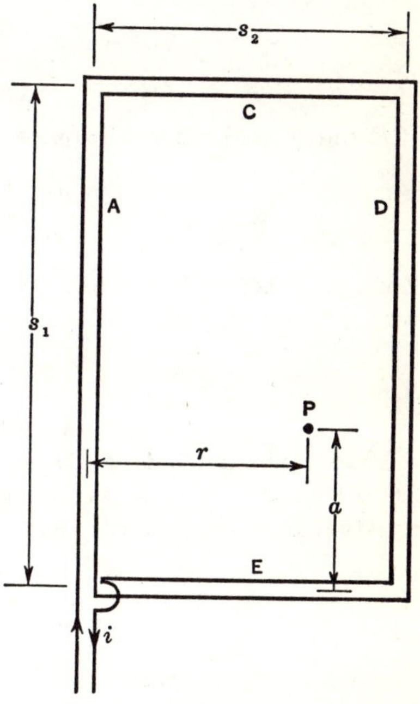

FIG. 10.16. Inductance of a rectangular coil.

$$\phi_0 = \frac{\mu_M i}{\pi}\left[\sqrt{s^2 + r^2} - r - s \ln\frac{s + \sqrt{s^2 + r^2}}{r}\right]_{r=R}^{r=D}$$

$$\phi_0 = \frac{\mu_M i}{\pi}\left[\sqrt{s^2 + D^2} - \sqrt{s^2 + R^2} + R - D + s \ln\frac{s + \sqrt{s^2 + R^2}}{R} - s \ln\frac{s + \sqrt{s^2 + D^2}}{D}\right] \tag{10.42}$$

This quantity gives all the flux external to the wires because all the flux lines which cross $AA'B'B$ curl around the wires and cross the produced plane of $AA'B'B$ extending out to infinity. Dividing equation 10.42 by the current gives the partial inductance of the two wires due to the external flux. To this quantity let us add twice the partial self-inductance of one wire alone (equation 10.43). This result gives the total inductance of the two-wire combination.

$$L = \frac{\mu_M}{\pi}\left(\sqrt{s^2 + D^2} - \sqrt{s^2 + R^2} + R - D + s \ln \frac{s + \sqrt{s^2 + R^2}}{R} - s \ln \frac{s + \sqrt{s^2 + D^2}}{D}\right) + \frac{\mu_W s}{4\pi} \quad (10.43)$$

If R and D are both small compared with s, equation 10.43 reduces to

$$L = \frac{\mu_M s}{\pi} \ln \frac{D}{R} + \frac{\mu_W s}{4\pi} \qquad \text{(See equation 10.31.)} \qquad (10.44)$$

Equation 10.44 holds for ordinary transmission lines. Further, if the wire is nonmagnetic, $\mu_W = \mu_0$, and

$$L = \frac{\mu_M s}{\pi} \ln \frac{D}{R} + \frac{\mu_0 s}{4\pi} \qquad (10.45)$$

10.15. Inductance of a Rectangle of n Turns Closely Wrapped. Consider two parallel wires of length s_1 as shown in Fig. 10.16. From equation 10.33 the intensity at P due to wire A alone is

$$H_{PA} = \frac{ni}{4\pi r}\left[\frac{s_1 - a}{\sqrt{(s_1 - a)^2 + r^2}} + \frac{a}{\sqrt{a^2 + r^2}}\right] \qquad (10.46)$$

The flux external to wire A contained in a length s_1 running from R, the radius of the equivalent wire group, out to infinity is

$$\phi_{OA} = \int_{r=R}^{r=\infty} dr \int_{a=0}^{a=s_1} B_{PA}\, da$$

Multiplying by n and dividing by i gives the partial self-inductance of wire A due to the flux external to the wire.

$$L_{AO} = \frac{n\phi_{AO}}{i} = \frac{n\mu_M}{i} \int_{r=R}^{r=\infty} dr \int_{a=0}^{a=s_1} H_{PA}\, da$$

$$= \frac{\mu_M n^2}{4\pi} \int \frac{dr}{r} \int \left[\frac{s_1 - a}{\sqrt{(s_1 - a)^2 + r^2}} + \frac{a}{\sqrt{a^2 + r^2}}\right] da$$

$$= \frac{\mu_M n^2}{4\pi} \int \frac{dr}{r} \left[-\sqrt{(s_1 - a)^2 + r^2} + \sqrt{a^2 + r^2} \right]_{a=0}^{a=s_1}$$

$$= \frac{\mu_M n^2}{4\pi} \int \frac{dr}{r} (-r + \sqrt{s_1^2 + r^2} + \sqrt{s_1^2 + r^2} - r)$$

$$L_{AO} = \frac{n^2 \mu_M}{2\pi} \int_{r=R}^{r=\infty} \left(\frac{\sqrt{s_1^2 + r^2}}{r} - 1 \right) dr \tag{10.47}$$

Similarly the mutual inductance of wire D with respect to wire A may be obtained by using the proper limits.

$$M_{DA} = \frac{n^2 \mu_M}{2\pi} \int_{r=s_2}^{r=\infty} \left(\frac{\sqrt{s_1^2 + r^2}}{r} - 1 \right) dr \tag{10.48}$$

If the length s_1 is large compared with the wire radius R, equation 10.35 for the partial self-inductance of a wire due to its internal flux holds.

$$L_{Ai} = \frac{\mu_W n s_1}{8\pi} \tag{10.49}$$

Combining these three inductances and doubling as in equation 10.42, the total inductance of wires D and A is

$$L_{AD} = 2(L_{Ai} + L_{AO} - M_{DA})$$

$$L_{AD} = \frac{\mu_W n s_1}{4\pi} + \frac{n^2 \mu_M}{\pi} \left[\int_{r=R}^{r=\infty} \left(\frac{\sqrt{s_1^2 + r^2}}{r} - 1 \right) dr - \int_{r=s_2}^{r=\infty} \left(\frac{\sqrt{s_1^2 + r^2}}{r} - 1 \right) dr \right]$$

$$L_{AD} = \frac{\mu_W n s_1}{4\pi} + \frac{n^2 \mu_M}{\pi} \int_{r=R}^{r=s_2} \left(\frac{\sqrt{s_1^2 + r^2}}{r} - 1 \right) dr$$

$$= \frac{\mu_W n s_1}{4\pi} + \frac{n^2 \mu_M}{\pi} \left[\sqrt{s_1^2 + r^2} - s_1 \ln \left(\frac{\sqrt{s_1^2 + r^2} + s_1}{r} \right) - r \right]_{r=R}^{r=s_2}$$

$$L_{AD} = \frac{\mu_W n s_1}{4\pi} + \frac{n^2 \mu_M}{\pi} \left[\sqrt{s_1^2 + s_2^2} - s_1 \ln \left(\frac{\sqrt{s_1^2 + s_2^2} + s_1}{s_2} \right) - s_2 - \sqrt{s_1^2 + R^2} + s_1 \ln \left(\frac{\sqrt{s_1^2 + R^2} + s_1}{R} \right) + R \right] \tag{10.50}$$

By interchanging s_1 and s_2 we obtain the total inductance of wires C and E.

$$L_{CE} = \frac{\mu_w n s_2}{4\pi} + \frac{n^2 \mu_M}{\pi}\left[\sqrt{s_1^2 + s_2^2} - s_2 \ln\left(\frac{\sqrt{s_1^2 + s_2^2} + s_2}{s_1}\right)\right.$$

$$\left. - s_1 - \sqrt{s_2^2 + R^2} + s_2 \ln\left(\frac{\sqrt{s_2^2 + R^2} + s_2}{R}\right) + R\right] \quad (10.51)$$

The total inductance of the rectangle is the sum

$$L = L_{AD} + L_{CE}$$

$$L = \frac{\mu_w n}{4\pi}(s_1 + s_2) + \frac{\mu_M n^2}{\pi}\left[2\sqrt{s_1^2 + s_2^2} - (s_1 + s_2) - \sqrt{s_1^2 + R^2}\right.$$

$$- \sqrt{s_2^2 + R^2} + 2R + s_1 \ln \frac{(\sqrt{s_1^2 + R^2} + s_1)s_2}{(\sqrt{s_1^2 + s_2^2} + s_1)R}$$

$$\left. + s_2 \ln \frac{(\sqrt{s_2^2 + R^2} + s_2)s_1}{(\sqrt{s_1^2 + s_2^2} + s_2)R}\right] \text{ henrys} \quad (10.52)$$

PROBLEMS

1. A circular coil of 5 turns closely wrapped, of radius 8 cm, carries a current of 4 amperes. Find the potential and potential gradient on the axis at a point 6 cm from the center. Find the solid angle subtended at the point. Repeat for the center of the coil. Compute the strength of the equivalent magnetic shell in air. Calculate the inductance if the wire radius is 0.1 cm.

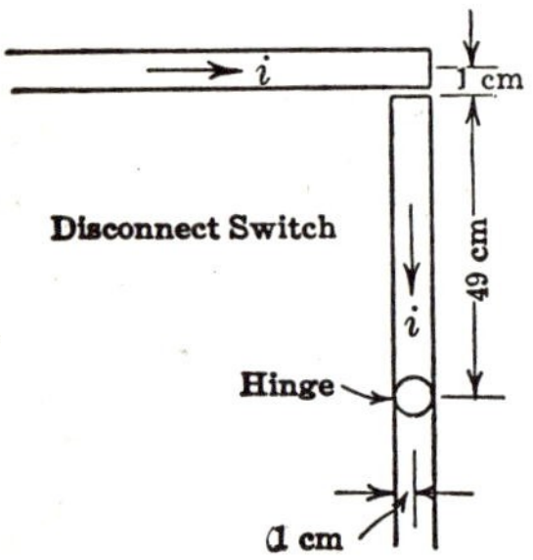

FIG. P6

2. Consider a solenoid of 1000 turns, carrying 1 ampere, 50 cm long and of 10-cm radius. Compute and draw the intensity along the axis. Now consider the coil to be infinitely long (also an infinite number of turns) and repeat. Compute the inductance on the assumption (in both cases) that the field is uniform across the center plane taken inside the coil and normal to the axis.

3. In the case of the infinitely long coil of Prob. 2, pass a wire carrying 5 amperes through the center of the coil and normal to the axis. Compute the force on the part of the wire inside the coil.

4. Compute the inductance of two thin concentric cylindrical wires of radii r_1 and r_2, considering only the flux between the tubes.

5. Two parallel wires, each of radius $R = 0.1$ cm, separated center distance $D = 10$ cm, carry oppositely directed currents of 2 amperes. Compute the force on each wire, and the inductance.

6. Consider the disconnect switch, shown in Fig. P6, which may carry as much as 30,000 amperes under short-circuit conditions. Compute the torque on the switch.

7. A wheel with conducting spokes of length R dips one spoke at a time into a mercury bath. A uniform field B runs parallel to the axle. Compute the torque on the wheel when one spoke carries current i.

8. Find the force on wire segment BC in Fig. P8 due to the infinite wire A.

9. A long, straight, vertical wire of radius 0.1 cm connects at one end by a flexible connection to a short straight section 10 cm long which makes 60° with the long section produced. The short section dips into a mercury bath. Compute the torque on the short section if the wires carry 3 amperes. Indicate its motion.

10. Consider two infinitely long, rigid, straight wires carrying currents, placed in parallel planes D centimeters apart. The projection of one wire upon the plane of the other is perpendicular to the other wire. The wires are free to move. Discuss the force action between them. What final position will they take, D remaining constant?

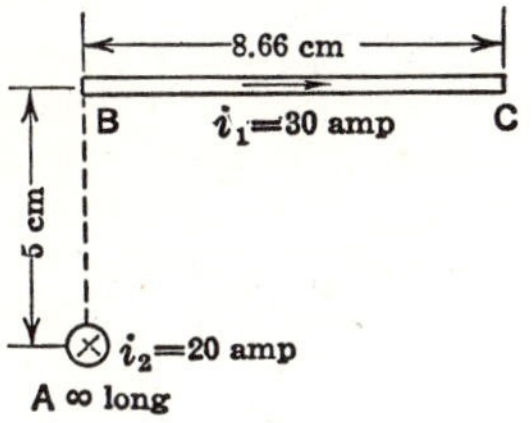

FIG. P8

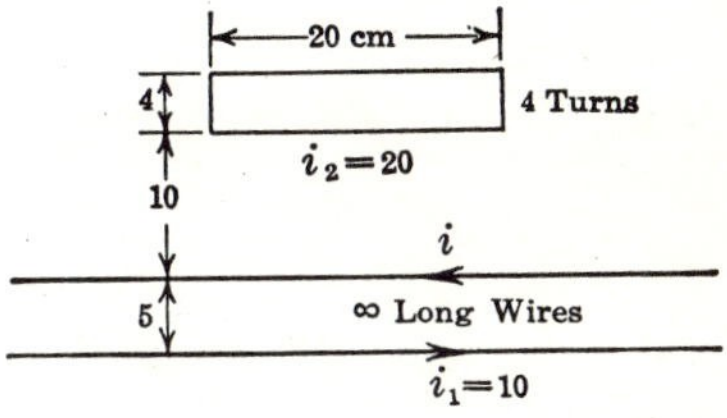

FIG. P13

11. Suppose the wires of Prob. 5 are 50 cm long. Compute the inductance of each wire, and the inductance of the pair.

12. Compute the inductance of the rectangle made by closing the ends of the wires of Prob. 11 with straight conductors.

13. Compute the mutual inductance of the rectangle with respect to the infinite line, shown in Fig. P13. Find the net force on the rectangle.

14. For the infinite concentric cylinders, shown in Fig. P14, carrying equal and opposite currents:

(*a*) Sketch the curve and derive the equation for the intensity.
(*b*) Sketch the curve and derive the equation for the flux density.
(*c*) Compute the inductance for s meters.

15. Compute the mutual inductance of two parallel lines, 1, 2 and 3, 4. Let d_{14} be the distance between the first and fourth wire, etc.

16. How would you build a high resistance with minimum inductance? What effect does capacitance have?

17. Compare the resistance and internal inductance ratios for a copper wire of radius 1 cm at frequencies of 100 and 10,000 cycles per second. Repeat for a copper wire of 0.1-cm radius at 100 and 1,000,000 cycles per second.

18. Two like circular coils of n turns and radius R are placed coaxially a distance S apart. S is large compared with R. Compute the mutual inductance.

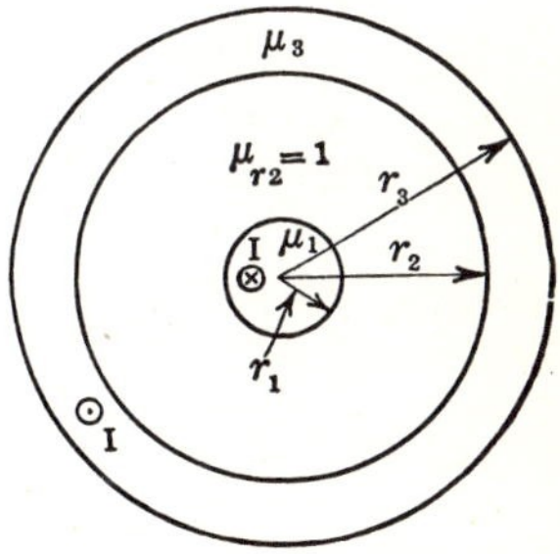

FIG. P14

19. A 4-meter diameter coil (No. 1) of 60 turns carries a current of 15 amperes. Find the magnetic flux density at the coil center. Place an 0.2-meter diameter coil

(No. 2) of 180 turns at the center of coil No. 1. Assuming that No. 1 sets up a uniform field in the neighborhood of coil No. 2, compute their mutual inductance. Coils lie in the same plane.

20. A solenoid of 200 turns is 20 cm long and 10 cm in diameter. At its center is a closely wrapped rectangular coil 6 cm square having 8 turns. The two coils have the same axis.

(*a*) Find the mutual inductance between these two coils, assuming that the flux density throughout the smaller one due to the larger one is uniform.

(*b*) If now the smaller one is turned so that it lies in a plane passing through the axis, find the torque tending to restore it to the coaxial position.

21. Consider a 1-turn square coil. The wire radius is 0.1 cm, the side of the square is 10 cm. If the current is 2 amperes, compute the potential and intensity at the center of the square. Compute the force on one side caused by the current in *one* of the adjacent sides.

22. Two round ($R = 1$) parallel copper bus bars, with center spacing $D = 20$ cm, carry currents in the same direction, $I_1 = 1000$, $I_2 = 2000$ amperes. Each bus is supported at 3-meter spans. Find the total force on each bus for one span.

23. An arc is struck straight across the ends of two parallel half infinite conductors, each of 1-cm radius and separated 10 cm, carrying 1000 amperes. Find the force on the arc.

24. A long straight copper wire, radius $R = 1$ cm, is surrounded concentrically with an iron pipe ($\mu_r = 1000$), radii $r_2 = 2$ and $r_3 = 2.5$ cm. Compute and plot the flux density for all values of r. For 1 meter of pipe length, compute the number of flux lines in the pipe if the current is 1000 amperes.

25. A current I of uniform density in a long straight wire of circular cross section, radius R, lies in its own internal magnetic field and is subject therefore to an inward radial force which tends to reduce the cross section. The phenomenon, known as the "pinch effect," can be seen readily in a mercury conductor. Prove that the expression for the mechanical "pressure" at any radius r is

$$p = \int_r^R BJ\,dr = \frac{\mu I^2}{4\pi^2 R^4}(R^2 - r^2) \quad \frac{\text{newtons}}{\text{square meter}}$$

26. Two coaxial circular turns, radii R_1 and R_2, separated distance S in air, have mutual inductance as follows:

If $S \gg R_1$ and R_2

$$M \cong \frac{\mu_0\pi}{2}\frac{(R_1R_2)^2}{(R_1^2 + S^2)^{3/2}}$$

If $S \ll R_1$ and R_2, and $R_1 \cong R_2$

$$M \cong \mu_0 R_1\left(\ln\frac{8R_1}{\sqrt{(R_1 - R_2)^2 + S^2}} - 2\right)$$

Compare results with those of Prob. 18.

CHAPTER 11

MAGNETIC PRODUCTION OF VOLTAGE

A voltage difference between any two points in an electric circuit is induced whenever the magnetic flux through the circuit is changing in time, or whenever a portion of the circuit is moved crosswise to the existing magnetic flux. Simultaneously an electric field intensity is created, which causes a current flow through the circuit provided that the circuit is composed of conducting material.

A. STATIONARY CONDUCTORS

11.1. Induction of Voltage in a Closed Circuit. Whenever the current in a wire changes in magnitude, there occurs also a change in the magnetic field strength and in the amount of energy stored in the field. Let us study the effect of a changing magnetic field upon a stationary conductor. See Fig. 11.1.

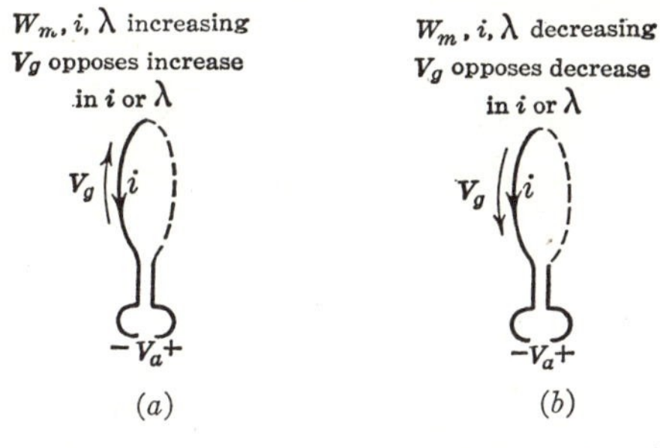

FIG. 11.1. Current-voltage relations in an inductive circuit. (a) Increasing current. (b) Decreasing current.

Consider a coil of conducting material of resistance R connected to a battery or any generator of electric potential difference V_a, constant or otherwise. A current i flows in the circuit, changing in time in a manner that is to be described. During a small time interval dt, $V_a i\, dt$ is the energy supplied to the circuit by the generator and $i^2R\, dt$ is the heat lost in the circuit resistance, leaving the difference between these two as the change in energy stored in the magnetic field dW_m. The last quantity may be expressed as the product of the current by the change of flux linkages $d\lambda$ occurring during the time interval dt.

Thus:

$$V_a i\,dt - i^2 R\,dt = dW_m = i\,d\lambda \tag{11.1}$$

$$V_a - iR = \frac{d\lambda}{dt}$$

or

$$i = \frac{V_a - \dfrac{d\lambda}{dt}}{R} = \frac{V_a + V_g}{R}$$

and

$$V_g = -\frac{d\lambda}{dt} = \frac{-\text{magnetic flux linkage change}}{\text{seconds}} = \text{volts} \tag{11.2}$$

This forces us to conclude that whenever the current or its accompanying flux linkages through a circuit change in time we shall find a voltage generated in the circuit equal in magnitude to the negative of the time rate of change of flux linkages.

When the current i is increasing, so also is λ. Then di/dt and $d\lambda/dt$ are positive and $V_g = -d\lambda/dt = -(d/dt)(Li) = -L(di/dt)$ is negative, where L is the inductance of the coil. This is shown in Fig. 11.1*a*. The converse [decreasing i and negative $(d\lambda/dt)$] is shown in Fig. 11.1*b*. Both cases are summed up in Lenz's law: The voltage generated in a circuit is equal in magnitude to the time rate of change of magnetic flux linkages and is so directed in the circuit as to oppose the change in the magnetic flux.

Rewriting equation 11.1,

$$V_a i - i^2 R = \frac{dW_m}{dt} = i\frac{d\lambda}{dt} = -iV_g \tag{11.3}$$

in which $V_a i$ = power supplied by the generator to the coil.
i^2R = power transformed into heat.
$-iV_g$ = power supplied to the magnetic field by the circuit.
$+iV_g$ = power supplied by the magnetic field to the circuit, as a consequence of the change in energy storage.

Equation 11.2 can be applied only to *closed circuits*. Except in a closed circuit it is difficult to give meaning to the phrase "time rate of change of flux linkages." This law of voltage induction is very general and may be used in a variety of situations. It is immaterial, for instance, whether the time rate of change of flux through the circuit is produced by:

(*a*) A changing current in the circuit itself.

(*b*) A changing current in an adjacent circuit, the flux of which links the first circuit.

(*c*) Change of shape (change of self-inductance) of the circuit when it is carrying a current.

(*d*) Change of position (change of mutual inductance) of the adjacent circuit when it is carrying a current.

(*e*) Movement of a permanent magnet in the neighborhood of the circuit.

In all these changes the essential requirement for voltage induction in a given closed circuit is simply that there must be a time rate of change of flux linking the circuit.

11.2. Voltage Generation in a Self-Inductance. The voltage induced in a closed circuit may be expressed in terms of the self-inductance of the circuit. For many purposes this method of expression is preferable to the one given in the preceding article, involving the time rate of change of flux linkages.

$$L = \frac{d\lambda}{di} \tag{11.4}$$

or

$$d\lambda = L\,di$$

but

$$-\frac{d\lambda}{dt} = +V_g = -L\frac{di}{dt} \tag{11.5}$$

From equation 11.5 it may be seen that the negative of the product of the self-inductance of a circuit and the time rate of change of current in the circuit gives the value of the voltage induced therein. The direction in the circuit is given by Lenz's law, stated in the preceding article. Rewriting equation 11.5 gives the inductance as

$$L = \frac{-V_g}{\left(\dfrac{di}{dt}\right)} = \frac{\text{volts}}{\left(\dfrac{\text{amperes}}{\text{second}}\right)} = \text{henrys} \tag{11.6}$$

Equation 11.6 may be used as a third definition of self-inductance to supplement the two other definitions given in Art. 9.24. The self-inductance of a circuit is the ratio of the negative of the induced voltage to the time rate of change of current in the circuit.

This method of defining the self-inductance is peculiarly useful in connection with the laboratory *measurement* of inductance; the definitions given in Art. 9.24 are better suited to its *calculation*.

11.3. Rise of Current in an Inductive Circuit. Every electric circuit has three properties: resistance, inductance, and capacitance.

In this development we shall consider the effect of capacitance to be negligibly small. A coil of wire is a circuit in which resistance and inductance predominate. As shown in Fig. 11.2*a*, let us consider a coil, of resistance R and inductance L, connected to a source of constant electric potential difference whose value is V_a.

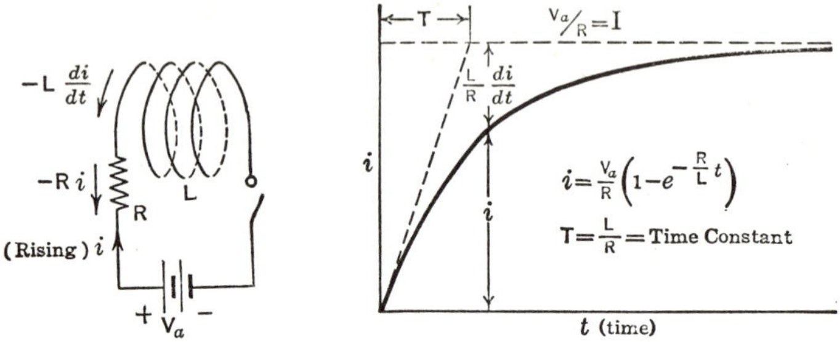

(*a*) (*b*)

FIG. 11.2. Current rise in an inductive curcuit.

On connecting the battery V_a, charges flow around the circuit giving current $+i$, which rises in value. In the resistance R there appears a back voltage $-Ri$ oppositely directed to the current flow. In the inductance L, with a rising current, there appears an induced voltage $-L(di/dt)$ which opposes the rise of current and hence is directed against the current flow. These induced voltages total $-Ri - L(di/dt)$. Consequently the applied battery potential V_a must be just equal and opposite to the induced voltages.

$$V_a = Ri + L\frac{di}{dt} \tag{11.7}$$

The solution of this equation for current as a function of time is simple, provided that L is a constant. This is the case whenever there is no magnetic material near the coil, and it is approximately true if magnetic material is present but the flux density is sufficiently low.

Then, assuming L is a constant,

$$\frac{di}{\left(\dfrac{V_a - Ri}{L}\right)} = dt$$

$$\frac{\left(-\dfrac{R}{L}\,di\right)}{-\dfrac{R}{L}\left(\dfrac{V_a - Ri}{L}\right)} = dt$$

Integrating we obtain

$$-\frac{L}{R}\ln\left(\frac{V_a - Ri}{L}\right) = t + \ln C \tag{11.8}$$

where ln C is the integration constant. To evaluate ln C let us consider the initial conditions. When time t is zero, i is also zero, if we count time from the instant when the battery is connected to the coil. Then

$$-\frac{L}{R}\ln\frac{V_a}{L} = \ln C$$

If this initial condition is put back into equation 11.8,

$$-\frac{L}{R}\ln\left(\frac{V_a - Ri}{L}\right) = t - \frac{L}{R}\ln\frac{V_a}{L}$$

$$\ln\frac{\left(\dfrac{V_a - Ri}{L}\right)}{\left(\dfrac{V_a}{L}\right)} = -\frac{R}{L}t$$

$$\frac{V_a - Ri}{V_a} = e^{-\frac{R}{L}t}$$

$$i = \frac{V_a}{R}(1 - e^{\frac{-R}{L}t}) \tag{11.9}$$

This is a logarithmic curve (see Fig. 11.2*b*); $i = 0$ when $t = 0$; $i = V_a/R$ when $t = \infty$. Theoretically the current never rises to its final value V_a/R, but practically this is usually accomplished in a rather short time.

The time constant T of this circuit is defined as the time that would be required for the current to reach its final value if the initial rate of change of current continued unchanged during this period. Then

$$\left(\frac{di}{dt}\right)_{t=0} T = I = \frac{V_a}{R} \tag{11.10}$$

but the initial rate of change of current is V_a/L, and, as shown in Fig. 11.2*b*,

$$T = \frac{L}{R} \tag{11.11}$$

Accompanying the current rise there is an increasing storage of energy in the magnetic field and a dissipation of energy in the form of heat due to the current flow through the resistance. If equation 11.7 is multiplied by $i\,dt$,

$$V_a i\,dt = Ri^2\,dt + Li\,di, \tag{11.12}$$

in which $V_a i\,dt$ = total energy supplied to the circuit.
$Ri^2\,dt$ = energy dissipated as heat.
$Li\,di$ = energy to storage in the magnetic field.

If we substitute the value of current $i = \frac{V_a}{R}(1 - e^{-\frac{R}{L}t})$ and integrate from the current starting time ($t = 0$) up to any time t_1, the heat loss is

$$H\Big]_0^{t_1} = \frac{V_a^2}{R}\left(t_1 + \frac{2L}{R}e^{\frac{-R}{L}t_1} - \frac{1}{2}\frac{L}{R}e^{\frac{-2R}{L}t_1} - \frac{2L}{R} + \frac{1}{2}\frac{L}{R}\right) \tag{11.13}$$

and the energy storage is

$$W\Big]_0^{t_1} = \frac{V_a^2 L}{R^2}\left[-e^{\frac{-R}{L}t_1} + \frac{1}{2}e^{\frac{-2R}{L}t_1} + \frac{1}{2}\right] \tag{11.14}$$

As t_1 increases in value the current rises more and more slowly, reaching its final value of V_a/R only when t_1 is infinite, when the heat loss is

$$H\Big]_0^{\infty} = \text{infinity} \tag{11.15}$$

and the energy storage in the magnetic field approaches a finite value

$$W = \frac{V_a^2 L}{2R^2} = \frac{1}{2}LI^2 \quad \text{joules} \tag{11.16}$$

where I is the final current value. But since L equals the flux linkages per unit of current,

$$W = \frac{\lambda I}{2} \quad \text{joules} \tag{11.17}$$

or

$$W = \frac{n\phi I}{2} \quad \text{joules} \tag{11.18}$$

if all the flux ϕ links all the turns n. Compare this with equation 9.78.

11.4. Decay of Current. In the preceding case, let us assume that the current and energy storage have for all practical purposes reached their final values. Now let the circuit be changed by connecting in resistance R_1, after which the battery is disconnected. See Fig. 11.3*a*. Without the battery the current cannot maintain its full value of V_a/R and therefore falls in value but maintains its direction in the circuit. The time rate of change of current develops a voltage $-L(di/dt)$ which opposes the change in current. Its direction there-

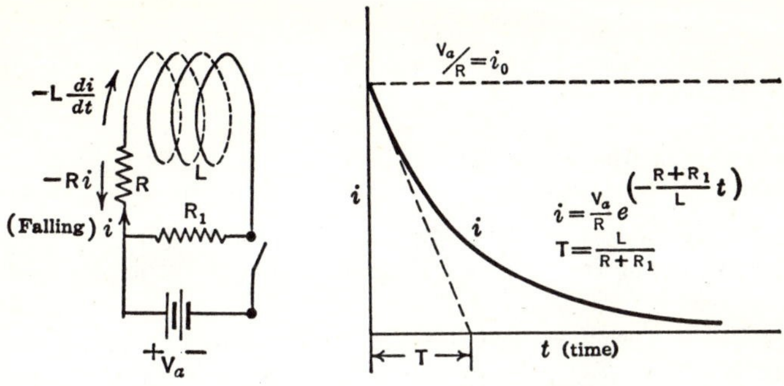

(a) (b)

FIG. 11.3. Current decay in an inductive circuit.

fore is such as to aid the current flow as shown. The magnitude of the current is

$$i = \frac{-L\dfrac{di}{dt}}{R + R_1} \tag{11.19}$$

$$\frac{di}{i} = -\frac{R + R_1}{L}\,dt$$

$$\ln i = -\frac{R + R_1}{L}t + \ln C$$

$$i = Ce^{-\frac{R+R_1}{L}t}$$

Measuring time from the instant when the circuit change was made, $i = V_a/R$ when $t = 0$, gives

$$C = \frac{V_a}{R}$$

Then

$$i = \frac{V_a}{R}e^{-\frac{R+R_1}{L}t} \tag{11.20}$$

as shown in Fig. 11.3*b*.

Let us now investigate the ratio of the voltage induced $-L(di/dt)$ in the coil to the original voltage V_a.

$$\text{Ratio} = \frac{-L\dfrac{di}{dt}}{V_a} = \frac{-L\dfrac{V_a}{R}\left[-\left(\dfrac{R+R_1}{L}\right)\right]e^{-\left(\frac{R+R_1}{L}\right)t}}{V_a}$$

$$\text{Ratio} = \left(\frac{R+R_1}{R}\right)e^{-\frac{R+R_1}{L}t} \tag{11.21}$$

The maximum value of this ratio is $(R + R_1)/R$, occurring when $t = 0$. At the instant the switch is opened there is induced for the moment a voltage greater than the battery potential in the ratio $(R + R_1)/R$, and of this voltage the fraction $R_1/(R + R_1)$ appears across the terminals of R_1.

If the circuit is short-circuited, $R_1 = 0$ and the induced potential equals V_a; if the circuit is merely broken, $R_1 = \infty$, and, theoretically at least, an infinite voltage is developed. Actually, long before this value is reached the induced voltage causes a breakdown of the dielectric between the open ends of the coil, and a spark passes across the ends. The spark lasts until all the original energy stored in the magnetic field ($W = \frac{1}{2}LI^2$) has dissipated itself in the resistance R of the coil and of the spark path. This illustrates the danger of suddenly breaking an inductive circuit in which a current is flowing.

11.5. Voltage Generation in a Mutual Inductance. Mutual inductance, as explained in Art. 9.22, exists only when there are two coils so placed in space that, if a current flows in one of them, some of the flux lines of the resulting magnetic field link the other coil. From equation 9.71,

$$\lambda_{21} = Mi_1 \tag{11.22}$$

If coil 1 carries current i_1, it sets up a flux ϕ_1 of which a part ϕ_{21} links the turns of coil 2, yielding flux linkages through coil 2 of λ_{21}.

Taking the negative of the time derivative of equation 11.22,

$$V_{g2} = -\frac{d\lambda_{21}}{dt} = -M\frac{di_1}{dt} \tag{11.23}$$

If the current in coil 1 is changing at the time rate di_1/dt, there is induced in coil 2 a voltage equal to the negative of the product of the time rate of change of current in coil 1 by the mutual inductance. Similarly, if the current is changing in coil 2, a voltage is induced in coil 1.

$$V_{g1} = -\frac{d\lambda_{12}}{dt} = -M\frac{di_2}{dt} \tag{11.24}$$

From the last two equations it is possible to develop a third definition of mutual inductance to supplement the two other definitions given in Art. 9.22.

The mutual inductance of two coils is equal to the voltage induced in either coil divided by the time rate of change of current in the other coil.

$$M = \frac{-V_{g1}}{\left(\frac{di_2}{dt}\right)} = \frac{-V_{g2}}{\left(\frac{di_1}{dt}\right)} = \frac{\text{volts}}{\left(\frac{\text{amperes}}{\text{second}}\right)} = \text{henrys} \qquad (11.25)$$

This method of defining mutual inductance is especially useful in connection with its *measurement* in the laboratory. The definitions given in Art. 9.22 are particularly adaptable to a *calculation* of mutual inductance.

11.6. Current–Voltage Relations in a Mutual Inductance. Let a source of electric potential V_a be connected to a coil having resistance R_1 and self-inductance L_1. Let a short-circuited second coil of resistance R_2 and self-inductance L_2 be magnetically connected to the first through their mutual inductance M.

The current–voltage relations in the two coils may be expressed by two equations, which hold simultaneously:

$$\left.\begin{aligned} V_a &= R_1 i_1 + L_1 \frac{di_1}{dt} + M \frac{di_2}{dt} \\ 0 &= R_2 i_2 + L_2 \frac{di_2}{dt} + M \frac{di_1}{dt} \end{aligned}\right\} \qquad (11.26)$$

These equations are true for any time variation of the applied voltage V_a. In particular, if $V_a = V_{am} \sin \omega t$, the usual alternating-current relations for a transformer may be obtained.

If the applied voltage V_a is constant, the currents have the time relations shown in Fig. 11.4. A *rising* current in coil 1 sets up an increasing flux, a part of which generates a potential in coil 2 which sends a current whose flux effect is in opposition to that of coil 1. During this period the coils repel each other. As the current in coil 1 reaches a steady state, i_2 approaches zero. On the break, while i_1 is *decreasing*, i_2 is reversed and now aids i_1 in its flux effect. The coils attract each other during this period. Current i_2 again disappears when i_1 becomes zero.

Assuming a *variable* potential V_a and *neglecting the resistances* of the coils, we may obtain a simple expression for the total inductance of coil 1 in the presence of short-circuited coil 2, connected magnetically by their mutual inductance. Equations 11.26 reduce as follows:

$$\left.\begin{aligned} V_a &= L_1 \frac{di_1}{dt} + M \frac{di_2}{dt} \\ 0 &= L_2 \frac{di_2}{dt} + M \frac{di_1}{dt} \end{aligned}\right\} \tag{11.27}$$

From the second of these equations

$$\frac{di_2}{dt} = -\frac{M}{L_2}\frac{di_1}{dt}$$

Hence

$$V_a = \left(L_1 - \frac{M^2}{L_2}\right)\frac{di_1}{dt}$$

and

$$\begin{aligned} L_{1\,\text{total}} &= \frac{V_a}{\left(\dfrac{di_1}{dt}\right)} = \left(L_1 - \frac{M^2}{L_2}\right) \\ &= L_1 - K^2 L_1 \\ &= L_1(1 - K^2) \qquad (R \text{ neglected}) \end{aligned} \tag{11.28}$$

in which K = coefficient of coupling ≤ 1. If $K = 1$, coil 2 completely eliminates any inductive effect for the combination.

See Art. 11.10 for the case in which the resistance of the coils is not neglected and the first coil is connected to a battery.

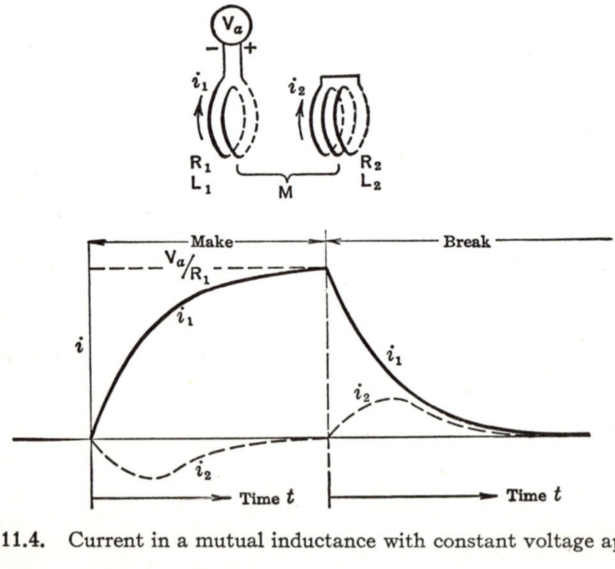

FIG. 11.4. Current in a mutual inductance with constant voltage applied.

11.7. Eddy Currents. In any conducting body, any closed path has an induced voltage or electric intensity which drives a current, if a magnetic flux through the path is changing in time. The time change in flux may be produced in several ways: a stationary flux changing its magnitude as in a transformer; a flux that varies in magnitude from point to point in space and is moved across the conducting body, as in the case of the passage of a magnet over a copper disk; or a movement of the conducting body through the flux that varies in its space distribution. Invariably the eddy currents so created are directed around the closed path so that by their magnetic effects they

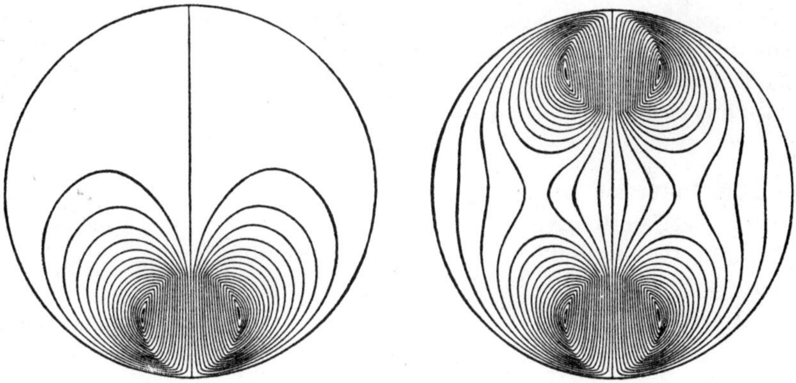

(a) (b)

FIG. 11.5. Eddy currents induced in a rotating disk. (From W. R. Smythe, *Electrical Engineering*, **61,** 681 (September 1942.) (*a*) Eddy currents induced by a circular magnet. (*b*) Eddy currents induced by two circular magnets with antiparallel flux.

tend to oppose the magnetic change which produced them, and, if the flux is moving relatively to the body in which eddy currents are induced, the body experiences a force or "drag" effect which tends to pull it along with the moving flux.

The parts of alternating-current apparatus that carry flux are laminated or composed of thin sheets so arranged that the plane of the sheet is parallel to the flux. In this way the eddy currents are forced to confine themselves for the most part to the individual sheets in which they are generated, and the power loss is reduced, inasmuch as it depends upon the square of the lamination thickness. (See Art. 13.10.) In some applications, wires are used instead of sheets, reducing the eddy current losses still further; and again, compressed-powder magnetic cores are practical, resulting in almost negligible eddy current losses.

An illustration of importance is that of a circular, metallic (usually aluminum) disk of a watthour meter or relay which is revolved over the stationary magnetic field of a permanent magnet. The eddy current distribution induced in the disk is illustrated in Fig. 11.5*a*. These currents, reacting with the permanent magnet field, give a retarding or braking torque on the disk proportional to the strength of the field of the permanent magnet ϕ, the disk rotational speed n, the conductivity of the disk σ, and to the effective radius R of the point where the field is applied.

$$\mathfrak{J} = k\phi n\sigma R \tag{11.29}$$

Equation 11.29 is valid only for low rotational speeds. At sufficient speed $\mathfrak{J}$ reaches a maximum and declines for greater speeds. Fig. 11.5*b* illustrates the eddy currents induced in the disk by two magnets with antiparallel flux of equal strength.

B. MOVING CONDUCTORS

11.8. Induction of Voltage in Moving Conductors. In the preceding section the voltage was induced in a closed conductor by changing the flux linkages through the conductor.

$$V_g = -\frac{d\lambda}{dt} \tag{11.30}$$

It is also possible to generate a voltage in any conductor, or even a portion of the conductor, by moving the conductor or the portion across

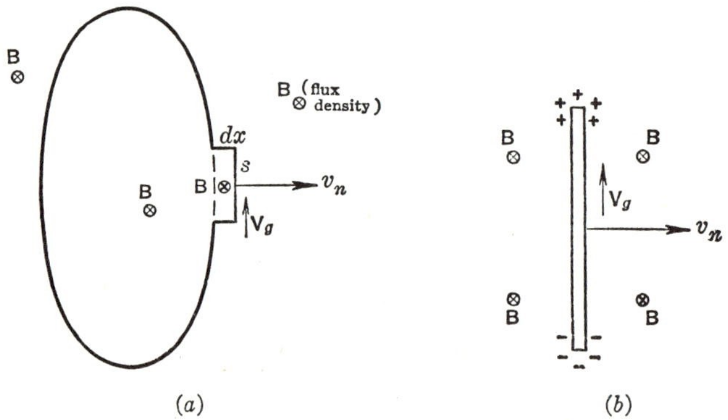

FIG. 11.6. Induction of voltage in a moving conductor.

a magnetic field. The formula that gives the magnitude and direction of the induced voltage may be derived from equation 11.30 by the aid of Fig. 11.6*a*.

Consider a closed conductor in the plane of the paper and a magnetic field of density B downward perpendicular to the plane of the paper. Now by permitting a small section of the conductor of length s to be moved a distance dx perpendicular both to the conductor and to the magnetic field with a velocity v_n in a time dt, the flux linking the circuit is changed by an amount $d\phi$, and therefore (from equation 11.30) a voltage must be induced in the conductor portion that is moved. The magnitude of the voltage is derived as follows:

$$V_g = -\frac{d\lambda}{dt} = -\frac{d\phi}{dt} \qquad (\text{as } n = 1 \text{ turn})$$

$$= -B\frac{dA}{dt} \qquad (\text{as } d\phi = B\,dA)$$

$$= -Bs\frac{dx}{dt} \qquad (\text{as } dA = s\,dx)$$

$$V_g = -Bsv_n \qquad \left(\text{as } \frac{dx}{dt} = v_n\right) \text{ volts} \qquad (11.31)$$

From this we see that a voltage is induced in a conductor whenever it is moved across a magnetic field and has a magnitude given by the product of the magnetic flux density B, the length of the wire s, and the component of the velocity normal to the conductor and to the magnetic field. Lenz's law gives the direction of the potential in the conductor. In Fig. 11.6*a*, inasmuch as the motion assumed increases the total quantity of flux linking the circuit, the generated voltage is directed upward as shown, so that it sends a current whose magnetic effect tends to oppose the increase of magnetic flux linking the circuit.

From this discussion it should be obvious that a voltage may be induced even in an open conductor when the latter is moved across a magnetic field. Equation 11.31 may be easily checked in the laboratory. A simple thumb rule for keeping straight the directions of this relation may be developed. See Fig. 11.6*b*. Pointing the fingers of the right hand in the direction of the flux, and holding the thumb in the direction of the normal velocity, the induced voltage separating the charges is directed along the conductor outward from the palm of the hand. If the conductor is not closed, its ends are charged with positive and negative electricity.

11.9. The Motor-Generator Principle. Consider a conductor, as shown in Fig. 11.7, placed normally to the magnetic flux. The combination may be made to give either motor or generator action, depending upon the manner in which the conductor is handled.

Motor Action. The combination becomes a motor (Fig. 11.7) if a current is passed through the conductor from an outside battery,

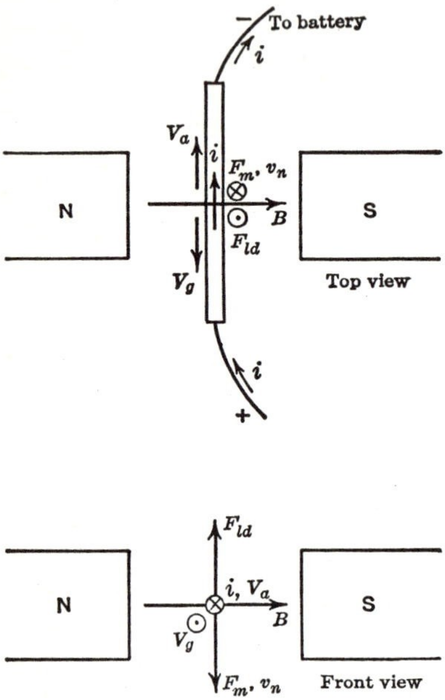

FIG. 11.7. The motor-generator. Motor action.

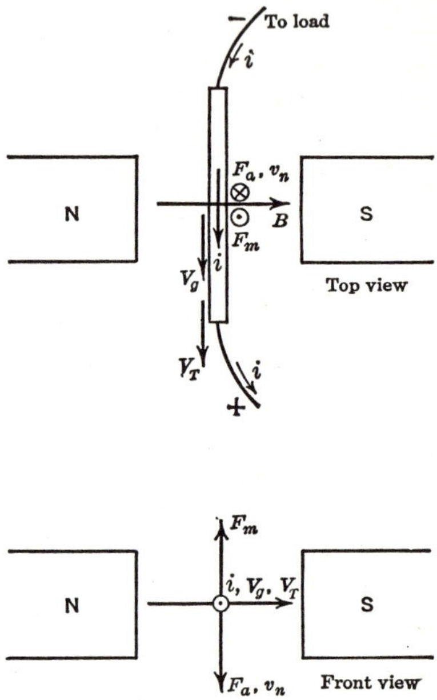

FIG. 11.8. The motor-generator. Generator action.

giving a force F_m on the conductor, which sets up motion and velocity v_n when F_m equals the back force F_{ld} of the mechanical load. As a consequence of the velocity, a voltage is generated in the conductor in opposition to the current flow. The voltage and power balances are

$$V_a = (-V_g) + Ri \tag{11.32}$$

$$V_a i = (-V_g)i + Ri^2 \tag{11.33}$$

$$\underset{\text{battery}}{\overset{\text{power}}{\text{from}}} = \underset{\text{work}}{\overset{\text{power to}}{\text{mechanical}}} + \underset{\text{heat}}{\overset{\text{power}}{\text{to}}}$$

Generator Action. The combination becomes a generator (Fig. 11.8) if the conductor is moved crosswise to the field with a velocity v_n, giving a generated voltage which sends a current through the conductor to an outside load. Assuming the same velocity direction as in the case of the motor, we now find that the generated voltage, current, and force of the conductor are reversed. The voltage and power balances are

$$V_T = (-V_g) - Ri \tag{11.34}$$

$$V_T i = (-V_g)i - Ri^2 \tag{11.35}$$

$$\underset{\text{external circuit}}{\overset{\text{power to}}{}} = \underset{\text{mechanical work}}{\overset{\text{power from}}{}} - \underset{\text{heat}}{\overset{\text{power to}}{}}$$

The current and the flux density set up a force F_m which must be equaled by the force F_a applied to drive the generator.

C. SPECIAL PROBLEMS

11.10. Battery Connected to Transformer Coupled Coils. The circuit is shown in Fig. 11.9. R_3 plays no role when the switch is closed,

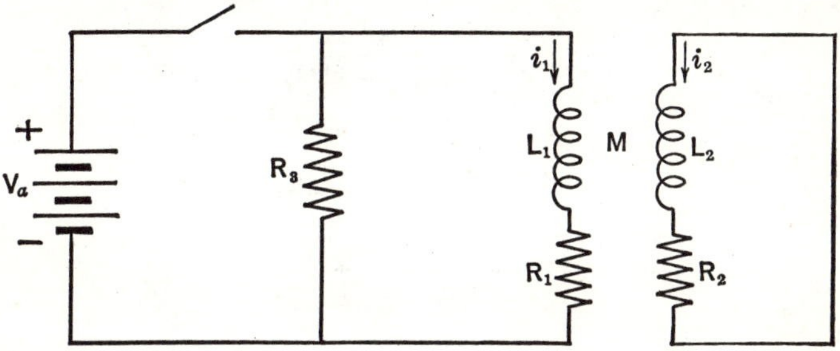

FIG. 11.9. Circuit for transformer coupled coils connected to a battery.

but on opening the switch it carries the discharge current i_1. The circuit equations, upon closing the switch, are:

$$V_a = R_1 i_1 + L_1 \frac{di_1}{dt} + M \frac{di_2}{dt} \tag{11.36}$$

$$0 = R_2 i_2 + L_2 \frac{di_2}{dt} + M \frac{di_1}{dt} \tag{11.37}$$

Solve equation 11.36 for $\dfrac{di_2}{dt}$ and differentiate for $\dfrac{d^2 i_2}{dt^2}$:

$$\left.\begin{aligned} M \frac{di_2}{dt} &= V_a - R_1 i_1 - L_1 \frac{di_1}{dt} \\ M \frac{d^2 i_2}{dt^2} &= -R_1 \frac{di_1}{dt} - L_1 \frac{d^2 i_1}{dt^2} \end{aligned}\right\} \tag{11.38}$$

Differentiate equation 11.37

$$0 = R_2 \frac{di_2}{dt} + L_2 \frac{d^2 i_2}{dt^2} + M \frac{d^2 i_1}{dt^2} \tag{11.39}$$

and substitute equation 11.38 into equation 11.39 and get

$$\left(\frac{L_1L_2 - M^2}{R_2}\right)\frac{d^2i_1}{dt^2} + \left(\frac{L_1R_2 + L_2R_1}{R_2}\right)\frac{di_1}{dt} + R_1i_1 = V_a \quad (11.40)$$

Differentiate equation 11.40, substitute $i_1 = e^{mt}$, and the equation for m becomes

$$\left(\frac{L_1L_2 - M^2}{R_2}\right)m^3 + \left(\frac{L_1R_2 + L_2R_1}{R_2}\right)m^2 + R_1m = 0$$

m is satisfied by three values,

$$m = 0, \qquad m = -\alpha' + \beta', \qquad m = -\alpha' - \beta'$$

where

$$\alpha' = \frac{L_1R_2 + L_2R_1}{2(L_1L_2 - M^2)}$$

$$\beta' = \sqrt{(\alpha')^2 - \frac{R_1R_2}{L_1L_2 - M^2}} \quad (11.41)$$

$$\alpha'^2 - \beta'^2 = \frac{R_1R_2}{L_1L_2 - M^2}$$

Then

$$i_1 = A_1 + e^{-\alpha' t}(A_2e^{+\beta' t} + A_3e^{-\beta' t}) \quad (11.42)$$

The secondary current i_2 is easily obtained by inspection of equations 11.36 and 11.37. Interchange the subscripts 1 and 2 in equation 11.40 and replace V_a by 0. Then

$$\left(\frac{L_1L_2 - M^2}{R_1}\right)\frac{d^2i_2}{dt^2} + \left(\frac{L_1R_2 + L_2R_1}{R_1}\right)\frac{di_2}{dt} + R_2i_2 = 0 \quad (11.43)$$

Substituting $i_2 = e^{mt}$, we obtain

$$i_2 = e^{-\alpha' t}(B_2e^{+\beta' t} - B_3e^{-\beta' t}) \quad (11.44)$$

where α' and β' have the same values as before.

The initial and final conditions determine the constants A_1, A_2, A_3, B_2, B_3.

$$\left.\begin{aligned} &\text{At } t = \infty,\ i_1 = A_1 = \frac{V_a}{R_1} \\ &\text{At } t = 0,\ i_1 = 0 = \frac{V_a}{R_1} + A_2 + A_3 \text{ or } A_2 + A_3 = -\frac{V_a}{R_1} \\ &\qquad\qquad i_2 = 0 = B_2 - B_3 \qquad \text{or } B_2 = B_3 = B \end{aligned}\right\} \quad (11.45)$$

Now we have constant A_3 in terms of A_2, and B_3 in terms of B_2; that is, there are two constants still to determine. Since $i_1 = i_2 = 0$ at $t = 0$ equations 11.36 and 11.37 may be written

$$\left.\begin{aligned} V_a &= L_1 \frac{di_1}{dt} + M \frac{di_2}{dt} \\ 0 &= L_2 \frac{di_2}{dt} + M \frac{di_1}{dt} \end{aligned}\right\} \quad \text{at } t = 0 \tag{11.46}$$

and

$$\left.\begin{aligned} \left(\frac{di_1}{dt}\right)_{t=0} &= V_a \frac{L_2}{L_1L_2 - M^2} = V_a \frac{L_2}{R_1R_2} (\alpha'^2 - \beta'^2) \\ \left(\frac{di_2}{dt}\right)_{t=0} &= -V_a \frac{M}{L_1L_2 - M^2} = -V_a \frac{M}{R_1R_2} (\alpha'^2 - \beta'^2) \end{aligned}\right\} \tag{11.47}$$

Differentiate equations 11.42 and 11.44, then set $t = 0$, insert the values from equations 11.45 and 11.47, and obtain

$$A_2(\beta' - \alpha') - A_3(\beta' + \alpha') = V_a \frac{L_2}{R_1R_2} (\alpha'^2 - \beta'^2) \tag{11.48}$$

$$B2\beta' = -V_a \frac{M}{R_1R_2} (\alpha'^2 - \beta'^2)$$

Combining equations 11.45 and 11.48, the constants become

$$\left.\begin{aligned} A_1 &= \frac{V_a}{R_1} \\ A_2 &= \frac{V_a}{R_1}\left[\frac{L_2}{R_2}(\alpha' - \beta') - 1\right]\frac{\alpha' + \beta'}{2\beta'} \\ A_3 &= -\frac{V_a}{R_1}\left\{1 + \left[\frac{L_2}{R_2}(\alpha' - \beta') - 1\right]\frac{\alpha' + \beta'}{2\beta'}\right\} \\ B = B_2 = B_3 &= -V_a \frac{M}{R_1R_2} \frac{\alpha'^2 - \beta'^2}{2\beta'} \end{aligned}\right\} \tag{11.49}$$

These values placed in equations 11.42 and 11.44 determine the two currents.

Current i_1 is shown in Fig. 11.10 for the circuit constants indicated for a pair of coils in which the coefficient of coupling is 0.9. i_1' is the current that would have flowed in coil 1 if the secondary had been

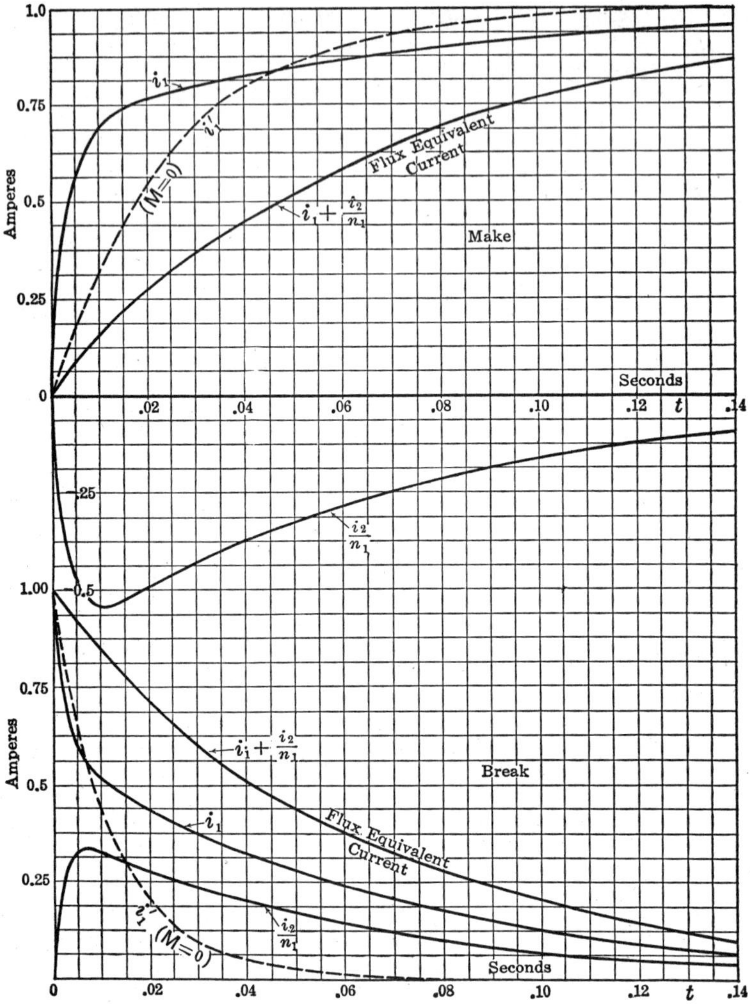

FIG. 11.10. Transformer coupled coils connected to a battery.

$L_1 = 2.5$ henrys	$R_1 = 100$ ohms	$n_1 = 5000$ turns
$L_2 = 10^{-7}$ henry	$R_2 = 2 \times 10^{-6}$ ohm	$n_2 = 1$ turn
$V_a = 100$ volts	$M = 0.45 \times 10^{-6}$ henry	$K = 0.9$
	$R_3 = 100$ ohms	

Make: $\alpha' = 157.9$ $\quad\beta' = 143.8$

$A_1 = 1 \quad A_2 = -0.3095 \quad A_3 = -0.6905 \quad B = -3290$

$i_1 = 1 - 0.3095e^{-14.1t} - 0.6905e^{-301.7t}$ ampere

$\frac{i_2}{n_1} = -0.658(e^{-14.1t} - e^{-301.7t})$

Break: $\alpha'' = 263$ $\quad\beta'' = 246.6$

$D_1 = 0.61 \quad D_2 = -0.39 \quad G = 1923$

$i_1 = 0.61e^{-16.4t} + 0.39e^{-509.6t}$ ampere

$\frac{i_2}{n_1} = 0.3846(e^{-16.4t} - e^{-509.6t})$

open circuited, that is, for $M = 0$. i_2 is the actual secondary current, but $i_2(n_2/n_1)$ is the *equivalent* current that would have to flow in coil 1 (but actually does not) to produce the magnetic effect of i_2.

The magnetic flux set up by the pair of coils of n_1 and n_2 turns is determined by the combined action of both n_1i_1 and n_2i_2. The total equivalent current of coil 1 for *flux-producing purposes*, but which does not actually flow, is

$$I_{1\,\text{equivalent}} = \frac{n_1i_1 + n_2i_2}{n_1} = i_1 + \frac{n_2}{n_1}i_2 \tag{11.50}$$

From the curves of Fig. 11.10 we note that, though current i_1 has risen much faster as a consequence of the coupling, the rise in flux which is proportional to $I_{1\,\text{equivalent}}$ is far slower.

For the limiting case of infinitely tight coupling, that is, with coefficient of coupling $K = 1$ and $M^2 = L_1L_2$, i_1 rises instantly to its final value V_a/R_1 and i_2 rises instantly to its peak value and then falls in simple exponential manner. The equivalent current and flux now rise only as fast as i_2 falls.

At the "break," upon opening the switch in Fig. 11.9, equation 11.36 must be modified by replacing R_1 by $R_1 + R_3$ and V_a by 0. Equation 11.37 still holds. The same change must then be made in equation 11.40 and in equation 11.43. Then

$$\left(\frac{L_1L_2 - M^2}{R_2}\right)\frac{d^2i_1}{dt^2} + \left[\frac{L_1R_2 + L_2(R_1 + R_3)}{R_2}\right]\frac{di_1}{dt} + (R_1 + R_3)i_1 = 0 \tag{11.51}$$

$$\left(\frac{L_1L_2 - M^2}{R_1 + R_3}\right)\frac{d^2i_2}{dt^2} + \left[\frac{L_1R_2 + L_2(R_1 + R_3)}{R_1 + R_3}\right]\frac{di_2}{dt} + R_2i_2 = 0 \tag{11.52}$$

Substituting $i = e^{mt}$ as usual we find solutions for both currents of the same form, but with different constants

$$i_1 = e^{-\alpha''t}(D_1e^{+\beta''t} - D_2e^{-\beta''t}) \tag{11.53}$$

$$i_2 = e^{-\alpha''t}(G_1e^{+\beta''t} - G_2e^{-\beta''t}) \tag{11.54}$$

in which

$$\alpha'' = \frac{L_1R_2 + L_2(R_1 + R_3)}{2(L_1L_2 - M^2)}$$

$$\beta'' = \sqrt{(\alpha'')^2 - \frac{(R_1 + R_3)R_2}{L_1L_2 - M^2}} \tag{11.55}$$

$$\alpha''^2 - \beta''^2 = \frac{(R_1 + R_3)R_2}{L_1L_2 - M^2}$$

Upon opening the switch, the initial current V_a/R_1 must continue to flow on through R_3, but it is sharply damped thereafter.

At $\qquad t = 0, \quad i_2 = 0 = G_1 - G_2 \quad \text{or} \quad G_1 = G_2 = G$

$$i_1 = \frac{V_a}{R_1} = D_1 - D_2$$

and at $t = 0$, from equation 11.36 modified to include R_3, we note that

$$L_1 \left(\frac{di_1}{dt}\right)_{t=0} + M \left(\frac{di_2}{dt}\right)_{t=0} = -\frac{R_3}{R_1} V_a \tag{11.56}$$

$$L_2 \left(\frac{di_2}{dt}\right)_{t=0} + M \left(\frac{di_1}{dt}\right)_{t=0} = 0 \tag{11.57}$$

Substituting into equations 11.56 and 11.57 the derivatives of equations 11.53 and 11.54, we obtain the values of the constants:

$$\left.\begin{aligned} D_1 &= \frac{V_a}{R_1}\left[1 - \frac{R_3L_2}{R_2(R_1 + R_3)}(\alpha'' - \beta'')\right]\frac{\alpha'' + \beta''}{2\beta''} \\ D_2 &= \frac{V_a}{R_1}\left\{\left[1 - \frac{R_3L_2}{R_2(R_1 + R_3)}(\alpha'' - \beta'')\right]\frac{\alpha'' + \beta''}{2\beta''} - 1\right\} \\ G &= \frac{V_a}{R_1}\,\frac{MR_3}{R_2(R_1 + R_3)}\,\frac{\alpha''^2 - \beta''^2}{2\beta''} \end{aligned}\right\} \tag{11.58}$$

For a discharge resistance of $R_3 = R_1 = 100$, the discharge currents that flow upon opening the switch have been plotted in Fig. 11.10. i_1' is the primary current that would have been present if M had been zero.

11.11. Time Delay Magnet. Consider a d-c magnet with an air gap. See Fig. 11.11. With an air gap in the magnetic circuit the inductances are practically constant for all values of current in the coils. Around the poles of the magnet we may put a short-circuited secondary winding or even copper cylinders; these are known as lag coils, and over the lag coils the primary is wound. The equations of the preceding article evidently apply to this case, and from a knowledge of the effective primary (coil 1) current and the flux we may compute the force of the magnet on its armature as a function of time. Equating this force to the force actually required to attract the armature gives the time delay of the armature action after either closing or opening the switch.

With the dimensions indicated in Fig. 11.11 the magnet will have approximately the constants used in calculating Fig. 11.10.

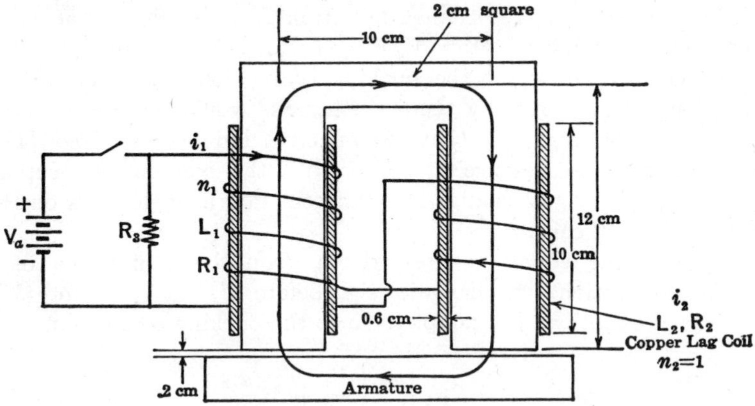

FIG. 11.11. Time delay magnet, n_1 = 5000 turns, No. 26 wire.

11.12. Coils in Parallel Connected to a Battery. Let us connect two coils in parallel and to a battery as shown in Fig. 11.12. If there were no mutual coupling the current in each branch, after closure of the switch, would rise independently of the other in an exponential manner according to the equations given in Art. 11.3.

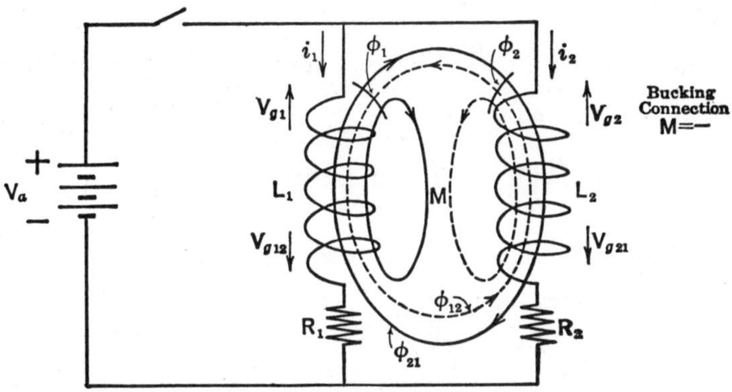

FIG. 11.12. Two coils in parallel, with mutual inductance. Bucking connection. Currents assumed positive after switch is closed.

But with mutual coupling the flux of each coil links the other partially and generates a voltage which modifies the rate of rise of current and the effective inductance of each coil. The final current values are, of course, independent of the coupling. The situation which prevails if there is a *bucking* connection is shown in Fig. 11.12. Consider coil 1. With current i_1 and its own flux ϕ_1 there is a generated voltage V_{g1} =

$-L_1(di_1/dt)$ *opposing* the change in current. The influence of coil 2 is effected through the voltage $V_{g12} = -M(di_2/dt)$ generated in coil 1. (M is itself negative for the bucking connection.) The current i_1 then changes more rapidly than it otherwise would if there were no coupling, or we may say that the effective inductance of coil 1 is decreased by the presence of coil 2 with the bucking connection. However, the flux will change at a different rate as it depends on the currents in both coils.

The following equations are written upon the assumption of a *boosting* connection. A change of sign before M, but not before M^2, in the final equations will adapt them to the bucking connection.

$$V_a = R_1 i_1 + L_1 \frac{di_1}{dt} + M \frac{di_2}{dt} \tag{11.59}$$

$$V_a = R_2 i_2 + L_2 \frac{di_2}{dt} + M \frac{di_1}{dt} \tag{11.60}$$

Differentiate equation 11.60:

$$0 = R_2 \frac{di_2}{dt} + L_2 \frac{d^2 i_2}{dt^2} + M \frac{d^2 i_1}{dt^2} \tag{11.61}$$

Solve equation 11.59 for $\frac{di_2}{dt}$, differentiate for $\frac{d^2 i_2}{dt^2}$ and substitute both in equation 11.61, and obtain

$$\left(\frac{L_1L_2 - M^2}{R_2}\right)\frac{d^2 i_1}{dt^2} + \left(\frac{L_2R_1 + L_1R_2}{R_2}\right)\frac{di_1}{dt} + R_1 i_1 = V_a \tag{11.62}$$

Interchanging subscripts gives the base equation for i_2.

$$\left(\frac{L_1L_2 - M^2}{R_1}\right)\frac{d^2 i_2}{dt^2} + \left(\frac{L_2R_1 + L_1R_2}{R_1}\right)\frac{di_2}{dt} + R_2 i_2 = V_a \tag{11.63}$$

Substituting $i_1 = e^{mt}$ in equation 11.62, following the usual procedure, and noting that $i_1 = V_a/R_1$ at $t = \infty$, the solution for i_1 takes the form:

$$i_1 = \frac{V_a}{R_1} + e^{-\alpha' t}(Be^{+\beta' t} + De^{-\beta' t}) \tag{11.64}$$

Similarly

$$i_2 = \frac{V_a}{R_2} + e^{-\alpha' t}(Ge^{+\beta' t} + He^{-\beta' t}) \tag{11.65}$$

in which

$$\alpha' = \frac{L_2R_1 + L_1R_2}{2(L_1L_2 - M^2)}$$

$$\beta' = \sqrt{\left(\frac{L_2R_1 + L_1R_2}{2(L_1L_2 - M^2)}\right)^2 - \frac{R_1R_2}{L_1L_2 - M^2}}$$

At $\qquad t = 0, \qquad i_1 = i_2 = 0,$

whence

$$\left.\begin{aligned} B + D &= -\frac{V_a}{R_1} \\ G + H &= -\frac{V_a}{R_2} \end{aligned}\right\} \tag{11.66}$$

Differentiate equations 11.64 and 11.65, and substitute in equation 11.59. Upon letting $t = 0$, two more equations for B, D, G, H are obtained:

$$\left.\begin{aligned} (\beta' - \alpha')B - (\beta' + \alpha')D &= V_a \frac{L_2 - M}{L_1L_2 - M^2} \\ (\beta' - \alpha')G - (\beta' + \alpha')H &= V_a \frac{L_1 - M}{L_1L_2 - M^2} \end{aligned}\right\} \tag{11.67}$$

Equations 11.66 and 11.67 yield

$$\left.\begin{aligned} B &= V_a\left[\frac{L_2 - M}{L_1L_2 - M^2}\left(\frac{1}{2\beta'}\right) - \frac{1}{R_1}\left(\frac{\beta' + \alpha'}{2\beta'}\right)\right] \\ D &= -V_a\left[\frac{L_2 - M}{L_1L_2 - M^2}\left(\frac{1}{2\beta'}\right) + \frac{1}{R_1}\left(\frac{\beta' - \alpha'}{2\beta'}\right)\right] \\ G &= V_a\left[\frac{L_1 - M}{L_1L_2 - M^2}\left(\frac{1}{2\beta'}\right) - \frac{1}{R_2}\left(\frac{\beta' + \alpha'}{2\beta'}\right)\right] \\ H &= -V_a\left[\frac{L_1 - M}{L_1L_2 - M^2}\left(\frac{1}{2\beta'}\right) + \frac{1}{R_2}\left(\frac{\beta' - \alpha'}{2\beta'}\right)\right] \end{aligned}\right\} \tag{11.68}$$

The values from equations 11.68 may be substituted back into equations 11.64 and 11.65 to give the expressions for current for the *boosting* connection (M is $+$).

These equations have been plotted in Fig. 11.13, both for the boosting and bucking connections. An interesting feature is that with the boosting connection the current through the larger self-inductance L_2

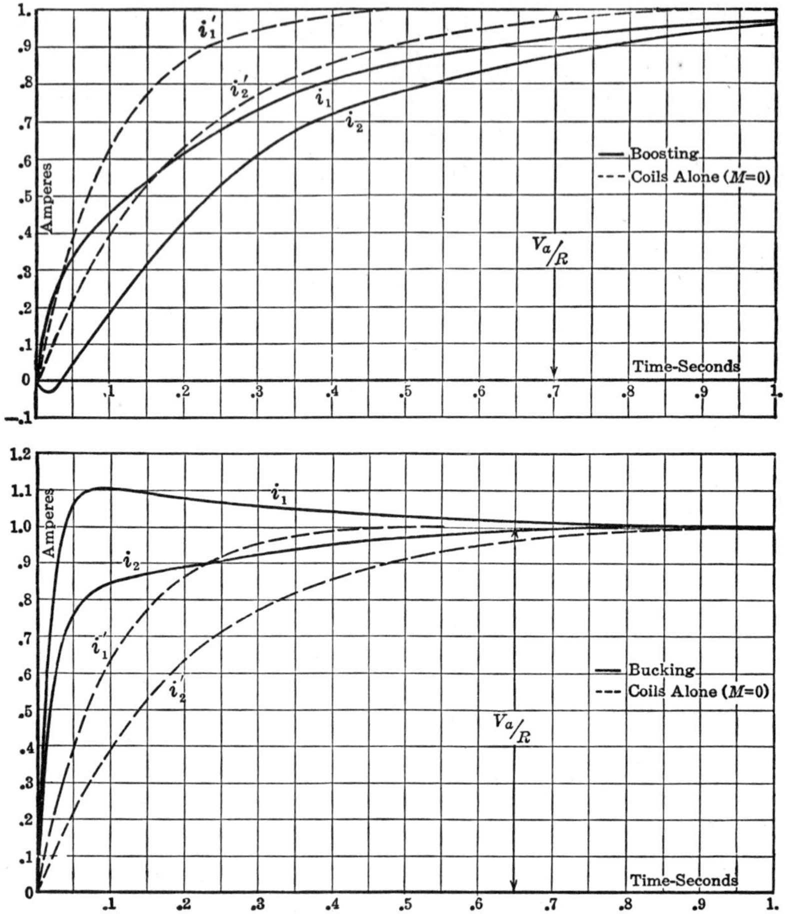

FIG. 11.13. Two coils in parallel connected to a battery, at make. See Fig. 11.12.

$L_1 = 1$ henry $\quad R_1 = 10$ ohms

$L_2 = 2$ henrys $\quad R_2 = 10$ ohms

$V_a = 10$ volts $\quad M = 1.225 \quad K = 0.866$

$L_1L_2 - M^2 = 0.5 \quad L_1 - M = -0.225 \quad L_2 - M = +0.775$

$\alpha' = 30 \quad \beta' = 26.46$

Boosting Connection: Take M positive.

$B = -0.775 \quad D = -0.225 \quad G = -1.153 \quad H = +0.153$

$i_1 = 1 - 0.775e^{-3.543t} - 0.225e^{-56.457t}$

$i_2 = 1 - 1.153e^{-3.543t} + 0.153e^{-56.457t}$

Bucking Connection: Take M negative.

$B = +0.153 \quad D = -1.153 \quad G = -0.225 \quad H = -0.775$

$i_1 = 1 + 0.153e^{-3.543t} - 1.153e^{-56.457t}$

$i_2 = 1 - 0.225e^{-3.543t} - 0.775e^{-56.457t}$

may be reversed momentarily, provided that $M > L_1$. If $M = L_1$, the current i_2 will start off with zero slope.

The equivalent inductance of the pair of coils is not a constant. At $t = 0$, the inductance is

$$L_{t=0} = \frac{V_a}{\left(\dfrac{di}{dt}\right)_{t=0}} = \frac{L_1L_2 - M^2}{L_1 - 2M + L_2} \tag{11.69}$$

which proves to be, for the case illustrated, 0.909 for the boosting connection and 0.0918 for the bucking connection. At $t = \infty$, when there is no rate of change of current, we take the equivalent inductance to be twice the energy stored divided by the square of the final current, or

$$L_{t=\infty} = \frac{2W}{I^2} = \frac{L_1R_2^2 + 2MR_1R_2 + L_2R_1^2}{(R_1 + R_2)^2} \tag{11.70}$$

For the boosting connection, this is 1.363, and 0.1375 for the bucking connection.

PROBLEMS

1. A battery of potential $V = 100$ volts is connected in series with a coil of resistance $R = 2$ ohms and self-inductance $L_1 = 0.12$ henry. A second coil of self-inductance $L_2 = 0.30$ henry remains open circuited but is inductively connected to the first coil. The coefficient of coupling $K = 0.8$. Find the value of the current, the rate of change of current in the first coil, and the induced potentials in both coils, at the instant when the switch is closed. Recompute these values after the elapse of a time interval equal to the time constant, and again after infinite time.

2. Resistance $R_1 = 10$ ohms is now connected across the terminals of the first coil in Prob. 1, and the switch to the battery is opened. Answer the questions asked in Prob. 1, but for this situation. How much of the induced voltage appears across R_1?

3. The current in a laboratory inductance reaches a final value of 4 amperes when 100 volts direct current is applied. The current reaches 95 per cent of its final value in 10 seconds. If 80 ohms is connected across its terminals, find the induced voltage in the coil at the instant of removal of the d-c line. Find the inductance. Assume permeability is constant.

4. Given a closely wrapped coil having an inductance of 0.4 henry and a resistance of 20 ohms. With 200 volts applied, it was found that it took the current a longer time than was desired to rise to 0.9 of its final value. In order to remedy this, the coil was split in two and the halves were separated until the total inductance was just enough to allow the current to rise to 0.9 of its final value in 0.8 the time it took previously. Find the coefficient of coupling between the two coil halves.

5. Two coils have a mutual inductance of 0.05 henry. Coil B has a self-inductance of 0.1 henry and is short-circuited. Coil A has a self-inductance of 0.2 henry.

(a) What is the total inductance of the combination measured at the terminals of coil A?

(b) If the current in coil A is decreasing at the rate of 100 amperes per second, what voltage is induced in coil B?

6. A battery $V = 100$ connects to a coil with $R_1 = 1$ and $L_1 = 1$. A second coil $L_2 = 4$ is short-circuited. $M = 1.5$. Find the time constant of the combination, as of the terminals of coil 1.

7. Two coils are wound on a transformer core. The primary has 100 turns with a self-inductance of 10^{-3} henry. The secondary has 300 turns. Assume no leakage flux. If the primary current is changing at the rate of 200,000 amperes per second, find the voltage induced in the secondary.

8. Two circular coils are hung by strings so that their common axis is horizontal. They are placed a few centimeters apart. Coil 2 is short-circuited. Coil 1 is connected through a switch to a battery. What happens when:

(a) The switch is closed?
(b) The switch is opened?

9. If an electric lamp is connected to an unloaded generator, the voltage across the lamp is suddenly decreased to zero at the instant of connection. Explain this. If the load is one lamp, and a second is connected, what happens to the voltage?

10. A watthour meter uses a brake composed of a thin aluminum disk revolving through a permanent magnet field. Derive the law of drag torque against disk speed.

11. How rapidly must a wire of 5-cm length be moved across a field of 1 weber per square meter in order to develop a potential of 0.01 volt between the ends?

12. In connection with the transformer coupled coils and time delay magnet discussed in this chapter, how could the rise of the flux equivalent current be slowed up? What physical changes in the magnet would be required?

13. With coils in parallel, mutually coupled, and boosting connection, explain why the current in one coil may be reversed when connected to a battery. What relations must exist among the circuit constants if there is to be no current reversal?

CHAPTER 12

ENERGY AND FORCES IN AN INDUCTANCE

An inductance carrying current stores energy in the magnetic field and experiences forces which tend to move it in directions that increase the inductance. The forces tend to expand a circular coil, cause two coils carrying current to attract each other if the currents are similarly directed or to repel each other if the currents are oppositely directed, and tend to turn one coil in the presence of the other. If the coils move, the inductance and energy storage change, and mechanical work is performed. The change in magnetic field generates voltages in the circuit which alter the current magnitudes and the power supplied by the battery or alternator, as well as the heat loss in the resistance.

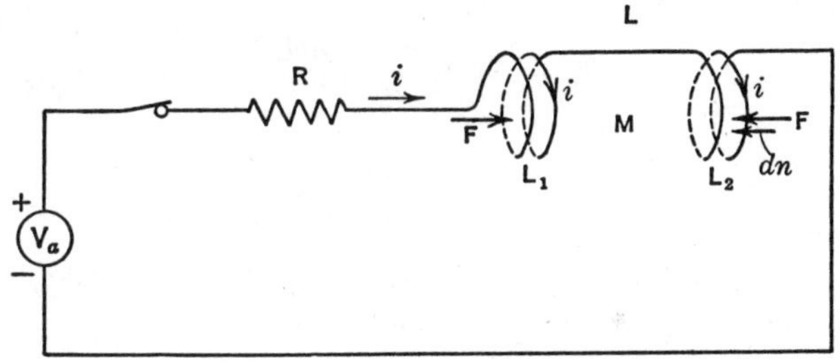

FIG. 12.1. Forces on an inductance.

The relations between these quantities are rather complicated in the general case, so we turn to a number of special cases:

I. Constant Current. The law of force is obtained by studying the energy changes taking place in a moving inductance in which the current is held constant.

II. Constant Flux Linkages, with R $= 0$. With no resistance in the circuit, no voltage is required to sustain a current and a fixed number of flux linkages. Any motion then occurs under the condition of constant flux linkages.

III. Energy Changes for a Finite Movement, with Constant Applied Voltage.

12.1. Constant Current. Let us consider an inductance L, as shown in Fig. 12.1, composed of its separate self-inductances, L_1 and L_2, and a mutual inductance M so connected that the current i flows in the same relative direction in the two parts and develops forces which are attractive. The force law is obtained by studying the changes which occur when the inductance moves a very small distance dn in time dt, under the assumption of *constant current*. The inductance is

$$L = L_1 + L_2 + 2M \tag{12.1}$$

and

$$\frac{dL}{dt} = +2\frac{dM}{dt} \tag{12.2}$$

The energy storage is

$$W_M = \tfrac{1}{2}Li^2 \tag{12.3}$$

and

$$\begin{aligned} dW_M &= Li\,di + \tfrac{1}{2}i^2\,dL \\ &= Li\,di + i^2\,dM \qquad \text{(for any current variation)} \\ &= i^2\,dM \qquad \text{(for constant current as } di = 0) \end{aligned} \tag{12.4}$$

The voltage relation for any current value, and a constant battery potential V_a, is

$$\begin{aligned} V_a - Ri &= \frac{d\lambda}{dt} = \frac{d(Li)}{dt} \\ &= L\frac{di}{dt} + i\frac{dL}{dt} \end{aligned}$$

In order to keep the current constant during the movement $(di/dt) = 0$, it is necessary to insert into the circuit an additional voltage

$$V' = -L\frac{di}{dt} \tag{12.5}$$

Then

$$\begin{aligned} V_a + V' - Ri &= +L\frac{di}{dt} - L\frac{di}{dt} + i\frac{dL}{dt} \\ &= V' = i\frac{dL}{dt} \qquad \left(\text{as } V_a - Ri = 0 \text{ always, if } \frac{di}{dt} = 0\right) \\ &= V' = i2\frac{dM}{dt} \end{aligned} \tag{12.6}$$

The energy relations are

$$\begin{aligned} V_a i\,dt - Ri^2\,dt &= i\,d\lambda = i\,d(Li) \\ &= Li\,di + i^2\,dL \end{aligned} \tag{12.7}$$

and, upon inserting V', they are

$$\begin{aligned} V_a i\,dt + V'i\,dt - Ri^2\,dt &= +Li\,di - Li\,di + i^2\,dL \\ = V'i\,dt \qquad &= i^2\,dL \qquad (\text{as } V_a - Ri = 0) \\ &= 2i^2\,dM \end{aligned} \tag{12.8}$$

But, from equation 12.4, this becomes

$$\begin{array}{ccccc} V'i\,dt & = & i^2\,dM & + & i^2\,dM \\ \left\{\begin{matrix}\text{energy}\\ \text{supplied}\\ \text{to the}\\ \text{inductance}\\ \text{by inserted}\\ \text{potential}\end{matrix}\right\} & = & \left\{\begin{matrix}\text{increase in}\\ \text{energy}\\ \text{storage}\end{matrix}\right\} & + & \left\{\begin{matrix}\text{external}\\ \text{mechanical}\\ \text{work}\\ \text{performed}\end{matrix}\right\} \\ dW_i & = & dW_m & + & d(MW) \end{array} \tag{12.9}$$

or

$$d(MW) = dW_M = \tfrac{1}{2}\,dW_i \tag{12.10}$$

Consequently the forces on the inductance are

$$F_n = +\frac{d(MW)}{dn} = +\frac{dW_m}{dn} \tag{12.11}$$

$$= +i^2\frac{dM}{dn} = +\frac{1}{2}i^2\frac{dL}{dn} \qquad (\text{with } i \text{ constant}) \tag{12.12}$$

The forces are readily computed by multiplying the space rate of change of mutual inductance by the square of the current, under the condition of assumed constant current. Reconnecting coil 2 so that the current flows in the reverse direction relative to coil 1, the force is repulsive but of the same magnitude.

12.2. Constant Flux Linkages, with $R = 0$. Under these conditions no voltage need be applied in order to set up a current. ($V_a = R = 0$.)

$$V_a - Ri = 0 = \frac{d\lambda}{dt} = \frac{d(Li)}{dt} \tag{12.13}$$

Hence

$$\lambda = \underset{\uparrow\downarrow}{Li} = \text{constant} \tag{12.14}$$

As the forces move the coils over either an infinitesimal or finite distance and the mutual and the total inductance increase, the current declines inversely.

The energy relations are

$$\begin{aligned} 0 &= i\,d\lambda = i\,d(Li) \\ &= Li\,di + i^2 2\,dM \\ &= \underbrace{Li\,di + i^2\,dM}_{} + i^2\,dM \\ 0 &= \quad dW_m \downarrow \quad + \quad d(MW) \uparrow \end{aligned} \tag{12.15}$$

Under these conditions the external mechanical work is performed at the expense (decline) of the energy storage.

12.3. Energy Changes for a Finite Movement, with Constant Applied Voltage and with Resistance Considered. (Fig. 12.1.) Under these conditions let us permit the inductance to move a distance s with a "smooth" motion that lasts a total time T. During the completed motion the inductance increases from L_0 to L_f, and the current drops below I and recovers again. But since the power to the coil is proportional to the current, and the power to heat loss is proportional to the square of the current, during the motion there is a difference between these two quantities that gives the power to the magnetic energy storage plus the mechanical power delivered externally. The variation of these and other quantities from time to time throughout the motion is shown in Fig. 12.2. The "smooth" motion mentioned above is one that is started with zero velocity (at $t = 0$) by the forces on the coil and is brought to zero velocity at the end ($v = 0$ at $t = T$) by suitable restraining forces exerted by the object which is receiving the mechanical work. Let

V_a = constant battery potential.
$I = V_a/R$ = constant current, except during the motion.
i = current during motion $< I$.
t = elapsed time.
T = total time of motion.
L = the total inductance, varying from L_0 to L_f.
q = the integrated charge, becoming ΔQ at $t = T$.
λ = instantaneous flux linkages.
$\Delta\lambda$ = the integrated change in flux linkages.
p = instantaneous power.
P = average power over time T.
$P_0 = V_aI$ = power from battery if i equaled I at all times.
$\mathcal{E}_0 = V_aIT$ = integrated energy from battery if i equalled I at all times.

$\mathcal{E} = V_a \int i\,dt = V_a q$ = integrated energy actually received from battery over time T.

$H = \int i^2 R\,dt$ = integrated energy lost in heat in R over time T.

ΔW_m = integrated change in energy storage over time T.

M.W. = integrated mechanical work performed over time T.

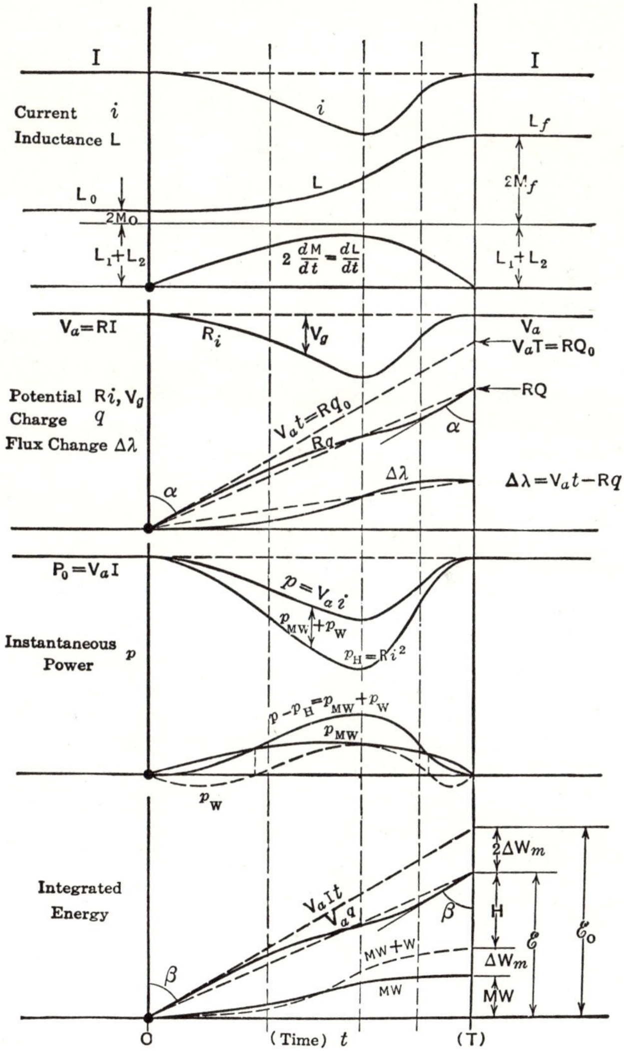

FIG. 12.2. Instantaneous changes during a "smooth" inductance movement.

Equation 12.16 gives the instantaneous value of the potential.

$$V_a - Ri = \frac{d\lambda}{dt} = \frac{d(Li)}{dt} = L\frac{di}{dt} + i\frac{dL}{dt} = L\frac{di}{dt} + 2i\frac{dM}{dt} \tag{12.16}$$

Integrated over the entire motion, it becomes

$$\int_{t=0}^{t=T} V_a\, dt - \int_{q=0}^{q=\Delta Q} Ri\, dt = \int_{\lambda=\lambda_1}^{\lambda=\lambda_2} d\lambda$$

or

$$V_aT - R\,\Delta Q = (\lambda_2 - \lambda_1) = \Delta\lambda \tag{12.17}$$

Multiplying by I, we get

$$V_aIT - RI\,\Delta Q = I\,\Delta\lambda = 2\,\Delta W_m \quad (\Delta W_m = \tfrac{1}{2}I\,\Delta\lambda)$$

$$\mathcal{E}_0 - V_a\,\Delta Q = 2\,\Delta W_m$$

$$\mathcal{E}_0 - \mathcal{E} = 2\,\Delta W_m \tag{12.18}$$

Consequently, the difference between the energy that would have flowed from the battery at constant $I = V_a/R$ and the energy that actually flows is twice the change in magnetic energy storage.

Similarly a relation between the heat loss in the resistance and the energy flow can be proved. Instantaneously,

$$\frac{dH}{d\mathcal{E}} = \frac{Ri^2\,dt}{V_ai\,dt} = \frac{Ri}{V_a} = \frac{Ri}{RI} = \frac{i}{I} = \frac{i\,dt}{I\,dt} = \frac{dQ}{dQ_0} \tag{12.19}$$

$$\frac{dH}{d\mathcal{E}} = \frac{V_a\,dQ}{V_a\,dQ_0} = \frac{d\mathcal{E}}{d\mathcal{E}_0} \tag{12.20}$$

Integrated over the whole motion

$$\frac{H}{\mathcal{E}} = \frac{\mathcal{E}}{\mathcal{E}_0} \tag{12.21}$$

In addition, the energy balance gives

$$\mathcal{E} = H + \Delta W_m + \text{M.W.} \tag{12.22}$$

Equations 12.18, 12.21, and 12.22 are three equations with three unknown quantities (H, $\mathcal{E}$, and M.W.) and are thus solvable, assuming $\mathcal{E}_0$ and ΔW_m as known quantities. The latter are known if the voltage, resistance, inductance change, and total time are known. Gathering these equations and rearranging, we have

$$\mathcal{E} = \mathcal{E}_0 - 2\,\Delta W_m \qquad \text{(giving } \mathcal{E}\text{)} \tag{12.23}$$

$$H = \frac{\mathcal{E}^2}{\mathcal{E}_0} \qquad \text{(giving } H\text{)} \tag{12.24}$$

$$\text{M.W.} = \mathcal{E} - H - \Delta W_m \qquad \text{(giving M.W.)} \tag{12.25}$$

Equations 12.18, 12.21, and 12.22 may be restated to advantage, using the simplification

$$m = \frac{\Delta\lambda}{V_a T} = \frac{I\,\Delta L}{IRT} = \frac{\Delta L}{RT} = \frac{2\,\Delta W_m}{\mathcal{E}_0} \tag{12.26}$$

Then

$$\frac{\Delta W_m}{\mathcal{E}_0} = \frac{\frac{1}{2}I\,\Delta\lambda}{V_a I T} = \frac{m}{2}$$

or

$$\Delta W_m = \mathcal{E}_0 \frac{m}{2} = V_a I \left(\frac{\Delta L}{R}\frac{1}{2}\right) \tag{12.27}$$

and

$$\mathcal{E} = \mathcal{E}_0 - 2\,\Delta W_m = \mathcal{E}_0(1 - m) = V_a I \left(T - \frac{\Delta L}{R}\right) \tag{12.28}$$

Also

$$H = \frac{\mathcal{E}^2}{\mathcal{E}_0} = \mathcal{E}_0(1 - m)^2 = V_a I \left[T\left(1 - \frac{\Delta L}{RT}\right)^2\right] \tag{12.29}$$

and

$$\begin{aligned} \text{M.W.} &= \mathcal{E} - H - \Delta W_m \\ &= \mathcal{E}_0\left(\frac{m}{2} - m^2\right) = V_a I \left(\frac{\Delta L}{2R} - \frac{(\Delta L)^2}{R^2}\frac{1}{T}\right) \end{aligned} \tag{12.30}$$

Dividing these energy equations by the time for the completed motion gives the average power relations.

$$P_0 = \frac{\mathcal{E}_0}{T} = \frac{V_a I T}{T} = V_a I \text{ and } m = \frac{\Delta L}{RT} \tag{12.31}$$

$$P = \frac{\mathcal{E}}{T} = P_0(1 - m) \tag{12.32}$$

$$P_H = \frac{H}{T} = P_0(1 - m)^2 \tag{12.33}$$

$$P_W = \frac{\Delta W_m}{T} = P_0\left(\frac{m}{2}\right) \tag{12.34}$$

$$P_{\text{M.W.}} = \frac{\text{M.W.}}{T} = P_0\left(\frac{m}{2} - m^2\right) \tag{12.35}$$

These relations are plotted against various values of m in Fig. 12.3. It is obvious that, if the coil is to perform external mechanical work,

$$m = \frac{\Delta L}{RT} \leqq 0.5 \tag{12.36}$$

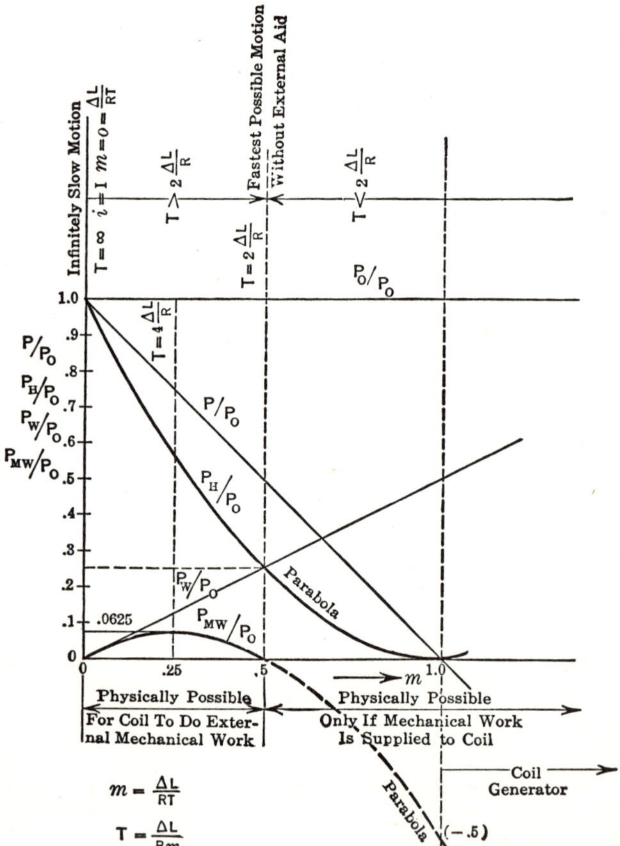

FIG. 12.3. Average power in a moving inductance.

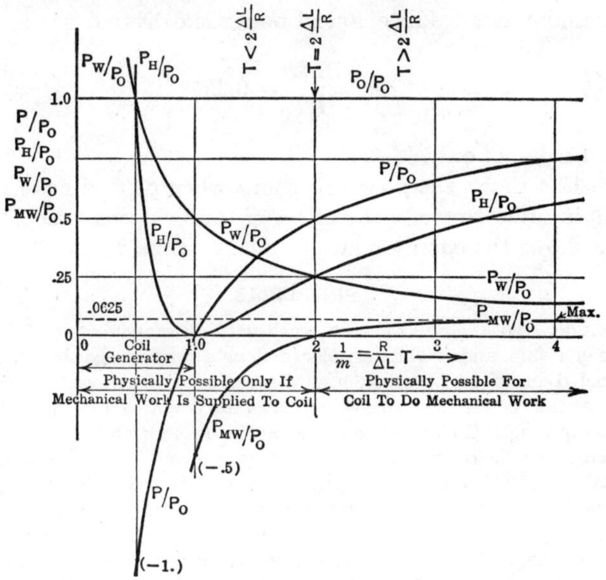

FIG. 12.4. Average power in a moving inductance.

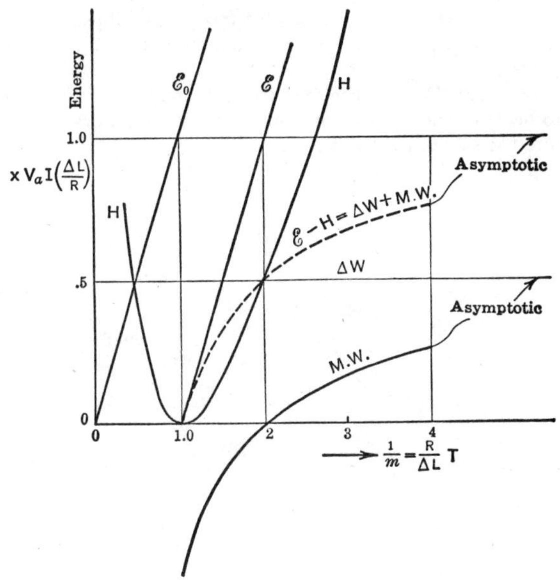

FIG. 12.5. Energy balance in an inductance with "smooth" motion.

The maximum average mechanical power is delivered when

$$m = \frac{\Delta L}{RT} = 0.25 \tag{12.37}$$

and has a value of only $\frac{1}{16} P_0$.

Figure 12.4 shows the power relations when plotted against $1/m = (R/\Delta L)T$ in order to study the effect of total elapsed time T directly; Fig. 12.5 shows the energy relations against time T.

PROBLEMS

1. Consider two like circular coils, each with an inductance of 0.25 henry, a resistance of 1 ohm, and 1000 turns, placed along a common axis in air. The coils are connected in series, boosting, and the current in each coil is 20 amperes. Assume that the coefficient of coupling follows the law $K = 1/(1 + n)$, in which n is the axial spacing. Compute the forces on the coils for any spacing.

If the current is held constant, compute the change of energy storage and the mechanical work performed in changing the spacing from 9 to 4 cm.

2. Wrap a coil of $n = 1000$ turns and 0.25 henry closely around the middle of a permanent magnet of pole strength equal to 10^{-4} weber. Short-circuit the coil and withdraw the magnet. Assuming the resistance of the coil to be zero, compute the magnitude of the current generated and the energy stored in the coil.

3. If the coils of Prob. 1 are permitted to move with a "smooth" motion and the spacing changes from 9 to 4 cm in T seconds, compute:

(*a*) The total time required for the fastest motion possible.

(*b*) The total time required for the motion which gives the maximum average mechanical power.

(*c*) Under condition (*b*), compute the average power supplied during the motion to the coil, to heat, to energy storage, and to mechanical work.

(*d*) Compute the same quantities for $T = \infty$.

PART III
THE FERROMAGNETIC FIELD

CHAPTER 13

FERROMAGNETISM

Most materials are, for all practical purposes, nonmagnetic with a relative permeability of unity. The magnetic materials may be divided into three groups, distinguished primarily by the magnitudes of their relative permeabilities, as shown in Table 9.1. The diamagnetic materials have relative permeabilities slightly smaller than unity, and for the paramagnetic materials the relative permeabilities are slightly greater than unity.

The ferromagnetic materials have relative permeabilities that, at the present time, reach as high as 1,000,000 in commercial specimens. This group is composed largely of cobalt, nickel, iron, and their alloys, and gadolinium. In addition, there are some strongly magnetic alloys (such as Heusler's alloy and some of the new permanent magnet materials) that are composed of metals which by themselves are only weakly magnetic. The ferromagnetic group is the only one of technical importance at the present time.

A. THE MAGNETIZATION CURVE

13.1. Rowland's Ring versus the Epstein Square. Direction of Rolling of Magnetic Sheets. The properties of the ferromagnetic materials are best exhibited by filling the core of Rowland's ring (see Art. 9.40) with circular sheets punched to the required size. The material is laminated to reduce the eddy current losses. The lamination thickness most commonly used is No. 29 gage, 0.014 inch, but the thickness may vary from No. 11 gage, 0.125 inch, to a thickness of 0.001 inch for use at frequencies considerably above the 60-cycle value commonly used in power applications. Magnetic properties of a ferromagnetic sample are dependent in no small degree on the relation of the direction of the flux density and the direction of rolling of the sheet. In general, the properties are more favorable in the direction of rolling. The ring sample, such as is required for the Rowland ring, gives an average result, averaged over all the rolling directions, which is applicable, for instance, to the material used in a motor or generator. Here the directions of the flux density and of the rolling vary throughout the flux path.

An alternative arrangement is to make the Rowland ring in the form of a square of laminated material in which the individual sheets of each side are carefully picked so that the flux density direction and that of the rolling are everywhere parallel. This arrangement can be tested in the Epstein Square apparatus, by using sheet sections with carefully squared ends and butt joints which close as completely as possible. This gives improved magnetic characteristics that are more truly representative of the core conditions in a transformer.

13.2. The Magnetization Curve. The magnetic flux density and the intensity are linearly related (constant permeability $\mu = B/H$) only when the flux passes through a nonferromagnetic medium. In a ferromagnetic medium, the flux density and the permeability depend upon the intensity in a manner that must be determined by experiment, and the relation of B and H expressed graphically is known as

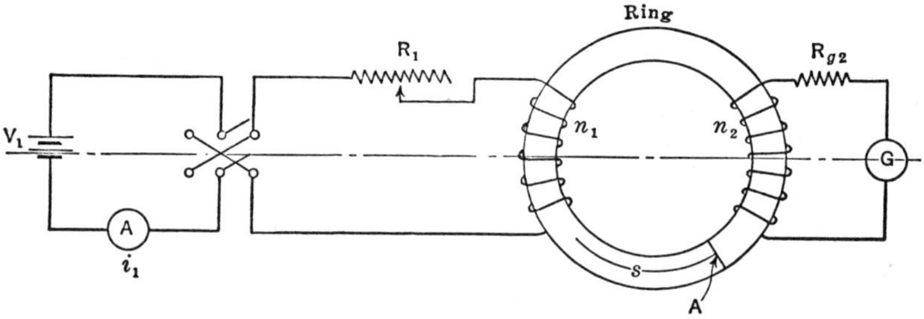

FIG. 13.1. Circuit for obtaining hysteresis loops and the magnetization curve.

the magnetization curve. The following article gives an experimental method for determining this curve.

13.3. The *B–H* Relation Experimentally Determined—Permeability. Consider a closed ring or square of magnetic material, of uniform cross section A, small compared with its circumferential length s, around which is wound a coil of n_1 turns uniformly spaced along the circumference. The coil has a resistance R_1 and is connected through an ammeter and reversing switch to a battery of constant potential V_1. See Fig. 13.1. In addition, let there be a second coil of n_2 turns wound uniformly along the ring and underneath the first coil, as close to the ring as possible. Connect a ballistic galvanometer to the terminals of the second coil. R_{g2} is the total resistance of this circuit.

Under these physical conditions, any current i_1 in the first coil creates a uniform field (practically) throughout the iron volume. The leakage flux is negligible. On the basis of this assumption, which is quite accurate, the magnetic field intensity H can be computed for any current i_1. The closed-path magnetic potential difference is

$$\underset{\text{closed path}}{\mathfrak{F}} = n_1 i_1 = Hs \quad \text{ampere-turns} \tag{13.1}$$

and

$$H = \frac{n_1 i_1}{s} \quad \frac{\text{newtons}}{\text{weber}} \quad \text{or} \quad \frac{\text{amperes}}{\text{meter}} \tag{13.2}$$

The addition of the second coil and ballistic galvanometer permits a measure of the flux density B accompanying the current i_1. During the period in which the current in the first coil is rising from zero to its final value i_1, the flux and the flux density are rising to their final values. The time rate of change of flux linkages $n_2(d\phi/dt)$ yields an electric voltage V_{g2} which sends a current i_2 through the galvanometer.

$$V_{g2} = n_2 \frac{d\phi}{dt} = R_{g2} i_2 = R_{g2} \frac{dQ_2}{dt} \tag{13.3}$$

Multiplying the second and fourth parts of this equation by dt, and integrating, we obtain

$$n_2 \cdot \Delta\phi = R_{g2} \cdot \Delta Q_2$$

$$\Delta\phi = \frac{R_{g2}}{n_2} \cdot \Delta Q_2 \tag{13.4}$$

Equation 13.4 shows that the total flux change in the ring is directly proportional to the amount of charge shifted around the galvanometer circuit. But with a ballistic galvanometer, the angular deflection θ of the first throw is directly proportional to the charge ΔQ_2 which passes through it, provided that the discharge is completed by the time the galvanometer moves through only a small portion of its total deflection. Consequently,

$$\theta = K \cdot \Delta Q_2 \tag{13.5}$$

The galvanometer constant K may be determined readily by discharging a known charge on a condenser through a resistance and the galvanometer, noting the deflection.

Then, finally, the change in flux is related to the deflection by the equation

$$\Delta\phi = \frac{R_{g2}}{n_2} \frac{\theta}{K} \tag{13.6}$$

and the change in flux density is

$$\Delta B = \frac{\Delta\phi}{A} = \frac{R_{g2}}{An_2} \frac{\theta}{K} \tag{13.7}$$

The magnetization curve is obtained by plotting B against the corresponding H, and the permeability is defined as $\mu = B/H$.

The *initial relative permeability* μ_{ri} is the value of relative permeability at the origin where B approaches zero.

There is, however, no one fixed B–H relation. Many relationships are possible, depending upon the previous magnetic history of the test sample and the method of test chosen. We shall consider two test methods: the step-by-step method, which yields the *rising B–H characteristic* and the *hysteresis loop;* and the alternating-current

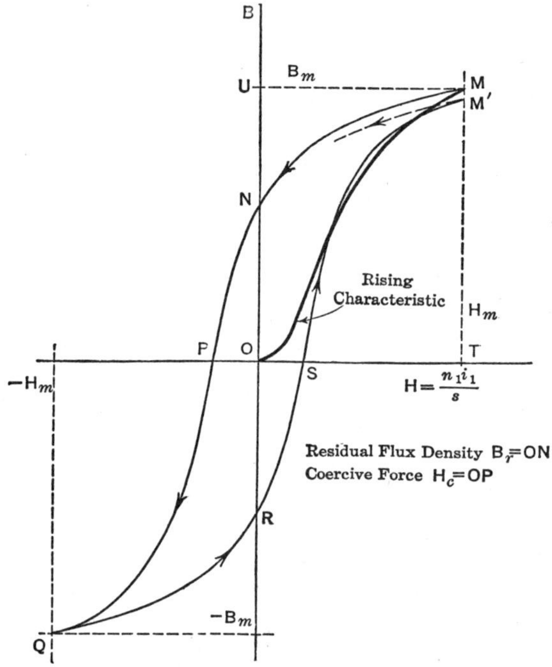

FIG. 13.2. The hysteresis loop and the rising characteristic.

oscillograph method, which yields the *alternating B–H characteristic* and the *hystero-eddy current loop*.

13.4. The Step-by-Step Method. Starting with a relatively high resistance R_1, as shown in Fig. 13.1, throw the switch to the right. As the current builds up to some final value i_1, the galvanometer deflection gives a measure of the corresponding final flux density. Then reduce the resistance R_1 to a new value. The current increases and the galvanometer throw registers the *increase* in flux density above the preceding value. Repeat this process again and again until some desired maximum current is reached. Plotting the computed values of B against H, the rising characteristic OM in Fig. 13.2 is obtained, provided that the magnetic sample used was originally in a demagnetized condition. If the magnetic sample had never before been magnetized, the resulting rising characteristic would have been not far different from the characteristic just described.

Continuing the process with declining currents, the curve *MN* is obtained. With reversed currents, the curve *NPQR* is obtained, and again using the original current direction gives the remaining curve portion *RSM'*. The point *M'* does not quite coincide with *M*, but repeating this cyclic variation between + and $-H_m$ a number of times gives a repeatable loop that is not far different from the first.

13.5. The Hysteresis Loop, Residual Flux Density, and Coercive Force. The loop shown in Fig. 13.2 is known as the *hysteresis loop*, and it shows graphically that the flux density characteristic of a declining current differs from that of a rising current. It will presently be shown (Art. 13.10) that, every time a piece of magnetic material is carried through the flux changes corresponding to the complete loop, there is a definite energy transformation to the heat form that tends to increase the temperature of the specimen. Were it not for this heat loss, the rising and falling characteristics of iron would be identical.

When the flux density is decreased along line *MN* to *N*, the specimen retains a certain degree of flux density *ON* known as its *residual flux density* B_r, even though the current and the magnetic intensity are zero. Reducing the flux density still further along line *NP* to *P*, by using a reverse current and a reverse intensity, the magnetic specimen reaches a condition of zero flux density. The value of the reverse intensity necessary to produce this condition is known as the *coercive force* H_c. Permanent magnets are made of materials having moderate values of residual flux density and high values of coercive force.

13.6. A-C Method for Obtaining Hystero-Eddy Current Loops. Figure 13.3 illustrates the details of a test circuit with which hystero-eddy current loops may be obtained. The alternating voltage is supplied by a sine-wave generator (especially designed to give a pure sinusoidal wave shape) which is connected to a frequency meter (F.M.), a voltmeter V_1, and to the magnetizing winding of the ring or square sample of iron through an ammeter and a variable resistance shunted by one of the elements of an oscillograph. The secondary circuit is connected to a voltmeter and to a second element of the oscillograph. Total resistance R_1 in the magnetizing circuit must be kept as low as possible so that the R_1i_1 drop, which is nonsinusoidal in shape, will not introduce a nonsinusoidal distortion of appreciable magnitude into the voltage applied to the magnetizing winding.

With this circuit we obtain simultaneous time oscillograms of the voltage V_2, which is related to the flux density B in the ring, and of the magnetizing current i_1, which is a direct measure of the intensity H. The flux varies sinusoidally in time since the induced voltage in the

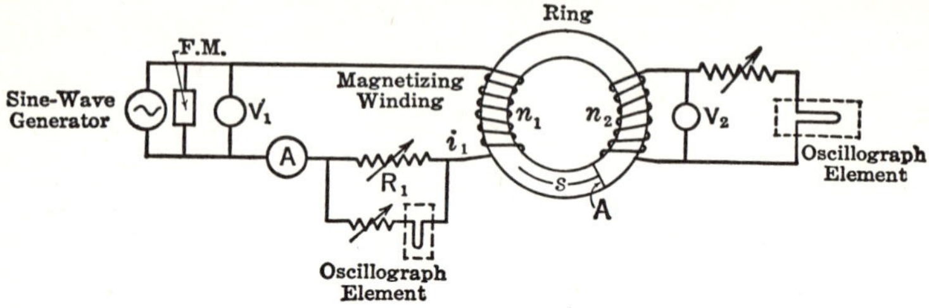

FIG. 13.3. Circuit for obtaining a-c hystero-eddy current loops.

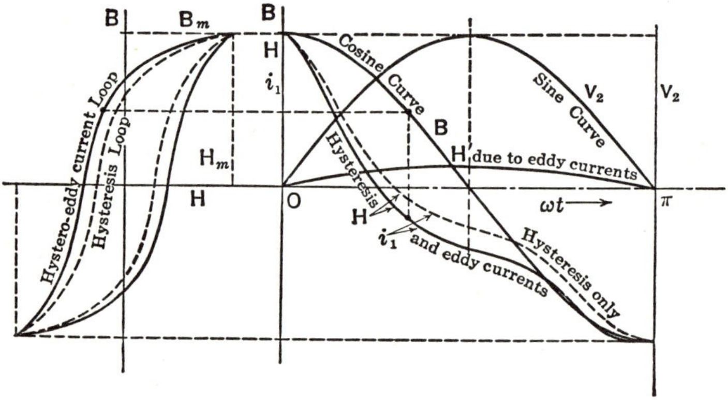

FIG. 13.4. Loop widening and current distortion with sine flux.

magnetizing winding must be sinusoidal to oppose the sine-wave generator voltage. Then if

$$\begin{aligned} \phi &= \phi_{\max} \cos \omega t \\ B &= B_{\max} \cos \omega t \end{aligned} \tag{13.8}$$

$$V_2 = -n_2 \frac{d\phi}{dt} = -n_2 A \frac{dB}{dt} = +n_2 A B_{\max} \omega \sin \omega t \tag{13.9}$$

Also

$$H = \frac{n_1 i_1}{s} \tag{13.10}$$

which indicates that H and i_1 are directly proportional to each other. But neither can be sinusoidal in time since the B–H loop prescribes a definite relationship between B and H and therefore between B and i_1. See Fig. 13.4.

The sinusoidal variation in flux density induces eddy voltages in the iron and eddy currents flow in phase with the voltages. The eddy currents try to reduce the flux density, but this tendency is overcome by an extra current flow in the primary winding and an additional value of H is imposed on the iron for a given value of B. This amounts to a widening of the hysteresis loop in the H direction for given values of B.

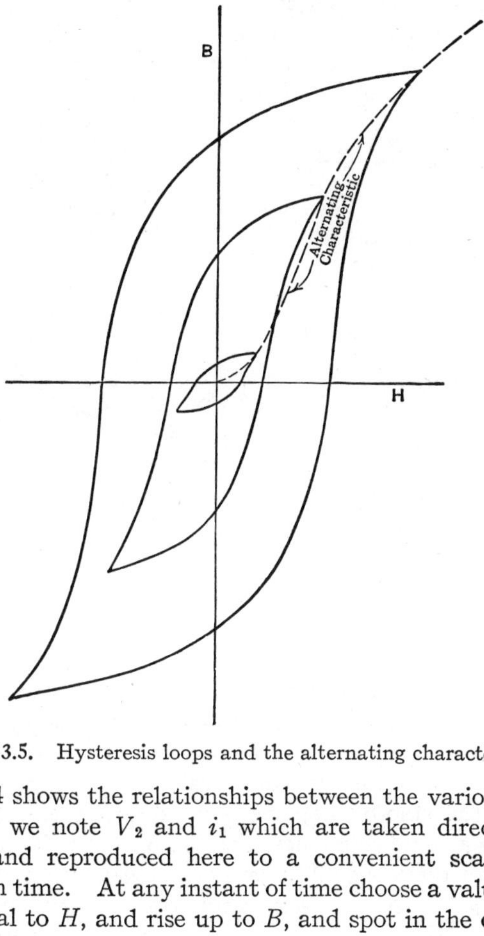

Fig. 13.5. Hysteresis loops and the alternating characteristic.

Figure 13.4 shows the relationships between the various quantities. To the right we note V_2 and i_1 which are taken directly from the oscillogram and reproduced here to a convenient scale. B is 90° ahead of V_2 in time. At any instant of time choose a value of i_1, which is proportional to H, and rise up to B, and spot in the corresponding point on the B–H hystero-eddy current loop. In this way the whole loop may be developed from the oscillogram.

Inside the hystero-eddy current loop lies the hysteresis loop proper, which would be obtained in this procedure only if there were no eddy currents. On the right in Fig. 13.4 in dashed lines is shown the current that would flow in the magnetizing winding under this condition. The amount of widening of the loop and of extra current required is directly proportional to the frequency (for low frequencies) so long as the flux density is uniform over the iron cross section. For ordinary iron at a frequency of 60 cycles per second the hystero-eddy current loop is approximately one and a half times as wide in H as the hysteresis loop proper. The widening in H caused by eddy currents varies sinusoidally in time and is in phase with V_2 as shown. This problem is discussed in more detail under core losses.

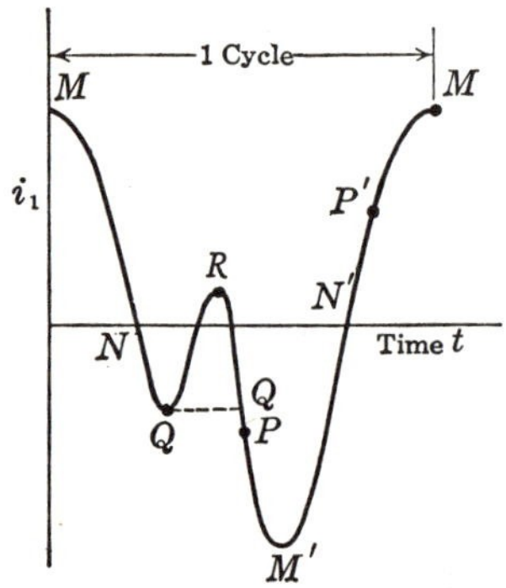

Fig. 13.6. Distorted magnetizing current. (See Fig. 13.7 for corresponding points.)

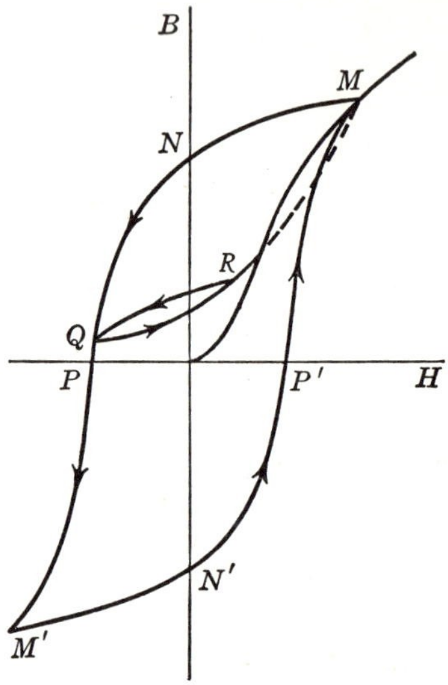

Fig. 13.7. Minor loop superimposed on major loop. (See Fig. 13.6 for corresponding points).

The hysteresis loop proper cannot be obtained directly, but two hystero-eddy current loops can be taken at two different frequencies, and, since the widening is proportional to frequency (for low frequencies), the results can be extrapolated to zero frequency. The hysteresis loop so obtained coincides very closely with the loop obtained by the step-by-step method except at low values of flux density where a slight difference usually appears. It should be noted that the tips of the loops all lie on the magnetization curve and that permeability $\mu = B_m/H_m$ applies only to that curve and to the tips and not at all to any other parts of the hysteresis loop. The magnetization curve derived from alternating current methods is known as the *alternating characteristic*. This is illustrated in Fig. 13.5.

13.7. Displaced Hysteresis Loops. If the magnetizing current flowing in coil 1 in Fig. 13.3 should have a wave shape with reversals as shown in Fig. 13.6, the hysteresis loop followed contains a minor or displaced loop of the type illustrated in Fig. 13.7. The arrows indicate the direction in which the loops are traced. The hysteresis loss per unit volume per cycle is now equal to the sum of the areas of the major and minor loops, measured in $B \times H$ units.

If, instead of reversing at point R, the current i_1 and H had continued increasing in the positive direction, the minor loop would have

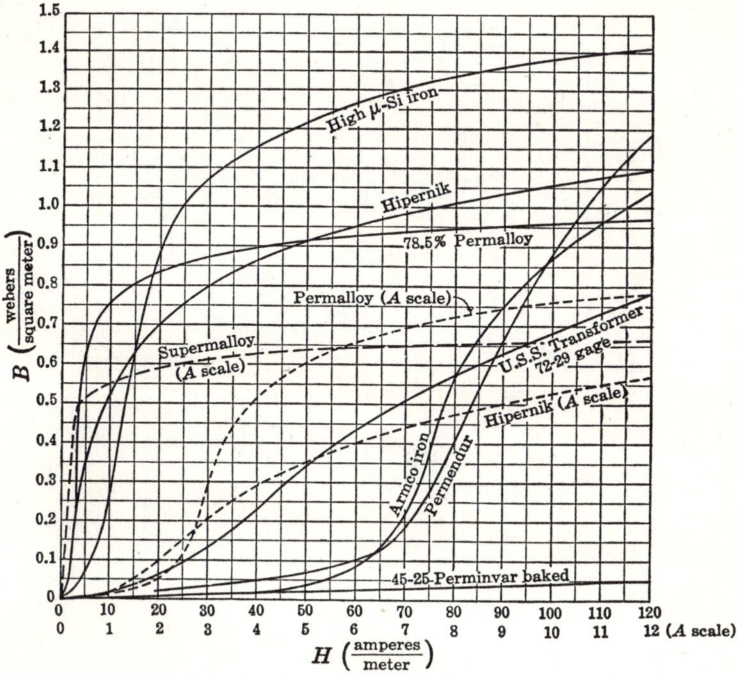

FIG. 13.8. Magnetization curves.

followed along the broken line RM back to the maximum point M on the magnetization curve.

13.8. Magnetization Curves and Hysteresis Loops. Magnetization (B versus H) curves for a wide variety of materials are given in Figs. 13.8 to 13.13 inclusive. The materials have been chosen to represent a wide spread in magnetic properties. At high values of H and B, shown in Fig. 13.11, $\beta = B - \mu_0 H$ also is plotted. β is called the *ferric flux density* and represents the flux density contribution of the magnetized particles of the material alone, while B, the total flux density, gives in addition the contribution of the magnetizing winding to the flux density. β therefore reaches a definite saturation value β_s

when the magnetized material has made its maximum contribution, whereas B continues to increase with H indefinitely. In the region of complete saturation any change in B is numerically equal to the corresponding change in $\mu_0 H$.

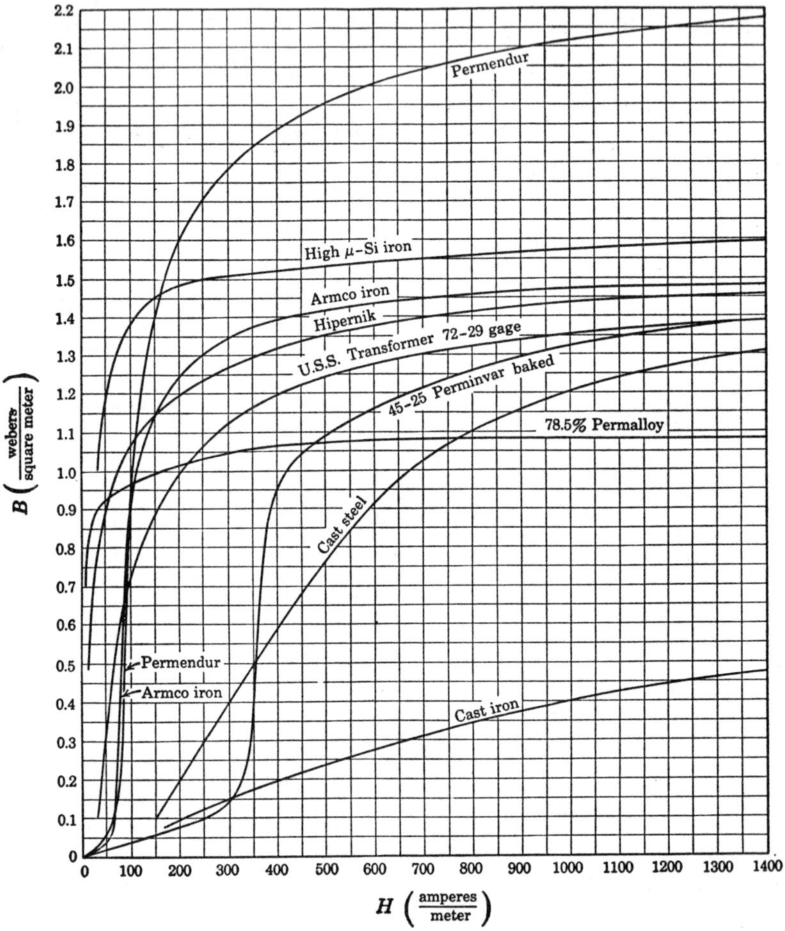

FIG. 13.9. Magnetization curves.

Permalloy with 78.5 per cent nickel and 21.5 per cent iron rises in relative permeability from an initial value $\mu_{ri} = 10{,}000$ with $B = 0$ to a maximum of 105,000 with $B = 0.51$. For higher values of B the permeability falls very rapidly and the material is no longer so good magnetically as many others. For this reason Permalloy must be used at the lower values of flux density if its especial qualities are to be utilized advantageously. *High permeability at low values of* H *implies a narrow hysteresis loop and low hysteresis loss.* Permalloy, in powdered form to reduce eddy current losses, is used extensively as a core material for inductance coils. In the form of a tape, it is used in submarine telegraph cables. Wrapped around the copper conductors it increases the inductance of the cable, which is necessary if messages are

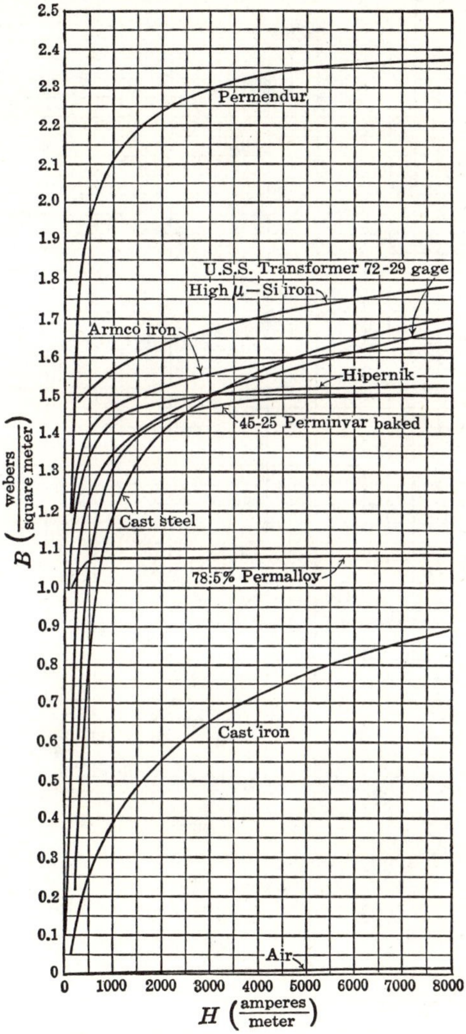

FIG. 13.10. Magnetization curves.

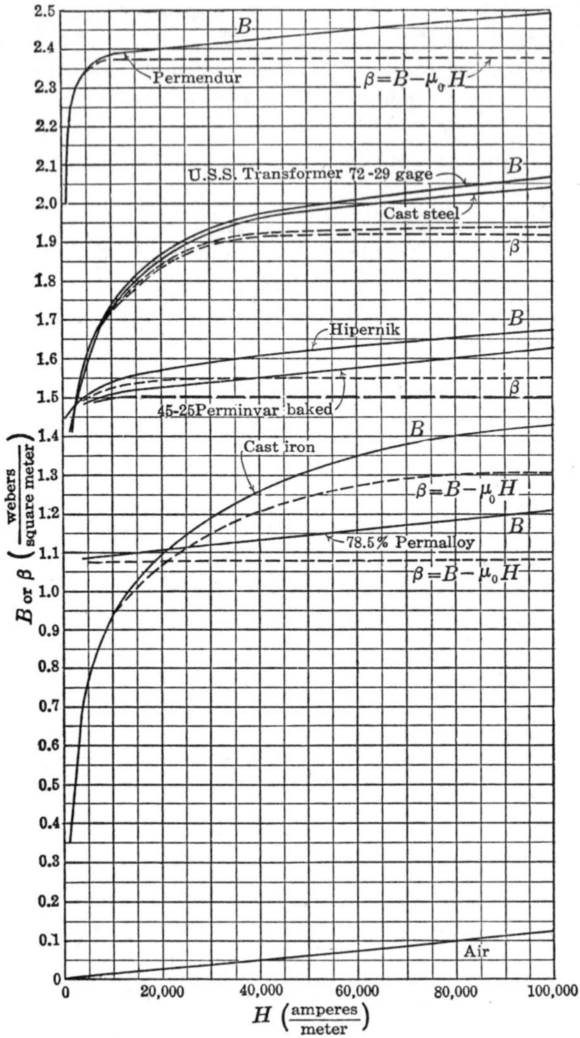

FIG. 13.11. Magnetization curves. (Note inclusion of the air line.)

to be sent with greatest speed. The qualities of Permalloy, as well as other magnetic materials, are to be realized only if the material has been given a very particular type of heat treatment.

Hipernik with 50 per cent nickel and 50 per cent iron has lower values of initial and maximum relative permeability, $\mu_{ri} = 5000$ and $\mu_{r\,\max} = 60{,}000$, but it may be used to advantage at higher flux densities than

Permalloy. Its saturation value of β_s is much higher, 1.55 compared with 1.08 for Permalloy. One of the most important applications of Hipernik is in the field of current transformer cores for instrumentation.

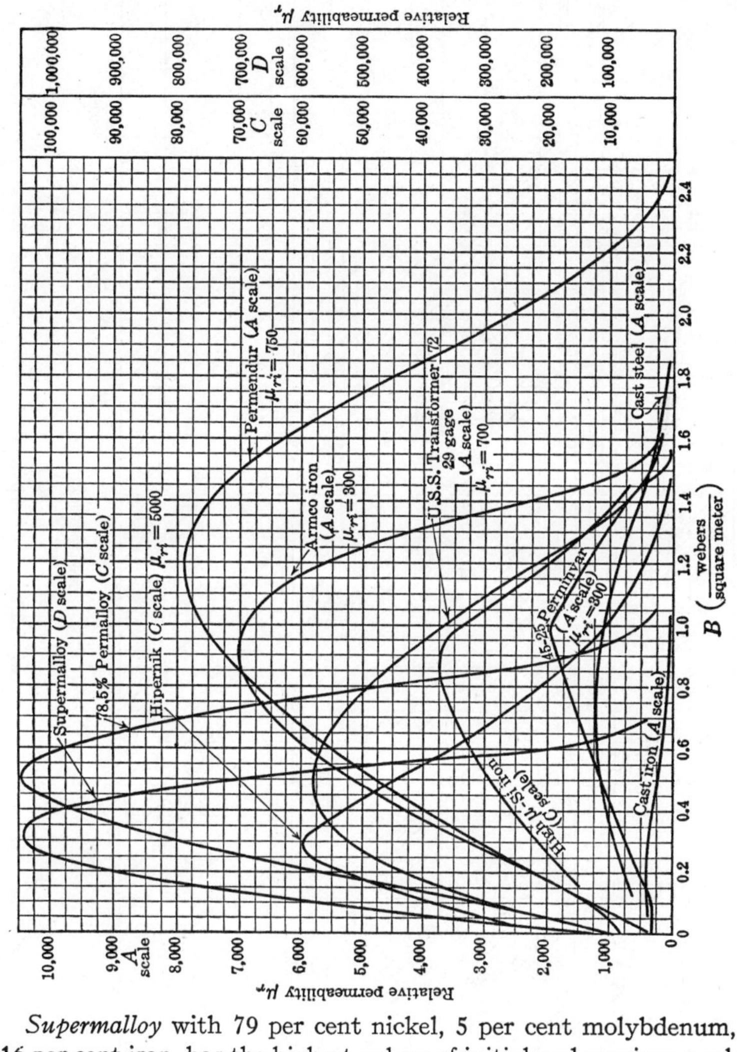

FIG. 13.12. μ versus B curves.

Supermalloy with 79 per cent nickel, 5 per cent molybdenum, and 16 per cent iron, has the highest values of initial and maximum relative permeabilities; $\mu_{ri} = 100{,}000$ and $\mu_{r\,\max} = 1{,}050{,}000$, occurring at $B = 0$ and 0.34 weber per square meter, respectively. The saturation value of $\beta_s = B_s - \mu_0 H$ is 0.8 weber per square meter. Supermalloy is used in the cores of special types of transformers.

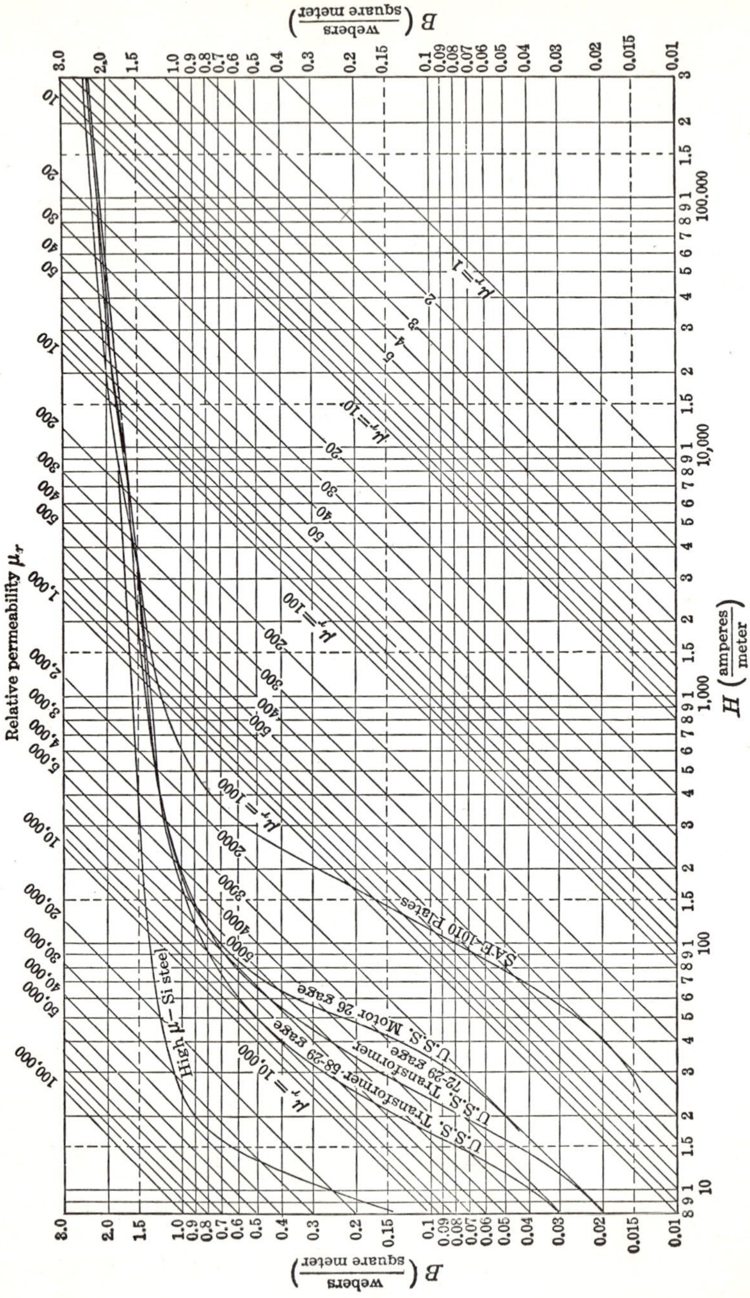

FIG. 13.13. Magnetization curves and relative permeability for commercial steel sheets.

Perminvar with 45 per cent nickel, 25 per cent cobalt, and 30 per cent iron, in contrast with the previous nickel alloys, has a low permeability for low flux densities, but with the special quality that the B

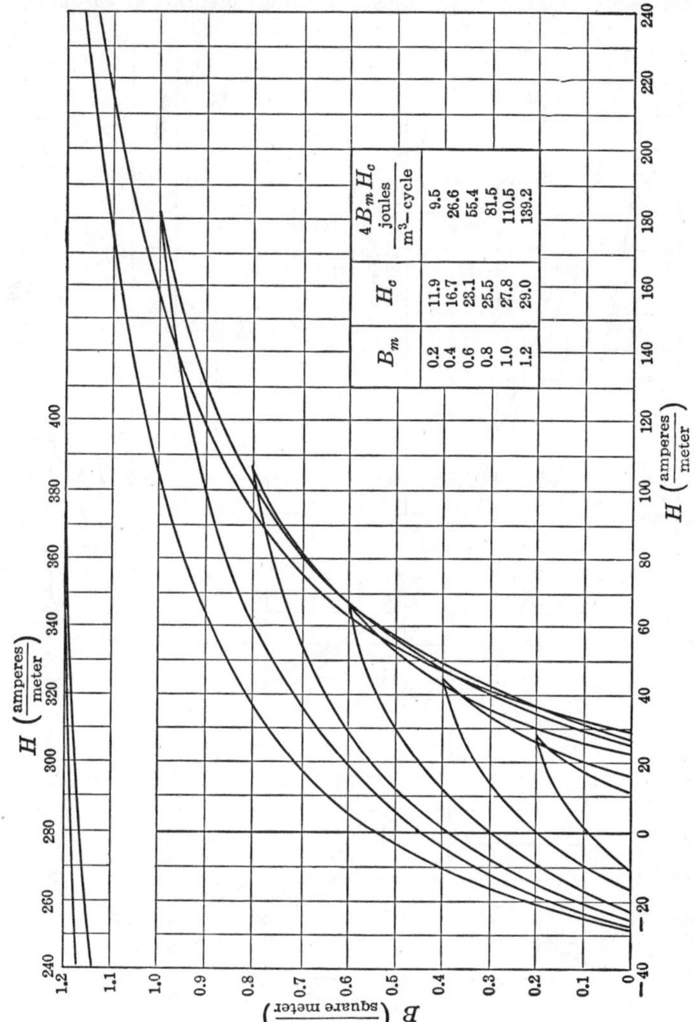

FIG. 13.14. Hysteresis loops for U.S.S. Transformer 58–29 gage. (For $B_{max} = 1.2$, $H_{max} = 380$.)

versus H curve is practically a straight line, and the permeability though low is constant, up to 0.08 weber per square meter. In this range the hysteresis losses are negligible and the flux density is directly proportional to the magnetizing current. This material is peculiarly adapted to the needs of communication apparatus in which the magnetic response should be directly proportional to the voice currents.

Permendur with 50 per cent cobalt and 50 per cent iron presents a strong contrast in its characteristics. Its high saturation value, $\beta_s =$

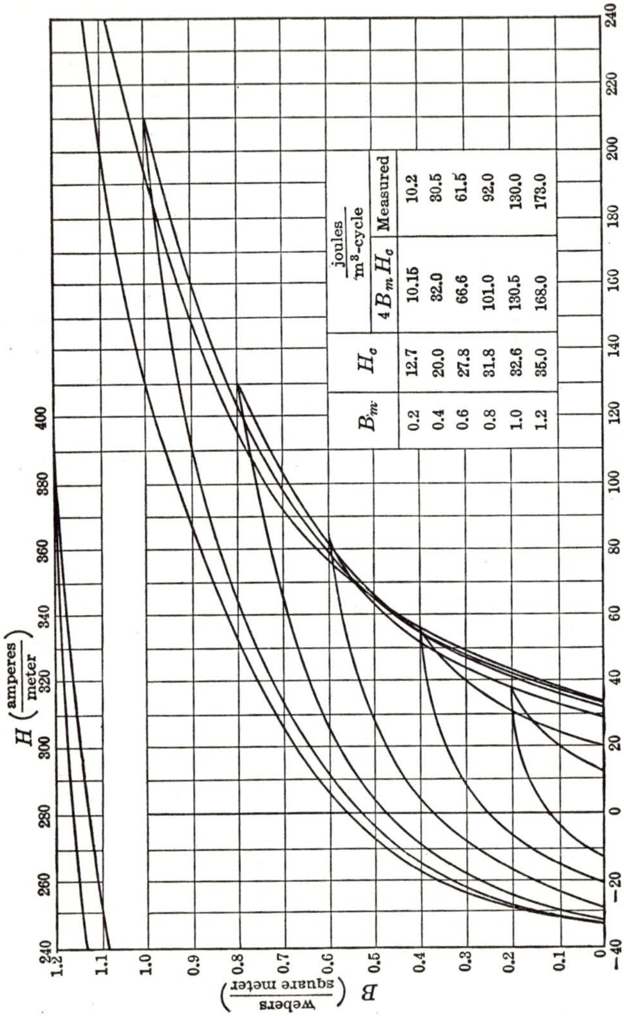

B_m	H_c	$\frac{\text{joules}}{\text{m}^3\text{-cycle}}$	
		$4B_mH_c$	Measured
0.2	12.7	10.15	10.2
0.4	20.0	32.0	30.5
0.6	27.3	66.6	61.5
0.8	31.8	101.0	92.0
1.0	32.6	130.5	130.0
1.2	35.0	168.0	173.0

FIG. 13.15. Hysteresis loops for U.S.S. Transformer 72–29 gage. (For $B_{\max} = 1.2$, $H_{\max} = 390$.)

2.37, and high permeability at high values of flux density indicate that Permendur is a useful component of the magnetic circuits of electromagnets and permanent magnets, where it is used for yokes and pole pieces. The 50–50 cobalt–iron composition of Permendur is rather brittle and cannot be rolled into thin sheets, but with the addition of about 2 per cent vanadium fabrication of thin sheets is possible. In this form the material is used in the thin diaphragms of telephone receivers.

Armco Iron, an iron in which the impurities have been reduced to very small values, is comparable in performance with ordinary sheet steel at the lower flux densities but continues to rise to a saturation

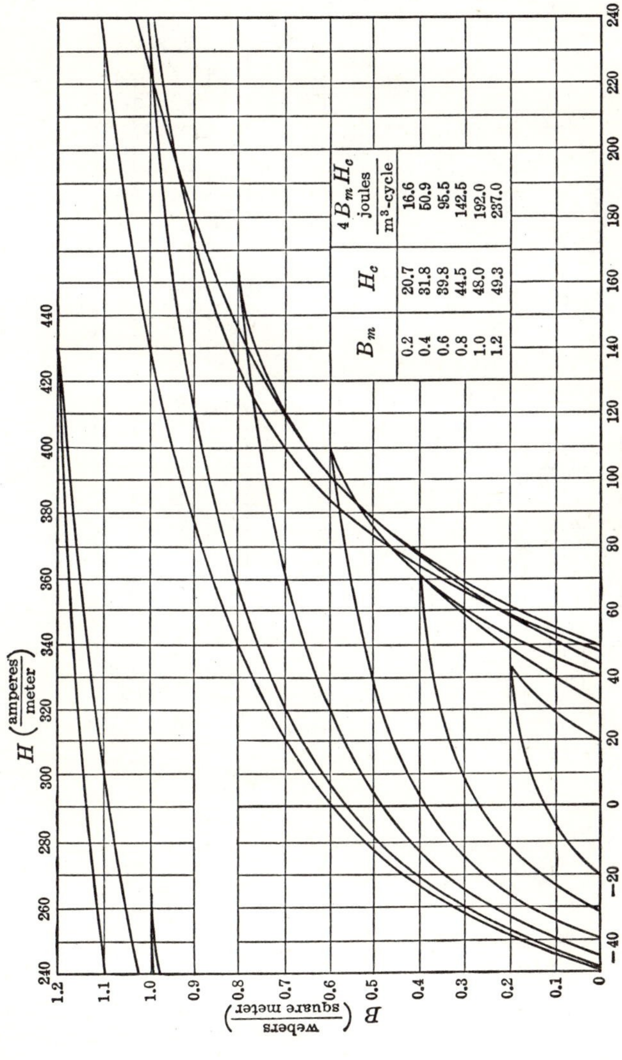

FIG. 13.16. Hysteresis loops for U.S.S. Motor grade—26 gage.

$\beta_s = 2.2$. For comparative purposes it may be noted, from Table 13.1, that iron containing only 0.02 per cent of impurities and heat-treated in a hydrogen-reducing atmosphere has attained a maximum relative permeability of 275,000.

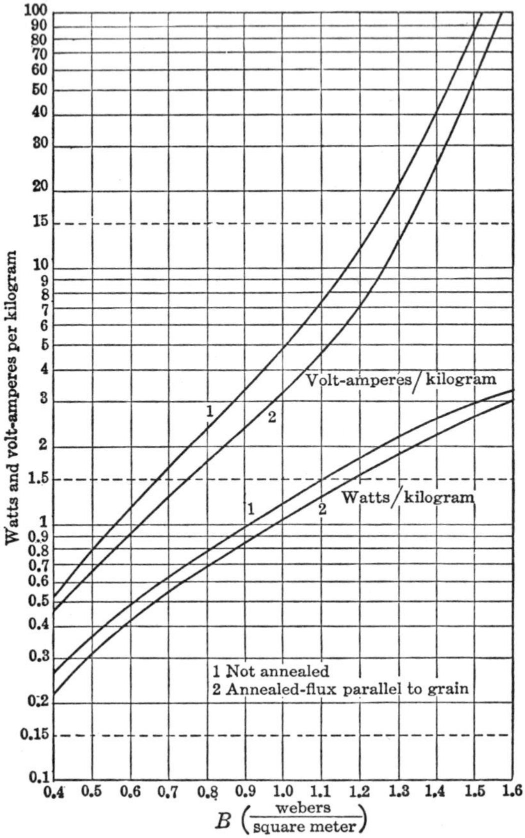

FIG. 13.17. 60-cycle core loss and volt-ampere curves for U.S.S. Transformer 58–29 gage.

Cast iron is a very poor material for magnetic purposes, but *cast steel* is nearly as good as electrical sheet steel, except in the lower flux densities; however, it cannot be laminated.

Electrical sheet steel is represented here by: U.S.S. Transformer 58–29 gage, 0.014 inch thick; U.S.S. Transformer 72–29 gage; U.S.S. Motor grade, 26 gage, 0.0185 inch thick; and U.S.S. SAE 1010 hot rolled low carbon steel plates. The designation of the transformer grades means that the guaranteed core loss at $B = 1$ weber per square meter, 60 cycles, will not exceed 0.58 and 0.72 watt per pound, respectively, as measured in a standard Epstein Test Square. For the Motor grade the value is 1.14 watts per pound.

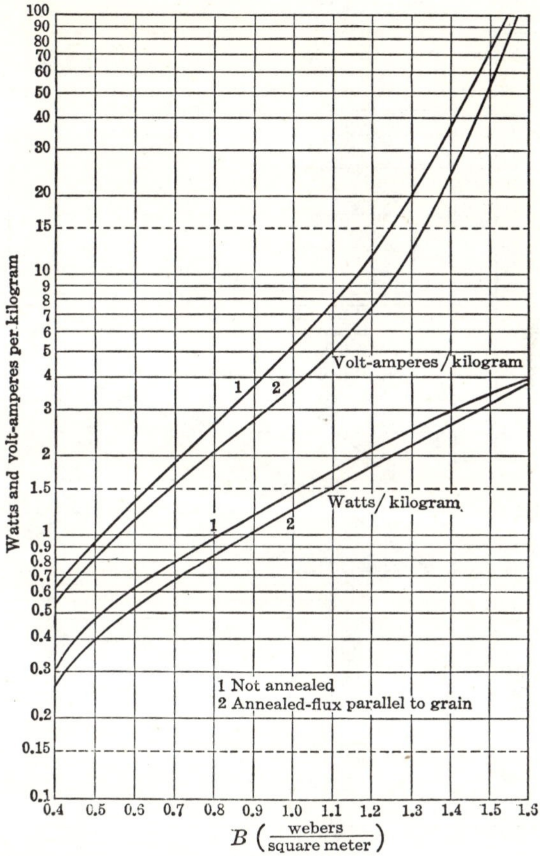

FIG. 13.18. 60-cycle core loss and volt-ampere curves for U.S.S. Transformer 72–29 gage.

Motor grade is used in motors, generators, small transformers, reactors, where medium valued core losses are permissible. SAE-1010 plates are used in the rotating pole structure of alternators and in the field yokes of d-c machines.

In addition, there are available high permeability silicon steels which, if properly annealed and used with the flux parallel to the direction of rolling, exhibit superior quality particularly in the medium and lower values of flux density.

The magnetization and relative permeability curves for these five materials are given in Fig. 13.13. Both properties are readily exhibited on one diagram by using log-log paper.

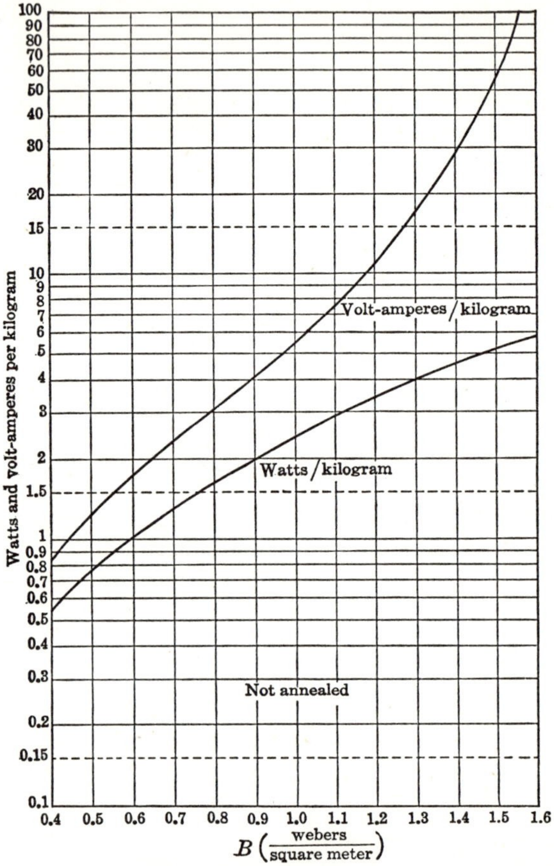

FIG. 13.19. 60-cycle core loss and volt-ampere curves for U.S.S. Motor grade—26 gage.

For U.S.S. Transformer 58 and 72 grades and Motor grade, the hysteresis loops are given in Figs. 13.14, 13.15, and 13.16. Volt-ampere curves and watts core-loss curves per kilogram for these materials are given in Figs. 13.17, 13.18, and 13.19, plotted against B. For the two transformer grades we note the improvement resulting from annealing and arranging the rolling direction parallel to the flux.

Figure 13.20 gives the variation of density, resistivity, and core loss of U. S. S. electrical sheets with silicon content. In general the effects of increasing silicon content are to decrease the core loss, both hysteresis and eddy losses, to decrease the saturation flux density β_s, to decrease the permeability at flux densities above 1.2 webers per square meter, but

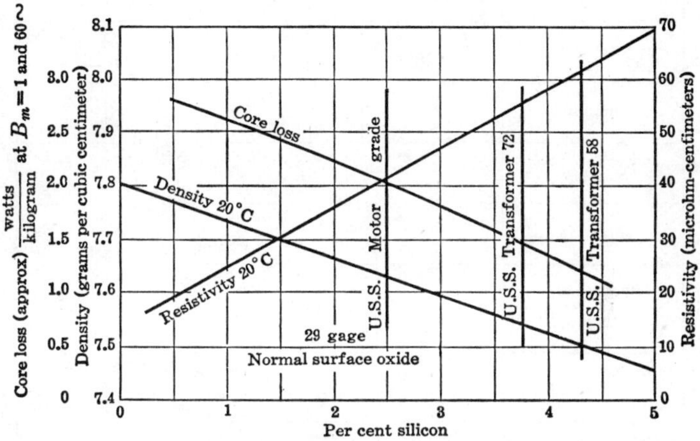

FIG. 13.20. Effect of silicon content on electrical steel sheets.

to increase the maximum permeability which occurs at lower values of flux density, and to increase the stiffness and hardness of the sheets.

Table 13.1 taken from the *Bell System Technical Journal* gives a wide variety of properties of certain "soft" magnetic materials, including some of those discussed in this chapter. "Soft" magnetic materials are those having low values of coercive force.

B. REDEFINITION OF FUNDAMENTAL CONCEPTS

13.9. Redefinition of Fundamental Concepts. In Chapter 9 the fundamental field concepts B and H are defined on the assumption that the magnetic flux set up by a coil is *linearly* related to the magnitude of the current passed through the coil. It follows from this that the magnetic flux density B (which depends on the total flux) is linearly related to the magnetic field intensity H (which depends on the magnetizing coil current), and the permeability $\mu = B/H$ is a constant—that is, is independent of the current magnitude.

In fact, no ferromagnetic material has a constant permeability. It is necessary to resort to experiment, as outlined in Art. 13.3, to determine B and H independently. Then, by definition, the permeability is the ratio of these independently determined values ($\mu = B/H$). It

TABLE 13.1

SOFT MAGNETIC MATERIALS IN SOLID OR SHEET FORM

Per Cent Composition						Raw Cost ¢/lb	Typical Anneal	Material	Initial Relative Permeability	Maximum Relative Permeability	Saturation	Hysteresis Loss at Saturation	Residual	Coercive Force	Resistivity	Curie Temperature	Density
Other	Mn	Mo	Ni	Co	Fe				μ_{ri}	μ_{rm}	$\beta_s = M_s$	W_∞	B_r	H_c	ρ	T_c	δ
											$\frac{webers}{m^2}$	$\frac{joules}{m^3}$	$\frac{webers}{m^2}$	$\frac{amp}{meter}$	microhm-cm		$\frac{gm}{cm^3}$
3 C, 2 Si	...	...	...	...	95	<1	800° (Centigrade)	Cast Iron	...	600	...	2000	0.53	366	30	...	...
	...	...	...	...	99.94	7	900° Pot	Magnetic Iron	250	5,500	2.15	500	1.3	79.5	10	770° C	7.88
	...	...	...	...	99.98	7	1480° H_2 + 880° H_2	Magnetic Iron H_2 Purified	25,000	275,000	2.15	30	1.36	4	10	770	7.88
0.5 Si	...	...	...	...	99.5	7	800° Pot	0.5 Si-Iron (Field)	250	3,700	2.10	450	1.28	63.5	18	760	7.7
4 Si	...	...	...	...	96	8	800° Pot	4 Si-Iron (Transformer)	400	6,700	2.0	350	1.2	40	60	690	7.5
9½ Si, 5½ Al	...	...	...	...	85	3	Cast	Sendust	30,000	120,000	1.0	10	0.5	4	80	...	7.1
0.2 Cu	...	...	99.0	...	0.4	35	1000° Pot	Nickel	110	600	0.61	300	0.36	270	8	360	8.85
	0.6	...	45	...	54.4	17	1100° Pot	45 Permalloy	2,700	23,000	1.65	120	0.8	23.8	45	440	8.17
	...	...	50	...	50	18	Long 1200° H_2	Hipernik	3,000	70,000	1.65	22	0.73	3.2	35	500	8.25
	0.6	...	78.5	...	20.9	28	1000° + Quench	78.5 Permalloy	9,000	105,000	1.07	20	0.6	4	16	580	8.60
5 Cu	1.0	...	74	...	20	27	900° + Quench	Mumetal	7,000	80,000	0.85	20	0.6	4	25	...	8.58
3.8 Cr	0.6	...	78.5	...	17.1	29	1000° Pot	3.8–78.5 Cr-Permalloy	10,000	40,000	0.8	20	0.45	4	65	455	8.56
	0.6	4	79	...	16.4	32	1000° Pot	4–79 Mo-Permalloy	22,000	72,000	0.85	20	0.5	4	55	460	8.72
15 Cu	1.0	3	71	...	10	29	1100° H_2	1040 Alloy	40,000	100,000	0.6	20	0.25	1.1	55	290	8.76
	...	12.5	80	...	7.5	40	800° H_2	12.5–80 Mo-Permalloy	9,000→1, over room temperature range					...	...	40	8.9
	...	...	...	99	...	136	1000° Pot	Cobalt	70	240	1.8	200	0.5	795	9	1120	8.9
	...	...	...	50	50	69	900° Pot	Permendur	800	5,000	2.4	1200	1.4	159	7	1000	8.3
2 V	...	...	...	49	49	73	800° Pot	2 V-Permendur	800	4,500	2.4	600	1.4	159	26	980	8.2
	0.6	...	45	25	29.4	50	1000° + 425° Bake	45–25 Perminvar	365	1,800	1.55	400	0.33	111	19	715	8.6
	0.6	...	70	7	22.4	35	1000° + 425° Bake	7–70 Perminvar	850	4,000	1.25	...	0.24	47.7	16	650	8.6
	0.6	7.5	45	25	21.9	57	1000° + 425° Bake	7.5–45–25 Mo-Perminvar	550	3,800	1.03	260	0.43	47.7	80	540	8.66
	...	5	79	...	16	...	1300° C, H_2	Supermalloy	100,000	1,050,000	0.8	...	...	0.32	60	400	8.87

Note: Values for μ_{ri}, μ_{rm}, W_∞, B_r, and H_c are subject to considerable variations, depending on purity of materials, composition, type of heat treatment, and final condition of mechanical stress.

Adapted from V. E. Legg, "Survey of Magnetic Materials and Applications in the Telephone System," *Bell System Technical Journal*, **18,** 438 (July 1939).

should be noted that this is not the slope of the magnetization curve. Figure 13.21 shows the usual B versus H curve as well as the μ versus H curve, which can be derived from it.

Table 13.2 lists the various basic magnetic field concepts. In one column the definitions are given for the condition of constant permeability, independent of field strength. The other column shows the revisions that must be made in order to take into account the dependence of the permeability upon the strength of the magnetizing current. Let us consider those definitions that are different under the two conditions of permeability.

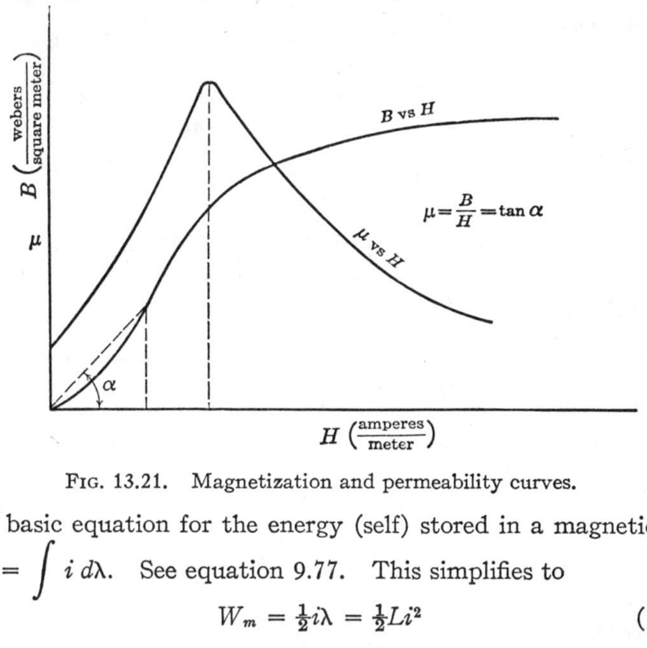

FIG. 13.21. Magnetization and permeability curves.

The basic equation for the energy (self) stored in a magnetic field is $W_m = \int i\,d\lambda$. See equation 9.77. This simplifies to

$$W_m = \tfrac{1}{2}i\lambda = \tfrac{1}{2}Li^2 \tag{13.11}$$

only in case the flux linkages are directly proportional to the current.

The definition for self-inductance based on the voltage-generating quality of a time-changing current is the same whether the permeability remains constant or otherwise.

$$L = \frac{-V_g}{\left(\dfrac{di}{dt}\right)} = \frac{\left(\dfrac{d\lambda}{dt}\right)}{\left(\dfrac{di}{dt}\right)} = \frac{d\lambda}{di} \tag{13.12}$$

With variable permeability, the flux linkages are not linearly related to the current, and the inductance cannot be a constant but must

TABLE 13.2

FUNDAMENTAL MAGNETIC CONCEPTS

COMPARISON FOR μ = CONSTANT AND μ = VARIABLE QUANTITY

	μ = constant	μ = variable i.e., depends on i or H or B
Coulomb's Law	$F = \dfrac{p_1 p_2}{4\pi\mu_0 r^2}$	
Unit Pole	$p = \sqrt{F 4\pi\mu_0 r^2}$	
Magnetic Intensity	$H = \dfrac{F}{p} = \dfrac{ni}{s}$	$H = \dfrac{ni}{s}$ from ring sample
Flux Density	$B = \mu H$	$B = \mu H = \dfrac{1}{nA}\int V_g\,dt$ from ring sample
Flux	$\phi = \iint_A B_n\,dA$	← Same
Permeability	$\mu = \dfrac{B}{H} = \dfrac{dB}{dH}$	$\mu = \dfrac{B}{H}$, not $\dfrac{dB}{dH}$
Magnetic Potential Difference	$\Delta U = \int_s - H\,ds\cos\alpha$	← Same
Closed Path MMF	$\oint_s d\mathfrak{F} = ni$ or 0	← Same
Energy (Self)	$W_m = \frac{1}{2}i\lambda$ $= \frac{1}{2}Li^2$	$^*W_m = \int i\,d\lambda$
Inductance (Self)	$L = \dfrac{\lambda}{i}$ = constant $= \dfrac{2W_m}{i^2}$ $= \dfrac{-V_g}{\left(\dfrac{di}{dt}\right)} = \dfrac{+\left(\dfrac{d\lambda}{dt}\right)}{\left(\dfrac{di}{dt}\right)}$	$L = \dfrac{d\lambda}{di}$ variable ← Same
Energy Density	$\omega_m = \dfrac{B^2}{2\mu} = \dfrac{BH}{2}$	$^*\omega_m = \int_0^B H\,dB$

* Energy supplied to the iron, only part of which is stored up, the remainder being transformed to heat.

TABLE 13.2. *Continued*

FUNDAMENTAL MAGNETIC CONCEPTS

COMPARISON FOR μ = CONSTANT AND μ = VARIABLE QUANTITY

	μ = constant	μ = variable i.e., depends on i or H or B
Reluctance	$\mathcal{R} = \int \frac{d\mathcal{F}}{\phi}$	← Same
Generated Voltage	$V_g = \frac{-d\lambda}{dt} = Bsv_n$	← Same
Energy (Mutual)	$W_m = \pm i_1\lambda_{12} = \pm i_2\lambda_{21} = \pm Mi_1i_2$	* $W_m = \pm \int i_1\, d\lambda_{12} = \pm \int i_2\, d\lambda_{21}$ providing μ is based on the combined flux density, not on the flux densities of the individual coil fields
Inductance (Mutual)	$M = \frac{\lambda_{12}}{i_2} = \frac{\lambda_{21}}{i_1} = \frac{W_m}{i_1i_2} = \frac{-V_{g1}}{\left(\frac{di_2}{dt}\right)} = \frac{-V_{g2}}{\left(\frac{di_1}{dt}\right)}$	$M = \frac{d\lambda_{12}}{di_2} = \frac{d\lambda_{21}}{di_1}$ with the same restriction as for mutual energy. M is variable; ← Same
Force on Current	$F_1 = B_{12}i_1s_1 \sin\theta_1$	← Same, except for the mutual energy restriction

* Energy supplied to the iron, only part of which is stored up, the remainder being transformed to heat.

depend on the current. Consequently, the two inductance definitions $L = \lambda/i$ and $L = 2W_m/i^2$ are valid only for constant permeability.

Consider the ring sample of Fig. 13.1. Since the total energy supplied to the magnetic field is $W_m = \int i\, d\lambda$, the expression for the energy density under the condition of variable permeability may be developed as follows:

$$H = \frac{n_1 i_1}{s} \quad \text{or} \quad i_1 = \frac{Hs}{n_1}$$

$$\lambda = n_1AB \quad \text{or} \quad d\lambda = n_1A\, dB$$

$$v = As \qquad \text{(volume)}$$

Then

$$W_m = \int i\,d\lambda = v \int_0^B H\,dB \tag{13.13}$$

and the energy density is

$$\omega_m = \int_0^B H\,dB \quad \frac{\text{joules}}{\text{cubic meter}} \tag{13.14}$$

In Fig. 13.2 the area $OMUO$ expressed in BH area units gives the energy supplied to 1 cubic meter of the ring when the iron is changed from its demagnetized condition to one of magnetic density OU. If there were no heat loss, all this energy would be stored in the magnetic field. As it is, the actual storage must be less than this value.

The mutual relations, mutual energy, mutual inductance, and the law of force must be treated in terms of infinitesimal changes in flux linkages accompanying infinitesimal changes in current, as in the case of self-energy and self-inductance. They also have the additional restriction that the permeability used in any of the relations depends, point by point in the field, upon the total flux density as determined by the *combined* field of the two or more coils or wires, and not on the flux densities of the separate coils or wires.

C. CORE LOSSES

13.10. Core Losses with Alternating Flux. When an alternating current of frequency f cycles per second is passed through coil 1 in Fig. 13.1, the magnetic intensity and the magnetic flux density in the ring alternate between positive and negative values $\pm H_m$ and $\pm B_m$ respectively, as shown in Fig. 13.22. As a consequence of the alternating flux, there are two losses or transformations of energy into the form of heat known as the *hysteresis loss* and the *eddy current loss.*

The hysteresis loss may be computed from the hysteresis loop. Traversing the loop in Fig. 13.22 from S to M, the areas 1 and 2 give the magnetic energy density supplied to the magnetic material at a maximum flux density B_m. From M to N, the area 2 gives the energy per cubic meter returned by the magnetic material to the electric circuit. From N to P, the area 3 represents the energy supplied to the magnetic material. On the other half of the loop, areas 4, 5, and 6 are proportional to the energy supplied to the magnetic material, but that corresponding to 5 is returned. The net result is that the loop area, measured in BH units, gives the energy per cycle per cubic meter that is supplied to the magnetic material and not returned, but is transformed to heat and raises the temperature of the material.

$$\frac{\text{Energy/cycle}}{\text{cubic meter}} = \text{area of hysteresis loop} \tag{13.15}$$

Assuming uniform flux density throughout the ring, the loss is proportional to the ring volume. Experiment has shown that the hysteresis

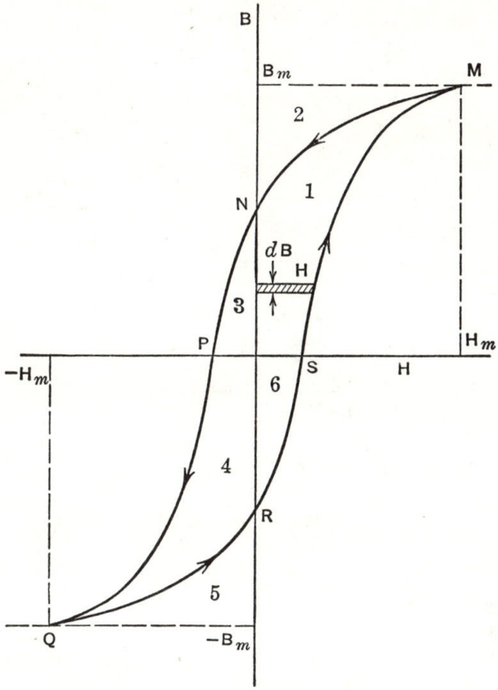

FIG. 13.22. Hysteresis loss is equal to the loop area. $\omega_m = \oint H\,dB \, \frac{\text{joules}}{\text{m}^3\text{-cycle}}$.

loss per cycle is independent of the time required to traverse the loop. Consequently the loss is proportional to the frequency. The total hysteresis loss per second then is

$$P_H = \text{volume} \cdot f \cdot \text{area of loop} \quad \frac{\text{joules}}{\text{second}} \text{ or watts} \tag{13.16}$$

As the maximum flux density B_m is increased, the area of the loop increases, causing an increase in the hysteresis loss.

An empirical rule for determining the hysteresis loss that is applicable for tip flux densities between 0.2 and 1.2 webers per square meter is

$$P_H = \text{volume} \cdot \eta \cdot f \cdot B_{\max}{}^x \quad \text{watts} \tag{13.17}$$

For use in this formula the coefficients η and x must be determined by experiment. A number of values of η and x are given in Table 13.3.

TABLE 13.3

HYSTERESIS AND EDDY CURRENT LOSS COEFFICIENTS

Material	Hysteresis		Eddy Current		
	η	x	ϵ	Resistivity ρ microhm-centimeter	
				Computed*	From D-C Test
U.S.S. Transf. 72...	130	1.582	0.747×10^7	16.5	56
Hipernik...........	25	1.75	0.745×10^7	16.5	45
Cast Steel..........	2500	1.6	....	....	..
Cast Iron..........	3750	1.6	....	....	30

* Computed from equation 13.39.

Another empirical rule for computing the hysteresis loss is

$$P_H = \text{volume} \cdot 4fH_{\text{coercive}}\, B_{\max} \quad \text{watts} \tag{13.18}$$

This rule is based on the approximation that the area of the hysteresis loop is equal to $4H_{\text{coercive}} \cdot B_{\max}$ which gives results which are somewhat too high at flux densities below 1.0 weber per square meter and gives results which are somewhat too low at the higher flux densities.

Silicon steel, containing 1.5 to 4 per cent silicon, possesses the very desirable quality of nonaging. Steels which age show a gradually increasing magnetic loss under prolonged heating. In the past, transformers with cores made of steel without the proper silcon content frequently showed losses increasing to the point where it became economical to replace them.

Conducting materials subject to changing magnetic fields are usually laminated in sheet form with the sheets placed parallel to the field. The changing field sets up eddy voltages and currents in the conducting material and causes a heat loss. The eddy voltages are set up principally along the sheet and perpendicular to the flux. If the flux is uniformly distributed over the cross section, the loss per sheet is proportional to the i^2R or $V_g{}^2G$ of each sheet, but the generated voltage is proportional to the time rate of change of flux, which depends on the frequency, maximum flux density, and thickness. The conductance G also depends on the thickness d.

The loss per sheet is proportional to V_g^2G

or

$$(fB_{\max}d)^2d$$

or

$$f^2B_{\max}^2d^3 \quad \text{watts} \tag{13.19}$$

but the loss per unit volume depends on the square of the thickness, as the number of sheets in a given volume varies inversely with the thickness. Consequently the total eddy current loss is

$$P_E = \text{volume} \cdot \epsilon f^2B_{\max}^2d^2 \quad \text{watts} \tag{13.20}$$

The eddy current coefficient ϵ is usually obtained experimentally. A number of values of ϵ are given in Table 13.3. A more detailed analysis of the eddy current losses is given in Art. 13.12.

The total core loss per unit volume, with uniform density of flux which varies sinusoidally in time, is

$$p_T = p_H + p_E = \eta B_m^x f + \epsilon B_m^2d^2f^2 \quad \frac{\text{watts}}{\text{cubic meter}} \tag{13.21}$$

Frequency f cycles per second appears only to the first power in the hysteresis term but to the second power in the eddy current term. Dividing by f gives the loss in somewhat different terms.

$$\frac{p_T}{f} = \eta B_m^x + \epsilon B_m^2d^2f \quad \frac{\text{joules/cubic meter}}{\text{cycle}} \tag{13.22}$$

In this form, per cycle rather than per second, the hysteresis loss is a constant, and the eddy current loss varies linearly with frequency.

When plotted in log-log coordinates, as shown in Fig. 13.23, both terms of equation 13.22 appear as straight lines, for the loss in joules per cubic meter per cycle against the maximum flux density. The slope of the hysteresis loss curve is x, that of the eddy current losses is 2. The hysteresis losses for U.S.S. Transformer 72–29 gage correspond with the hysteresis loops of Fig. 13.15. The total loss for this material, on test, was 186.7 joules per cubic meter per cycle at $f = 60$ and $B_m = 1$. The hysteresis value is taken 130, leaving 56.7 for eddy current losses at 60 cycles, 1 weber per square meter. This method determines η, x, and ϵ, for which values are given in Table 13.3.

For 29 gage, $d = 0.014$ inch $\equiv 0.03556$ centimeter. The losses for Hipernik are analyzed in the same manner.

The losses per cycle as a function of frequency, shown in Fig. 13.24, are for a flux density of 1 weber per square meter. It should be noted

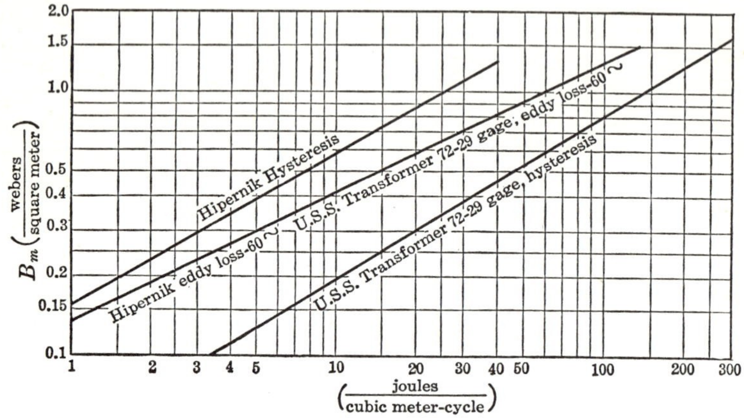

FIG. 13.23. 60-cycle hysteresis and eddy current losses as a function of flux density.

U.S.S. Transformer 72–29 gage	$\eta = 130$	$x = 1.582$	$\epsilon = 0.747 \times 10^7$
Hipernik	$\eta = 25$	$x = 1.75$	$\epsilon = 0.745 \times 10^7$

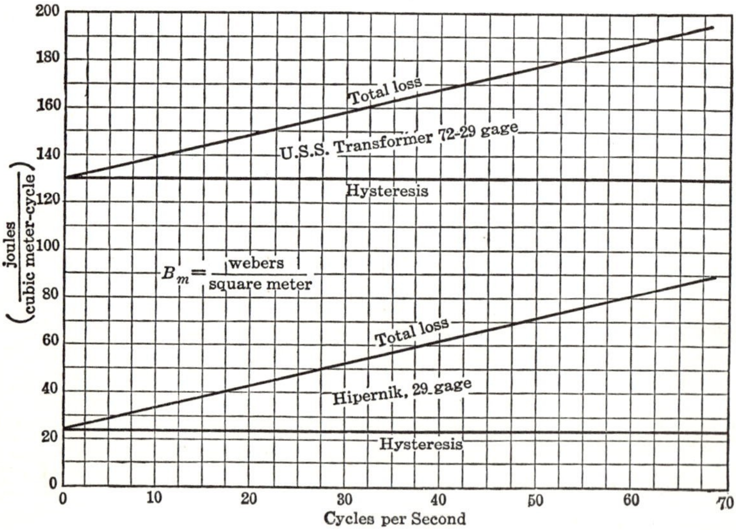

FIG. 13.24. Hysteresis and eddy current losses at $B_m = 1$ as a function of frequency.

that the hysteresis loss for Hipernik is about one-fifth that for silicon iron, indicating a hysteresis loop that is correspondingly narrow, but the eddy current losses are not greatly different.

D. SPECIAL PROBLEMS

13.11. The Closed Ring—Differential Permeability. Consider a closed ring of magnetic material, of uniform cross section A small compared with its mean length s, around which is wound a coil of n_1 turns carrying current i_1, as shown in Fig. 13.25. Unless the coil is uniformly spaced around the ring there is bound to be a leakage flux

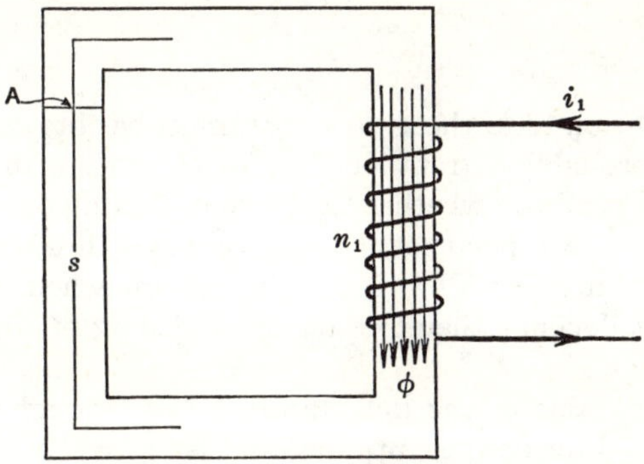

FIG. 13.25. Field of a closed ring.

which traverses a path that exists partly in the air. Nevertheless, for simplicity of solution, the assumption is usually made that the flux ϕ forms a uniform field of length s and area A entirely constrained to the iron. Under this assumption the closed-path MMF is

$$\mathfrak{F}_{\text{closed path}} = n_1 i_1 = Hs \quad \text{amperes} \tag{13.23}$$

and the magnetic intensity is

$$H = \frac{n_1 i_1}{s} \quad \frac{\text{amperes}}{\text{meter}} \tag{13.24}$$

The flux density B is obtained for any value of H by referring to the magnetization curve for the magnetic material composing the ring. The permeability for the value of flux density used is

$$\mu = \frac{B}{H} \tag{13.25}$$

and the total flux is

$$\phi = BA \quad \text{webers} \tag{13.26}$$

The inductance is

$$L = \frac{d\lambda}{di_1} = n_1 \frac{d\phi}{di_1}$$

$$= n_1 \frac{A\,dB}{\left(\frac{s}{n_1}\right) dH}$$

$$= \frac{n_1^2 A}{s} \frac{dB}{dH} \tag{13.27}$$

For use in equation 13.27 the ratio dB/dH must be obtained graphically from the hystero-eddy current loop. dB/dH varies with each value of current and depends on whether the current is rising or falling. With alternating voltage applied the inductance varies throughout the cycle from a maximum when $B = 0$ to a minimum when B is receding from B_{max}. It depends therefore on the widening of the loop caused by eddy currents.

The average value of the inductance, taken throughout the cycle, is the value used for design purposes, and is

$$L_{\text{average}} = \frac{n_1^2 A}{s} \left(\frac{dB}{dH}\right)_{\text{average}} = \frac{n_1^2 A}{s} \frac{B_m}{H_m} = \frac{n_1^2 A}{s} \mu \tag{13.28}$$

which is a function only of the BH values at the tip of the loop and depends therefore on the magnetization curve and not upon the shape of the hystero-eddy current loop. The average inductance varies with the permeability.

The slope of the hystero-eddy current loop is called the *differential permeability*,

$$\mu_d = \frac{dB}{dH} \tag{13.29}$$

So defined, its variation throughout a cycle determines the distortion of the magnetizing current with a sinusoidal variation of the flux.

The reluctance $\mathcal{R}$ is obtained by taking the ratio of the closed-path magnetic potential to the total flux set up.

$$\mathcal{R} = \frac{\mathcal{F}}{\phi} = \frac{n_1 i_1}{\phi} \tag{13.30}$$

The reluctance is not a constant but varies with the flux density, which in turn depends on the permeability.

13.12. Relation of Eddy Current Losses and Loop Widening. Figure 13.26 illustrates an iron lamination of width g, thickness d, and length s arranged to form a continuous iron and magnetic path. Magnetic flux lines, set up by an external coil with n turns and current i, lie along the path length s. The flux is alternating with frequency f cycles per second. For thin laminations ($d = 0.03556$ centimeter for No. 29 gage) we may assume, to a first approximation, that the flux density over the iron cross section is uniformly distributed. Let us study the generation of electric voltage around the elementary path dx, dy, under these conditions.

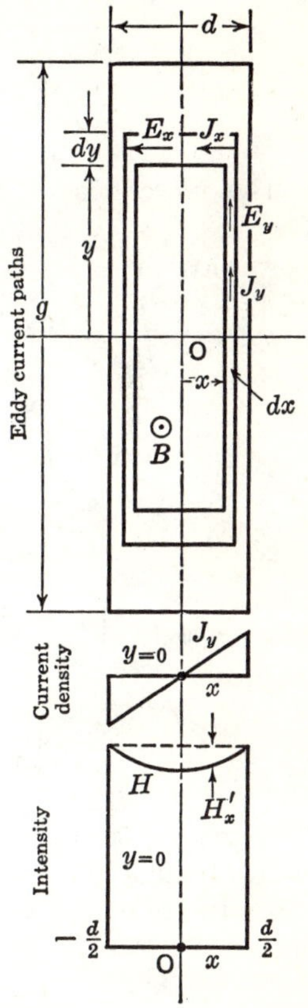

FIG. 13.26. Eddy current paths and magnetic intensity in laminations.

The voltage is generated by the time rate of change of magnetic flux, but it must be equal to the electric intensity integrated over the path.

$$V_g = -\frac{d\phi}{dt} = -4xy\frac{dB}{dt} = 4xE_x + 4yE_y \tag{13.31}$$

The elementary paths dx and dy must be related to each other so that the dx path integrated from 0 to $d/2$ corresponds to the dy path integrated from 0 to $g/2$, or

$$\frac{y}{x} = \frac{dy}{dx} = \frac{g}{d} \tag{13.32}$$

For one meter depth of iron (in the s direction) the electric flux density E_y times the normal area 1 dx is equal to the electric flux density E_x times the normal area 1 dy. This holds similarly for the eddy current density J times area, since $J = \sigma E$, in which σ is the conductivity.

$$\left.\begin{aligned} E_y 1\,dx &= E_x 1\,dy \\ J_y 1\,dx &= J_x 1\,dy \\ J_x &= J_y\left(\frac{d}{g}\right) \end{aligned}\right\} \tag{13.33}$$

Hence, with substitutions,

$$\left.\begin{aligned} V_g &= -4xy\frac{dB}{dt} = 4yE_y\left[1+\left(\frac{d}{g}\right)^2\right] \\ J_y &= \sigma E_y = -\frac{\sigma x}{\left[1+\left(\frac{d}{g}\right)^2\right]}\frac{dB}{dt} \\ J_x &= J_y\left(\frac{d}{g}\right) \end{aligned}\right\} \tag{13.34}$$

The power loss due to eddy currents per cubic meter, by Ohm's law, is $JE = J^2/\sigma$, and for the whole sheet the time average of the power loss is

$$\begin{aligned} \overline{P_E} &= \int_{\text{vol}} \frac{\overline{J^2}}{\sigma}\, d\,(\text{volume}) = \frac{1}{\sigma}\int \overline{J_y{}^2}\, 4y \cdot dx \cdot s + \overline{J_x{}^2}\, 4x \cdot dy \cdot s \\ &= \frac{4s}{\sigma}\left(\frac{g}{d}\right)\int_{x=0}^{x=d/2} \overline{J_y{}^2}\left[1+\left(\frac{d}{g}\right)^2\right] x\, dx \\ &= 4\sigma s\,\frac{\left(\frac{g}{d}\right)}{\left[1+\left(\frac{d}{g}\right)^2\right]}\overline{\left(\frac{dB}{dt}\right)^2}\int_{x=0}^{x=d/2} x^3\, dx \\ &= (sgd)\,\frac{\sigma d^2}{16\left[1+\left(\frac{d}{g}\right)^2\right]}\overline{\left(\frac{dB}{dt}\right)^2}\ \frac{\text{watts}}{\text{sheet}} \end{aligned} \tag{13.35}$$

The heavy bar over $\overline{P_E}$, $\overline{(dB/dt)^2}$, and $\overline{J^2}$ indicates the time average of each of these quantities. Dividing by the volume of the sheet, sgd, gives the time and volume average eddy current power loss per cubic meter

$$\overline{p_E} = \frac{\sigma d^2}{16\left[1+\left(\frac{d}{g}\right)^2\right]}\overline{\left(\frac{dB}{dt}\right)^2}\ \frac{\text{watts}}{\text{m}^3} \tag{13.36}$$

in which the factor $1 + (d/g)^2$ is very nearly equal to one for the (d/g) values used in practice.

If the flux density varies sinusoidally in time

$$B = B_m \cos \omega t \quad \text{and} \quad \frac{dB}{dt} = -\omega B_m \sin \omega t$$

$$\overline{\left(\frac{dB}{dt}\right)^2} = \omega^2 B_m{}^2\, \overline{\sin^2 \omega t} = \frac{\omega^2 B_m{}^2}{2} = 2\pi^2 f^2 B_m{}^2 \tag{13.37}$$

since the time average of $\sin^2 \omega t = \frac{1}{2}$. Then equation 13.36 becomes

$$\overline{p_E} = \frac{\sigma\pi^2 B_m{}^2 f^2 d^2}{8} \quad \frac{\text{watts}}{\text{cubic meter}} \tag{13.38}$$

Comparing this equation with equation 13.21 we see that the eddy current coefficient is

$$\epsilon = \frac{\sigma\pi^2}{8} \tag{13.39}$$

The resistivity is $\rho = 1/\sigma = \pi^2/8\epsilon$ in ohm-meters. Multiplying by 10^8 puts the result in microhm-centimeters. Computed values for U.S.S. Transformer 72 and Hipernik are given in Table 13.3 and may be compared with values obtained by direct-current test methods.

The only treatment of this subject the author has ever seen gives a 6 in place of the 8 in equation 13.39. The use of 6 is believed to be an error which arises in the assumption that the eddy currents flow all the way up one side of the lamination and down all the way on the other side, rather than in the rectangular path assumed here.

Returning to the general case, not necessarily sinusoidal, we compute the demagnetizing intensity in the iron resulting from the eddy current flow. Around the periphery of the center rectangular core, $2x$ by $2y$, the field set up by the eddy currents between $x = x$ and $x = d/2$ and between $y = y$ and $y = g/2$ is called a demagnetizing field H' because it modifies the main field created by the external magnetizing winding. As a result the current in the winding must be increased to an amount to compensate for this demagnetizing influence with the consequent widening of the hysteresis loop. H' is directed along the path s, perpendicular to the cross section xy.

The total closed-path magnetomotive force around the closed path s due to this effect is

$$\mathfrak{F}_{\text{closed path}} = H's = \int_{x=x}^{x=d/2} J_y \, dx \cdot s$$

$$H' = \frac{\sigma}{\left[1 + \left(\frac{d}{g}\right)^2\right]} \left(\frac{dB}{dt}\right)\left(\frac{d^2}{8} - \frac{x^2}{2}\right) \tag{13.40}$$

We obtain the space average of H' over the cross section by integration of the product of H' by the elementary area $4(y\,dx + x\,dy) = 8(g/d)x\,dx$ and division by the total area gd. Then

$$H'_{\text{average}} = \frac{1}{gd} \frac{\sigma}{\left[1 + \left(\frac{d}{g}\right)^2\right]} \left(\frac{dB}{dt}\right) \int_{x=0}^{x=d/2} \left(\frac{d^2}{8} - \frac{x^2}{2}\right) \left(8 \frac{g}{d} x\, dx\right)$$

$$= \frac{1}{16} \frac{\sigma d^2}{\left[1 + \left(\frac{d}{g}\right)^2\right]} \left(\frac{dB}{dt}\right) \tag{13.41}$$

If we again assume a sinusoidal time variation for the flux density,

$$B = B_m \cos \omega t \quad \text{and} \quad \frac{dB}{dt} = -\omega B_m \sin \omega t$$

Then the peak value of dB/dt (maximum instantaneous value) occurs when $\sin \omega t = 1$ and $B = 0$, and the peak value of average H' is

$$H'_{\substack{\text{average}\\ \text{peak}}} = \frac{1}{16} \frac{\sigma d^2}{\left[1 + \left(\frac{d}{g}\right)^2\right]} \omega B_m = \frac{\pi}{8} \frac{\sigma d^2 f B_m}{\left[1 + \left(\frac{d}{g}\right)^2\right]} \tag{13.42}$$

The demagnetizing field is 90° out of phase with the main flux but is in phase with the supply voltage, and, by virtue of the fact that it requires additional current flow in the magnetizing winding, it calls for a power input just sufficient to supply the eddy current losses. See Fig. 13.4. This peak value of H'_{average} is, of course, the maximum widening of the hystero-eddy current loop.

With sinusoidal flux density the ratio of the peak value of H'_{average}, corresponding to the maximum widening of the loop, to the time average of the eddy current losses takes a simple form.

$$\frac{H'_{\substack{\text{average}\\ \text{peak}}}}{\overline{p_E}} = \frac{1}{\pi f B_m} \tag{13.43}$$

The ratio involves only the frequency and the maximum flux density. Using this ratio and the measured value of eddy current loss, it is possible to compute the widening of the loop. This method is independent of the conductivity σ but does assume that the latter is constant over the cross section. In silicon iron particularly, this condition may not be fulfilled owing to both crystal structure and surface strains. The conductivity of the layers near the surface may be changed by the mechanical work done on the sheet. Since it is here that the maximum current density occurs, the result will likely be that the loop is actually considerably wider than predicted from the foregoing formula. An ordinary silicon iron sample on test showed a widening 40 per cent greater than predicted.

13.13. Ring and Air Gap. In Fig. 13.27 is shown a ring identical with that of Fig. 13.25, except that a parallel-faced air gap of length s_a has been cut perpendicular to the flux direction. The simplest assumption concerning the flux distribution is that all the flux lines follow the iron and cross the air gap without refraction. Under this assumption, the flux densities in the iron and in the air are equal.

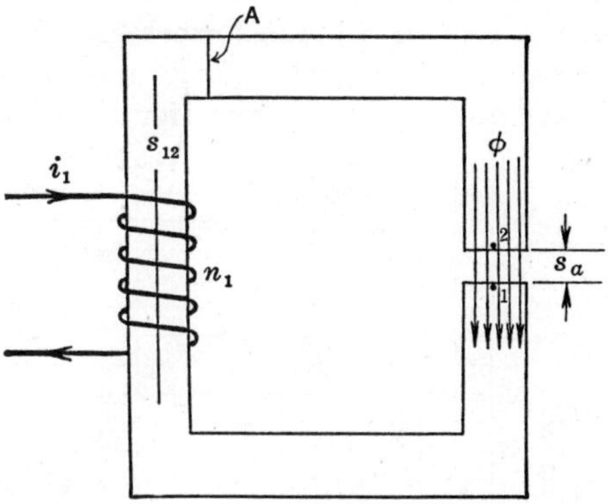

FIG. 13.27. Field of a ring and air gap.

$$B = B_{12} = B_a = \frac{\phi}{A} \tag{13.44}$$

The closed-path MMF is

$$\underset{\text{closed path}}{\mathfrak{F}} = n_1 i_1 = H_{12}s_{12} + H_a s_a \tag{13.45}$$

$$= B\left(\frac{s_{12}}{\mu_{12}} + \frac{s_a}{\mu_0}\right) \tag{13.46}$$

$$= \frac{\phi}{A}\left(\frac{s_{12}}{\mu_{12}} + \frac{s_a}{\mu_0}\right) \tag{13.47}$$

Any one of these last three equations must be solved by cut-and-try methods, because each contains three variables (i_1, H_{12}, and H_a; i_1, B, and μ_{12}; i_1, ϕ, and μ_{12}). Perhaps the simplest procedure is to assume the flux density B, get the corresponding intensity H_{12} from Figs. 13.8 to 13.13, and solve the first equation for i_1, repeating again and again until the desired current is reached.

Equation 13.45 may be rewritten in the form

$$n_1 i_1 = \left(H_{12} + H_a \frac{s_a}{s_{12}}\right) s_{12} = H_Q s_{12} \tag{13.48}$$

in which H_Q is the equivalent H for the combination of iron and air gap in series. H_Q is likely to be several times as large as H_{12} even for an air gap ratio s_a/s_{12} as small as 0.001. The ratio of H_Q to H_{12} is equal to the ratio of the current required with and without an air gap.

Figure 13.28 illustrates the tremendous influence which an air gap plays in magnetizing an open iron circuit. To the left is plotted the magnetization curve and hysteresis loop for $B_m = 1.0$ weber per square meter for U.S.S. Transformer 72. The line OG is the magnetization curve of the air gap. The sum of these in H gives the magnetization curve and hysteresis loop (for $B_m = 1$) for the iron plus the air gap.

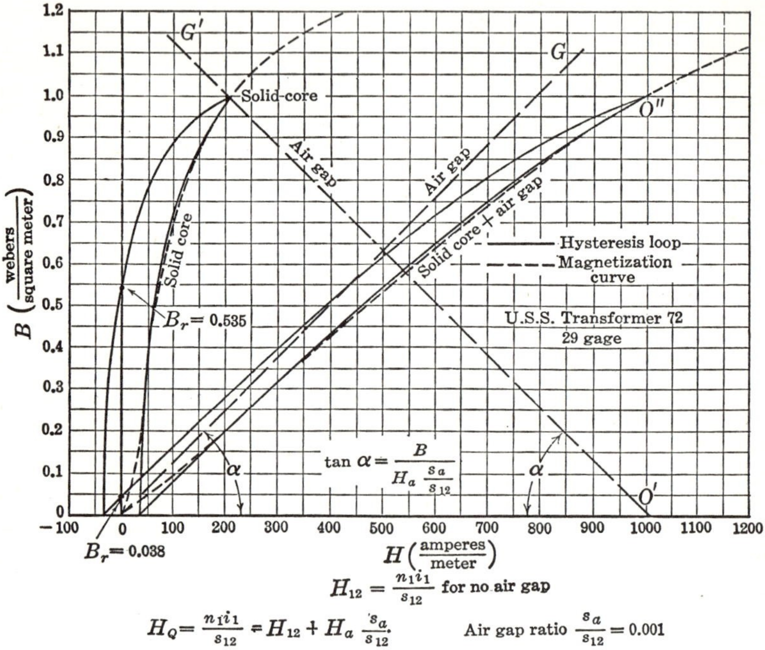

FIG. 13.28. Effect of air gap ratio on magnetization of iron and air gap.

This requires a replot of the magnetization curve and hysteresis loop for each air gap ratio s_a/s_{12}. The same result may be achieved more easily by plotting the line $O'G'$ at the angle α through the flux density desired on the solid core magnetization curve as shown. The required H_Q is given by the value at O' as well as at O''. Angle α is determined from the relation

$$\tan \alpha = \frac{B}{H_a \dfrac{s_a}{s_{12}}} \tag{13.49}$$

In the case illustrated $H_Q = 1000$ and $H_{12} = 210$. This means that the ratio of currents required with and without an air gap is $1000/210 = 4.76$. For a larger air gap ratio the current ratio will be even greater.

Upon removal of the magnetizing current, the remanent flux density with no air gap would (in this case) be 0.535, and with the air gap ratio indicated it would fall to a value of 0.038. This effect is quite evident in motors and generators where the remanent magnetism appears to be quite low valued. As the field current is increased the flux density starts at the remanent point (corresponding to 0.038) and builds up to join the solid core plus air gap curve.

13.14. Total Pressure at an Iron–Air Boundary. The pressure at an iron–air boundary is normal to the surface and has the value given in equation 8.40 in Art. 8.12, if B, H, and μ are substituted for D, E, and ϵ respectively. Then

$$p_{\text{boundary}} = \frac{1}{2}\left[H_{1T}{}^2(\mu_1 - \mu_2) + B_{1N}{}^2\left(\frac{1}{\mu_2} - \frac{1}{\mu_1}\right)\right] \quad \frac{\text{newtons}}{\text{square meter}} \tag{13.50}$$

urging medium 1 toward medium 2 if $\mu_1 > \mu_2$.

For iron, $\mu_1 \gg \mu_0$ ordinarily
For air, $\mu_2 = \mu_0$
Also $H_{1T} = H_{2T}$
$B_{1N} = B_{2N}$

Consequently equation 13.50 may be reduced to

$$\begin{aligned} p_{\text{boundary}} &= \frac{1}{2}\left(H_{2T}{}^2\mu_1 + \frac{B_{2N}{}^2}{\mu_2}\right) \\ &= \frac{1}{2}\left(\frac{B_{2T}{}^2\mu_1}{\mu_2{}^2} + \frac{B_{2N}{}^2}{\mu_2}\right) \end{aligned} \tag{13.51}$$

But with the flux on the air side practically normal to the surface, $B_{2T} \ll B_{2N}$, and equation 13.51 reduces further to

$$p_{\text{boundary}} = \frac{1}{2}\frac{B_2{}^2}{\mu_2} = \omega_{m\,\text{air}} \quad \frac{\text{newtons}}{\text{square meter}} \tag{13.52}$$

The pressure on the iron–air boundary is equal practically to the energy density in the air and tends to urge the iron surface normally toward the air region.

13.15. Parallel Magnetic Flux Paths. An infinite number of variations in magnetic path arrangements is possible. One such variation with parallel paths for the flux is shown in Fig. 13.29. With the currents directed as shown, the probability is that flux will be directed upward in branches 1 and 2, adding together to give a downward flux in branch 3. (But this need not be true; if the ampere-turns of coil 1 are considerably greater than those of coil 2, the first coil might send the flux down the center leg as well as down branch 3.) Perhaps the best procedure is to assume the direction of the flux everywhere. The H's will differ wherever the flux densities differ, whether due to changes in cross-sectional area or to the conjunction of two or more flux paths. Cut-and-try solutions are almost inevitable.

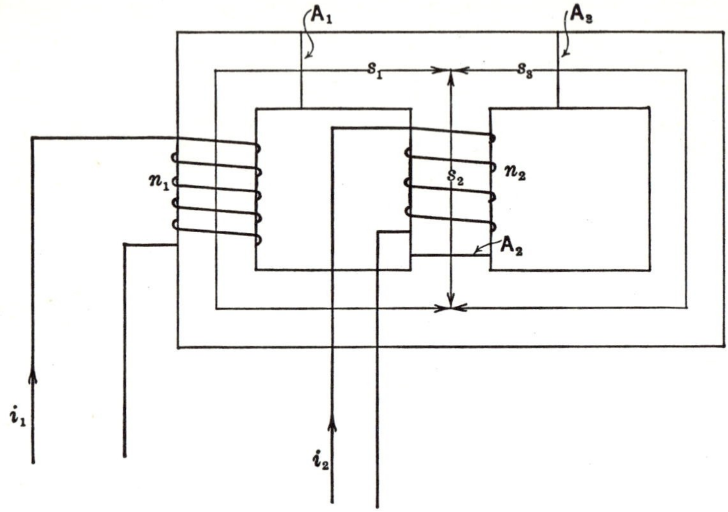

FIG. 13.29. Parallel magnetic flux paths.

A basic formula that governs the solution is

$$\Sigma ni = \Sigma Hs \tag{13.53}$$

taken along any path, but preferably along one or more of the paths assumed for the flux. In addition,

$$\phi = BA \tag{13.54}$$

must be used, and the magnetization curve controls the relation of B to H at all points in the iron. Negative values of a part of the solution indicates that the flux direction in that part was incorrectly chosen. A second solution with new flux directions may be required.

13.16. Displaced Hysteresis Loops—Incremental Permeability. When ordinary alternating current is passed through a coil enclosing an iron core, the H and B values swing equal distances positively and negatively on succeeding half cycles, as shown in the major loop of Fig. 13.30. However, in many applications (radio chokes, d-c reactors, etc.) a direct component of current is superposed in addition to the alternating component, with the result that the B–H variation follows a minor loop ab.

The upper tip of the minor loop (point a) lies on the magnetization curve at a point given by the sum of the H_{dc} due to the direct current and H_{ac} due to the alternating current, at the instant when the latter has reached its maximum value in the same sense as H_{dc}, taken around the magnetic circuit. There is, of course, a slight widening of the

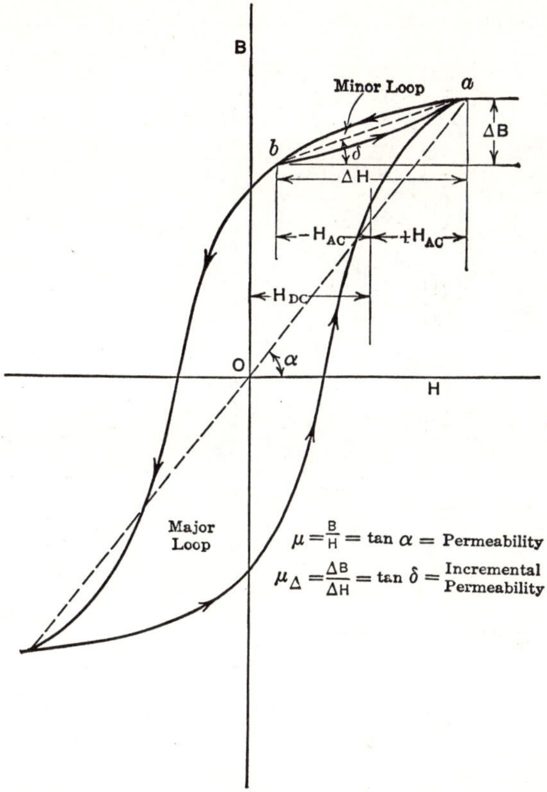

FIG. 13.30. Displaced hysteresis loop and incremental permeability.

minor loop in the H direction due to eddy currents, varying from zero at points a and b to a maximum in between, but this may be neglected, in a first approximation, for the purposes of this article, since the widening is proportional to the change in flux $\Delta B/2$ which is usually smaller than B_{max}.

With sinusoidal flux density both the thickness of the minor hysteresis loop and the eddy current widening contribute to a distortion of the magnetizing current from the sine form, but this effect is likely to be much smaller in this case than one in which the whole hysteresis loop is involved.

H_{dc} is the component of intensity set up by the direct current, and $\pm H_{ac}$ is the alternating variation added by the alternating current, so that the intensity varies between H_a and H_b (ΔH) as the flux density varies from B_a to B_b (ΔB), following the minor loop in the direction indicated by the arrows. For conditions of this kind let us define a

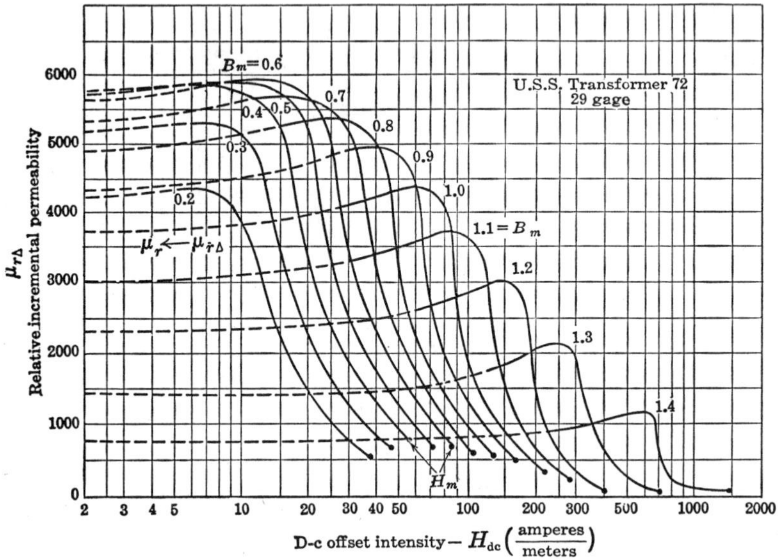

FIG. 13.31. Relative incremental permeability versus H_{dc} with parameter B_m.

new type of permeability, known as the *incremental permeability*, given by the ratio of the change in B to the change in H.

The two permeabilities are

$$\mu = \frac{B}{H} = \tan \alpha = \text{permeability} \tag{13.55}$$

$$\mu_\Delta = \frac{\Delta B}{\Delta H} = \tan \delta = \text{incremental permeability} \tag{13.56}$$

The former or ordinary permeability is useful only under conditions that take the iron through the ordinary hysteresis loop. The latter is used when the iron is carried through a displaced loop.

The relative incremental permeability for U.S.S. Transformer 72–29 gage has been computed from the hysteresis loops of Fig. 13.15 and is shown in Fig. 13.31, in which $\mu_{r\Delta}$ is plotted against the magnitude of the d-c offset, H_{dc}, with the maximum values of flux density B_m considered as parameters. The hysteresis loops are given for even values of B_m. $\mu_{r\Delta}$ for odd values of B_m has been interpolated. As the direct current is decreased to zero, H_{dc} approaches zero, and, for a fixed maximum value B_m, $\mu_{r\Delta}$ approaches μ_r, ΔB in the limit becomes $2B_m$,

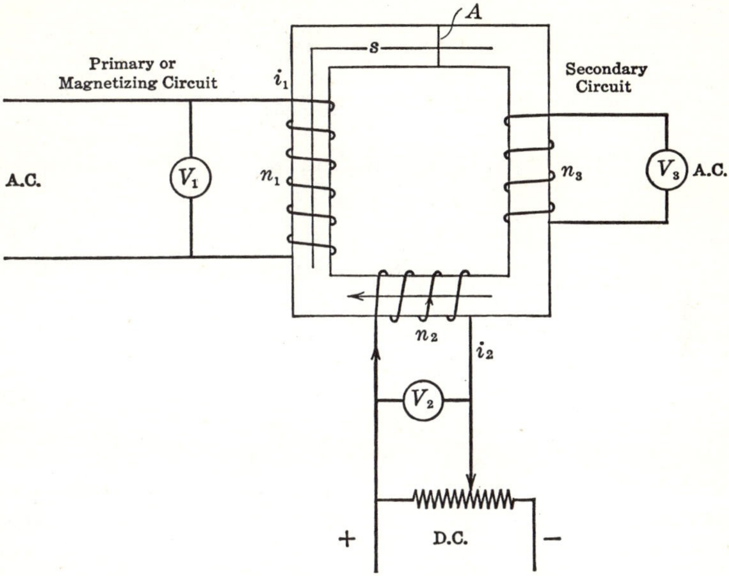

FIG. 13.32. Circuit for closed ring with d-c offset.

and ΔH becomes $2H_m$. At the other extreme H_{dc} approaches H_m and then ΔB and ΔH both approach zero.

An interesting article on this subject is: R. F. Edgar, "Loss Characteristics of Silicon Steel at 60 Cycles with D-C Excitation," *Trans. A.I.E.E.*, **52,** 721 (1933).

13.17. Closed Ring with D-C Offset—Figure-Eight Reactor. The circuit for which the incremental permeability is useful is illustrated in Fig. 13.32. An alternating sinusoidal voltage V_1 is connected to the primary winding of n_1 turns and a direct voltage V_2 through a potentiometer to coil 2; let coil 3 be connected to an a-c voltmeter.

When the magnetizing effect of the primary a-c coil has the same sense as that of the d-c coil the total H in the magnetic path is the sum of the two, or

$$H_{max} = H_{dc} + H_{ac} = \frac{n_2 i_2 + n_1 i_{1\,max}}{s} \tag{13.57}$$

Corresponding to H_{max} is the flux density B_{max}, to be found from the magnetization curve for the iron.

One half-cycle later the alternating current has reversed and H reaches its minimum value.

$$H_{min} = H_{dc} - H_{ac} = \frac{n_2 i_2 - n_1 i_{1\,max}}{s} \tag{13.58}$$

The corresponding flux density is found, not on the magnetization curve, but along the hysteresis loop which has B_m for its upper tip. From equations 13.57 and 13.58 we see that

$$H_{dc} = \frac{H_{max} + H_{min}}{2} = \frac{n_2 i_2}{s} \tag{13.59}$$

Also

$$\Delta H = H_{max} - H_{min} = 2H_{ac} = \frac{2n_1 i_{1\,max}}{s} \tag{13.60}$$

From the hysteresis loop ΔB may be found and

$$\mu_\Delta = \frac{\Delta B}{\Delta H} = \mu_0 \mu_{r\Delta} \tag{13.61}$$

or, using Fig. 13.31, $\mu_{r\Delta}$ may be found if H_{dc} and B_m are known.

In proceeding around the minor loop ΔB is the total swing of magnetic flux density. $\pm \Delta B/2$ is the swing or change from its midposition. The electric potential generated in coil 3 is

$$V_{g3} = -n_3 A \frac{dB}{dt} = -n_3 A \omega \frac{\Delta B}{2} \sin \omega t \tag{13.62}$$

and the applied potential to coil 1, out of phase by 180°, is

$$V_1 = +n_1 A \frac{dB}{dt} = +n_1 A \omega \frac{\Delta B}{2} \sin \omega t \tag{13.63}$$

The average inductance of the circuit as viewed from the primary input is equal to the incremental change in flux linkages divided by the incremental change in current, or

$$\begin{aligned} L_{1\,average} &= \frac{n_1 \Delta\phi_1}{\Delta i_1} \\ &= \frac{n_1 A \, \Delta B/2}{\Delta H s/(2n_1)} = \frac{n_1^2 A}{s} \mu_\Delta \end{aligned} \tag{13.64}$$

With large values of H_{dc} the average inductance falls sharply with the fall in μ_Δ.

If a load is connected to coil 3, additional current flows in that coil and the additional current required in coil 1, 180° out of phase, is

$$i_1' = \frac{n_3 i_3'}{n_1} \tag{13.65}$$

There is a serious objection to the arrangement shown in Fig. 13.32. The alternating flux density $\Delta B/2$ induces a voltage in the d-c winding.

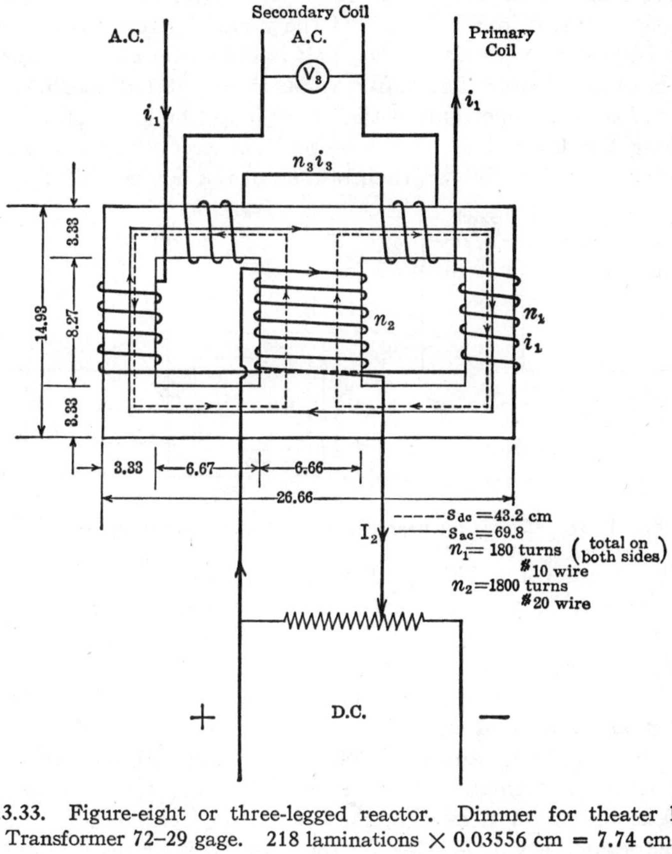

FIG. 13.33. Figure-eight or three-legged reactor. Dimmer for theater lights. U.S.S. Transformer 72–29 gage. 218 laminations × 0.03556 cm = 7.74 cm deep.

This may be of considerable value because the d-c coil is usually wound with a large number of turns n_2 and is likely to create an undesirable fluctuation in the d-c current with repercussions on the a-c performance. This feature is avoided by using a figure-eight or three-legged reactor of a design shown in Fig. 13.33. The iron cross section is the same throughout except in the center leg where it is double to accommodate both branches of the d-c flux (shown in dashed lines) with the same value of H_{dc} in all parts of the magnetic circuit.

The primary a-c coil is wound in two equal parts on the outer legs so connected that the a-c flux (shown by a solid line) flows around the outer path and does not penetrate the center leg. At some instant when the alternating current in coil 1 is at its maximum and the flux is directed as shown, the right half of the device is operating at the upper tip, point a in Fig. 13.30, of the minor hysteresis loop on the magnetization curve, and the left half is operating at the lower end, point b, of the loop. Thereafter while the right half declines from point a along the upper half of the minor loop, the left half is increasing along the lower half of the loop. The secondary a-c coil 3 is connected as shown, half on each branch of the device.

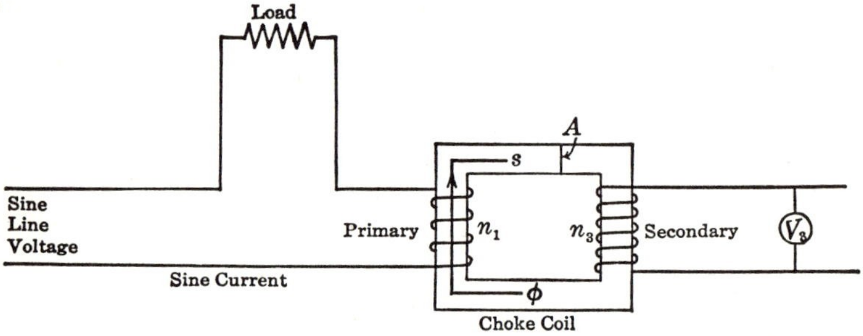

FIG. 13.34. Circuit for generation of peaked alternating voltages.

13.18. Dimmer for Theater Lights. The three-legged reactor of Fig. 13.33 may be operated as a dimmer for theater lights. For this application coil 3 is not required. Coil 1 is connected to a 120-volt 60-cycle line through a bank of lamps in series. Coil 2 is connected through the potentiometer to a 115-volt d-c source. With the coils n_1 and n_2 as shown and with the iron dimensions indicated, the device will control a bank of lamps of 1000 watts rating with a maximum d-c demand of approximately 1 ampere. With no direct current the inductance is high and most of the alternating voltage is used across the reactor and very little is left for the lamps. With maximum direct current the inductance, due to low μ_Δ, is low, the drop across the reactor is small, and the lamps light to high brilliance.

13.19. Generation of Peaked A-C Voltage. Alternating voltages which are not sinusoidal but instead have high sharp peaks are desired for certain applications. A peaked voltage of short duration might be applied to a neon (or mercury) lamp to give short sharp flashes

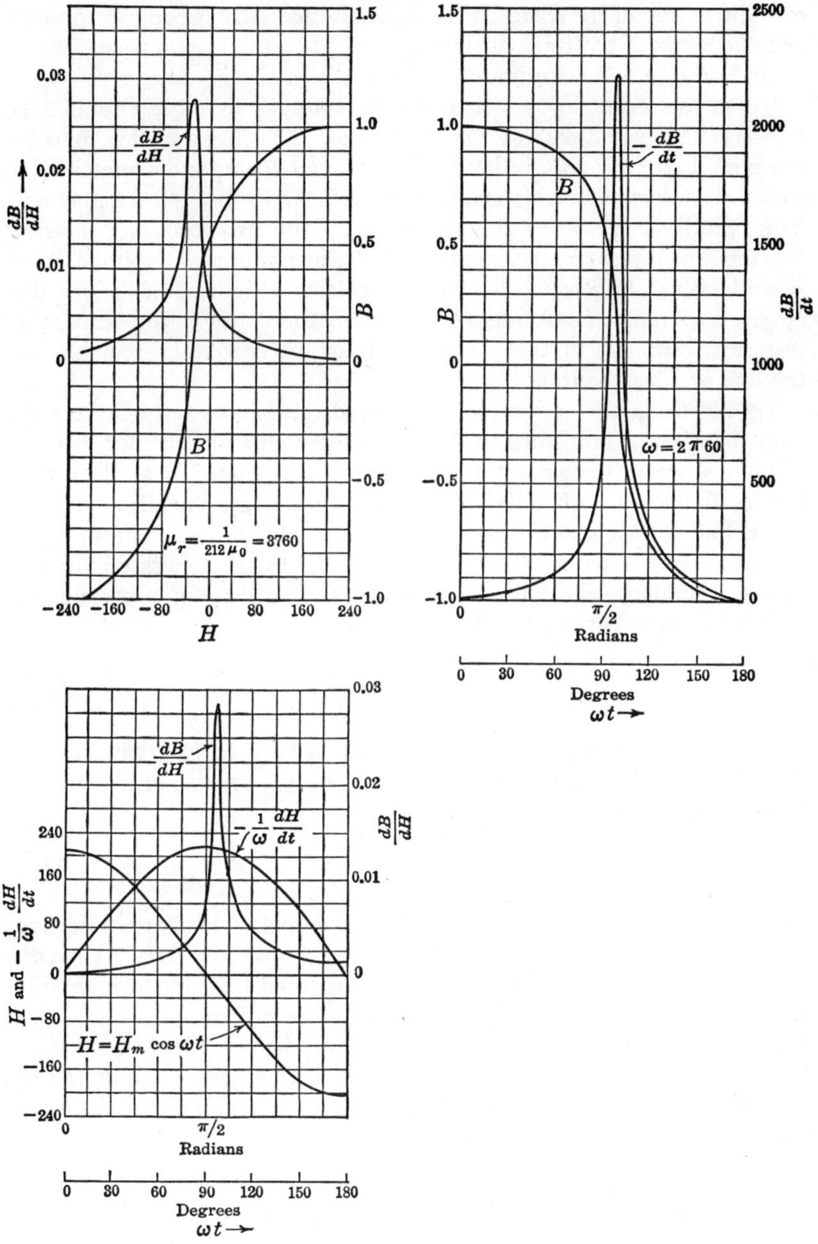

FIG. 13.35. 60-cycle peaked voltage generated by circuit of Fig. 13.34.

of light at regularly spaced time intervals for stroboscopic or photographic work. A voltage of this type might be used to puncture paper wound on a moving drum and thus serve as a timing device.

Up to this point in this chapter sine voltages have been applied to various magnetic devices with iron cores. The flux and flux density are bound to vary sinusoidally, while the magnetizing current is distorted. If, however, a sinusoidal current is *forced* through the magnetizing winding, H varies sinusoidally and the flux and flux density are distorted and by their rates of change induce a voltage in a secondary winding which is sharply peaked. Figure 13.34 gives the circuit required. To obtain a sine current the voltage drop across the primary terminals of the choke coil must be a small percentage of the line and load voltages.

To illustrate the magnitudes involved, one-half of the hysteresis loop for U.S.S. Transformer 72 iron, with a maximum flux density of $B_m = 1$ weber per square meter, is plotted in Fig. 13.35 at the upper left-hand corner. dB/dH has been evaluated graphically. At the left H and dH/dt, assumed sinusoidal, have been drawn. From these curves, B and $-dB/dt$, which is proportional to the generated voltage in the secondary coil, have been drawn as functions of time and are shown at the upper right. At any instant

$$\frac{dB}{dt} = \frac{dB}{dH} \cdot \frac{dH}{dt} \tag{13.66}$$

and the separate values in the product can be read from the curves. A highly peaked voltage is obtained. The voltage generated in the secondary is

$$V_3 = -n_3 A \frac{dB}{dt} \tag{13.67}$$

The peak might have been made much higher in magnitude and narrower in its time width by using larger values of primary current, H, and B_m. In this discussion we have ignored the influence of eddy currents by considering the hysteresis loop proper. Had we used the hystero-eddy current loop instead, the eddy currents would have been included automatically, and a somewhat higher value of H and primary magnetizing current would have been required for the same generated voltage.

E. PHYSICAL BASIS OF FERROMAGNETISM

13.20. Physical Basis of Ferromagnetism. The ferromagnetic materials of technical importance are iron, cobalt, nickel, and their alloys. All these are solid materials having definite types of crystalline structure such as body-centered cubic (Fe), hexagonal (Co, if slowly cooled), and face-centered cubic (Ni, and Co if quenched). In each crystal there are certain principal directions depending on the arrangement of the atoms in the crystal. These directions, for a *single crystal* of iron, are illustrated in Fig. 13.36 and are numbered according to the Miller system of indices for crystals. Figure 13.36 gives the magnetization curves for magnetization of the single crystal in the 100, 110, and 111 directions. In this order magnetization is "easy," "medium," or "hard."

Any ordinary ferromagnetic material is composed of a large number of crystals, and each crystal is composed of a large number of smaller

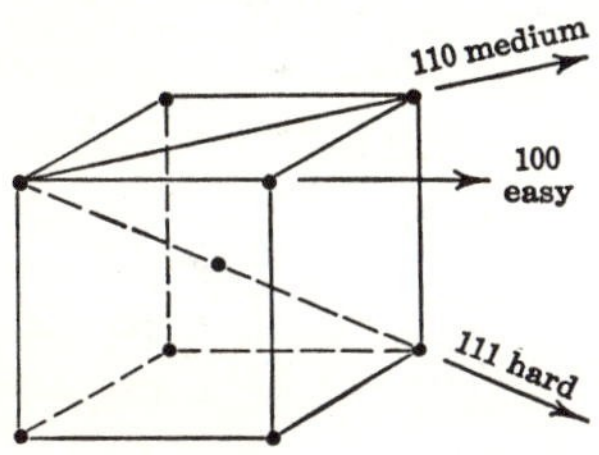

(*a*)

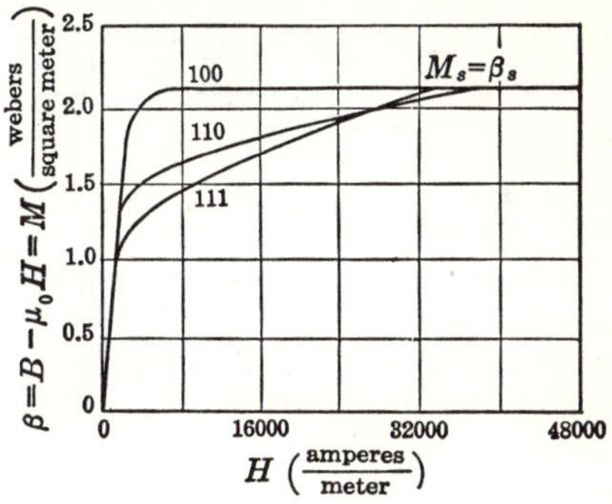

(*b*)

FIG. 13.36. Iron single crystal structure and magnetization curves.

units called domains. A cubic centimeter of iron, for instance, containing about 10^{23} molecules, is divided into perhaps 10,000 single crystals, each divided into perhaps 100,000 domains. Approximately, there are 10^{19} molecules in a crystal and 10^{14} molecules in a domain. If these units were cubic, the linear dimensions would be about 0.050 centimeter for a crystal and 0.001 centimeter for a domain.

According to modern theory the ferromagnetic state is ultimately traceable to the cooperative behavior of certain of the outermost electrons (probably two) in the atoms of ferromagnetic materials, particularly with respect to the cooperative behavior of the electron spins. In this theory the electron spins are "locked" together in groups with their axes pointing in one or another of the "easy" crystalline directions. In each domain the spins are pointed in one "easy" direction which we shall call the parallel direction, or in the reverse "easy" direction which we shall call the antiparallel direction. At $T = 0°$ absolute, all the spins in one domain are pointed in the parallel direction and the domain is thus "spontaneously magnetized" to the maximum degree, without the assistance of any external field H. The domain is then completely saturated, with a magnetic moment per unit volume of $M_0 = \beta_0$.

A ferromagnetic material in bulk form may appear nonmagnetized simply because there are as many domains with magnetization pointing one way as another, giving an average external effect of zero magnetization.

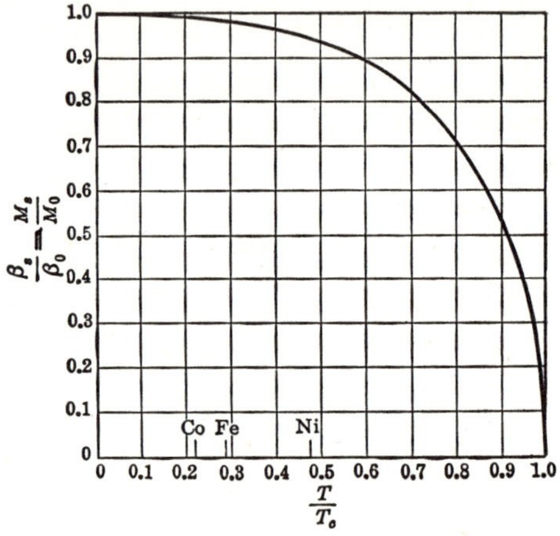

FIG. 13.37. Domain saturation as a function of temperature.
Saturation for Ni, Fe, and Co marked for $T = 300°$ absolute

Curie temperature T_c { Nickel 630° absolute; Iron 1043° absolute; Cobalt 1385° absolute

At higher temperatures (say 300° absolute) a small percentage of the spins reverse and are pointed in the "antiparallel" direction, giving a somewhat decreased saturation and decreased magnetic moment per unit volume $M_s = \beta_s$. With increasing temperature more spins reverse and $M_s = \beta_s$ declines toward zero at the Curie temperature T_c. Above T_c the material is paramagnetic, not ferromagnetic. The equation for M_s/M_0 takes the form

$$\frac{\beta_s}{\beta_0} = \frac{M_s}{M_0} = \tanh \frac{\left(\frac{M_s}{M_0}\right)}{\left(\frac{T}{T_c}\right)} \tag{13.68}$$

and is plotted in Fig. 13.37. It may be noted that the M_s/M_0 values for cobalt, iron, and nickel, at 300° absolute, are about 99 per cent, 98 per cent, and 96 per cent.

According to this theory the domain is the fundamental magnetized unit in bulk material. It is inherently magnetized to saturation, depending only on temperature, without the aid of any external magnetic field. $\beta_s = M_s$ is readily obtained as the saturation value of ferric induction on the magnetization curve for temperature T.

Any piece of bulk ferromagnetic material appears nonmagnetized simply because it is composed of a large number of domains, each completely magnetized, whose directions of magnetization are randomly distributed. Our next problem is to consider the influence of an externally applied magnetic field H on orienting the domain magnetization in the H direction and producing a bulk magnetization as given by the magnetization curve and hysteresis loops, obtained from a sample in the form of Rowland's ring (Arts. 9.40 and 13.3).

Starting with a nonmagnetized sample the current is increased in steps, giving the initial curve shown in Fig. 13.38. This curve has four parts, while the hysteresis loop, obtained thereafter, has three parts, each related to different physical processes.

In Part I it is believed that the random orientation of magnetization of the domains is upset, upon applying external field H, by a "crumbling" or translation of the domains with magnetization not parallel to H and a corresponding increase in the size of the domains magnetized parallel to H. This is illustrated in Fig. 13.39.

In Part II magnetization proceeds much more strongly, not smoothly as indicated on the initial curve, but rather in a series of jerky steps, as indicated in the magnified detail in Fig. 13.38. This is known as the Barkhausen effect. This part of the curve is believed to correspond to a jerky rotation of entire domains from "easy" crystal directions to other "easy" directions more nearly parallel to the applied field H. See Fig. 13.40.

Part III represents the approach toward saturation $\beta_s = M_s$ which occurs in Part IV. In Part III it is believed that the domain magnetic spins are rotated gradually from the "easy" crystal direction most parallel to H to complete parallelism with H. Part IV represents the completion of this process and saturation $M_s = \beta_s$ to the extent permitted by the temperature.

The hysteresis loop is composed of Parts II, III, IV. In each of these the process is similar to that occurring along the initial curve.

In concluding this paragraph we shall make a rough calculation of the saturation moment per cubic meter $M_s = \beta_s \cong M_0 = \beta_0$ for iron. From equation 9.147 we note that the magnetic moment per spin is 1.164×10^{-29} weber-meter. Multiplying this by 2 spins per atom and by 8.55×10^{28} atoms per cubic meter gives $M_0 = \beta_0 \cong 2$

webers per square meter, which is the correct order of magnitude for saturation for ferromagnetic materials.

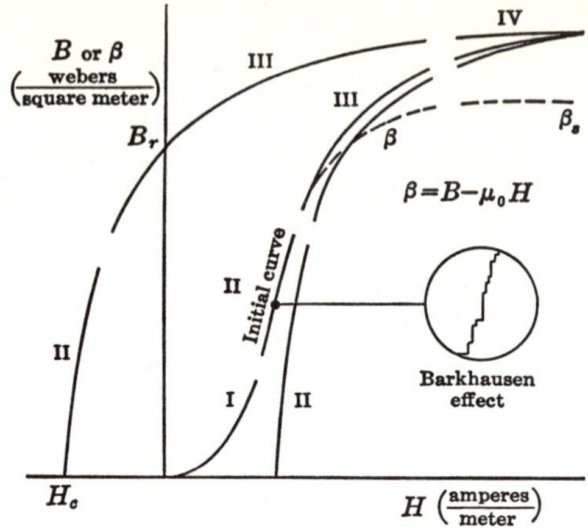

FIG. 13.38. Characteristics of the technical magnetization curve and hysteresis loop.

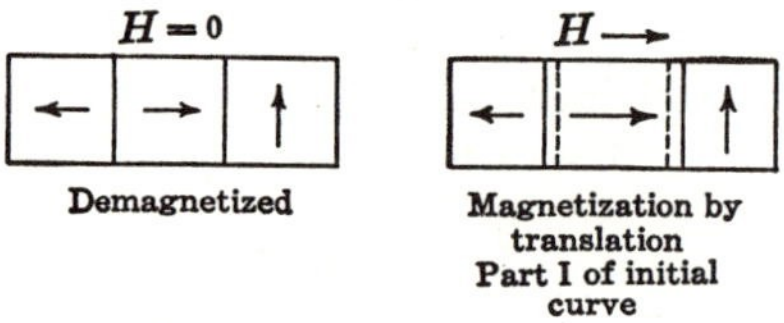

FIG. 13.39. Magnetization by translation (Part I).

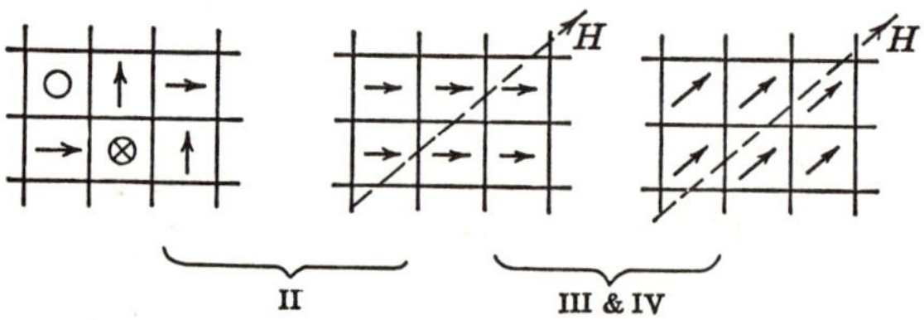

FIG. 13.40. Magnetization by rotation (Parts II, III, IV).

An excellent review of this and allied subjects is given in an article entitled "Magnetism," by R. M. Bozorth, *Rev. Mod. Phys.*, **19,** 29 (January 1947).

PROBLEMS

1. Discuss the relative merits of making $B = H = 0$ in an iron sample by four methods: heating above the Curie temperature T_c and annealing; slow removal of the sample from an alternating field; slow reduction of magnetizing current to zero; and adjustment of magnetizing current so that point R in Fig. 13.7 is made to coincide with the origin.

2. Assume that the core of Fig. 13.25 is U.S.S. Transformer 72–29 gage, and that $s = 25$ cm, $A = 5$ sq cm, $n_1 = 100$ turns, and $i_1 = 0.4225$ ampere. Compute the intensity, flux density, flux, permeability (relative), and reluctance. If the current above is the alternating peak value, compute the average inductance of the ring.

Compute the maximum value of the inductance at some instant in the cycle, if, at $B = 0$, the slope of the hysteresis loop is equal to the slope of the hystero-eddy current loop.

Compute the maximum value that the average inductance may attain. What values of B, H, μ, and i_1 are required?

How much hysteresis energy is required to accomplish a reversal of magnetization at the current peak of 0.4225 ampere?

3. Following Fig. 13.27, let $s_{12} = 25$ cm, $s_a = 0.1$ cm, $A = 5$ sq cm, $n_1 = 100$ turns, $i_1 = 0.4225$ ampere. Compute H, B, ϕ, μ_r, and $\mathcal{R}$. How much current i_1 is necessary in order to bring the flux density up to the value computed in Prob. 2? Of the total number of ampere-turns, how many are necessary to send the flux across the gap? Compute the mechanical pull on the air gap in both cases.

Compute the equivalent H for the iron and air gap in series. Estimate the value of the residual flux density upon removal of the magnetizing current.

4. Place a second coil of $n_2 = 300$ turns on the ring of Prob. 2 and let it carry a direct current of 0.1842 ampere. Compute the incremental permeability and compare it with the ordinary permeability taken at the extreme of the alternating-current swing. Sketch the shape of the minor hysteresis loop described in one cycle.

Estimate the hysteresis and eddy-current losses per cycle. Compare with the values for Prob. 2 (frequency 60 cycles per second).

5. See Prob. 4 and Fig. 13.32. Add coil $n_3 = 200$ turns. Compute the voltage induced in coil 3, assuming that the flux varies sinusoidally 60 times per second.

6. Assuming that the MMF drop in the iron is negligible compared with that in the air gap, prove that a reactor has a minimum power factor when the coil copper loss is equal to the core iron loss.

7. Following Fig. P7, compute H, B, ϕ, μ, and $\mathcal{R}$.

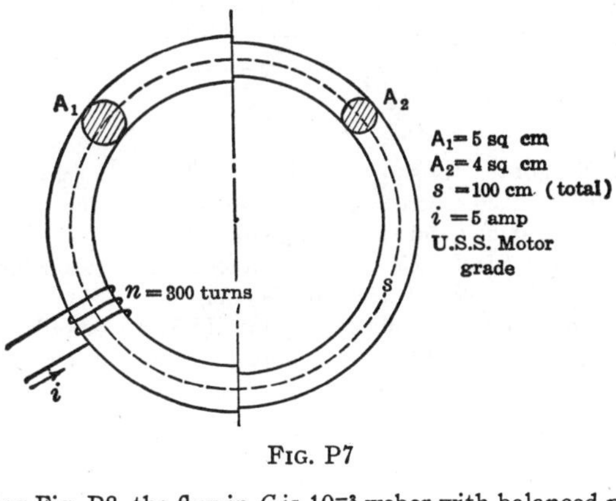

FIG. P7

8. Following Fig. P8, the flux in C is 10^{-3} weber with balanced pulls on the two gaps. Compute the area required in section D and the amperes required for balanced pull. Must the current be larger or smaller than this amount in order to operate the relay?

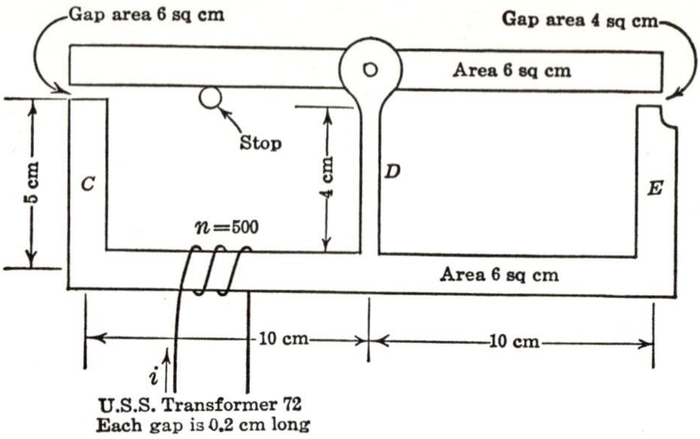

FIG. P8

9. In Fig. 13.29, let $s_1 = s_3 = 15$ cm, $s_2 = 5$ cm, $A_1 = A_2 = A_3 = 5$ sq cm. The material is U.S.S. Transformer 72. Compute B, H, ϕ, and μ in the three legs, if $n_1 = 200$ turns, $i_1 = 1$ ampere, $n_2 = 50$ turns, $i_2 = 0.5$ ampere. Repeat for current i_2 reversed.

10. Compute the current required if the air gap flux density in Fig. P10 is 1 weber per square meter (U.S.S. Transformer 72).

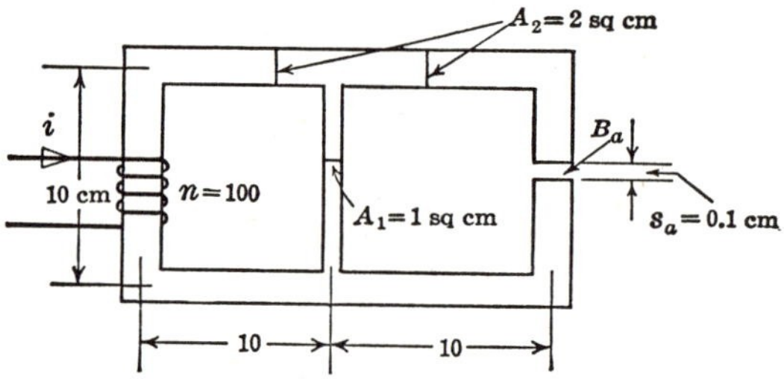

FIG. P10

11. In the figure-eight reactor shown in Fig. P11 let $n = 100$, $i = 1$ ampere (peak value), $N = 1000$, $I = 0.2$ ampere. Compute ϕ, B, and H in each leg. Compute the incremental permeability and the average inductance.

Disconnect the d-c control. Compute the average inductance and the flux in the center leg, if the alternating current is unchanged.

12. Test values for the theater dimmer of Fig. 13.33 are as follows: $V_{\text{line}} = 120$ volts 60 cycles, $V_1 = 55$ volts dimmer drop, $I_1 = 10$ amperes rms, $I_2 = 0.65$ ampere. The load is purely resistive, and we assume that the dimmer drop is entirely reactive. Compute the incremental permeability. The result lies beyond the range of the curves of Fig. 13.31. Approximately, however, for large values of H_{dc} and values of ΔH which are not too great, the minor hysteresis loop lies along the magnetization curve. With this assistance, compute $H_{\max}$ and $H_{\min}$.

13. In Fig. P13 let $A = 3$ sq cm, $s_1 = 1.5$ cm, $s_0 = 0.2$ cm, $s_2 = 0.83$ cm, $s_g =$

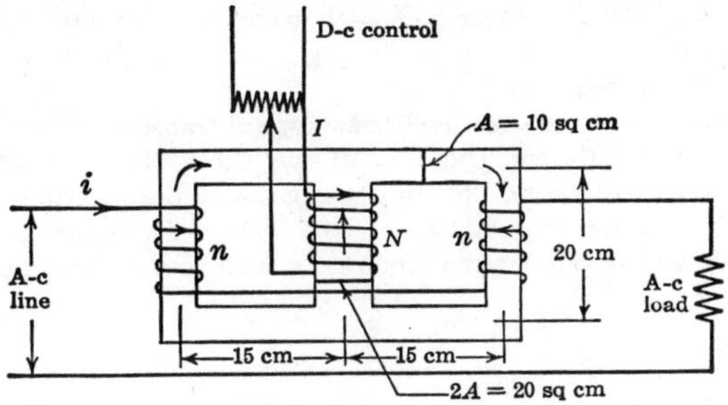

FIG. P11. U.S.S. Transformer 72-29 gage.

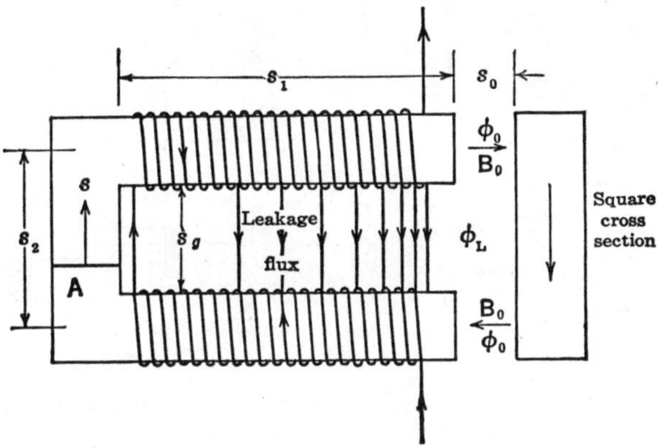

FIG. P13

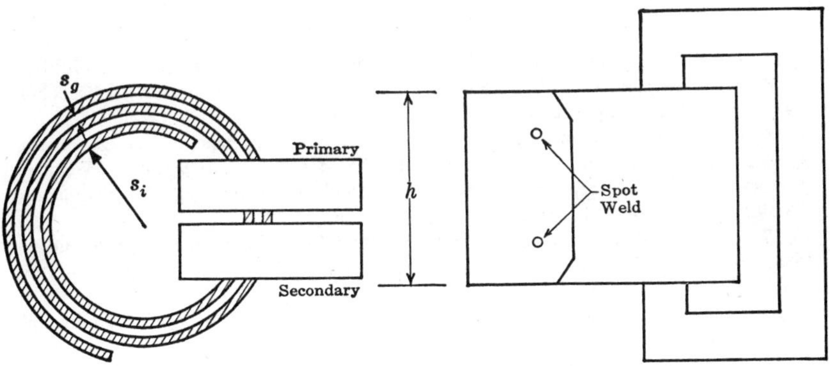

FIG. P14

0.5 cm, and $\phi_0 = 0.0003$ weber. $B_0 = 1$ weber per square meter. Assume $\mu_r = 3750$ everywhere. Compute the leakage flux density and total leakage flux ϕ_L. Compute the ni required.

14. Figure P14 illustrates a relatively new type of transformer core construction. Let $s_i = 5$ cm, $s_o = 0.01$ cm, and $h = 10$ cm; U.S.S. Transformer 72–29 gage; 150 turns of iron in the core. Sketch the paths of the flux lines. If the flux density around the inner circumference is 1.2 webers per square meter, evaluate the flux density at the outer circumference, remembering that the permeability varies with the flux density.

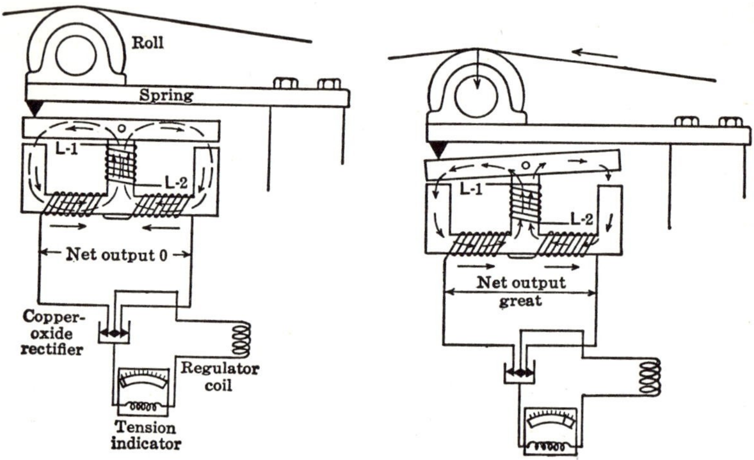

(*a*) With zero tension in the strip, the air gaps are equal and, therefore, the flux due to the exciting winding is the same in the two secondaries.

(*b*) When the strip presses downward or is under tension, the movable armature is deflected down toward the left in proportion to the tension on the strip.

FIG. P15. Diagram of tension control for cold-strip rolling.

15. Figure P15 gives an application of a magnetic device, taken from F. Mohler, "Tension Control in Cold Strip Rolling," *Iron and Steel Engineer*, **13**, 19 (October 1936). Explain its operation.

16. Calculate the hysteresis and eddy current coefficients for U.S.S. Motor grade, using $B = 1$ and $B = 0.6$ on Fig. 13.19. $d = 0.0185$ inch thick. What is the percentage of hysteresis loss at $B = 1$? Calculate the resistivity from equation 13.39 and compare with the test value given in Fig. 13.20.

17. Compute the saturation magnetic moment per cubic meter for Permendur. Assuming 9×10^{28} atoms per cubic meter, compute the moment per atom. How many active electron spins are there per atom?

18. Compute the maximum pressure on a Permendur surface when the material is magnetized to saturation.

19. Hipernik is frequently used in the cores of current transformers to improve operating characteristics. Discuss the improvements to be expected in ratio and phase angle errors.

CHAPTER 14

PERMANENT MAGNETS

Permanent magnets for a long time have been indispensable elements in many types of instruments and machines—voltmeters and ammeters for direct-current uses, galvanometers, watthour meters, magnetos, etc. But the development during the last few years of new permanent magnet alloys has altered the picture considerably. Magnets can now be made with shorter iron sections and longer air gaps, and greater quantities of energy stored in the gaps are available. In addition,

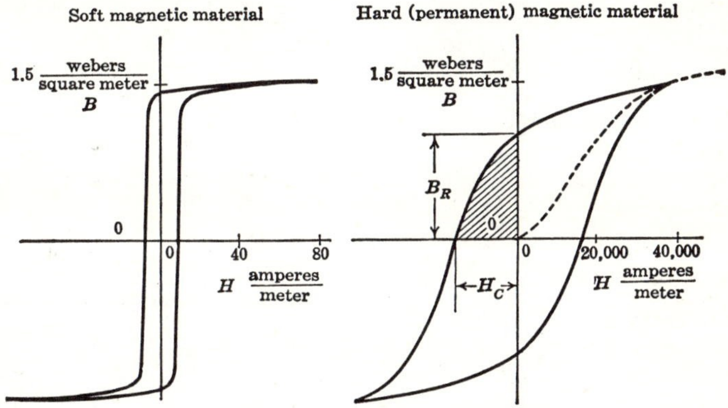

FIG. 14.1. Comparison of soft and hard magnetic materials. From Ellis and Schumacher, "A Survey of Magnetic Materials in Relation to Structure," *Bell System Technical Journal*, **14**, 8 (1935).

permanent magnets are replacing electromagnets, eliminating the need for magnetizing currents, in various applications—on mechanical-type oscillographs, magnetos, etc. Unfortunately the new alloys are frequently rather costly, and designs based on their use must utilize the materials as effectively as possible. Furthermore, many of the new alloys are so hard that the magnets must be forged or cast to shape or sintered.

Permanent magnet materials are said to be "magnetically hard," that is, they require high values of current and H_i to saturate and have moderately good values of remanent flux density B_r and high values of coercive force H_c. Figure 14.1 illustrates the magnitudes of these quantities for hard and soft magnetic materials. In general, the higher the residual flux density, the smaller the cross section of the magnet required; the higher the coercive force, the shorter the magnet may be made.

14.1. Fundamental Principles. A permanent magnet, to be useful, must have an air gap; but in order to develop the fundamental principles we will first consider a closed ring sample of the magnet material,

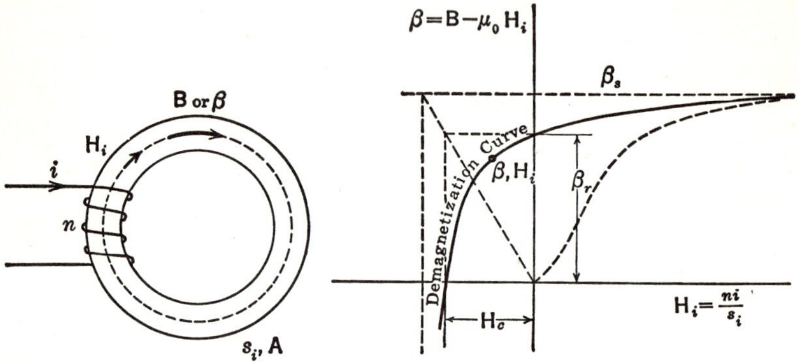

FIG. 14.2. Closed ring sample and demagnetization curve.

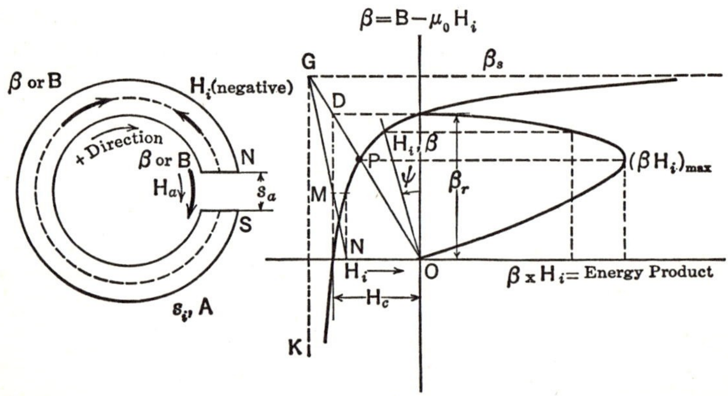

FIG. 14.3. Permanent magnet with air gap, and energy products.

magnetize it to saturation, then introduce an air gap. Following Fig. 14.2 let us magnetize the closed ring sample to saturation, in which condition the ferric induction $\beta = \beta_s = B_s - \mu_0 H_i$ has reached its limiting value of the order of 1.6 webers per square meter at values of H_i which may run up to tens of thousands with correspondingly large values of current. In this condition the vectors H_i, B, and β lie in the same direction around the ring. The equations that hold are

$$\left.\begin{aligned} B_s - \mu_0 H_i &= \beta_s = M_s \\ H_i s_i &= ni \end{aligned}\right\} \tag{14.1}$$

where M_s is the magnetic moment per cubic meter of the material. For $\beta_s = 1.6$ webers per square meter, $M_s = 1.6$ weber-meters per cubic meter. If H_i, for example, were 80,000 amperes per meter, and $s_i = 0.15$ meters, ni would be 12,000 ampere-turns. For $n = 500$ turns, $i = 24$ amperes.

As the current is reduced to zero, H_i becomes zero, and

$$\beta = \beta_r = B_r = M_r \tag{14.2}$$

the remanent value. The β versus H_i relation obtained from test, shown in Fig. 14.2, may be pushed to the left of the remanent point by either of two methods: The first is to apply a reverse current and set up a reverse H_i while β is still positive but reduced in magnitude; the second method is the one which we shall now follow. The portion of the curve in which B or β is positive and H_i is negative is known as the *demagnetization curve* and is the portion of chief interest in the application of permanent magnets.

With the current zero, imagine the magnet to be sawed through and the gap lengthened to s_a. Ignoring any leakage flux, the flux density B is the same in both iron and gap but is reduced in magnitude to an amount to be determined presently. At the gap faces, magnetic poles of opposite signs appear, as shown in Fig. 14.3. In the air gap B is equal to $\mu_0 H_a$, but in the iron H_i now has a negative value with respect to B. The equations are

$$B = \mu_0 H_a \tag{14.3}$$

$$ni = 0 = H_i s_i + H_a s_a \tag{14.4}$$

$$H_i = -\frac{H_a s_a}{s_i} = -\frac{\mathfrak{F}_{\text{gap}}}{s_i} \tag{14.5}$$

$$\beta = M = B - \mu_0 H_i = |B| + |\mu_0 H_i| \tag{14.6}$$

Equation 14.5 gives only the relation between H_a and H_i. Rewriting equation 14.5,

$$-\frac{H_a}{H_i} = -\frac{B}{\mu_0 H_i} = \frac{s_i}{s_a} \cong -\frac{\beta}{\mu_0 H_i} \tag{14.7}$$

since β is approximately equal to B in this range. Assuming s_i and s_a known, it is possible to lay off a straight line through the origin at angle ψ with the β axis so that

$$\tan \psi = -\frac{\mu_0 H_i}{\beta} = \frac{s_a}{s_i} \tag{14.8}$$

and where this line crosses the demagnetization curve of the magnet is the "working point" which gives the operating β and H_i values for the given magnet with an air gap.

The air gap reluctance, ignoring leakage, is

$$\mathcal{R}_{\text{gap}} = \frac{\mathfrak{F}_{\text{gap}}}{\phi} = \frac{H_a s_a}{BA} = \frac{s_a}{\mu_0 A} = -\frac{H_i}{B}\frac{s_i}{A} \cong \frac{1}{\mu_0}(\tan\psi)\frac{s_i}{A} \tag{14.9}$$

and the energy stored in the gap is

$$W_{\text{gap}} = \frac{\mu_0 H_a^2}{2} s_a A = \frac{1}{2}\mathfrak{F}_{\text{gap}}\phi = \frac{\mu_0 H_a^2 A^2}{2}\frac{s_a}{A} = \frac{\phi^2}{2}\mathcal{R}_{\text{gap}}$$
$$= -\tfrac{1}{2}(H_i B)(s_i A) \cong -\tfrac{1}{2}(H_i\beta)(s_i A) \tag{14.10}$$

From this we note that the energy in the *gap* is proportional to the $(H_i\beta)$ product at the working point and to the volume of the *iron*.

The $(H_i\beta)$ or energy products for various working points are plotted to the right of the origin in Fig. 14.3. The energy product is perhaps the best all-around measure of the effectiveness of a given magnet. The most economical use of expensive magnet material, for a given stored gap energy, occurs when the energy product is a maximum and the magnet volume is a minimum. This optimum condition is obtained in a design which places the working point at P, in Fig. 14.3, where the line OD crosses the demagnetization curve. D is located as shown where the perpendiculars from the axes meet. For maximum gap energy, for a given iron volume, $\tan\psi = \mu_0 H_c/\beta_r$. Energy products for ordinary materials may be a few hundred thousands but may run up to five million for the new alloys.

Watson[1] has shown that the demagnetization curve for a large class of "magnetically hard" materials is a rectangular hyperbola, provided that the materials have been properly heat-treated to make homogeneous substances. According to this rule H (same as H_i) and β are related by the equation

$$\frac{H + H_c}{\beta} = a + b(H + H_c) \tag{14.11}$$

or

$$\beta = \frac{(H + H_c)}{a + b(H + H_c)} \tag{14.12}$$

in which H_c is taken as a positive value, H as negative in the demagnetization range, and a and b are constants. When H is positive and

[1] E. A. Watson, *J. Inst. of Elec. Engs*, **61**, 641 (1923).

large compared with H_c

$$\beta = \beta_s \cong \frac{1}{b} \tag{14.13}$$

and since

$$\frac{d\beta}{dH} = \frac{a}{\left[a + \frac{1}{\beta_s}(H + H_c)\right]^2} \tag{14.14}$$

we note that, when $H = -H_c$, $\beta = 0$, and

$$\left(\frac{d\beta}{dH}\right)_{\beta=0} = \frac{1}{a} \tag{14.15}$$

Hence b is the reciprocal of the saturation ferric induction, and a is the reciprocal of the slope of the demagnetization curve at the H axis.

These equations can be put into rectangular hyperbola form by shifting the axes in Fig. 14.3 from origin at O to origin at G, which is the point where the line OD produced cuts β_s. Using the substitutions

$$\left.\begin{aligned} r &= \frac{\beta_s}{\beta_r} \\ \beta' &= \beta_s - \beta \\ H' &= H_c\frac{\beta_s}{\beta_r} + H = H_c r + H \\ a &= \frac{H_c}{\beta_r}\frac{r-1}{r} \end{aligned}\right\} \tag{14.16}$$

Equation 14.12 transforms into

$$\beta'H' = \beta_s H_c(r - 1) = \text{constant} \tag{14.17}$$

a rectangular hyperbola with axes $\beta_s G$ and GK. Only three points are necessary to determine the hyperbola, for instance β_s, β_r, and H_c, or *any other set of three points*. In practice, it would be better to use β_r, H_c, and any other point on the demagnetization curve rather than β_s.

The demagnetization curve is easily determined graphically. From G draw a straight line to any point N on the H axis, crossing the vertical line through D at M. Where a horizontal line through M meets a vertical line through N is a point on the curve.

Watson's law, when applied to large positive values of H, so that $H \gg H_c$, reduces to Lamont's law (sometimes called Fröhlich's law), given by

$$\frac{H}{\beta} = a_1 + bH \tag{14.18}$$

in which $b = 1/\beta_s$ and $a_1 = (d\beta/dH)_{H=0}$.

A sample set of figures will help to indicate the magnitudes involved. Given a magnetic material whose $(\beta H)_{\max} = 2560$ joules per cubic meter at $\beta = 0.8$ weber per square meter, $H = 3200$ amperes per meter, to be formed into a magnet to supply $\phi = 0.00016$ weber over an air gap $s_a = 0.0025$ meter. Then $A = \phi/\beta = 0.00016/0.8 = 0.0002$ square meter, $\tan \psi = \mu_0 H/\beta = (4\pi 3200 \times 10^{-7})/0.8 \cong 0.005$, and $s_i = s_a/\tan \psi = 0.0025/0.005 = 0.5$ meter.

If Alnico V is used to produce the same flux $\phi = 0.00016$ weber over the same air gap $s_a = 0.0025$ meter, the calculation indicates that a much shorter magnet is required. $(\beta H)_{\max} = 36{,}000$ joules per cubic meter at $\beta = 0.97$ weber per square meter and $H = 36{,}600$ amperes per meter. $A = \phi/\beta = 0.00016/0.97 = 0.000165$ square meter, $\tan \psi = \mu_0 H/\beta = (4\pi 36{,}600 \times 10^{-7})/0.97 = 0.0475$ and $s_i = s_a/\tan \psi = 0.0025/0.0475 = 0.0527$ meter. If the magnet were operated at any working point other than the one corresponding to maximum βH, the iron length required would be still greater. The new alloys with greatly increased values of $\tan \psi$ require much shorter lengths.

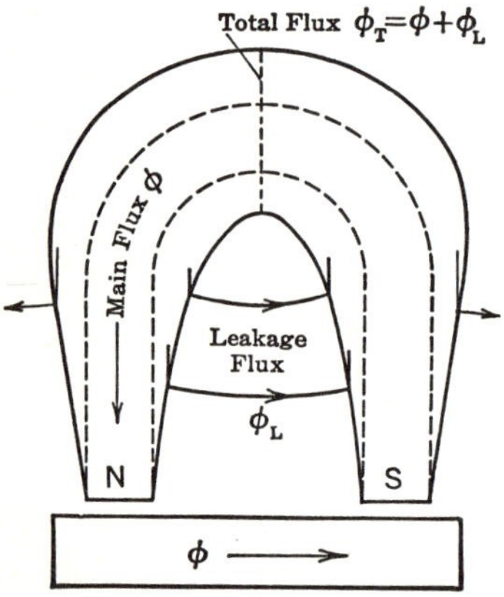

FIG. 14.4. Permanent magnet design to provide for leakage flux.

So far leakage flux has been ignored, but it may well have a value as large as that of the useful flux. Figure 14.4 shows a design in which the cross section of the magnet has been increased to provide for leakage, leaving the center core for the useful flux. The equations previously given apply, of course, to the useful flux.

14.2. Permanent Magnet Materials. Prior to 1929, magnets were always composed of carbon steels in which the magnetic hardness was due to the carbides contained. The earliest type, with 0.7 to 1.0 per cent carbon, had a coercive force of about 4000 amperes per meter, requiring a long magnet. The βH product was about 1600 joules. Early in the century 6 per cent tungsten was added and β_r, H_c, and $(\beta H)_{\max}$ were improved materially at a considerable increase in cost. See Tables 14.1 and 14.2. During World War I it was found that 3 per cent chromium instead of the tungsten produced much the same improvement at a greatly reduced cost. The first big development came in 1916 with the announcement by Dr. Honda of his K.S. magnet steel, containing among other elements from 15 to 42 per cent cobalt. The addition of cobalt quadrupled the coercive force (16,000 amperes per meter) with only a slight reduction of β_r, and raised $(\beta H)_{max}$ to 8000 joules. However, the high cost of cobalt has retarded the use of this material. All the materials so far mentioned depend, for their magnetic hardness, upon the influence of the various alloy elements in controlling the carbide effect.

In 1930 it was found that excellent magnetic qualities could be obtained with carbon-free iron alloys, containing, in various proportions, tungsten, molybdenum, tantalum, beryllium, and cobalt. These alloys require special solution and aging heat treatments known as precipitation hardening, which is quite different in its metallurgical effect from the more usual quench hardening. For some of these expensive alloys H_c approaches 20,000 amperes per meter and $(\beta H)_{max}$ is approximately 8000 joules.

The next step in advance, also 1930, came with the announcement of Dr. Mishima's nickel-aluminum M.K. steel, which, though it contains iron, is not truly a steel as it does not contain carbon. The addition of cobalt, and cobalt and copper, have led to alloys which are most favored at the present time, for example, Alnico. H_c has been boosted to 40,000 amperes per meter and $(\beta H)_{max}$ to 11,200 joules; the cost, in spite of the improved performance, has been reduced very materially.

Nickel-aluminum type magnets are not malleable and so are to be obtained only in cast or sintered (powdered, compressed, and heated) forms. In practice it is often desirable to use short straight sections of nickel-aluminum alloy and complete the magnetic path with soft iron. Recently a small Alnico magnet, suitably combined with a soft iron extension of adequate design, has succeeded in lifting an iron object whose weight was 4500 times that of the magnet. To accomplish this the flux is forced to cross the air gap several times. Each additional crossing increases the tractive force of the magnet.

The Alnico alloys appear to be far superior to tungsten and cobalt magnet steels with respect to their ability to withstand high temperatures, strong demagnetizing fields, and vibrations.

The copper-nickel-iron alloy Magnetoflex (von Auwers and Neumann, 1935) is distinguished by the fact that it can be forged, drilled, tapped, and machined, and its magnetic properties are greatly enhanced by cold working.

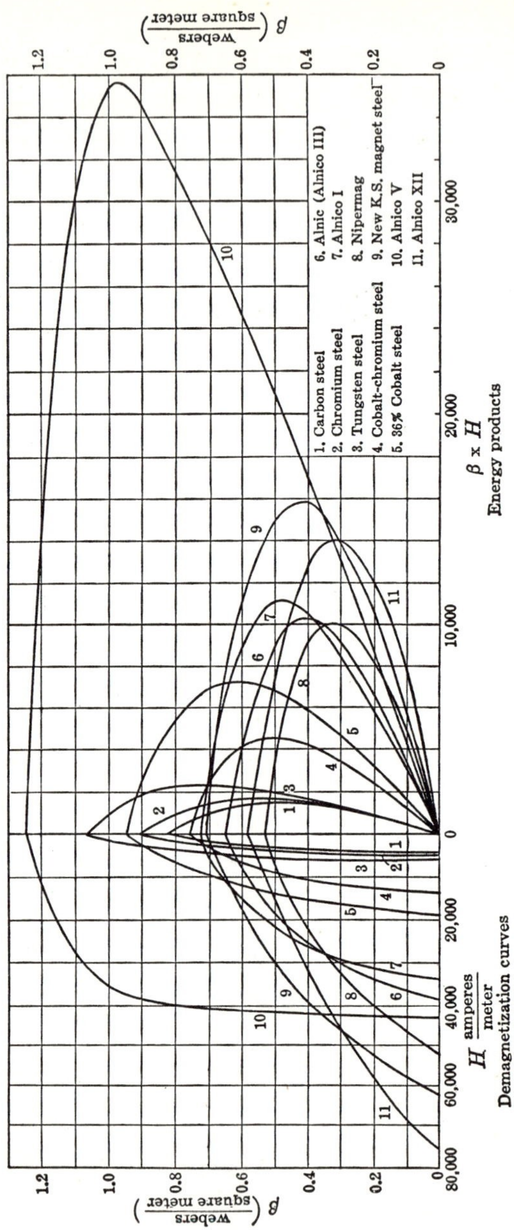

FIG. 14.5. Demagnetization curves and energy products for various permanent magnet materials. [From International Nickel Company, *Nickel Alloy Steels*, Section IV, Data Sheet No. 2, November 1938. Alnico V data from the *G.E. Review*, **45**, 573 (October 1942).]

TABLE 14.1

REPRESENTATIVE MAGNETIC PROPERTIES OF SOME PERMANENT MAGNET MATERIALS

Curve	Name	Coercive Force, amperes/meter	Residual Induction, webers/square meter	Energy Product $(\beta H)_{max}$, joules per cubic meter	Specific Gravity, gm/cc
1	1% carbon steel	4,000	0.82	1,600	7.7
2	3% chromium steel	4,800	0.9	1,840	7.7
3	6% tungsten steel	5,600	1.06	2,480	8.0
	Copper-nickel-iron alloy (60% Cu, 20% Ni, 20% Fe) cast	32,000	0.34	3,760	....
4	15% cobalt, 10% chromium steel	14,800	0.77	4,800	7.8
	18% tungsten, 15% cobalt steel	12,000	1.15	5,600	....
5	36% cobalt steel	19,200	0.96	7,360	8.1
	Kato's oxide magnet	48,000	0.35	8,240	4.0
	19% molybdenum, 12% cobalt steel	24,000	0.84	8,320	8.3
	Copper-nickel-iron alloy (60% Cu, 20% Ni, 20% Fe) cold rolled	36,800	0.528	8,560	8.6
	42% cobalt steel	18,400	1.06	8,640	8.1
8	Nipermag (32% Ni, 12% Al + Ti)	52,000	0.53	10,000	7.0
6	Alnic (Alnico III) (25% Ni, 12% Al)	39,200	0.65	10,400	6.9
	Alnico IV (28% Ni, 12% Al, 5% Co)	48,000	0.62	10,400	7.0
7	Alnico I (20% Ni, 12% Al, 5% Co)	36,400	0.72	11,200	7.0
	Alnico II (17% Ni, 10% Al, 12.5% Co, 6% Cu)	43,200	0.72	13,200	7.1
9	New K.S. magnet steel (27% Co, 18% Ni, 7% Ti, 3.7% Al)	62,800	0.715	16,240	7.3
	Platinum-iron alloy (77.8% Pt, 22.2% Fe)	125,600	0.538	24,560	13.5
	Platinum-cobalt alloy (76.7% Pt, 23.3% Co) cast	122,400	0.64	28,400	14.6
	Platinum-cobalt alloy (76.7% Pt, 23.3% Co) quenched	212,000	0.453	30,160	14.6
10	Alnico V	44,000	1.25	36,000	....
11	Alnico XII	76,000	0.58	14,000	....

Adapted from International Nickel Company, *Nickel Alloy Steels*, Section IV, Data Sheet No. 2, November 1938.

TABLE 14.2

Permanent Magnet Steels

Per Cent Composition					Raw Cost ¢/lb	Typical Heat Treatment		Material	Initial Relative Permeability	Reversible Relative Permeability	Saturation	Residual	Coercive Force	Energy Product	B for $(\beta H)_{max}$	Resistivity	Curie Temperature	Density
Other	Cr	Ni	Co	Fe		Quench	Aging		μ_{ri}		β_s	B_r	H_c	$(\beta \cdot H)_{max}$		ρ	T_c	δ
											webers/sq meter	webers/sq meter	amperes/meter	joules/cu meter	webers/sq meter	microhm-centimeter		gm/cm[1]
0.6 C, 0.8 Mn	...	...	....	98.8	0.8	800° C Water	.....	Manganese	75	....	2.1	1.0	4,000	0.16×10^4	0.69	20	750° C	7.8
0.6 C, 0.4 Mn	1	...	....	98	0.9	800° Oil	.....	1 % Chrome	....	....	....	0.98	4,000	0.16	0.69	....	....	....
1 C, 0.4 Mn	3	...	....	96	1.1	840° Oil	.....	3 % Chrome	....	31	....	0.97	5,200	0.24	0.61	38	....	7.7
5 W, 1 C	...	...	....	94	9	840° H_2O	.....	5 % Tungsten	....	32	....	1.08	4,800	0.24	0.7	30	....	8.0
7 W, 0.5 Mn	3.5	...	36	52	62	940° Oil	.....	Cobalt (Honda KS)	7	9.4	1.9	0.95	17,600	0.72	0.6	27	700	8.3
6.7 Ti, 3.7 Al	...	18	27	45	53	Cast	650°	Honda, New KS	3	3.8	....	0.71	62,400	1.6	0.41	65	....	7.3
13 Al	...	29	....	58	14	1200° Oil	600°	Mishima	4	5	1.16	0.6	44,000	1.12	0.35	60	750	7.1
14 Al	...	25	....	60	12	1200° Oil	600°	Oerstit 500	....	5	....	0.65	40,000	1.12	0.44	60	....	....
12 Al	...	20	5	63	17	1200° Oil	600°	Alnico I	....	4	....	0.73	34,400	1.12	0.45	60	....	7.0
10 Al, 6 Cu	...	17	12.5	54.5	26	1200° Oil	600°	Alnico II	....	4	....	0.72	43,200	1.28	0.44	....	....	7.1
17 Mo	...	...	12	71	33	1300° Oil	700°	Remalloy	8	12	1.7	1.05	20,000	0.88	0.65	45	780	8.4
60 Cu	...	20	....	20	13	1000° Oil	600°	Magnetoflex[1]	3	3	0.5	0.34	31,200	0.4	0.18	....	....	....
41 Cu	...	24	35	....	60	1050° Oil	600°	Magnetoflex[2]	4	4	0.86	0.53	35,200	0.8	0.34	38	850	8.7
77 Pt			23	....	$400	1200° Oil		77 Platinum—cobalt	1.1	1.1	....	0.45	208,000	3.04	0.25	50	....	....
2 Fe_2O_3 + 1 Fe_3O_4 + 1 Co_2O_3			....	....	25¢	950° Vacuum + magnetize at 500°		Oxide, compressed	1.7	1.7	....	0.18	48,000	....	....	10^{12}	350	3.8

[1] Newmann, Bücher, and Reinboth, *Z. f. Metallkunde* **29,** 173 (1937).
[2] Dannöhl and Neumann, *Z. f. Metallkunde* **30,** 217 (1938).
Note: Values for μ_{ri}, B_r, H_c, and $(\beta \cdot H)$max are subject to considerable variations, depending on purity of materials, composition, and heat treatment.
From V. E. Legg, "Survey of Magnetic Materials and Applications in the Telephone System," *Bell System Technical Journal*, **18,** (July 1939).

Platinum-cobalt alloys far outshine in performance any magnets previously developed, but the cost per pound is some 40,000 times that of an ordinary carbon-steel magnet.

Tables 14.1 and 14.2 give, in condensed form, pertinent information about these materials. Figure 14.5 illustrates the demagnetization curves and βH products for the materials listed in Table 14.1 for corresponding curve numbers.

In addition, new materials have recently appeared, designed to meet special magnet requirements. Among these are Cunife (60% Cu, 20% Ni, 20% Fe), Cunico (45% Cu, 25% Ni, 30% Co), Comol (12% Co, 15% Mo, 73% Fe), Silmanol (86% Ag, 8.8% Mn, 4.7% Al), and Vectolite, made of cobalt and iron-oxide powders, pressed to shape and sintered.

The following references contain interesting information concerning the application of permanent magnets: Hornfeck and Edgar, "The Output and Optimum Design of Permanent Magnets Subjected to Demagnetizing Forces," *A.I.E.E. Trans. Supplement*, **59,** 1017 (1940); Wey, "The Shielding of Permanent Magnets from Transient Magnetic Fields," *Elec. Eng.*, **60,** 875 (September 1941).

CHAPTER 15

IMAGING AND MAPPING MAGNETIC FIELDS

The principles of field mapping as applied to electric fields (Chapter 7) may be carried over and applied directly to the study of magnetic fields that lie outside regions containing magnetic material and outside conducting regions carrying current. Let us consider, then, three types of magnetic regions through which magnetic flux may pass—air (or any other nonmagnetic medium, $\mu = \mu_0$); iron (or any other magnetic material of high permeability) in which the permeability is *assumed* constant; and copper (or any other nonmagnetic conducting body) carrying a current, which is assumed to be a steady current of uniform current density in order to eliminate the very difficult problem of skin effect.

15.1. Mapping in the Air Region. In this region the field is considerably simplified by assuming that the magnetic flux approaches all air-iron surfaces normally. *Viewed from the air side*, the iron–air surface may be regarded as a magnetic equipotential surface. Actually, the flux is refracted in crossing from air to iron, but the permeability ratio is so high that, even for very large angles from normal in the iron (α_2 in Fig. 15.1), the angles in air α_1 are very small, according to the boundary rule

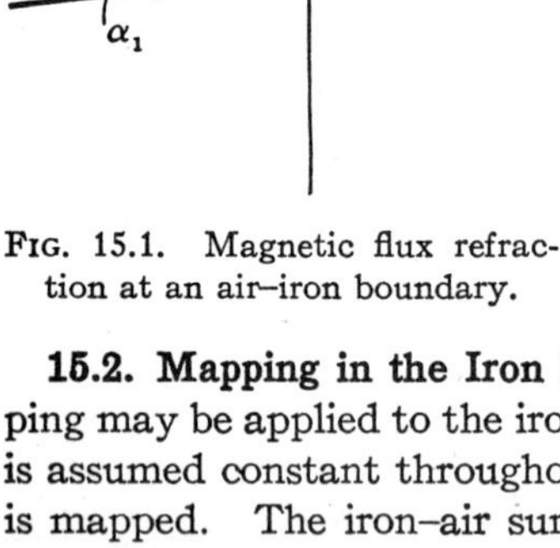

FIG. 15.1. Magnetic flux refraction at an air–iron boundary.

$$\frac{\tan \alpha_2}{\tan \alpha_1} = \frac{\mu_2}{\mu_1} \gg 1 \qquad (15.1)$$

This is equivalent to treating the iron as a medium of infinite permeability.

15.2. Mapping in the Iron Region. The usual rules for field mapping may be applied to the iron region, provided that the permeability is assumed constant throughout the portion of iron in which the field is mapped. The iron–air surface may *not* be considered as an equipotential from the point of view of the iron region.

15.3. Mapping in the Copper Region (Carrying Current). There is a line in the copper region, roughly paralleling the current flow, along which the magnetic intensity is zero. This line is known as the "core" or "kernel." The magnetic flux lies in closed curves surrounding the kernel, as shown in Fig. 15.2. The location of the kernel is usually rather difficult to determine inasmuch as it depends not only

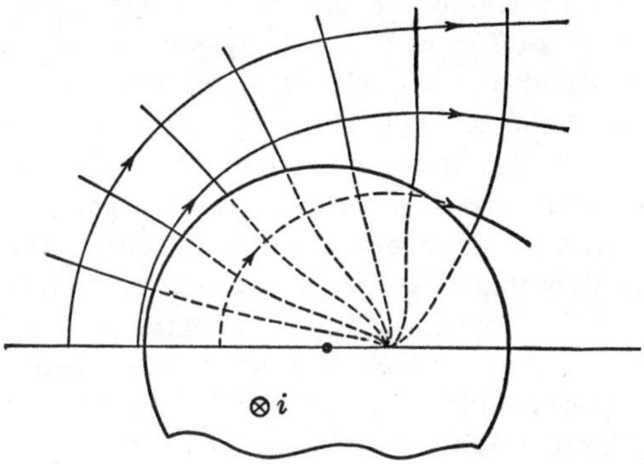

FIG. 15.2. Field about a kernel. (See Fig. 15.13.)

on the conductor shape but also on the proximity of other conductors carrying current and of magnetic materials. In general, in any one conductor, the kernel is shifted toward nearby magnetic materials, toward other conductors carrying current in the same direction as the first, and away from conductors carrying current in the opposite direction.

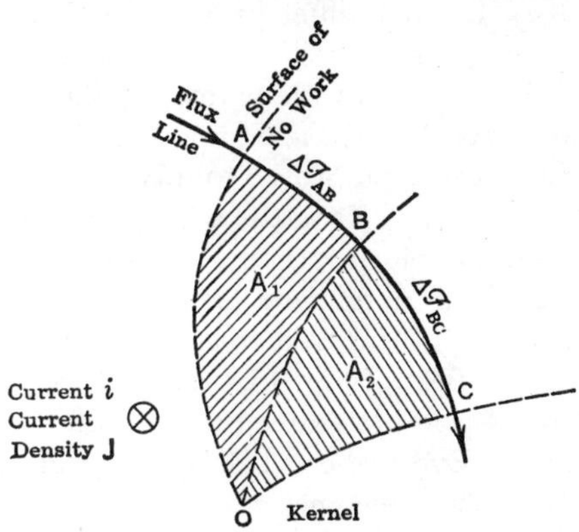

FIG. 15.3. Field inside a conductor.

The position of the kernel may be located by graphical methods. If, as shown in Fig. 15.3, surfaces normal to the flux are drawn from the kernel O and a flux line ABC is drawn perpendicular to them, the difference of magnetomotive force (MMF) from A to B taken along the flux line is to the MMF difference from B to C as the ratio of areas, A_1 to A_2, provided that the current density is constant over A_1 and A_2. The total closed-path MMF difference around $OABO$ is the same as the MMF between A and B.

$$\frac{\Delta\mathfrak{F}_{AB}}{\Delta\mathfrak{F}_{BC}} = \frac{JA_1}{JA_2} = \frac{A_1}{A_2} \tag{15.2}$$

Of course this relation does not locate the position of the kernel, but it can be combined with the knowledge of the position of the equipotential surfaces outside the copper to fix the former with fair accuracy.

Inside the copper conductor the surfaces normal to the flux lines are called "surfaces of no work." They join on to the outside equipotential surfaces at the copper–air boundary. At the copper–air surface the lines of flux are not refracted, as the permeabilities of both media are equal, but it can be shown that the flux lines have different radii of curvature at adjacent points in the two media.

The points just brought out are illustrated in the special cases following, which combine the methods of imaging and field mapping. The material of Chapters 5 and 7 on imaging and field mapping in electrostatic fields may be carried over and applied directly to the solution of magnetic fields outside regions carrying current.

15.4. Two Parallel Equal Currents. *I. With the Same Direction.* Figure 15.4. This case is symmetrical about the plane through the currents and about the normal bisecting plane. These planes are equipotential surfaces and cross at a singular line that also contains the center of gravity. That portion of the plane of the wires lying outside the wires may be given two values of potential ($-i/2$ and $+i/2$), which differ from each other by i. The potential of the plane between the wires is zero, as also is that of the center bisecting plane.

Far out from the wires, the flux lines form circles about the center of gravity, and the equipotential surfaces, for equal changes in potential, are asymptotic to planes making equal angles at the center of gravity. Near the wires, the field is most readily determined by field-mapping methods, aided by the knowledge that surfaces of equal potential increments leave each wire at equal angles when the wires are small compared with their separation.

The field of Fig. 15.4 remains unaltered on the air side of the boundary if iron of assumed infinite permeability is placed:

(1) Everywhere below the mid-bisecting plane. The lower current then becomes the *image* of the upper one, for the purpose of getting the field on the *air* side of the boundary when one wire in air is brought near a plane iron surface. See Art. 15.10 for the case in which the iron is assumed not to have infinite permeability.

(2) Everywhere to one side of the plane of the wires. This is the case of two equal currents of the same direction imbedded in the plane iron surface.

(3) Everywhere to one side of the plane of the wires and below the mid-bisecting plane. This is the case of a wire imbedded in one air–iron face of a 90° air corner bounded by 270° of iron.

II. With Opposite Directions. Figure 15.5. This is the case of the transmission line solved analytically in Art. 10.8. Field mapping also furnishes an easy solution. All the flux lines are circles, and the equipotential surfaces are portions of circular cylinders. Any equipotential leaving the wire at angle α with the plane of wires crosses the

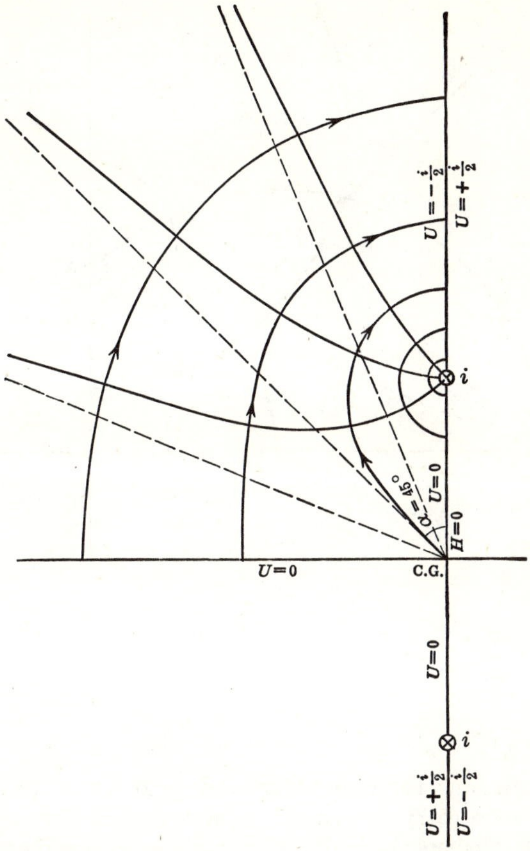

FIG. 15.4. Field of two like and equal parallel currents.

mid-bisecting plane at a point determined by drawing a construction line from the wire at angle $\alpha/2$. Far out from the wires, the 45° line CD is a construction symmetry lying halfway between the mid-bisecting plane of flux lines and the plane of wires equipotential. The flux lines and equipotentials each across CD at 45°, far out from the wires.

If iron of infinite permeability is placed everywhere to the right of the plane of wires, Fig. 15.5 gives the correct field on the air side for two equal currents of opposite direction imbedded in the plane iron surface.

15.5. Four Equal Currents in Rectangular Formation. *I. With Same Directions.* Figure 15.6 shows the field for this configuration of

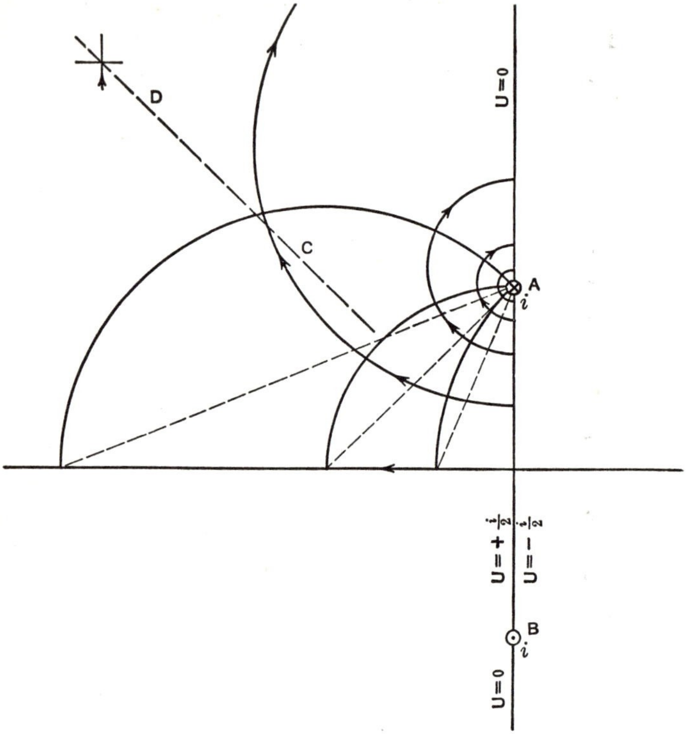

FIG. 15.5. Field of two unlike and equal parallel currents.

currents in air, and the upper left-hand quadrant is the correct field in air for a wire carrying current parallel to two plane iron surfaces meeting at 90°, where the wire is placed distances a and b from the two surfaces.

II. With Opposite Directions. Figure 15.7 shows the field when oppositely directed currents lie in rectangular formation. The lower half of the figure may be replaced by iron of infinite permeability without disturbing the air portion of the field.

15.6. Wires inside a Closed Iron Conductor. Figure 15.8 shows the electric and the magnetic fields about two wires placed inside a *closed* iron volume. The upper and lower closing ends of the iron are not shown. The fields are given with both like and opposite polarities of electric and magnetic fields.

15.7. Wires Lying Midway between Parallel Iron Surfaces. In this case the electric fields are practically the same as in Fig. 15.8*a* and *c*. The magnetic fields are changed in a number of respects,

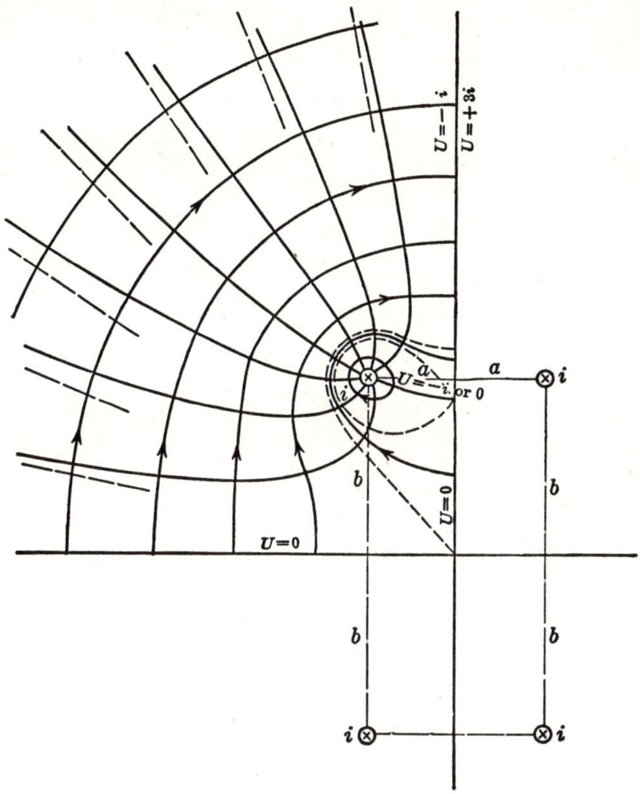

FIG. 15.6. Field of four equal currents. (Same direction.)

because without an end closing section of iron, the flux must return across the air gap between the iron faces. Two singular lines now appear on each iron face.

15.8. Field of Three Line Currents. The field of a conductor and its return circuit with current evenly divided and symmetrically placed are shown in Fig. 15.9. Figure 15.10 shows the field of the same current distribution placed midway between two iron planes. The right-hand half of the field may be replaced by iron. Figures 15.11

and 15.12 illustrate the same field except that the return currents are split in a three to one ratio.

15.9. Wires of Large Cross-Sectional Area. Two parallel circular wires which have radii R and are spaced center distance D carry equal

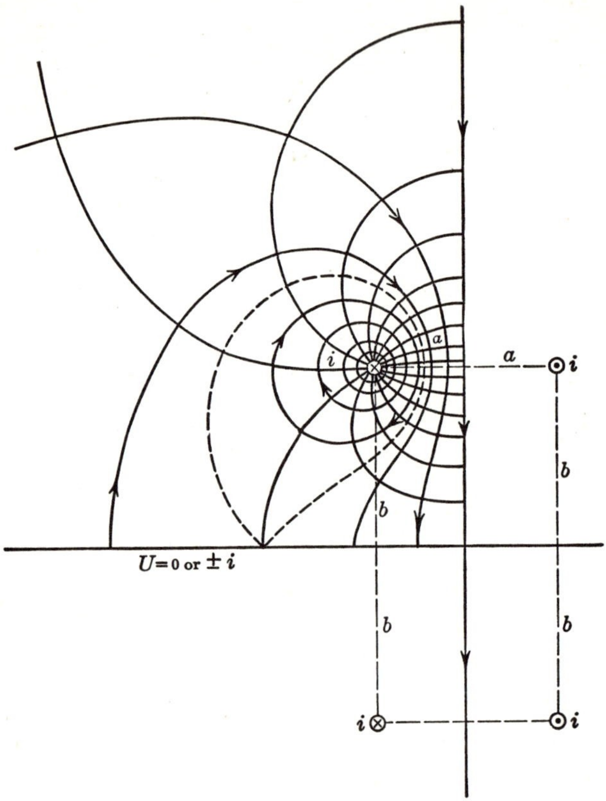

FIG. 15.7. Field of four equal currents. (Opposite directions.)

currents in the same direction, as shown in Fig. 15.13. Outside the wires, the field is the same as that given by two equal currents of small cross section (filaments of current) lying at the wire centers. Inside the wires, the field requires special study.

Along the plane of wire centers and inward from them a distance Δ lie the kernels, or lines of zero intensity. The kernels are located easily in this case. Let r be any distance from a wire center along the plane of centers but inside the wire. Then

$$H_{\text{axis}} = \frac{i}{2\pi R^2} r - \frac{i}{2\pi(D - r)} = 0 \tag{15.3}$$

or

$$\Delta = r = \frac{D}{2} - \sqrt{\left(\frac{D}{2}\right)^2 - R^2} \tag{15.4}$$

Inside the wire, the surfaces of no work emanate from the kernel in such a way that they divide the wire cross section into equal areas

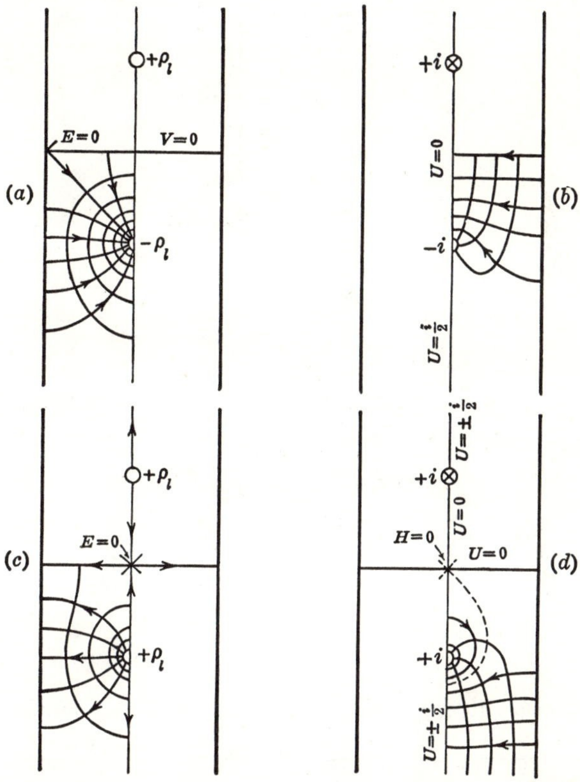

FIG. 15.8. Fields inside a closed iron conductor. (*a*) Electric field of two lines with opposite charges. (*b*) Magnetic field of two lines with opposite currents. (*c*) Electric field of two lines with like charges. (*d*) Magnetic field of two lines with like currents.

and at the wire surface meet equipotential surfaces having equal potential differences. The lines of flux form a series of closed lines about the kernel. They are everywhere normal to the equipotentials and surfaces of no work and cross from the wire to the air without refraction. The right half of Fig. 15.13 may be replaced with iron of infinite permeability.

In the special case that the two wires touch, or that one wire is placed against an iron plane, the kernel is drawn over until it lies at

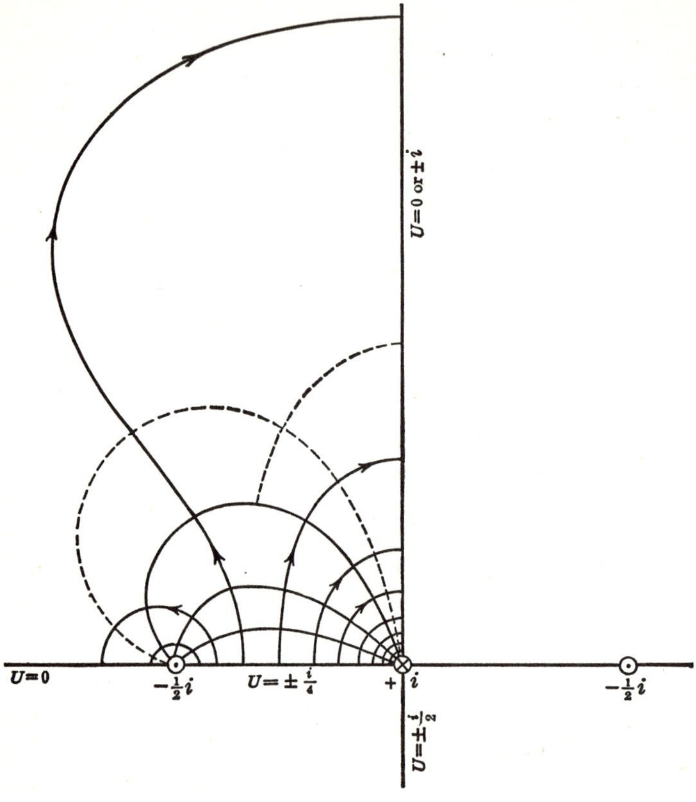

Fig. 15.9. Field of three line currents, symmetrically arranged.

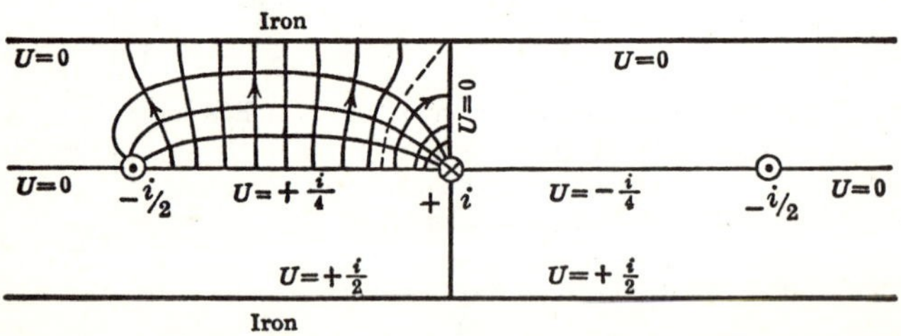

Fig. 15.10. Field of three line currents midway between iron surfaces.

the intersection of the two wires, or of the wire and the iron, as shown in Fig. 15.14.

On the basis of this discussion it is possible to say that the kernel is drawn away from the center toward another wire carrying a current in the same direction, and toward any adjacent body of iron, and is

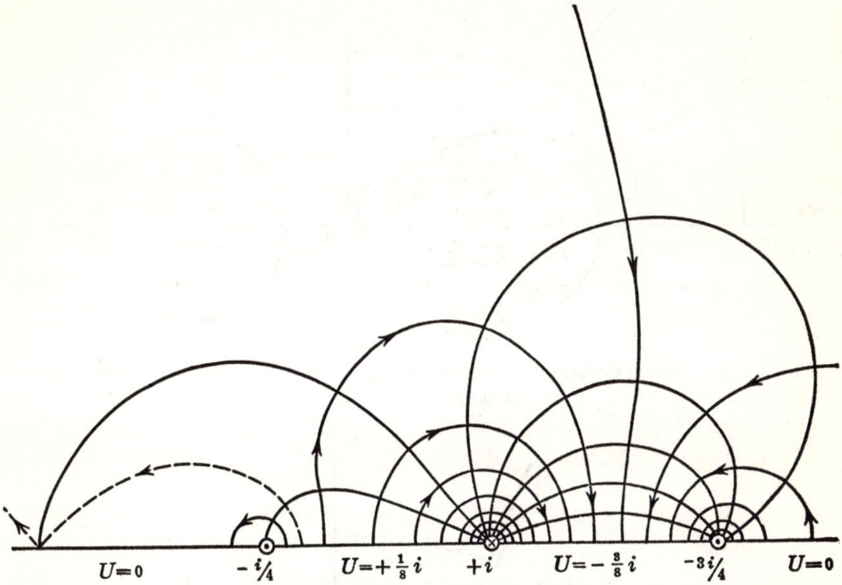

FIG. 15.11. Field of three line currents.

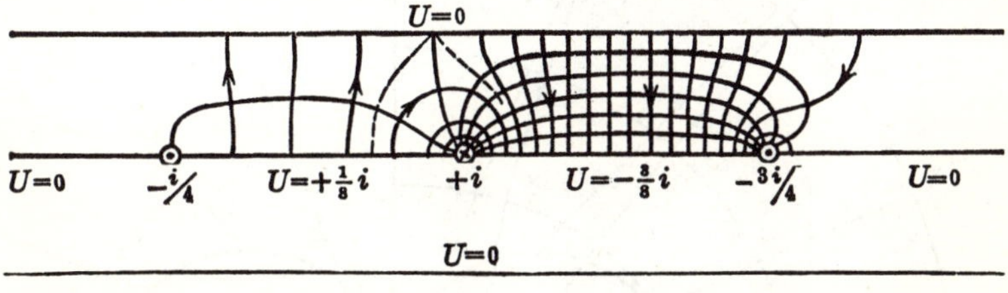

FIG. 15.12. Field of three line currents midway between iron surfaces.

pushed away from the center by a second wire carrying current in the opposite direction. See, for comparison, Fig. 10.10.

15.10. Current Parallel to a Plane Air-Iron Boundary. *I. Current in Air.* The method of imaging with dielectrics, explained in Art. 5.13, may be used, with modification, to determine the field of a current placed in air parallel to a plane surface of iron, for the case in which the iron permeability is finite. See Figs. 15.15 and 15.16.

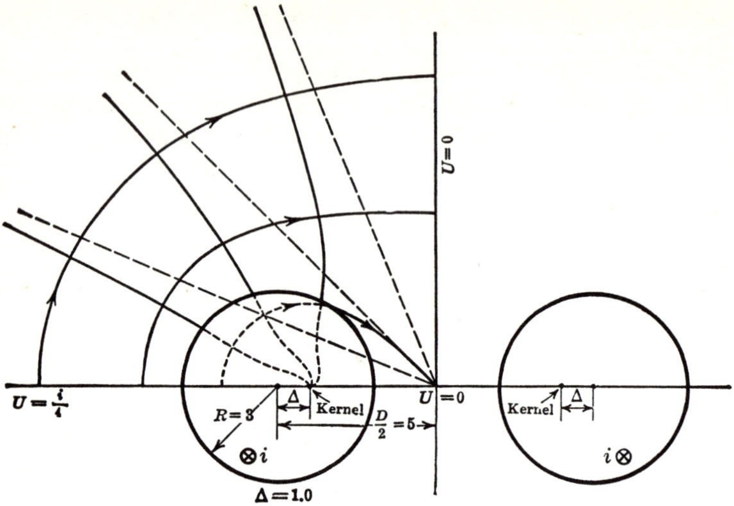

FIG. 15.13. Magnetic field of two large parallel wires. Currents in same direction. (Compare with Fig. 10.10.)

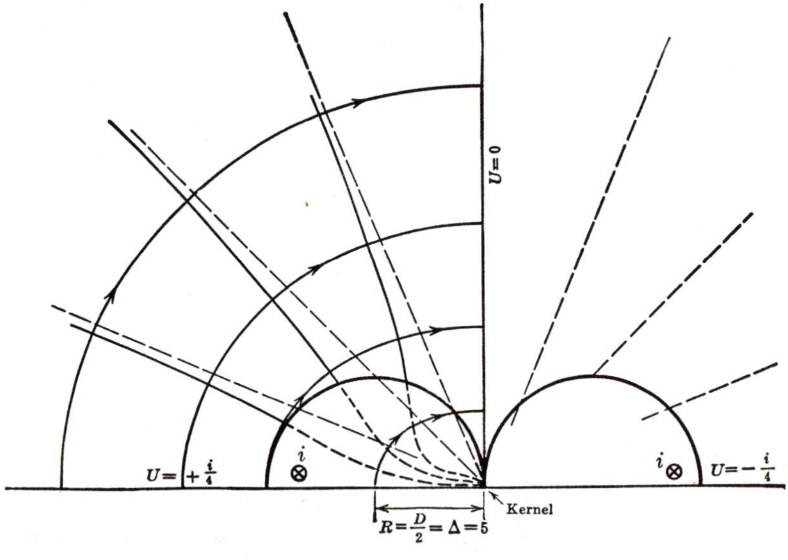

FIG. 15.14. Field of two like currents—touching.

Let h equal the distance of the wire from the iron. Place image i_2 inside the iron at distance h and image i_1 at the wire itself. The air field is determined by wire i and image i_2, assumed to be acting in a medium of permeability μ_1 everywhere. The iron field is given by wire i and image i_1, assumed to be acting in a medium of permeability μ_2 everywhere. The flux lines in the *iron* form a system of circles

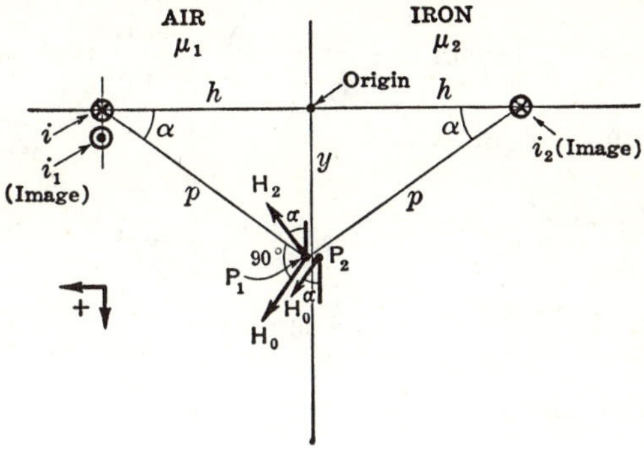

FIG. 15.15. Current (i) in air near a plane iron surface.

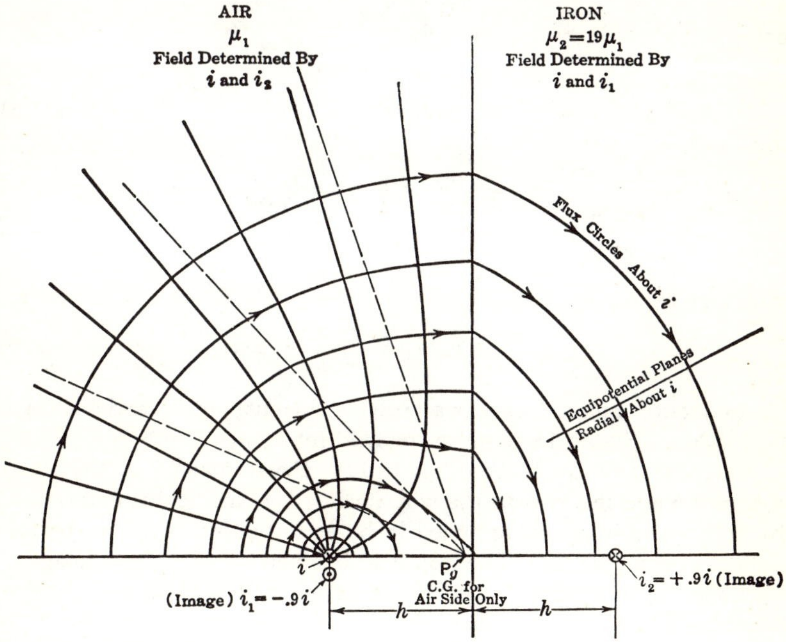

FIG. 15.16. Current (i) in air near a plane iron surface.

concentric about the wire. In the following equations, the currents in the wire and both images are assumed to carry positive currents (downward perpendicular to the paper).

At Point P_1 (Fig. 15.15), use i and i_2 in μ_1.

$$H_{1T} = \frac{i}{2\pi p}\cos\alpha - \frac{i_2}{2\pi p}\cos\alpha = \frac{h}{2\pi p^2}(i - i_2)$$

$$B_{1N} = \frac{\mu_1}{2\pi}\left(\frac{i}{p}\sin\alpha + \frac{i_2}{p}\sin\alpha\right) = \frac{y\mu_1}{2\pi p^2}(i + i_2)$$

At point P_2, use i and i_1, in μ_2.

$$H_{2T} = \frac{i}{2\pi p}\cos\alpha + \frac{i_1}{2\pi p}\cos\alpha = \frac{h}{2\pi p^2}(i + i_1)$$

$$B_{2N} = \frac{\mu_2}{2\pi}\left(\frac{i}{p}\sin\alpha + \frac{i_1}{p}\sin\alpha\right) = \frac{y\mu_2}{2\pi p^2}(i + i_1)$$

The boundary conditions demand that

$$H_{1T} = H_{2T} = \frac{h}{2\pi p^2}(i - i_2) = \frac{h}{2\pi p^2}(i + i_1)$$

or

$$i - i_2 = i + i_1$$

$$i_1 = -i_2 \tag{15.5}$$

and that

$$B_{1N} = B_{2N} = \frac{y\mu_1}{2\pi p^2}(i + i_2) = \frac{y\mu_2}{2\pi p^2}(i + i_1)$$

or

$$\mu_1(i + i_2) = \mu_2(i + i_1)$$

which gives

$$i_2 = i\,\frac{\mu_2 - \mu_1}{\mu_2 + \mu_1} = -i_1 \tag{15.6}$$

Consequently image i_2 carries a positive current, but i_1 carries a negative value. Figure 15.16 gives the field for $\mu_2/\mu_1 = 19$.

II. Current in Iron. See Fig. 15.17. Place image i_1 in the air at distance h and image i_2 at the wire itself. The air field is determined by wire i and image i_2, assumed to be acting in a medium of permeability μ_1 everywhere. The iron field is given by wire i and image i_1, assumed to be acting in a medium of permeability μ_2 everywhere. The flux lines in the *air* form a system of circles concentric about the wire. In the following equations, the currents in the wire and both images are assumed to carry positive currents.

At point P_1 (Fig. 15.17), use i and i_2 in μ_1.

$$H_{1T} = -\frac{(i + i_2)}{2\pi p} \cos \alpha = -\frac{h}{2\pi p^2}(i + i_2)$$

$$B_{1N} = +\mu_1 \frac{(i + i_2)}{2\pi p} \sin \alpha = +\frac{y\mu_1}{2\pi p^2}(i + i_2)$$

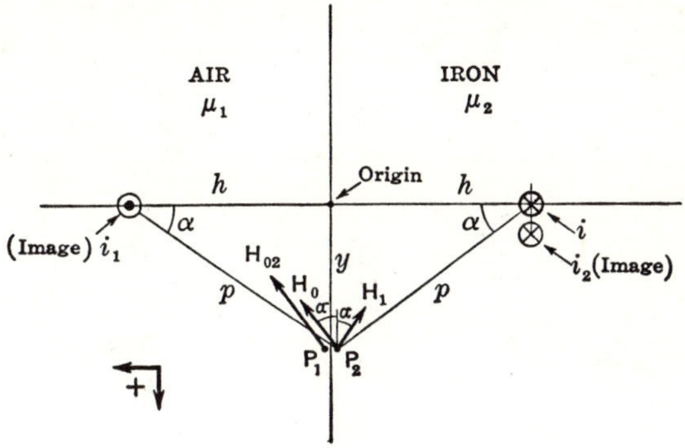

FIG. 15.17. Current (i) in iron near a plane air surface.

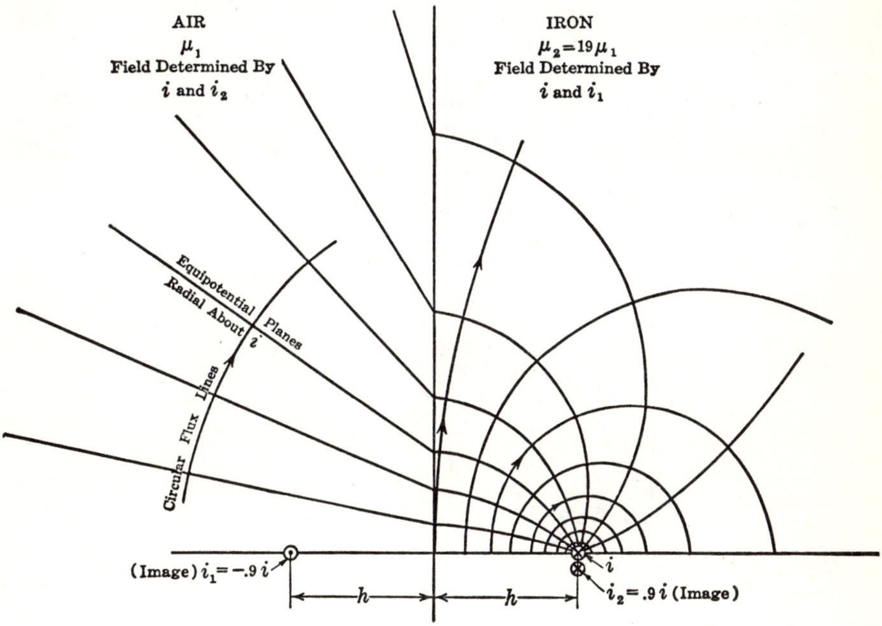

FIG. 15.18. Current (i) in iron near a plane air surface.

At point P_2, use i and i_1, in μ_2.

$$H_{2T} = -\frac{i}{2\pi p}\cos\alpha + \frac{i_1}{2\pi p}\cos\alpha = \frac{h}{2\pi p^2}(-i + i_1)$$

$$B_{2N} = +\frac{\mu_2}{2\pi}\left(\frac{i}{p}\sin\alpha + \frac{i_1}{p}\sin\alpha\right) = \frac{y\mu_2}{2\pi p^2}(i + i_1)$$

The boundary conditions demand that

$$H_{1T} = H_{2T} = -\frac{h}{2\pi p^2}(i + i_2) = +\frac{h}{2\pi p^2}(-i + i_1)$$

or

$$i + i_2 = i - i_1$$

$$i_1 = -i_2 \tag{15.7}$$

and that

$$B_{1N} = B_{2N} = +\frac{y\mu_1}{2\pi p^2}(i + i_2) = +\frac{y\mu_2}{2\pi p^2}(i + i_1)$$

or

$$\mu_1(i + i_2) = \mu_2(i + i_1)$$

which gives

$$i_2 = \frac{\mu_2 - \mu_1}{\mu_2 + \mu_1}i = -i_1 \tag{15.8}$$

Consequently image i_2 carries a positive current, but i_1 carries a negative value. Figure 15.18 gives the field for $\mu_2/\mu_1 = 19$.

PROBLEMS

1. Map the field in the air region shown in Fig. P1. If the flux density at the center of the air gap is 0.1 weber per square meter, compute the total flux running from A to B for 1-cm depth taken normal to the paper. Compare this value with the total flux figured on the assumption that the flux density is 0.1 throughout the gap and that there is no leakage flux.

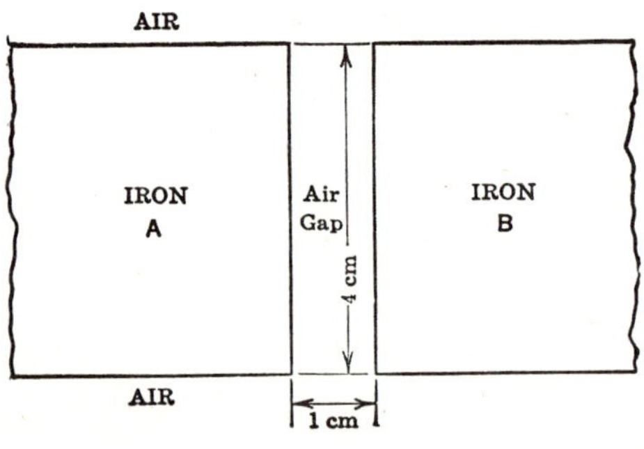

FIG. P1

2. Compute the force on the iron surface of the air gap in Prob. 1. At the corners the flux density approaches an infinite value, but the force remains finite.

3. Sketch the magnetic field inside and outside a rectangular bus bar 6 cm by 2 cm. Bus bars are made of copper, but if iron were used what would be the effect on the flux distribution and magnitude?

4. Following Fig. 10.10, locate the kernels if the two wires are almost touching.

5. Sketch the fields of Figs. 15.6 and 15.7, if a equals b.

6. Find the values of the images required in Figs. 15.15 to 15.18 if the iron has a relative permeability of 99; of 999; of infinity. What is the influence of the permeability on the shape of the field?

7. Compute the flux density at any height y, inside the current or outside, in Fig. P7.

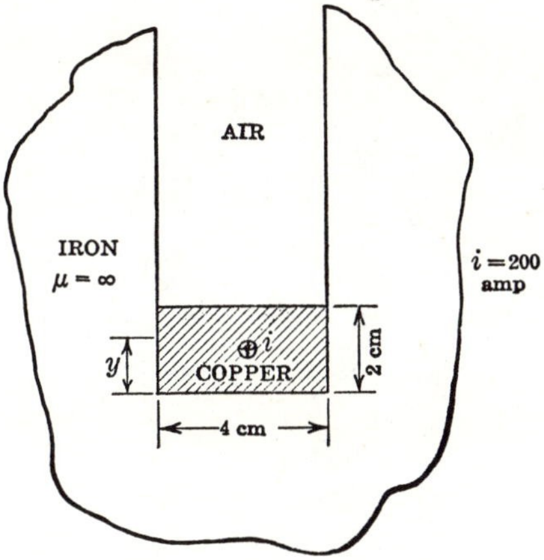

FIG. P7

8. Two wires lie midway between two parallel iron surfaces. Sketch the magnetic fields for currents in the same direction; in opposite directions.

PART IV

COMBINED ELECTRIC AND MAGNETIC FIELDS

In this portion of the text, consideration is given to some of the elementary interactions that occur whenever electric and magnetic fields are simultaneously distributed throughout the same space.

CHAPTER 16

INTERACTIONS OF ELECTRIC AND MAGNETIC FIELDS

In electrical engineering practice it is convenient to divide circuits containing both inductance and capacitance into two classes—those containing the inductance and the capacitance in different locations, leading to the idea of "lumped" constants; and those in which the inductance and capacitance are intermixed and "distributed" over a considerable space. A coil and a condenser illustrate the first class, and a transmission line the second.

16.1. Charging a Condenser at Constant Potential through an Inductance and a Resistance. Article 4.22 gives the equations for the time-current relations which result when a battery of constant potential is connected through a resistance to a condenser; Art. 11.3 gives the similar relations for a resistance connected to an inductance. We now complete the series by developing the equations which describe the charging of a condenser connected through an inductance and a resistance to a battery of constant potential.

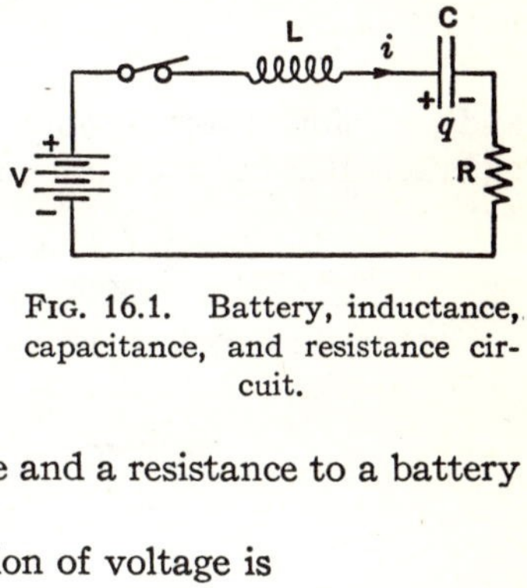

FIG. 16.1. Battery, inductance, capacitance, and resistance circuit.

Following Fig. 16.1, the circuit equation of voltage is

$$V = L\frac{di}{dt} + Ri + \frac{1}{C}\int i\,dt$$

$$V = L\frac{d^2q}{dt^2} + R\frac{dq}{dt} + \frac{1}{C}q \tag{16.1}$$

Hence we must look for a time function of q which has the same form as its first and second derivatives. This suggests the exponential function. Differentiating equation 16.1 gives an equation that does not contain V.

$$0 = L\frac{d^3q}{dt^3} + R\frac{d^2q}{dt^2} + \frac{1}{C}\frac{dq}{dt} \tag{16.2}$$

Let us try

$$q = Ae^{mt} \tag{16.3}$$

where A and m are constants to be determined by substitution in equations 16.1 and 16.2, and by reference to the initial conditions of the circuit at the instant when the switch is closed.

From equation 16.2

$$\left(Lm^2 + Rm + \frac{1}{C}\right) m = 0 \tag{16.4}$$

and we see that equation 16.3 is the type of solution if equation 16.4 is satisfied.

Thus m takes the three values:

$$\left.\begin{aligned} m_1 &= 0 \\ m_2 &= -\frac{R}{2L} + \sqrt{\left(\frac{R}{2L}\right)^2 - \frac{1}{LC}} \\ m_3 &= -\frac{R}{2L} - \sqrt{\left(\frac{R}{2L}\right)^2 - \frac{1}{LC}} \end{aligned}\right\} \tag{16.5}$$

and the solution for q is obtained by adding three terms of the form of equation 16.3 but with different values of the constants.

$$q = A_1 + e^{-\frac{R}{2L}t}\left(A_2e^{+\sqrt{\left(\frac{R}{2L}\right)^2 - \frac{1}{LC}}\,t} + A_3e^{-\sqrt{\left(\frac{R}{2L}\right)^2 - \frac{1}{LC}}\,t}\right) \tag{16.6}$$

For ease in handling, let us use the abbreviations

$$\alpha = +\frac{R}{2L} \qquad \text{(Damping constant)}$$

$$\beta = +\sqrt{\left(\frac{R}{2L}\right)^2 - \frac{1}{LC}}$$

Then

$$q = A_1 + e^{-\alpha t}(A_2e^{+\beta t} + A_3e^{-\beta t}) \tag{16.7}$$

and the current

$$i = \frac{dq}{dt} = e^{-\alpha t}[(\beta - \alpha)A_2e^{+\beta t} - (\beta + \alpha)A_3e^{-\beta t}] \tag{16.8}$$

After a long time ($t = \infty$) the condenser is fully charged to its final value CV and the current has dropped to zero. The first condition gives the value of A_1.

$$A_1 = CV \tag{16.9}$$

The constants A_2 and A_3 cannot be determined until it is known whether β is real, zero, or imaginary. The behavior of the circuit, in fact, depends markedly upon the value of β. We will now obtain three equations for q and three for i corresponding to the choice for β.

Case I. $\beta = \sqrt{\left(\frac{R}{2L}\right)^2 - \frac{1}{LC}}$ *is real. Overdamped Condition.*

$$\frac{R^2}{4L^2} > \frac{1}{LC} \quad \text{or} \quad R > 2\sqrt{\frac{L}{C}} = R_c \tag{16.10}$$

$R_c = 2\sqrt{L/C}$ is called the *critical damping resistance*. Equations 16.7 and 16.8 are already in suitable form for this case. The constants A_2 and A_3 are easily determined from the initial conditions.

We assume, at the moment when the switch is closed, that $t = 0$ and the condenser is uncharged, $q = 0$; and, since inductance is present, the current must be zero.

$$\text{At } t = 0 \qquad q = 0 = CV + A_2 + A_3$$
$$i = 0 = (\beta - \alpha)A_2 - (\beta + \alpha)A_3$$

from which we get

$$A_2 = -CV\frac{\alpha + \beta}{2\beta} \quad \text{and} \quad A_3 = +CV\frac{\alpha - \beta}{2\beta}$$

and

$$q = CV\left[1 + e^{-\alpha t}\left(-\frac{\alpha + \beta}{2\beta}e^{+\beta t} + \frac{\alpha - \beta}{2\beta}e^{-\beta t}\right)\right] \tag{16.11}$$

$$i = CV\left(\frac{\alpha^2 - \beta^2}{2\beta}\right)e^{-\alpha t}(e^{+\beta t} - e^{-\beta t}) \tag{16.12}$$

During discharge, that is with R, L, and C connected together,

$$q = -CVe^{-\alpha t}\left(-\frac{\alpha + \beta}{2\beta}e^{+\beta t} + \frac{\alpha - \beta}{2\beta}e^{-\beta t}\right) \tag{16.13}$$

$$i = -CV\left(\frac{\alpha^2 - \beta^2}{2\beta}\right)e^{-\alpha t}(e^{+\beta t} - e^{-\beta t}) \tag{16.14}$$

since, $\quad$ at $t = \infty \quad q = 0 = A_1$

at $t = 0 \quad q = CV = A_2 + A_3$

$$i = 0 = (\beta - \alpha)A_2 - (\beta + \alpha)A_3$$

The curves for q and i are plotted as curve I in Fig. 16.2 for a value of resistance which is twice the critical damping resistance.

Case II. $\beta = j\sqrt{\frac{1}{LC} - \left(\frac{R}{2L}\right)^2} = j\gamma$ *is imaginary. Underdamped or Oscillatory Condition.*

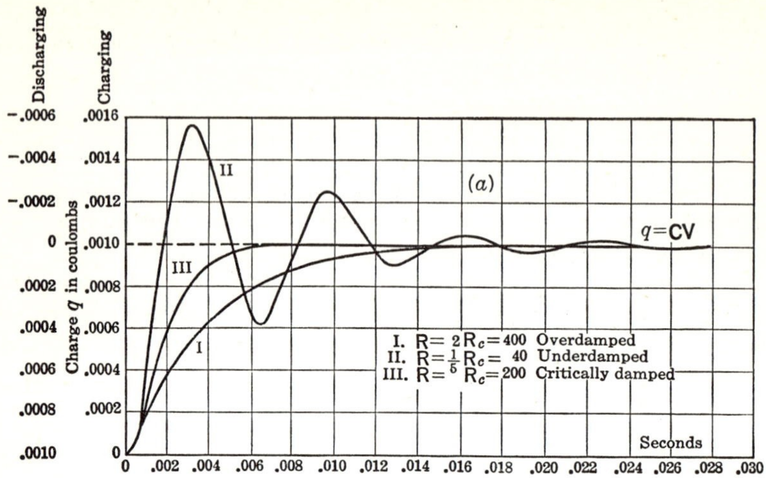

(a) Charge on the condenser.

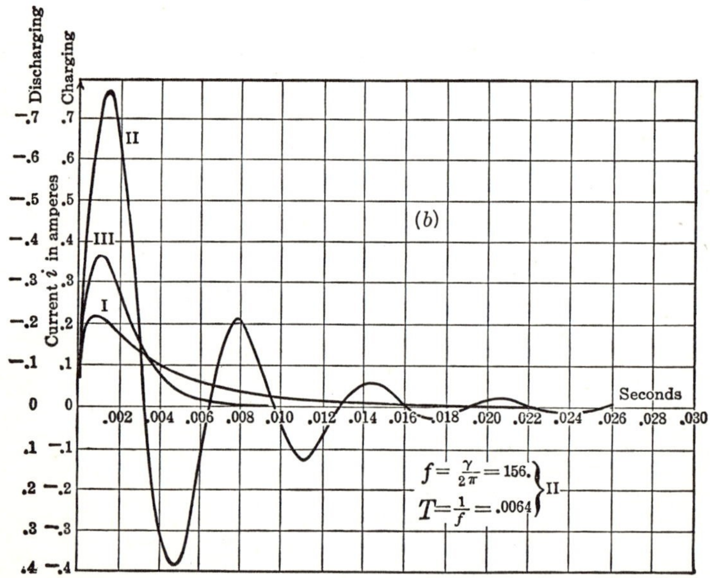

(b) Current in the circuit.

FIG. 16.2. Charge and discharge curves for V, R, L, C circuit. $V = 100$ volts, $L = 0.1$ henry, $C = 10^{-5}$ farad, $R_c = 2\sqrt{L/C} = 200$ ohms.

$$\frac{R^2}{4L^2} < \frac{1}{LC} \quad \text{or} \quad R < 2\sqrt{\frac{L}{C}} = R_c \tag{16.15}$$

In this case, the solutions for q and i take imaginary exponential forms which are sines and cosines, and we may expect the instantaneous values to oscillate about their final values. For $\beta = j\gamma$, from equations 16.11 and 16.12, using the equivalent values

$$e^{+j\gamma t} = \cos \gamma t + j \sin \gamma t$$

$$e^{-j\gamma t} = \cos \gamma t - j \sin \gamma t$$

we find the corresponding charging equations for the oscillatory case to be

$$q = CV\left[1 - e^{-\alpha t}\left(\cos \gamma t + \frac{\alpha}{\gamma} \sin \gamma t\right)\right] \tag{16.16}$$

$$i = CVe^{-\alpha t}\frac{\alpha^2 + \gamma^2}{\gamma} \sin \gamma t \tag{16.17}$$

These equations are plotted as curve II in Fig. 16.2 for a value of resistance one-fifth of the critical damping value. It should be noted that the condenser charge, and consequently its voltage, may rise to a value nearly double the battery potential. The current is a damped sine wave with frequency

$$f = \frac{\gamma}{2\pi} = \frac{1}{2\pi}\sqrt{\frac{1}{LC} - \left(\frac{R}{2L}\right)^2} \tag{16.18}$$

The equations for the oscillatory discharge are

$$q = +CVe^{-\alpha t}\left(\cos \gamma t + \frac{\alpha}{\gamma} \sin \gamma t\right) \tag{16.19}$$

$$i = -CVe^{-\alpha t}\frac{\alpha^2 + \gamma^2}{\gamma} \sin \gamma t \tag{16.20}$$

Case III. $\beta = \gamma = 0$. *Critically Damped Condition.*

$$\frac{R^2}{4L^2} = \frac{1}{LC} \quad \text{or} \quad R = 2\sqrt{\frac{L}{C}} = R_c \tag{16.21}$$

The critically damped condition is the border line between the two previous cases, in which the circuit just fails to oscillate. With critical damping the charge on the condenser is built up to its final value most rapidly. See Fig. 16.2, curve III.

The equations for this case are obtained from either equations 16.11 and 16.12 by permitting β to approach zero, or equations 16.16 and 16.17 as γ approaches zero. Choosing the latter pair, note that, as $\gamma \to 0$,

$$\cos \gamma t \to 1, \qquad \frac{\alpha}{\gamma} \sin \gamma t = \alpha t \left(\frac{\sin \gamma t}{\gamma t} \right)_{\gamma \to 0} \to \alpha t$$

Then

$$q = CV[1 - e^{-\alpha t}(1 + \alpha t)] \tag{16.22}$$

$$i = CV\alpha^2 t e^{-\alpha t} \tag{16.23}$$

During discharge

$$q = +CVe^{-\alpha t}(1 + \alpha t) \tag{16.24}$$

$$i = -CV\alpha^2 t e^{-\alpha t} \tag{16.25}$$

Many examples of this type of behavior are to be found in nature. Consider a mass M, moving with velocity v, attached to a vertical spring, acting under the force of gravity F. If R is the resistance per unit velocity (due to moving through air or any other medium) and K is the elasticity or force per unit length required to stretch the spring and s is the spring deflection from its unstressed length, the equation of motion of the system is

$$F = M \frac{dv}{dt} + Rv + K \int v \, dt$$

$$= M \frac{d^2 s}{dt^2} + R \frac{ds}{dt} + Ks \tag{16.26}$$

Comparing this with equations 16.1 and 16.2 it is seen that similar solutions are to be expected. The same forms will describe the swing of a galvanometer under a constant applied torque.

16.2. Steep Wave-Front Transient Voltages. Surges. Time Lag. For testing insulators it is desirable to use transient voltages which rise in value rapidly to predetermined peaks or maxima. The circuit of Fig. 16.3 serves this purpose. A condenser is charged through a rectifier from a controllable a-c source. When the condenser voltage has attained a certain value the gap breaks down and a current flows through the C, L, R circuit. For small values of L the circuit is nonoscillatory and equation 16.14 for current is valid. The current discharge and the voltage drop across R have the general form of curve I in Fig. 16.2*b*. The voltage drop over R may be applied to an insulator placed in the test cup.

Figure 16.4 shows the transient voltages for $C = \frac{1}{8}$ microfarad,

R = 1000 ohms, and for several different values of L, with the condenser initially charged to 77 kilovolts. With larger values of L the peak voltage drops somewhat. However, if the insulator in the test cup breaks down before the peak is reached, the voltage across the

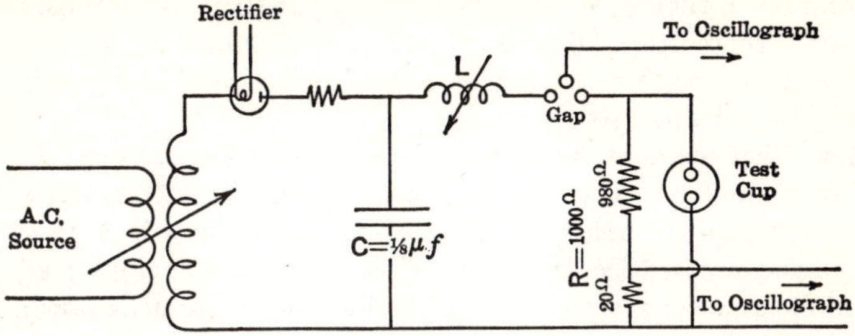

FIG. 16.3. Circuit for producing steep wave-front voltages.

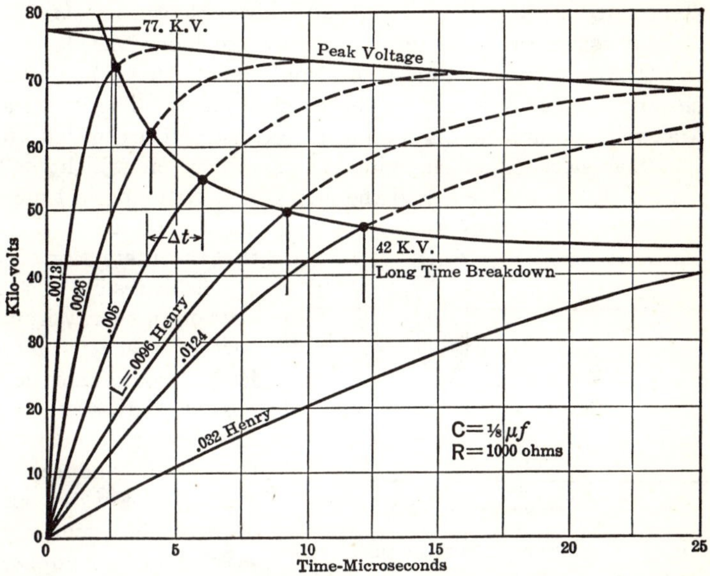

FIG. 16.4. Transient voltages applied to insulator in Fig. 16.3. Test data obtained by Dr. W. H. Bixby.

test cup will drop suddenly practically to zero. The heavy dots on the transient voltage curves indicate test values of breakdown for liquid carbon tetrachloride (CCl_4) in the test cup between 1-inch diameter spheres with a spacing of 0.0635 centimeter.

Δt in Fig. 16.4 gives the time lag of the insulation which is about 2 to 2.5 microseconds. With slowly applied voltages the time lag is hardly apparent, but it is revealed clearly when the rate of application is high. In the latter case the voltage applied overshoots the long-time breakdown value. The time lag is particularly important in connection with lightning surges which travel down a transmission line and reach their peak values in a matter of 10 microseconds, rapidly declining thereafter. For surges of this character it is clear that insulation can resist much higher voltages than would be possible with slowly rising transients.

16.3. Displacement Current. The magnetic relations that are developed in Part II are based on the assumption that the current which sets up the magnetic field is one that flows in a *closed path* around the circuit (See Art. 4.3). But if a condenser is inserted in the circuit it is no longer possible to consider the current as flowing in a closed path, if current is conceived merely as the movement of positive charge and electrons. James Clerk Maxwell surmounted this difficulty by assuming that the electric current *always* flows in a *closed path.* Under this assumption the term "current" includes the idea of any electric field changing in time (the displacement current) in addition to the simple concept of electric charge movement. The justification for Maxwell's assumption is found in its consequences, which are, in brief, that an electric field changing in time is a current which sets up a magnetic field about itself, and the latter, also changing in time, sets up the electric field, and that the interaction of these two fields results in a propagation of energy from one part of space to another. Radio communication is possible only by the transfer of energy from one place to another and serves as ample proof of the validity of Maxwell's point of view.

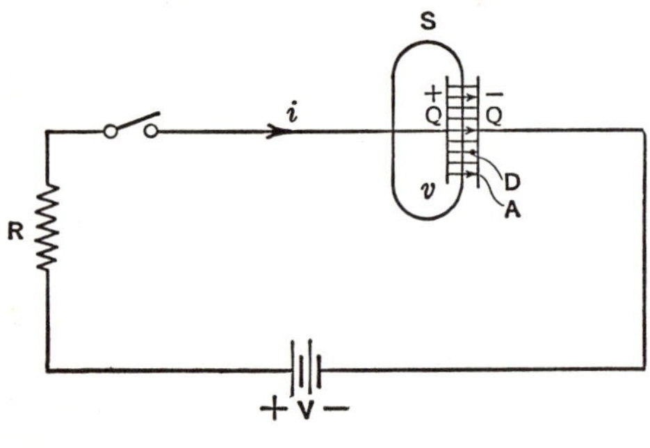

FIG. 16.5. Displacement current in a condenser.

In order to compute the value of the displacement current density J let us consider, for simplicity, a parallel plate condenser of area A, separation s, and dielectric constant ϵ connected through a switch and a resistance R to a source of potential V, as shown in Fig. 16.5. Consider also a closed surface S (enclosing volume v) which passes through the center of the condenser and back across one of the wires leading to the condenser.

The charging current i which flows through the wire into v after the switch is closed must, by Maxwell's assumption, be equal to the current which flows out of volume v across that part of surface S that lies between the two plates.

$$i = \frac{dQ}{dt} = A\frac{d\sigma}{dt} = A\frac{\partial D}{\partial t}$$

$$= A\frac{\partial D}{\partial t} \quad \text{amperes}$$

and

$$J = \frac{i}{A} = \frac{\partial D}{\partial t} \quad \frac{\text{amperes}}{\text{square meter}} \tag{16.27}$$

The displacement current density J at a point is a vector quantity whose direction and magnitude are equal to the time rate of change of the electric flux density.

Remembering that D may be written in terms of the polarization and intensity vectors, we get:

$$D = P + \epsilon_0 E \tag{16.28}$$

$$J = \frac{\partial D}{\partial t} = \frac{\partial P}{\partial t} + \epsilon_0\frac{\partial E}{\partial t} \quad \frac{\text{amperes}}{\text{square meter}} \tag{16.29}$$

The displacement current density is the sum of a current density due to changing polarization $\partial P/\partial t$, which is an actual charge movement or bound charge shift in the dielectric, and a current density $\epsilon_0(\partial E/\partial t)$ which occurs even in free space and is sometimes called the ethereal current density.

16.4. Propagation of Energy through the Medium. Poynting's Vector. When two conductors carry currents equal in magnitude and opposite in direction and have a voltage difference, they form a transmission line that guides a flow of energy from one end of the line to the other. The power or time rate of energy transfer across any surface normal to the line is the product of the potential difference and the current.

$$P = VI \quad \frac{\text{joules}}{\text{second}} \text{ or watts} \tag{16.30}$$

But in addition to the above point of view which regards energy transfer as a property of the transmission *circuit*, it is possible to consider energy transfer as a property of the electric and magnetic *fields* which surround the wires. In this view, energy is propagated from the generator to the load throughout the insulating medium along lines formed by the intersection of the magnetic and electric equipotential surfaces. These energy flow lines are necessarily perpendicular at any point in the field to the electric and magnetic intensity vectors E and H at that point.

Following Fig. 16.6*b*, let us consider a small parallelogram at some field point, whose sides (da and db) are formed by the magnetic and the

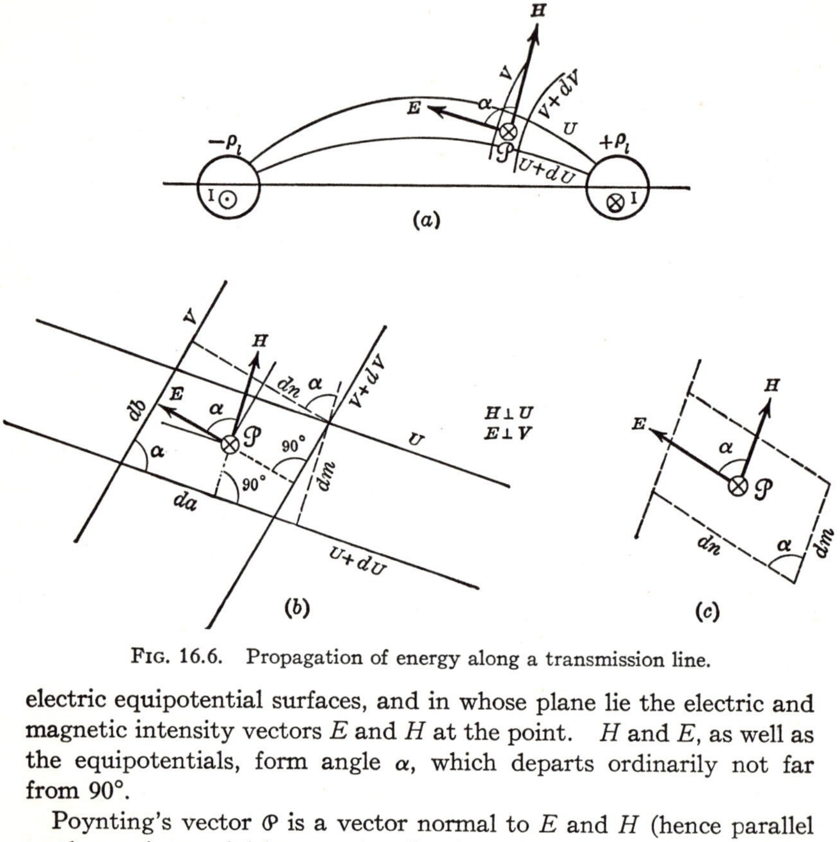

FIG. 16.6. Propagation of energy along a transmission line.

electric equipotential surfaces, and in whose plane lie the electric and magnetic intensity vectors E and H at the point. H and E, as well as the equipotentials, form angle α, which departs ordinarily not far from 90°.

Poynting's vector $\mathcal{P}$ is a vector normal to E and H (hence parallel to the equipotential intersection lines) in the right-hand screw order $EH\mathcal{P}$. Poynting's vector gives the energy per square meter per second flowing normally across the small parallelogram and is

$$\mathcal{P} = EH \sin \alpha \quad \frac{\text{watts}}{\text{square meter}} \tag{16.31}$$

The proof of this statement is obtained by integrating over the entire surface made of adjacent small parallelograms but excluding the wire cross-sectional areas, and by showing that the result yields power flow from the generator to the load as given by equation 16.30.

$$\begin{aligned} d \text{ power} &= \mathcal{P}\, dA = \mathcal{P}\, da\, db \sin \alpha \\ &= \mathcal{P} \frac{dm\, dn}{\sin \alpha} \\ &= (EH \sin \alpha) \frac{dm\, dn}{\sin \alpha} \\ &= EH\, dm\, dn \end{aligned}$$

but

$$E = -\frac{dV}{dn} \quad \text{and} \quad H = -\frac{dU}{dm}$$

Hence

$$\begin{aligned} d \text{ power} &= \frac{dV}{dn} \frac{dU}{dm}\, dm\, dn \\ &= dV \cdot dU \end{aligned}$$

By extending the small parallelograms over the entire area, exclusive of the wire cross sections, we get the total power flowing along the wires as the product of the electric difference of potential between the wires and the closed-path magnetomotive force.

$$\begin{aligned} \text{Power} &= \int_{V_1}^{V_2} dV \int_0^{U=\mathfrak{F}} dU \\ &= \Delta V \cdot \mathfrak{F} \\ &= \Delta V \cdot I \quad \text{watts} \end{aligned} \tag{16.32}$$

which is equal to equation 16.30.

Let us consider the flow of energy along a transmission line, composed of wires having resistance. Figure 4.7 shows the electric field of a line carrying current. Some of the electric equipotentials pass through both the battery and the load; others converge toward the surfaces of the wires. The magnetic equipotential surfaces are portions of circular cylinders which include both wire centers. The energy from the battery flows along the intersections of the magnetic and the electric equipotentials. At some point just off the wire surface Poynting's vector is directed at an angle to the surface. The tangential component of the vector gives the energy per unit area per second flowing *along* the wire, and the normal component gives the energy flow *into* the wire that is transformed to heat according to the I^2R relation.

If the wires are assumed to have zero resistance, the electric equipotentials run directly from the battery to the load, Poynting's vector along the wire surface is entirely tangential, and no energy is directed into the wire.

16.5. Circuits with Distributed Constants. Guided Waves. When the inductance and the capacitance are "distributed" as in a long transmission line or a long cable, rather than "lumped" as in a coil and condenser, the potential and current are propagated as waves and are guided along the line or cable with a very high velocity that, theoretically at least, may equal the velocity of light.

Consider a transmission line or cable, in which L and C are the inductance and capacitance per meter of length, and in which resistance

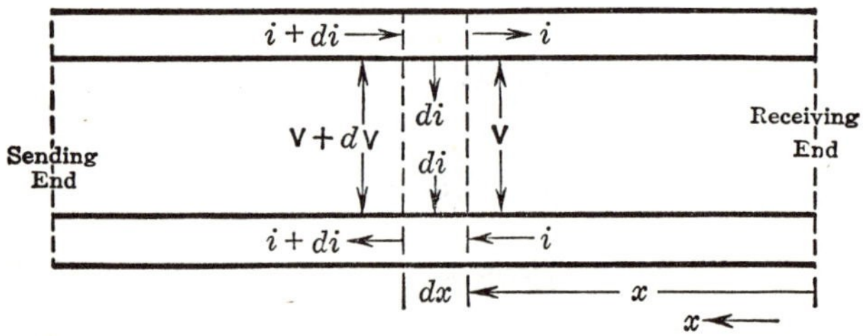

FIG. 16.7. Current and voltage on a long line.

R is assumed zero. Let x, in Fig. 16.7, be the distance measured from the receiving end of the line where a load may be connected. At the sending end of the line a battery or generator may be connected. For a small length of line dx the capacitance is $C\,dx$ and the inductance $L\,dx$.

Under transient conditions, the potential difference between the wires is V on the receiving side of dx and $V + dV$ on the sending side. The current in the line wires is i on the receiving side of dx and $i + di$ on the other side. The current difference di gives a transient *excess* charge flow into the left side of dx on the upper wire, or it may be considered as a Maxwellian displacement current flowing from the upper wire to the lower across the dx condenser region.

The relations between V and i and time t are obtained as follows:

$$\partial i = (C\,\partial x)\frac{\partial V}{\partial t} \qquad \partial V = (L\,\partial x)\frac{\partial i}{\partial t}$$

$$\frac{\partial i}{\partial x} = C\frac{\partial V}{\partial t} \qquad \frac{\partial V}{\partial x} = L\frac{\partial i}{\partial t}$$

$$L\frac{\partial^2 i}{\partial x\,\partial t} = LC\frac{\partial^2 V}{\partial t^2} \qquad C\frac{d^2 V}{\partial x\,\partial t} = LC\frac{\partial^2 i}{\partial t^2}$$

$$\frac{\partial^2 i}{\partial x^2} = C\frac{\partial^2 V}{\partial x\,\partial t} \qquad \frac{\partial^2 V}{\partial x^2} = L\frac{\partial^2 i}{\partial x\,\partial t}$$

$$\frac{\partial^2 i}{\partial x^2} = LC\frac{\partial^2 i}{\partial t^2} \qquad \frac{\partial^2 V}{\partial x^2} = LC\frac{\partial^2 V}{\partial t^2} \tag{16.33}$$

The inductance–capacitance product for the region outside of two parallel wires (per meter) is

$$LC = \frac{\mu}{\pi}\ln\frac{D}{R}\cdot\frac{\pi\epsilon}{\ln\frac{D}{R}} = \mu\epsilon = \frac{1}{v^2} \tag{16.34}$$

and for the region between two concentric cylinders (per meter) it is

$$LC = \frac{\mu}{2\pi}\ln\frac{r_2}{r_1}\cdot\frac{2\pi\epsilon}{\ln\frac{r_2}{r_1}} = \mu\epsilon = \frac{1}{v^2} \tag{16.35}$$

Consequently equations 16.33 may be rewritten as

$$\frac{\partial^2 i}{\partial x^2} = \frac{1}{v^2}\frac{\partial^2 i}{\partial t^2} \quad \text{and} \quad \frac{\partial^2 V}{\partial x^2} = \frac{1}{v^2}\frac{\partial^2 V}{\partial t^2} \tag{16.36}$$

These are well known as wave equations, indicating that the voltage and the current are propagated as waves along the x direction with a velocity of $1/\sqrt{\mu\epsilon} = v$. For instance, the potential equation

$$\frac{\partial^2 V}{\partial x^2} = \frac{1}{v^2}\frac{\partial^2 V}{\partial t^2} \tag{16.37}$$

is solved by any function of the variable z, where

$$z = (x + vt) \quad \text{or} \quad (x - vt) \tag{16.38}$$

and

$$V = \phi_1(x + vt) \quad \text{or} \quad \phi_2(x - vt) \tag{16.39}$$

This means that any voltage or charge concentration found at any given instant at a position x on the wire will be found at a later time at a new position in undiminished magnitude, having moved to the new position with a constant velocity v. It is possible that the voltage may divide into two waves,

ϕ_1 with velocity $-v$

ϕ_2 with velocity $+v$

moving in opposite directions along the line. The current equation may be treated similarly.

As a particular case, consider a battery of constant potential V connected through a switch at the sending end of the line. Let the switch be closed for a brief period and opened again after time Δt.

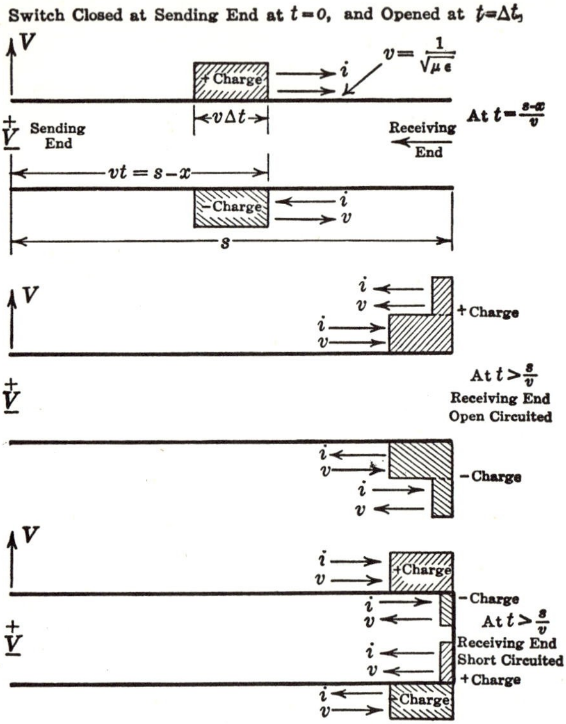

FIG. 16.8. Charge movement on a long line.

The voltage or charge concentration of constant magnitude is propagated as a rectangular wave of height V and length $v\Delta t$ along the line toward the receiving end, as shown in Fig. 16.8. There is current only in that part of the line occupied by charge.

At the receiving end of the line, we may distinguish three conditions:

I. Line Open-Circuited at Receiving End. In this case, when the voltage wave reaches the receiving end at $t = s/v$, the potential is completely reflected and starts back toward the sending end, reversing the current thereafter. In this way the current at the receiving end is always zero, but the potential doubles until the reflection is completed, at $t = (s/v) + \Delta t$, returning again to zero after the wave has passed.

The voltage of the upper line is always positive, but the current reverses.

II. Line Short-Circuited at Receiving End. In this case, when the voltage wave reaches the receiving end at $t = s/v$, each charge continues on around the circuit, thus reversing the voltages on the two wires, but keeping the current always in the same direction around the circuit. Under these conditions there is always zero voltage at the receiving end but there is a double valued current during the interval of wave passage from $t = s/v$ to $t = (s/v) + \Delta t$.

Any load at the receiving end lies in between the two extreme conditions just described and permits partial reflection and partial passage. If the wire has resistance or if there is a leakage conductance from wire to wire, the wave equation 16.36 must be replaced by more complicated expressions. The effect of resistance is to absorb energy from the wave and to transform it into heat, which results in a reduced voltage and wave velocity and a distortion of the wave shape from its original rectangular character.

III. Surge Impedance. Reflectionless Termination. The surge impedance of the line is the quantity which must be multiplied into the current to give the potential on the line. L and C are the line constants per meter, ρ_l the charge per meter, and from equation 16.34 the velocity of the wave is $v = 1/\sqrt{LC}$. Then

$$V = \frac{\rho_l}{C} = \frac{i}{vC} = \frac{\sqrt{LC}}{C}\,i = \sqrt{\frac{L}{C}}\,i \tag{16.40}$$

and the surge impedance of the line is $\sqrt{L/C}$. This quantity has the dimensions of a resistance. If, then, a resistance of value $R = \sqrt{L/C}$ is connected to the end of the line, the wave will disappear without reflection and the energy is lost in the resistance. This property is used in communication circuits and in circuit connections to cathode-ray oscillographs where reflections are troublesome.

16.6. Distortionless Line. The square wave of voltage and current described in the preceding paragraph will travel down the line without attenuation (diminished magnitude) or change in wave shape only if there are no losses in the line or in its insulating dielectric. Applying the analytical method of Fourier analysis, the square wave may be considered to be broken into an infinite number of sinusoidal waves of differing amplitudes and frequencies. So long as the line losses are zero, all these frequencies travel with the same velocity, $v = 1/\sqrt{\mu\epsilon} = c/\sqrt{\mu_r\epsilon_r}$, and the wave shape is maintained.

But since the line wires have resistance R and the insulation has a leakage conductance G, per meter of line, there are losses at the expense of the energy stored in the wave, which now attenuates as it moves. Moreover, under these conditions the various frequencies travel at different velocities and the wave changes its shape as it travels along the line. There is one exception to the latter statement. If

$$RC = LG \tag{16.41}$$

all the frequencies travel at the same velocity and the wave shape is unchanged, though it is attenuated. This is the condition for a "distortionless line."

Distortionless conditions are particularly important if speed in submarine telegraphy is to be obtained. If the waves of successive signals (dots and dashes) are distorted, the signals must be spaced farther apart along the line to avoid overlapping and confusion of signals, or, what amounts to the same thing, a smaller number of signals per second can be carried over the cable. Cables built in the past have usually had an LG product less than the RC product. A corrective measure would be to increase the inductance until equation 16.41 is satisfied. This approach has been used by the Bell Telephone Laboratories. They wind a layer of high permeability Permalloy tape over the conductor of the cable and thereby increase the inductance to the required value.

16.7. Field of a Charge Moving with Uniform Velocity. An electric charge is far too small to be observed directly, but, if we make reasonable assumptions about the distribution of the electricity in the charge, results of value are obtained. Let us assume that the elementary charge (electron or proton) is a sphere of electricity, uniformly distributed, which moves in a straight line with constant velocity v, small compared with the velocity of light c (say one-tenth or less), and that the total energy of the charge is the sum of the energies stored in its electric and magnetic fields. The former is potential energy and is independent of the velocity (for low velocities), and the latter is kinetic energy with a value $\frac{1}{2}m_0v^2$. The behavior of electrons in cathode-ray tubes and oscillographs provides fairly direct proof of the essential correctness of the point of view just outlined.

Figure 16.9 illustrates a charge $(+e)$ moving along the axis with velocity v. From Ampère's rule (equation 9.95) the magnetic intensity at point P is

$$H = \frac{ev \sin \theta}{4\pi r^2} = \epsilon Ev \sin \theta = Dv \sin \theta \tag{16.42}$$

a vector directed circumferentially around the axis. ev takes the place of the current element is. The kinetic energy of the moving charge

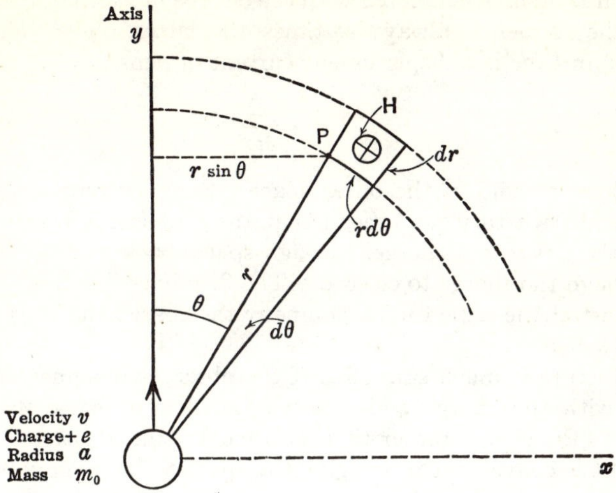

FIG. 16.9. Charge moving with velocity v.

is equal to the energy stored in the magnetic field outside the radius a, or

$$\frac{1}{2} m_0 v^2 = \iiint_{\text{vol}} \frac{\mu H^2}{2} d(\text{volume}) \tag{16.43}$$

As a volume element we choose the annular ring of radius $r \sin \theta$ and cross section $dr \cdot r\, d\theta$. Then

$$d(\text{vol}) = 2\pi r \sin \theta \cdot dr \cdot r\, d\theta$$

and

$$\frac{1}{2} m_0 v^2 = \frac{\mu e^2 v^2}{16\pi} \int_{r=a}^{\infty} \frac{dr}{r^2} \int_{\theta=0}^{\pi} \sin^3 \theta\, d\theta$$

$$= \frac{\mu e^2 v^2}{12\pi a} \tag{16.44}$$

or

$$a = \frac{\mu e^2}{6\pi m_0} \tag{16.45}$$

The charge radius is therefore inversely proportional to its mass. For an electron, $e = 1.602 \times 10^{-19}$ coulomb, $m_0 = 9.1 \times 10^{-31}$ kilogram, and $a = 1.885 \times 10^{-15}$ meter. For a proton, the mass is 1836.5 m_0 and $a = 1.025 \times 10^{-18}$ meter.

As the charge (say positive) moves along the axis, the electric and magnetic fields at any point in space are continually changing. The change in H is small compared with the change in E, since, by equation 16.42, the former is always ϵv times the latter. The value of ϵE changing in time is a displacement current of density

$$J = \epsilon \frac{\partial E}{\partial t} \tag{16.46}$$

which flows around in the outer space. Since current always flows in a closed path this must mean that the flow is out from the front face of the advancing charge, through space, back into the rear face, and through the charge to closure. The flow lines "circling" through space must at the same time accompany the charge and thus "spiral" through space.

This picture is much simplified if the observer imagines himself to "ride" with the charge and observe the flow at some fixed point relative to the charge rather than at a fixed point in space. For this purpose it is convenient to imagine the space to be moving backward relative to the charge with velocity $-v$. At point P, fixed in position relatively to the charge, with coordinates r and θ or x and y, as shown in Fig. 16.9, the current density is

$$J = \epsilon \frac{\partial E}{\partial t} = \epsilon \frac{\partial E}{\partial y} \frac{\partial y}{\partial t} = -\epsilon v \frac{\partial E}{\partial y} \tag{16.47}$$

in which y is measured along the axis of motion and is computed at any point relative to the charge, not to the backward moving space.

E, a radial vector equal to $e/(4\pi\epsilon r^2)$, can be written in terms of its x and y components,

$$\begin{aligned} E = E_x i + E_y j &= \frac{1}{4\pi\epsilon}\left(\frac{e}{r^2}\sin\theta\right) i + \frac{1}{4\pi\epsilon}\left(\frac{e}{r^2}\cos\theta\right) j \\ &= \frac{1}{4\pi\epsilon}\frac{ex}{(x^2+y^2)^{3/2}}\, i + \frac{1}{4\pi\epsilon}\frac{ey}{(x^2+y^2)^{3/2}}\, j \end{aligned} \tag{16.48}$$

where i and j are unit vectors in the x and y directions, respectively. The displacement current density is

$$\begin{aligned} J = -\epsilon v \frac{\partial E}{\partial y} &= -\epsilon v \left(\frac{\partial E_x}{\partial y} i + \frac{\partial E_y}{\partial y} j\right) \\ &= -\frac{ev}{4\pi}\left[\frac{-3xy}{(x^2+y^2)^{5/2}}\, i + \frac{x^2 - 2y^2}{(x^2+y^2)^{5/2}}\, j\right] \end{aligned} \tag{16.49}$$

This vector is indicated in Fig. 16.10.

The equation of the *shape* of the displacement current flow lines is not difficult to obtain. The vector J is tangent to the flow lines, and thus the slope of the flow lines is

$$\frac{dy}{dx} = \frac{J_y}{J_x} = \frac{x^2 - 2y^2}{-3xy} \tag{16.50}$$

or

$$(x^2 - 2y^2)\,dx = -3xy\,dy \tag{16.51}$$

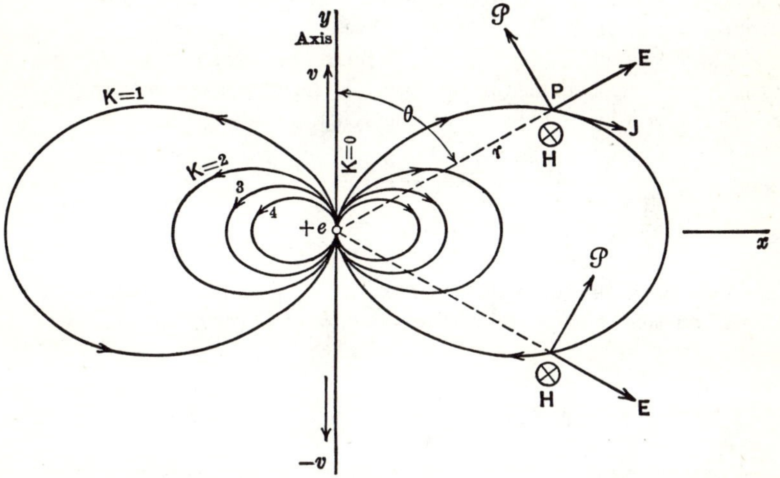

FIG. 16.10. Displacement current-density field about a charge moving with constant velocity v.

This is a homogeneous differential equation, solved by substitution of

$$y = xp \quad \text{and} \quad dy = x\,dp + p\,dx \tag{16.52}$$

and after a few algebraic steps we get

$$\frac{dx}{x} = -\frac{3}{2}\left(\frac{2p\,dp}{1 + p^2}\right) \tag{16.53}$$

which becomes

$$\ln x = -\tfrac{3}{2}\ln(1 + p^2) - \ln K \tag{16.54}$$

with K an arbitrary constant. Then

$$Kx = (1 + p^2)^{-3/2} = \left[1 + \left(\frac{y}{x}\right)^2\right]^{-3/2} \tag{16.55}$$

In polar coordinates, with $y = r\cos\theta$ and $x = r\sin\theta$, the equation of the current density flow lines becomes

$$r = \frac{1}{K} \sin^2 \theta \tag{16.56}$$

and is plotted in Fig. 16.10 for $K = 0, 1, 2, 3,$ and 4.

The flow of energy through the field takes place along lines tangent at any point to Poynting's vector $\mathcal{P}$, which is perpendicular to the electric and magnetic vectors E and H at the point. The energy, for a charge moving in a straight line with constant velocity, simply flows around in the field without gain or loss.

At high velocities (say $v/c > 0.1$) the mass of the charge, and in fact ordinary mass, becomes a function of the velocity according to the relation given by the theory of relativity,

$$m = \frac{m_0}{\sqrt{1 - \left(\frac{v}{c}\right)^2}} \tag{16.57}$$

in which m_0 is the low velocity mass or "rest" mass. As the velocity of the particle v approaches the velocity of light c, the mass and its

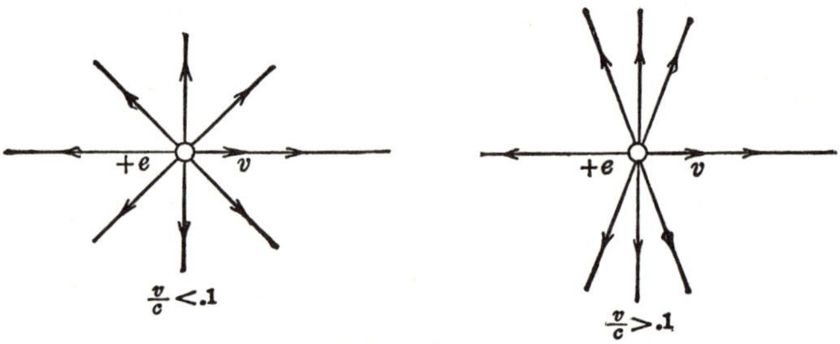

FIG. 16.11. Electric fields of a charge at low and high velocities.

energy approach infinite value. We cannot therefore expect to find charges or masses moving as rapidly as light waves. Electrons emitted from radioactive materials have been observed with velocities equal to 0.98 c. At high constant velocities the electric flux lines about a charge, while still straight lines, are located most densely near the equatorial plane and most sparsely before and behind the charge. This is shown in Fig. 16.11.

16.8. Radiation from an Accelerated Charge. A charge which is accelerated is accompanied by an increasing current, and the magnetic field energy must therefore increase also. At first the magnetic field in the neighborhood of the charge increases in strength, and with the passage of time it is propagated outward through the surrounding space with a velocity equal to $1/\sqrt{\mu\epsilon} = c/\sqrt{\mu_r\epsilon_r}$. The increase in total energy has, of course, been supplied through the mechanism of the force responsible for the acceleration of the charge. With a reverse force the charge is decelerated and the magnetic energy must be propagated from the outlying regions inward toward the charge and become reabsorbed by the system creating the force.

During acceleration and deceleration the magnetic field energy does not simply flow outward and again inward. During *each* of these periods there is an extra quantity of energy *propagated outward* and completely detached from the field of the charge. This energy is *radiated* and is permanently lost to this charge and its field, both upon acceleration and deceleration.

The problem of computing the radiant energy in terms of the acceleration (or deceleration) is really one of considerable difficulty, but the correct answer can be obtained with elementary mathematical tools if simplifying approximations are used judiciously.[1] The result indicates that the power radiated by a charge q with an acceleration or deceleration g is

$$P_r = \frac{dW}{dt} = \frac{2}{3}\frac{q^2 a^2}{c^3} \ \frac{\text{joules}}{\text{second}} \text{ or watts} \tag{16.58}$$

A charge moving in a straight line with constant velocity does not therefore radiate energy. If, however, the charge moves in a circular (or other shape) path, even at a constant speed, it should radiate energy because here there is acceleration, but of direction in space rather than of velocity. This conflicts with one of the basic tenets of Bohr's theory of the atom which assumes that an electron rotating in an atomic orbit is in a "stationary state" and can continue to rotate indefinitely without radiation of energy. Nevertheless, each point of view seems correct in its own sphere.

16.9. Radiation from an Oscillating Doublet. Two equal charges of opposite sign $\pm q$ spaced a distance apart $2h$, that is, distance h from their center point, constitute an electric doublet. $p = q2h$ is the moment of the doublet, a vector along the doublet axis directed from $-q$ to $+q$. If the charges in the doublet oscillate sinusoidally f times per second, moving up and down the doublet axis with equal and opposite velocities, the system is in a continuous state of acceleration and deceleration and must therefore continuously radiate energy.

[1] Richtmyer, *Introduction to Modern Physics*, Second Edition, McGraw-Hill, p. 131. Jeans, *Electricity and Magnetism*, Fifth Edition, Columbia University Press, 1925, p. 592.

Let h_m be the maximum displacement of each charge from its center position. Then the instantaneous position, velocity, and acceleration of each charge are given by

$$h = h_m \sin 2\pi ft = h_m \sin \omega t \tag{16.59}$$

$$v = \frac{dh}{dt} = \omega h_m \cos \omega t \tag{16.60}$$

$$g = \frac{dv}{dt} = \frac{d^2h}{dt^2} = -\omega^2 h_m \sin \omega t \tag{16.61}$$

in which $\omega = 2\pi f$ radians per second.

Equation 16.58 gives the instantaneous radiation for one charge. Here there are two, which doubles the charge and quadruples the power radiated. Then

$$P_r = \frac{dW}{dt} = \frac{80q^2g^2}{c^2} = \frac{80q^2\omega^4 h_m{}^2 \sin^2 \omega t}{c^2} \quad \text{watts} \tag{16.62}$$

is the instantaneous radiation of the doublet. Since the time average of $\sin^2 \omega t$ is $\frac{1}{2}$, the average time rate of radiation is

$$\overline{P_r} = \frac{40q^2\omega^4 h_m{}^2}{c^2} = \frac{640\pi^4 q^2 h_m{}^2 f^4}{c^2} \quad \text{watts} \tag{16.63}$$

This result can be written in terms of an equivalent sinusoidal current length. Let i equal the instantaneous value of a current flowing with uniform strength over distance s which is taken at the origin and along the doublet axis. Then

$$is = sI_m \cos \omega t = 2qv = 2q\omega h_m \cos \omega t$$

$$sI_m = 2q\omega h_m \tag{16.64}$$

and the time average of radiation is

$$\overline{P_r} = \frac{10s^2 I_m{}^2 \omega^2}{c^2} = \frac{20s^2 I_{\text{eff}}{}^2 \omega^2}{c^2} \quad \text{watts} \tag{16.65}$$

since the square of the effective current is one-half the square of the maximum value. This may also be expressed in free-space wave length $\lambda_0 = c/f$, giving

$$\overline{P_r} = \frac{80\pi^2 s^2 I_{\text{eff}}{}^2}{\lambda_0{}^2} \quad \text{watts} \tag{16.66}$$

The radiation resistance is

$$R_r = \frac{\overline{P_r}}{I_{\text{eff}}{}^2} = \frac{80\pi^2 s^2}{\lambda_0{}^2} \quad \text{ohms} \tag{16.67}$$

An oscillating current radiates energy with a time rate proportional to the product of the squares of the current, its length, and its frequency. The radiation from a 60-cycle transmission line is insignificant compared with the amount of power delivered along the line to a load. At high frequencies the proportion is increased greatly.

The average time rate of radiation given above is the total amount passed over a spherical surface described about the doublet or current element and far enough out so that the sphere radius r is large compared with the free-space wave length λ_0 of the radiation.

$$r \gg \lambda_0 = \frac{c}{f} = \frac{2\pi c}{\omega} \tag{16.68}$$

The radiation is *not* distributed uniformly over the sphere surface. It can be proved that the radial component of Poynting's vector is equal to

$$\begin{aligned} \mathcal{P}_{\text{radial}} &= \frac{30 q^2 h_m{}^2 \omega^4}{\pi c^2} \frac{\sin^2 \theta}{r^2} \sin^2 \omega \left(t - \frac{r}{c}\right) \\ &= \frac{15 s^2 I_m{}^2 \omega^2}{2\pi c^2} \frac{\sin^2 \theta}{r^2} \sin^2 \omega \left(t - \frac{r}{c}\right) \\ &= \frac{30\pi s^2 I_m{}^2}{\lambda_0{}^2} \cdot \frac{\sin^2 \theta}{r^2} \sin^2 \frac{2\pi}{\lambda_0} (ct - r) \ \frac{\text{watts}}{\text{square meter}} \end{aligned} \tag{16.69}$$

for $r \gg \lambda_0 = c/f = 2\pi c/\omega$, and the time average of the integral of this quantity over the spherical surface reduces to the value given in equation 16.65. Since the radial component of Poynting's vector varies with $\sin^2 \theta$, it is obvious that the energy radiated varies from zero in the direction of the doublet or current axis to a maximum along the equatorial plane.

Since the equations for the electric and magnetic vectors are rather complicated, even far out where $r \gg \lambda_0$, the forms will be given without proof.

$$E_\theta = \frac{60\pi s I_m \sin \theta}{\lambda_0 r} \sin \omega \left(t - \frac{r}{c}\right) \tag{16.70}$$

$$H_\phi = \frac{E_\theta}{120\pi} \tag{16.71}$$

The electric flux lines have been plotted in Fig. 16.12. They lie on the surfaces of peculiarly shaped coaxial toroids which move outward with velocity c and expend in size as they move. *Here the electric flux lines are closed lines which are no longer attached to charges, a condition which is possible in a radiation field but not in statics.* The magnetic flux lines everywhere lie in circles about the doublet axis.

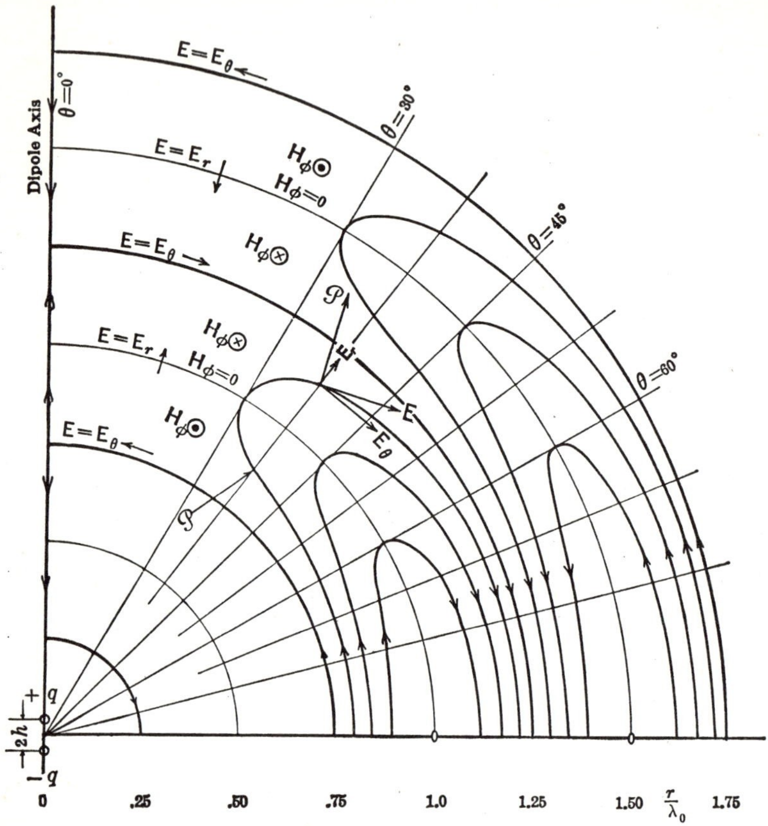

FIG. 16.12. Electric flux lines in the radiation field of an oscillating doublet, for $r \gg \lambda_0$. The scale of the figure is not given in true proportion. As drawn, only 1.75 wave lengths are shown. Actually the electric flux lines would have the shapes indicated only at radial distances where $r \gg \lambda_0$.

The manner in which the electric flux lines of the radiation field are detached from the oscillating charges is interesting. Suppose in the first place that the doublet is not oscillating and the two charges $\pm q$ are spaced the full distance $2h_m$ apart. The field is, of course, the electrostatic field of Fig. 16.13a, distributed over all space but with the higher concentrations of energy in the regions close to the doublet. Now suppose the doublet charges are moved together at the origin, with no spacing, in a time corresponding to one-quarter of an oscillating period $t = T/4 = 1/(4f)$. Electrostatically the field should have disappeared because the charges cancel, but actually the energy must flow back to the origin from all over space at velocity c. This requires time. During the movement a magnetic field is created and energy in this form is stored in the field. Upon cessation of motion this energy

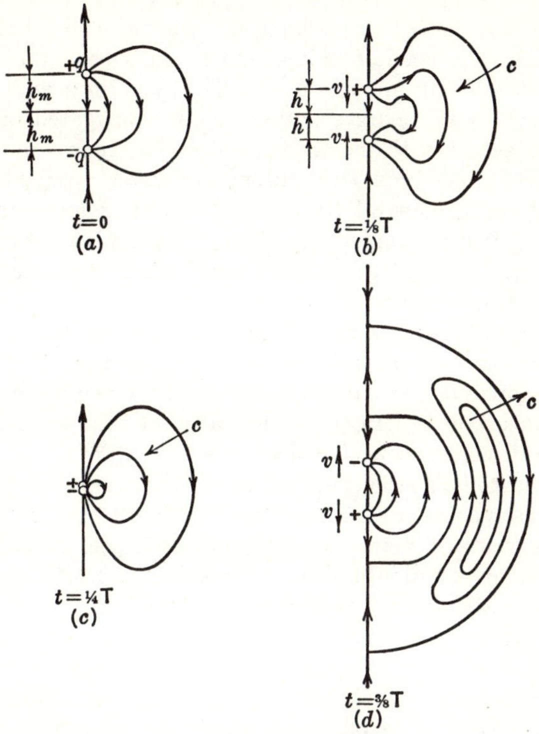

FIG. 16.13. Oscillating doublet, showing method of detachment of flux lines.

also must flow back to the origin with velocity c. If now the doublet in another quarter cycle is given its full displacement, but with the charges in reversed positions, energy in both electric and magnetic forms is again propagated outward before the incoming energy corresponding to the preceding position has actually arrived. With sinusoidal doublet motion there must therefore be a continuous radiation of energy over and above the amount required to establish the steady-state field.

For illustration, suppose the doublet is oscillating at a frequency of $f = 10^6$ cycles per second. During a full period, $T = 1/f = 10^{-6}$

second, the doublet makes a complete oscillation, and for steady-state conditions the field must be back again at the same value everywhere. During this time, however, energy in the field could flow only over a distance $s = cT = 3 \times 10^8 \times 10^{-6} = 3 \times 10^2$ meters. Energy in the outlying regions can merely shuffle back and forth, but with each shuffle a relatively small amount is radiated permanently. Figure 16.13 illustrates qualitatively the progress during the charge motion discussed in the preceding paragraph. As the charges reverse positions some of the electric flux lines cross themselves and a new group in the reverse direction is formed. The flux lines which cross are detached from their charges and are now propagated outward permanently, together with the energy which they represent.

The current in a broadcast antenna radiates energy in a similar manner. Since in general the current varies in strength along the antenna wire, it is convenient to consider the equivalent current to be a series of doublet oscillators of different strengths placed at various positions along the wire.

16.10. The Electromagnetic Equations. Seven equations, commonly known also as Maxwell's equations, summarize the fundamental principles expressing the nature and behavior of electromagnetic fields, whether static or dynamic. Practically all the material, except quantum and thermal considerations, previously presented in this book can be shown to be special applications of these general equations.

For the convenience of readers familiar with the forms of vector analysis, as well as others, the equations are expressed in both vector and rectangular coordinate forms. Reference is made to the original equations where first used in the text.

To solve the equations it may be necessary in practice, if not in theory, to assume that the material "constants," ϵ, μ, σ, are true constants independent of field strength and time or frequency. For alternating fields, operating at a particular frequency, alternating steady-state solutions may be obtainable by using the value of the "constants" corresponding to that frequency.

The equations are as follows:

(1) *Equation 1.44.*

$$D = \epsilon E \tag{16.72}$$

relates the two electric vectors D and E.

(2) *Equation 9.10.*

$$B = \mu H \tag{16.73}$$

relates the two magnetic vectors B and H.

(3) *Equation 6.5.*

$$\frac{d\psi_{\text{out}}^{\text{net}}}{dv} = \operatorname{div} D = \rho_{\text{net}} = \rho_+ - \rho_- \tag{16.74}$$

$$= \frac{\partial D_x}{\partial x} + \frac{\partial D_y}{\partial y} + \frac{\partial D_z}{\partial z} \tag{16.75}$$

$$= \epsilon\left(\frac{\partial E_x}{\partial x} + \frac{\partial E_y}{\partial y} + \frac{\partial E_z}{\partial z}\right) \tag{16.76}$$

Equations 16.74 and 16.75 are general and state that the net outward electric flux per unit volume is equal to the net charge per unit volume. Equation 16.76 holds for ϵ equal to a constant only.

For static electric fields only, and ϵ = constant,

$$E_x = -\frac{\partial V}{\partial x}, \quad E_y = -\frac{\partial V}{\partial y}, \quad E_z = -\frac{\partial V}{\partial z} \tag{16.77}$$

and equation 16.76 reduces to the Poisson equation

$$\nabla^2 V = \frac{\partial^2 V}{\partial x^2} + \frac{\partial^2 V}{\partial y^2} + \frac{\partial^2 V}{\partial z^2} = -\frac{\rho_{\text{net}}}{\epsilon} = -\frac{1}{\epsilon}\frac{d\psi_{\text{out}}^{\text{net}}}{dv} \tag{16.78}$$

or, for $\rho_{\text{net}} = 0$, to the Laplace equation

$$\nabla^2 V = \frac{\partial^2 V}{\partial x^2} + \frac{\partial^2 V}{\partial y^2} + \frac{\partial^2 V}{\partial z^2} = 0 \tag{16.79}$$

V is the electrostatic potential. See equations 6.8 and 6.9.

(4) *Equation 9.15.*

$$\frac{d\phi_{\text{out}}^{\text{net}}}{dv} = \operatorname{div} B = 0 \tag{16.80}$$

$$= \frac{\partial B_x}{\partial x} + \frac{\partial B_y}{\partial y} + \frac{\partial B_z}{\partial z} = 0 \tag{16.81}$$

$$= \mu\left(\frac{\partial H_x}{\partial x} + \frac{\partial H_y}{\partial y} + \frac{\partial H_z}{\partial z}\right) = 0 \tag{16.82}$$

Equations 16.80 and 16.81 express the fact that magnetic flux lines are always closed lines and that, therefore, the net outward flux per unit volume is always zero. Equation 16.82 holds for μ = constant only.

For static magnetic fields only, μ = constant, and in non-current-bearing regions only

$$H_x = -\frac{\partial U}{\partial x}, \quad H_y = -\frac{\partial U}{\partial y}, \quad H_z = -\frac{\partial U}{\partial z} \tag{16.83}$$

Equation 16.82 then reduces to the Laplace form

$$\nabla^2 U = \frac{\partial^2 U}{\partial x^2} + \frac{\partial^2 U}{\partial y^2} + \frac{\partial^2 U}{\partial z^2} = 0 \tag{16.84}$$

U is the magnetostatic scalar potential. See equations 9.26, 9.27, and 9.28.

(5) *Based on equation 11.2.* This equation states that the voltage generated in a closed line path is the negative time derivative of the magnetic flux linking the path.

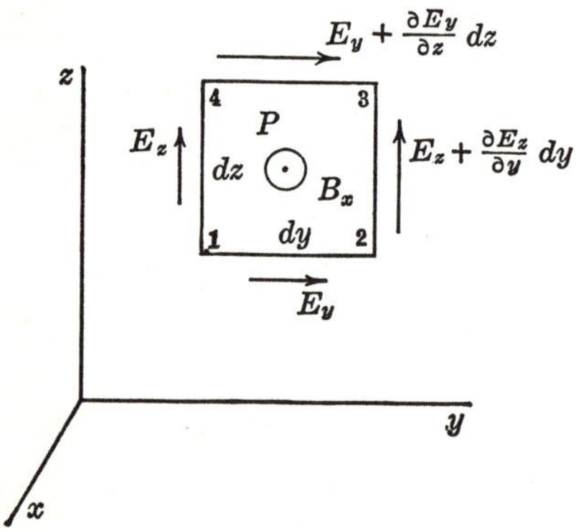

FIG. 16.14. Calculation of generated voltage around a closed path.

$$V_g = -\frac{d\phi}{dt} \tag{16.85}$$

This is to be restated in differential equation form for a small path and area.

At point P in Fig. 16.14 we assume a time-changing magnetic flux density vector B, whose x component is B_x. Let us consider a square plane area $dy\,dz$ (or 1–2–3–4–1), parallel to the y–z reference plane, including the point P. The integral of the electric intensity components E_y and E_z around the path is equal to the generated voltage, and B_x integrated over the normal area $dy\,dz$ is equal to the magnetic flux $d\phi$ crossing the area.

Then, over paths 1–2 and 3–4 the voltage generated is

$$(E_y)(+dy) + \left(E_y + \frac{\partial E_y}{\partial z} dz\right)(-dy) = -\frac{\partial E_y}{\partial z} dy\,dz \tag{16.86}$$

and over paths 2–3 and 4–1, it is

$$\left(E_z + \frac{\partial E_z}{\partial y} dy\right)(+dz) + (E_z)(-dz) = +\frac{\partial E_z}{\partial y} dy\,dz \tag{16.87}$$

The total closed path generated voltage equals

$$\left(\frac{\partial E_z}{\partial y} - \frac{\partial E_y}{\partial z}\right)(dydz) \tag{16.88}$$

which is equal to the negative time rate of change of magnetic flux across the area $dy\,dz$, or

$$-\frac{\partial B_x}{\partial t}(dy\,dz) \tag{16.89}$$

Per unit area (here $dy\,dz$, which has an x-directed normal)

$$\text{curl}_x\,E = \frac{\partial E_z}{\partial y} - \frac{\partial E_y}{\partial z} = -\frac{\partial B_x}{\partial t} \tag{16.90}$$

By cyclically permuting x, y, and z, we get the corresponding values on a y-directed area ($dz\,dx$) and on a z-directed area ($dx\,dy$), passed through point P.

$$\text{curl}_y\,E = \frac{\partial E_x}{\partial z} - \frac{\partial E_z}{\partial x} = -\frac{\partial B_y}{\partial t} \tag{16.91}$$

$$\text{curl}_z\,E = \frac{\partial E_y}{\partial x} - \frac{\partial E_x}{\partial y} = -\frac{\partial B_z}{\partial t} \tag{16.92}$$

The partial derivatives of E given above are the three components of a vector, curl E. If the area is oriented with its normal in such a direction as to give the *maximum* generated voltage per unit area, the value obtained is called curl E and its direction is that of the area normal.

For constant μ, $\partial B/\partial t$ may be replaced by $\mu(\partial H/\partial t)$. For a static field, $\partial B/\partial t = 0$ and the curl E components are all zero. In this event, E is derived from the electrostatic potential V by the relation $E = -\text{grad}\,V$. (See equation 1.62.)

(6) *Based on equations 4.10 and 4.11, and on equation 9.4.* From the first two we get the current density vector

$$J = \begin{cases} \sigma E + \dfrac{\partial D}{\partial t} & (16.93) \\[2ex] (\rho v)_{\pm} + \dfrac{\partial D}{\partial t} & (16.94) \end{cases}$$

The first form is used for a conducting region and the second for a region containing space charge. From equation 9.4 we note that the integration of vector H over a closed line path is equal to the current I crossing an area bounded by the path. (The sign is changed because the path element ds is here taken parallel to H, not to $-H$.)

The procedure followed now is identical with that used in (5) above and Fig. 16.14. Replace B_x with J_x and the components of E with those of H. Then

$$\text{curl}_x H = \frac{\partial H_z}{\partial y} - \frac{\partial H_y}{\partial z} = J_x \tag{16.95}$$

$$\text{curl}_y H = \frac{\partial H_x}{\partial z} - \frac{\partial H_z}{\partial x} = J_y \tag{16.96}$$

$$\text{curl}_z H = \frac{\partial H_y}{\partial x} - \frac{\partial H_x}{\partial y} = J_z \tag{16.97}$$

in which J_x, J_y, and J_z are the components of either equation 16.93 or 16.94.

For a static field $\partial D/\partial t = 0$. For regions with zero conductivity $\sigma E = 0$, and for regions free from space charge $(\rho v) = 0$. For regions where $J = 0$, H is derivable from the magnetic scalar potential U by the process $H = -\text{grad } U$, applicable therefore only to static fields in regions without conductivity or space charge.

(7) *Equation 9.100.* In any field, static or dynamic, the force on charge q is the vectorial summation

$$F = qE \oplus qv_n B \tag{16.98}$$

This equation was not originally included by Maxwell but was added as one of the fundamental group by Lorentz in about 1900 in modernizing the theory to include the electron as a fundamental physical unit of charge.

The solution of electromagnetic field problems, including those given in this text, require in one form or another the solution of one or more of the seven fundamental equations given above. If the problem has several materials, with differing values of ϵ, μ, σ, separate solutions are required for each medium, and the solutions must be joined mathematically by the boundary conditions appropriate at each interface.

16.11. Plane Wave Solution of Maxwell's Equations. We shall conclude with one particular solution of Maxwell's equations. The simplest solution of the dynamic equations is that for a plane wave. We here consider the wave to move with velocity v in the x direction through a perfect dielectric without conductivity $(\sigma = 0)$. For this case we take the electric field to be y directed everywhere and with equal value all over any particular y–z plane, $E = E_y$ only, though it will of necessity vary with x and with time. We assume $\epsilon = \epsilon_0\epsilon_r =$ constant and $\mu = \mu_0\mu_r =$ constant. (In any actual dielectric $\mu_r = 1$.)

For $\mu =$ constant, equation 16.92 yields

$$\frac{\partial E_y}{\partial x} = -\mu \frac{\partial H_z}{\partial t} \tag{16.99}$$

whence H_z is z-directed and constant over any y–z plane, though it also is a function only of x and t. Combining equations 16.99 and 16.96, and taking $J_y = \epsilon(\partial E_y/\partial t)$, we get

$$-\frac{\partial H_z}{\partial x} = J_y = \epsilon\frac{\partial E_y}{\partial t} \tag{16.100}$$

Differentiating equation 16.99 partially with respect to x and equation 16.100 partially with respect to t, and combining, gives

$$\frac{\partial^2 E_y}{\partial x^2} = \mu\epsilon\frac{\partial^2 E_y}{\partial t^2} \tag{16.101}$$

Similarly,

$$\frac{\partial^2 H_z}{\partial x^2} = \mu\epsilon\frac{\partial^2 H_z}{\partial t^2} \tag{16.102}$$

Equations 16.101 and 16.102 are wave equations and are solved by an infinite number of functions of the same type,

$$E_y = \phi_1\left(t - \frac{x}{v}\right) \tag{16.103}$$

or

$$E_y = \phi_2\left(t + \frac{x}{v}\right) \tag{16.104}$$

Similarly

$$H_z = \phi_3\left(t - \frac{x}{v}\right) \tag{16.105}$$

and

$$H_z = \phi_4\left(t + \frac{x}{v}\right) \tag{16.106}$$

in which the wave (or phase) velocity

$$v = \frac{1}{\sqrt{\mu\epsilon}} = \frac{c}{\sqrt{\mu_r\epsilon_r}}\ \frac{\text{meters}}{\text{second}} \tag{16.107}$$

Here $c \cong 3 \times 10^8$ meters per second is the wave velocity in free space.

Equations 16.103 and 16.105 prescribe similar waves of E_y and H_z moving with velocity v in the $+x$ direction, while equations 16.104 and 16.106 give similar waves moving in the $-x$ direction.

The wave shape, or the wave distribution in space, is obtained by taking $t = 0$, giving for $+x$ directed waves

$$E_y = \phi_1(x) \tag{16.108}$$

$$H_z = \phi_3(x) \tag{16.109}$$

Similarly, for the other pair moving in the $-x$ direction,

$$E_y = \phi_2(x) \tag{16.110}$$

$$H_z = \phi_4(x) \tag{16.111}$$

The variation of E_y and H_z in time, at a particular point in space, has a similar shape.

To satisfy equation 16.99, it is seen that $\phi_1\left(t - \frac{x}{v}\right)$ must equal $\mu v \phi_3\left(t - \frac{x}{v}\right)$, and $\phi_2\left(t + \frac{x}{v}\right)$ must equal $\mu v \phi_4\left(t + \frac{x}{v}\right)$, or that

$$E_y = \mu v H_z = \sqrt{\frac{\mu}{\epsilon}}\, H_z \tag{16.112}$$

$$= \sqrt{\frac{\mu_0}{\epsilon_0}}\sqrt{\frac{\mu_r}{\epsilon_r}}\, H_z$$

$$\cong 120\pi \sqrt{\frac{\mu_r}{\epsilon_r}}\, H_z \tag{16.113}$$

The ratio

$$\frac{E_y}{H_z} = \sqrt{\frac{\mu}{\epsilon}} \cong 120\pi \sqrt{\frac{\mu_r}{\epsilon_r}} \quad \frac{\text{volts/meter}}{\text{amp/meter}} = \text{ohms} \tag{16.114}$$

is called the impedance of the dielectric for this type of wave. The impedance of free space is

$$\frac{E_{y0}}{H_{z0}} = \sqrt{\frac{\mu_0}{\epsilon_0}} \cong 120\pi \cong 376.7 \quad \text{ohms} \tag{16.115}$$

For a single-frequency (monochromatic) type wave, moving in the $+x$ direction, we take

$$E_y = (E_y)_{\max} \cos \omega \left(t - \frac{x}{v}\right) \tag{16.116}$$

$$H_z = \sqrt{\frac{\epsilon}{\mu}}\, (E_y)_{\max} \cos \omega \left(t - \frac{x}{v}\right) \tag{16.117}$$

with $\omega = 2\pi f$. In terms of wave length $\lambda = v/f = 2\pi v/\omega$,

$$E_y = (E_y)_{\max} \cos \frac{2\pi}{\lambda}\, (vt - x) \tag{16.118}$$

$$H_z = \sqrt{\frac{\epsilon}{\mu}}\, (E_y)_{\max} \cos \frac{2\pi}{\lambda}\, (vt - x) \tag{16.119}$$

The Poynting vector $\mathcal{P}_x$, normal to E_y and H_z, is a double frequency function given by

$$\mathcal{P}_x = E_y H_z = \sqrt{\frac{\epsilon}{\mu}} (E_y)_{\max}^2 \cos^2 \frac{2\pi}{\lambda} (vt - x) \quad \frac{\text{watts}}{\text{square meter}} \tag{16.120}$$

Equations 16.118 and 16.120 are illustrated in Fig. 16.15. H_z is similar to E_y but is $\sqrt{\epsilon/\mu}$ times as large and is z directed.

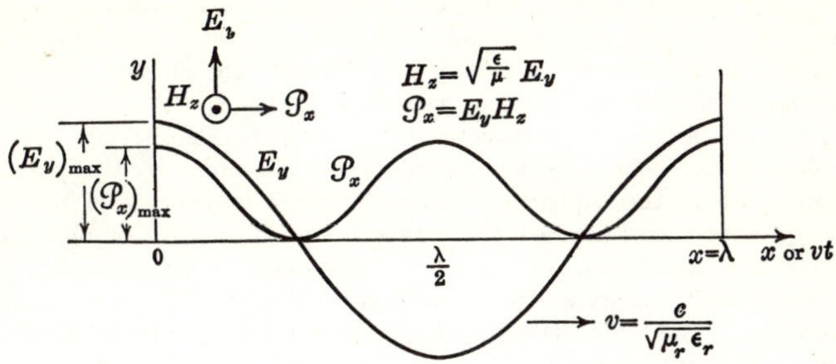

FIG. 16.15. Plane wave in a perfect dielectric.

The energy density in both electric and magnetic field forms are equal at any position but are distributed nonuniformly throughout space.

$$\omega_e = \frac{\epsilon E_y^2}{2} = \frac{\epsilon}{2} (E_y)_{\max}^2 \cos^2 \omega \left(t - \frac{x}{v}\right) \tag{16.121}$$

$$\omega_m = \frac{\mu H_z^2}{2} = \frac{\mu}{2} (H_z)_{\max}^2 \cos^2 \omega \left(t - \frac{x}{v}\right) \tag{16.122}$$

$$= \omega_e \quad \frac{\text{joules}}{\text{cubic meter}} \tag{16.123}$$

The total energy density

$$\omega = \omega_e + \omega_m = \epsilon (E_y)_{\max}^2 \cos^2 \omega \left(t - \frac{x}{v}\right) \tag{16.124}$$

Comparing equations 16.124 and 16.120, we note that

$$\mathcal{P}_x = \omega \frac{1}{\sqrt{\mu\epsilon}} = \omega v \quad \frac{\text{watts}}{\text{square meter}} \tag{16.125}$$

so that Poynting's vector is equal to the product of the total energy density and the wave velocity.

PROBLEMS

1. What size inductance is required with a 0.0001-μf condenser in order to produce oscillations of one million cycles per second? What is the value of the critical damping resistance? If a resistance equal to ⅕ of the critical damping resistance is inserted in series, compute the new value of frequency. In the time required for one cycle, by what percentage is the current peak reduced?

2. What changes should be made in the circuit of Fig. 16.3 to produce a wave front which rises to its peak with great rapidity and then falls very slowly? In this case breakdown might take place on the "tail" of the wave instead of on the wave front.

3. Parallel plate condenser, $s = 0.2$ cm, $A = 100$ sq cm (circular), $\epsilon_r = 3$. The potential is rising 3000 volts per second. Compute the displacement current density, and the magnetic field intensity at the edge of the plates.

4. A concentric cylindrical cable with wire radius $r_1 = 1$ cm, inner sheath radius $r_2 = 5$ cm, and relative dielectric constant $\epsilon_r = 4$, supports a potential of 60,000 volts and carries a current of 100 amperes. Compute by two methods the power flowing along the cable. Compute the surge impedance of the cable as determined by the fluxes between the wire and sheath only. Compute the wave velocity, ignoring any losses.

5. In Prob. 4 if the sending-end switch is closed for 0.01 second, how long is the block of charge on the line? How much charge is contained in the block? What current is carried by the block?

6. An electron moves in a straight line with a velocity equal to 0.1 that of light. At a point 10 cm away in the equatorial plane, compute E, H, and the displacement current density. Compute the ratio of the relativistic mass to the rest mass.

7. A current of 1 ampere maximum value, of length 1 cm, is alternating one million times per second. Compute the time average of the energy radiated and the wave length of the radiation.

8. High-potential ignition circuits for oil burners have proved to be troublesome sources of radio interference. Explain this phenomenon and suggest some possible methods of improvement.

9. Explain why a condenser, shunted around a switch in a circuit containing resistance and inductance is largely effective in eliminating sparking when the switch is opened.

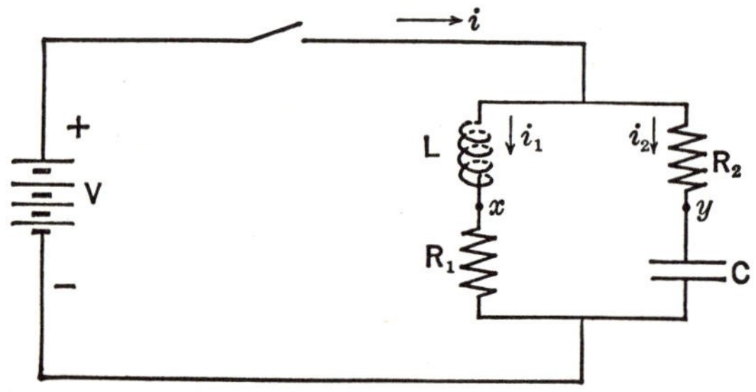

FIG. P10

10. Write the expressions for current i_1 and i_2 in the two branches of the circuit shown in Fig. P10. If now R_1 is set equal to R_2 and each is made equal to $\sqrt{L/C}$, show that the potential difference between points x and y is zero for all values of V and time t. Show that $i = i_1 + i_2 = \dfrac{V}{\sqrt{L/C}}$ for all values of t.

APPENDIX I

TABLE OF UNITS, SYMBOLS, AND DIMENSIONS

Gathered together in tabular form are the units used throughout the text, together with their symbols, unit names, defining equations, and dimensional formulas based on *LMTQ*.

Listed also are the cgs equivalents, in both cgs sub-systems, emu, and esu. The number of units in each sub-system required to equal one mks unit of the same quantity are tabulated also. These equivalents should be particularly useful during the present transition period of changing from cgs to mks units.

APPENDIX I

Table of Units, Symbols, and Dimensions

	Quantity	Symbol	mks Unit Rationalized	Defining Equation	Dimensional Formula Exponents of: L	M	T	Q	cgs emu	No. of emu / No. of mks	cgs esu	No. of esu / No. of mks	No. of esu / No. of emu
1	Length	L	m		1	0	0	0	cm	10^2	cm	10^2	1
2	Area	A	m^2	$A = L^2$	2	0	0	0	cm^2	10^4	cm^2	10^4	1
3	Volume	v	m^3	$v = L^3$	3	0	0	0	cm^3	10^6	cm^3	10^6	1
4	Mass	M, m	kilogram		0	1	0	0	gram	10^3	gram	10^3	1
5	Time	T, t	second		0	0	1	0	second	1	second	1	1
6	Velocity	v	m/sec	$v = L/T$	1	0	−1	0	cm/sec	10^2	cm/sec	10^2	1
7	Acceleration	a	m/sec^2	$a = L/T^2$	1	0	−2	0	cm/sec^2	10^2	cm/sec^2	10^2	1
8	Force	F	newton	$F = Ma$	1	1	−2	0	dyne	10^5	dyne	10^5	1
9	Energy	W	joule	$W = FL$	2	1	−2	0	erg	10^7	erg	10^7	1
10	Power	P	watt	$P = W/T$	2	1	−3	0	erg/sec	10^7	erg/sec	10^7	1
11	Charge	Q, q	coulomb	$F = Q^2/(4\pi\epsilon_0 L^2)$	0	0	0	1	abcoulomb	10^{-1}	statcoulomb	$10c$	$100c$
12	Dielectric constant of free space	ϵ_0	farad/m	$\epsilon_0 = 1/(\mu_0 c^2)$	−3	−1	2	2			1	$4\pi c^2/10^7$	
13	Dielectric constant	ϵ	farad/m		−3	−1	2	2					
14	relative	ϵ_r	numeric	$\epsilon_r = \epsilon/\epsilon_0$	0	0	0	0				1	
	Charge density												
15	volume	ρ	$coulomb/m^3$	$\rho = Q/v$	−3	0	0	1	$abcoulomb/cm^3$	10^{-7}	$statcoulomb/cm^3$	$c/10^5$	$100c$
16	surface	ρ_s	$coulomb/m^2$	$\rho_s = Q/A$	−2	0	0	1	$abcoulomb/cm^2$	10^{-5}	$statcoulomb/cm^2$	$c/10^3$	$100c$
17	line	ρl	coulomb/m	$\rho l = Q/L$	−1	0	0	1	abcoulomb/cm	10^{-3}	statcoulomb/cm	$c/10$	$100c$
18	Electric intensity	E	volt/m	$E = F/Q = -V/L$	1	1	−2	−1	abvolt/cm	10^6	statvolt/cm	$10^4/c$	$1/(100c)$
19	Electric flux density	D	$coulomb/m^2$	$D = \epsilon E = \psi/A$	−2	0	0	1		$4\pi/10^5$		$4\pi c/10^3$	$100c$
20	Electric flux	ψ	coulomb	$\psi = DA$	0	0	0	1		$4\pi/10$		$4\pi 10c$	$100c$
21	Electric potential	V	volt	$V = -EL$	2	1	−2	−1	abvolt	10^8	statvolt	$10^6/c$	$1/(100c)$
22	EMF	V_g	volt	$V_g = -d\phi/dt$	2	1	−2	−1	abvolt	10^8	statvolt	$10^6/c$	$1/(100c)$
23	Capacitance	C	farad	$C = Q/V$	−2	−1	2	2	abfarad	10^{-9}	statfarad	$c^2/10^5$	$(100c)^2$
24	Current	I, i	ampere	$I = Q/T$	0	0	−1	1	abampere	10^{-1}	statampere	$10c$	$100c$
25	Current density	J	$ampere/m^2$	$J = I/A$	−2	0	−1	1	$abampere/cm^2$	10^{-5}	$statampere/cm^2$	$c/10^3$	$100c$
26	Resistance	R	ohm	$R = V/I$	2	1	−1	−2	abohm	10^9	statohm	$10^5/c^2$	$1/(100c)^2$
27	Resistivity	ρ	ohm-m	$\rho = RA/L$	3	1	−1	−2	abohm-cm	10^{11}	statohm-cm	$10^7/c^2$	$1/(100c)^2$
28	Conductance	G	mho	$G = 1/R$	−2	−1	1	2	abmho	10^{-9}	statmho	$c^2/10^5$	$(100c)^2$
29	Conductivity	σ	mho/m	$\sigma = 1/\rho = J/E$	−3	−1	1	2	abmho/cm	10^{-11}	statmho/cm	$c^2/10^7$	$(100c)^2$
30	Electric polarization	P	$coulomb/m^2$	$P = D - \epsilon_0 E = \rho L$	−2	0	0	1	$abcoulomb/cm^2$	10^{-5}	$statcoulomb/cm^2$	$c/10^3$	$100c$
31	Electric susceptibility	χ_e	farad/m	$\chi_e = P/E = \epsilon_0(\epsilon_r - 1)$	−3	−1	2	2			1	$4\pi c^2/10^7$	
32	Electric dipole moment	m_e	coulomb-m	$m_e = QL$	1	0	0	1			statcoulomb-cm	10^3c	
33	Electric energy density	ω_e	$joule/m^3$	$\omega_e = DE/2$	−1	1	−2	0	erg/cm^3	1	erg/cm^3	10	1

APPENDIX I. *Continued.*

Table of Units, Symbols, and Dimensions

	Quantity	Symbol	mks Unit Rationalized	Defining Equation	Dimensional Formula Exponents of L	M	T	Q	cgs emu	No. of emu / No. of mks	cgs esu	No. of esu / No. of mks	No. of esu / No. of emu
34	Permeability of free space	μ_0	henry/m	$\mu_0 = 4\pi/10^7$	1	1	0	−2		$10^7/4\pi$			
35	Permeability	μ	henry/m	$\mu = B/H$	1	1	0	−2					
36	relative	μ_r	numeric	$\mu_r = \mu/\mu_0$	0	0	0	0		1			
37	Magnetic pole	p	weber	$p = A(B - B_0)$	2	1	−1	−1	pole = maxwell/4π	$10^8/4\pi$			
38	Magnetic moment	m	weber-m	$m = pL$	3	1	−1	−1	pole-cm	$10^{10}/4\pi$			
39	Magnetic intensity	H	ampere/m or newton/weber	$H = U/L$ or F/p	−1	0	−1	1	oersted or gilbert/cm	$4\pi/10^3$			
40	Magnetic flux density	B	weber/m²	$B = \mu H = \phi/A$	0	1	−1	−1	gauss or maxwell/cm²	10^4			
41	Magnetic flux	ϕ	weber	$\phi = BA = V_gT$	2	1	−1	−1	maxwell	10^8			
42	Magnetic potential	U	ampere	$U = \mathfrak{F} = HL$	0	0	−1	1	gilbert	$4\pi/10$			
43	MMF	$\mathfrak{F}$	ampere	$\mathfrak{F} = I$	0	0	−1	1	gilbert	$4\pi/10$			
44	Intensity of magnetization	M	weber/m²	$M = B - B_0 = m/L^3$	0	1	−1	−1	pole/cm² or gauss/4π	$10^4/4\pi$			
	Inductance												
45	self	L	henry	$L = \phi/I$	2	1	0	−2	abhenry	10^9		$10^5/c^2$	$1/(100c)^2$
46	mutual	M	henry	$M = \phi/I = W/I^2$	2	1	0	−2	abhenry	10^9		$10^5/c^2$	$1/(100c)^2$
47	Reluctance	$\mathcal{R}$	ampere/weber	$\mathcal{R} = \mathfrak{F}/\phi$	−2	−1	0	2					
48	Reluctivity	ν	meter/henry	$\nu = 1/\mu$	−1	−1	0	2					
49	Permeance	$\mathcal{P}$	weber/amp	$\mathcal{P} = 1/\mathcal{R}$	2	1	0	−2					
50	Permittivity	μ	henry/meter	$\mu = 1/\nu$	1	1	0	−2					
51	EMF	V_g	volt	$V_g = -d\phi/dt$	2	1	−2	−1	abvolt	10^8	statvolt	$10^6/c$	$1/(100c)$
52	Poynting's vector	$\mathcal{P}$	watts/m²	$\mathcal{P} = EH$	0	1	−3	0	abwatt/cm²	10^3	statwatt/cm²	10^3	1
53	Magnetic energy density	ω_m	joule/m³	$\omega_m = HB/2$	−1	1	−2	0	erg/cm³	10	erg/cm³	10	1
54	Magnetic susceptibility	χ_m	henry/m	$\chi_m = M/H = \mu_0(\mu_r - 1)$	1	1	0	−2	henry/m	$10^7/4\pi$			

$\mu_0 = 4\pi/10^7$ henrys/meter. For $c = 2.998 \times 10^8$ meters/sec, $\epsilon_0 = 1/\mu_0 c^2 = 10^7/(4\pi c^2) = 8.854 \times 10^{-12}$ farad/meter
For $c \cong 3 \times 10^8$ meters/sec, $\epsilon_0 \cong 1/(36\pi 10^9)$ farad/meter
$c^2 = 8.988 \times 10^{16} \cong 9 \times 10^{16}$

APPENDIX II

ELECTRONS IN ATOMS

The material presented in Arts. 1.1 to 1.7 inclusive has been given primarily for the purpose of acquainting the student with the magnitudes involved: size, mass, velocity, energy, spacing, electric charge of the electrical particles which constitute the atom; and with some of the molecular magnitudes encountered in gaseous, liquid, and solid states.

It is assumed in Art. 1.6 that the hydrogen atom can exist in any one of a series of "stationary states" of energy, and in each of these states the electron travels in a circular orbit about the nucleus. For $n = 1$ the electron travels in the smallest circle, and the atom is in its normal state of lowest total energy (kinetic plus potential energy). See equations 1.17 and 1.21. Higher values of n indicate larger circles and higher values of total energy, values which are less negative and which correspond to possible excited states of the atom.

In addition to energy, each circular orbit has associated with it a quantity called *angular momentum,* which is the product of the momentum of the electron (m_0v) by the orbital radius (a). Rewriting equation 1.15 in the form

$$m_0v \cdot a = k\left(\frac{h}{2\pi}\right) \tag{II.1}$$

it is seen that angular momentum occurs only in integral multiples of $h/2\pi$. In this equation integer k has been used in place of n so that n may be reserved as a measure of the total energy of the orbit, whereas k is a measure of its angular momentum. n is the *total quantum number* and k is the *azimuthal quantum number.* For circular orbits, $n = k$. Angular momentum is a vector quantity; its direction is taken normal to the plane of the orbit according to the right-hand screw rule.

Another possible type of orbit is the elliptic orbit. In this type the electron travels an elliptic path about the nucleus which is located at one of the foci of the ellipse. In general there are several possible elliptic orbits, any one of which may be used instead of a particular circular orbit. The analysis of elliptic orbits, while not particularly difficult, is somewhat too long for our present purpose. The analysis shows, first, that the total energy of an electron in an elliptic orbit is the same as that for a circular orbit, if the major axis of the ellipse is equal to the diameter of the circle and the same value of n is associated with both orbits. For $n = k$, the orbit is circular. The analysis shows, in the second place, that k can take integral values less than n, down to one for its lowest value, and for these values the orbits are elliptical and possess values of angular momentum proportional to k. The ratio of the minor to the major axis is equal to the ratio of k to n.

Stationary states of electrons in orbits, as outlined above, cannot be observed directly. Only *changes* in stationary states are observable as the atom emits or absorbs energy (see equations 1.23 and 1.24) and leaves its traces in the form of spectral lines with measurable wave lengths. From these spectral data one must deduce the nature of the stationary states themselves.

Spectral studies have shown that the foregoing picture of the hydrogen atom is only approximately correct. A better approximation is obtained if it is assumed that the angular momentum of the orbits is one unit, $h/2\pi$, too large. Angular momentum, instead of being $k(h/2\pi)$, is equal to $(k - 1)(h/2\pi) =$

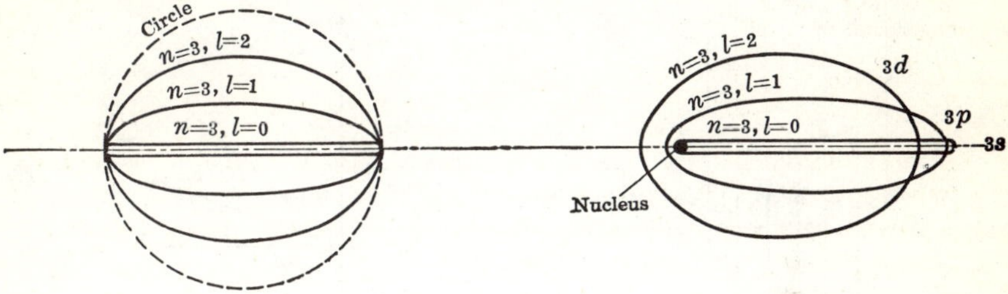

(a)
Relative shapes of orbits

(b)
Special arrangement of orbits about nucleus

FIG. II.1. Electron orbits for $n = 3$.

(Total energy $\sim \frac{1}{n^2}$. Angular momentum $\sim l$.)

$l(h/2\pi)$. Whereas k could assume all integral values from $k = n$ to $k = 1$, l can assume all integral values from $l = n - 1$ to $l = 0$. l is called the ***orbital quantum number***. With this change circular orbits are no longer possible, and in the case of $l = 0$ there is no orbital angular momentum at all, the motion of the electron is entirely radial, and *apparently* the electron and nucleus should collide. This catastrophe is actually avoided but only at the price of giving up this simple mechanistic and approximate picture of the atomic state in favor of a "wave-mechanical" atom which refuses to yield to pictorial treatment but in compensation is more accurate mathematically. Electron orbits for $n = 3$, and $l = 2, 1, 0$ are illustrated in Fig. II.1. Electrons for which $l = 0, 1, 2, 3 \ldots$, etc., are referred to as $s, p, d, f \ldots$ types, respectively.

An electron spins about its own axis and has therefore additional angular momentum over and above its orbital value. The ***spin quantum number*** is $s = \frac{1}{2}$, which means that the unit of spin angular momentum is one-half as large as the unit of orbital angular momentum, that is, $\frac{1}{2}(h/2\pi)$. Spin angular momentum may be represented by a vector along the axis of spin. Unlike quantum numbers n and l, s has only one value, $\frac{1}{2}$.

Every atom is pervaded by a magnetic field, created perhaps by the movement of electrons in the same or adjacent atoms, or imposed on the atom by external sources of magnetic field. The direction of the magnetic field forms a unique direction in space about which the vectors of angular momentum may take certain definite orientations only. In other words the electron orbits and spin axes can be oriented in space in only a few definite directions relative to the direction of the magnetic field.

If an electron orbit has l units of angular momentum, the directions which the vector can assume relative to the magnetic field are those, and those only, in which the projections of the vector on the magnetic direction are integral numbers of units, m_l, of angular momentum. m_l is called the *magnetic orbital*

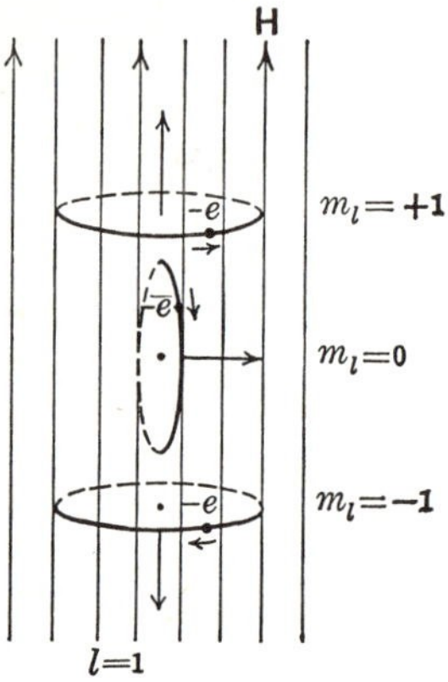

FIG. II.2. Orientation of orbits in a magnetic field for $l = 1$.

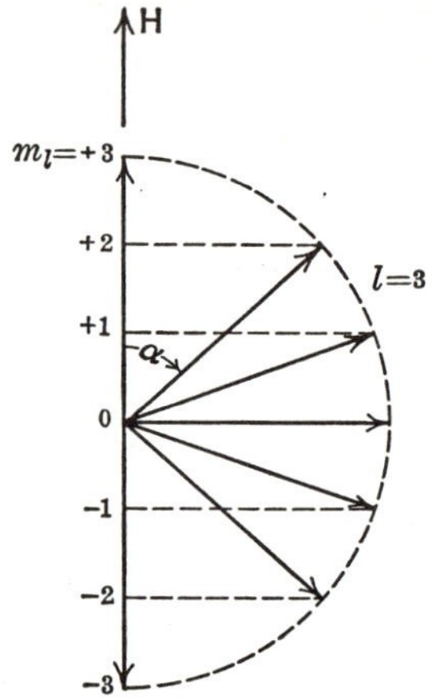

FIG. II.3. Orientation of angular momentum vector for $l = 3$.

quantum number. For illustration, if $l = 3$, then m_l can take the values 3, 2, 1, 0, −1, −2, −3. In general, m_l can assume $2l + 1$ possible values and this number of orbital orientations is possible, for each value of l. Figure II.2 illustrates the possible orientations for $l = 1$; and Fig. II.3, for $l = 3$, gives the same information in a different manner. The total energy of an orbit changes slightly with its orientation, a correction term which can be added to the energy represented by the total quantum number n. Positive values of m_l correspond to larger values of energy than do negative values.

In a magnetic field the axis of spin aligns itself either parallel or antiparallel with the field. These two possibilities are recognized by assigning a *magnetic spin quantum number* $m_s = \pm\frac{1}{2}$, since $s = \frac{1}{2}$. The positive value is taken when the spin axis, by the right-hand rule, is parallel to the field; this represents a slightly higher energy state than the negative value.

Each electron in an atom subject to a magnetic field intensity H may therefore be characterized by four quantum numbers:

1. Total quantum number. $n = 1, 2, 3, 4, \ldots, n.$
   Total energy, to the first approximation, is proportional to $1/n^2$.
   Size of orbit is proportional to n^2.
2. Orbital quantum number. $l = \begin{cases} 0, 1, 2, 3, \ldots, (n-1). \\ s, p, d, f, \ldots \end{cases}$
   Orbital angular momentum $= l(h/2\pi)$.

The larger the value of $n - l$, the narrower the orbital ellipse proves to be.

Total energy is independent of the value of l.

3. Magnetic orbital quantum number. $m_l = -l, -(l-1), \ldots, 0, \ldots, (l-1), l$.

   $m_l/l = \cos\alpha$, where α = angle between vectors $l(h/2\pi)$ and H.

   m_l can assume one of $2l + 1$ possible values.

   Total energy depends to a small degree upon m_l.

4. Magnetic spin quantum number. $m_s = \pm\frac{1}{2}$.

   Spin angular momentum $= m_s(h/2\pi)$, parallel or antiparallel with the field H.

   Total energy depends to a very small degree upon m_s.

The various electrons in an atom are not completely independent of each other, and quantum numbers cannot be assigned arbitrarily to each electron. Instead, the assignment of quantum numbers is subject to the restrictions of the exclusion principle. According to this rule, no two electrons in an atom can have all four quantum numbers, (n, l, m_l, m_s), alike. In Table II.1 are listed all the possible combinations of electronic quantum numbers for $n = 1, 2, 3$.

According to this scheme electrons may be arranged in shells about the nucleus. The number of the shell is the same as the total quantum number n, higher numbers corresponding to shells of larger size which can hold larger numbers of electrons. In the first shell, $n = 1$, $l = m_l = 0$, two s-type electrons only, with opposite spins, can be accommodated, and the shell is said to be closed. These two electrons may be characterized as $1s^2$, which means first shell, s-type electrons, two in number.

In the second shell, $n = 2$, $l = 0$ or 1. For $l = 0$ and $m_l = 0$, there may be not more than two electrons with opposite spins, characterized as $2s^2$, and this closes a sub-shell. For $l = 1$, m_l can take values $+1, 0, -1$, and for each of these orbital orientations there may be two electrons with opposite spins or six in all, characterized as $2p^6$. This closes another sub-shell. The whole second shell is composed of the two sub-shells and the eight electrons are characterized as $2s^2 2p^6$. In the third shell there may be up to eighteen electrons, split into three sub-shell groups, two for s, 6 for p, and 10 for d, characterized as $3s^2 3p^6 3d^{10}$.

With this background we may proceed to build up Table 1.2, which gives the electronic configurations for the first 36 atoms, each in its normal or unexcited state of lowest energy. For hydrogen, with atomic number 1, the nucleus carries charge $+e = +1.602 \times 10^{-19}$ coulomb, and the electron with an equal negative charge moves in a radial orbit without orbital angular momentum since $l = 0$, but has angular momentum due to its spin. For this electron, in its normal state of lowest energy, the n, l configuration, as shown in Table 1.2, is 1, 0, or $1s$. For helium, with atomic number 2, the nuclear charge is $+2e$ and its two electrons can go into the first shell, $n = 1$, without orbital angular momentum, $l = 0$, but with the electron spins oppositely directed. Thus there are two electrons for which n, l equals 1, 0 or $1s$, and this closes the first shell.

Lithium, atomic number 3, has a nuclear charge $+3e$, and three electrons. The first two electrons go into the first shell, $n = 1$ and $l = 0$. Since the first shell is closed the third electron enters the second shell as an s-type electron, $l = 0$, as this is the lowest possible energy state for that shell. With beryllium, atomic number 4, the fourth electron also enters the second shell as an s-type electron but with a spin opposite to that of the third electron. This closes the s-type sub-shell of the second shell. The fifth electron, required for

TABLE II.1

Possible Combinations of Electronic Quantum Numbers in atoms

Number of Shell	n	$l \leqq (n - 1)$	m_l	m_s	Orbital Type of Electrons = Type of Sub-shell	Number of Electrons in Sub-shell $= 2(2l + 1)$	Number of Electrons in Shell $= 2n^2$	Shell and Sub-shell Characterization
1	1	0	0	$+\frac{1}{2}$ $-\frac{1}{2}$	s	2	2	$1s^2$
2	2	0	0	$+\frac{1}{2}$ $-\frac{1}{2}$	s	2		
		1	+1	$+\frac{1}{2}$ $-\frac{1}{2}$	p	6	8	$2s^2 2p^6$
			0	$+\frac{1}{2}$ $-\frac{1}{2}$				
			−1	$+\frac{1}{2}$ $-\frac{1}{2}$				
3	3	0	0	$+\frac{1}{2}$ $-\frac{1}{2}$	s	2		
		1	+1	$+\frac{1}{2}$ $-\frac{1}{2}$	p	6		
			0	$+\frac{1}{2}$ $-\frac{1}{2}$				
			−1	$+\frac{1}{2}$ $-\frac{1}{2}$				
		2	+2	$+\frac{1}{2}$ $-\frac{1}{2}$	d	10	18	$3s^2 3p^6 3d^{10}$
			+1	$+\frac{1}{2}$ $-\frac{1}{2}$				
			0	$+\frac{1}{2}$ $-\frac{1}{2}$				
			−1	$+\frac{1}{2}$ $-\frac{1}{2}$				
			−2	$+\frac{1}{2}$ $-\frac{1}{2}$				

boron, enters the second shell but is a p-type electron with $l = 1$. The process of building on p-type electrons, with one unit of orbital angular momentum, continues up to neon, which has two s-type electrons in the first shell, two s-type electrons in a sub-shell of the second shell, and 6 p-type electrons in another closed sub-shell of the second shell. With neon the second shell, containing in all eight electrons, is closed. This process continues for the next eight atoms, from sodium to argon, but, as shown in Table 1.2, with the electrons beyond the neon core being taken into the third shell, and of these the first two are s-type and the remainder p-type.

The first break in regularity in the building-up process occurs with potassium, atomic number 19. In this atom the nineteenth electron enters the fourth shell as an s-type electron instead of entering the third shell as a d-type electron, $l = 2$, because the former possibility proves to be state of energy lower than that of the latter possibility. The third shell is not closed until copper, atomic number 29, is reached, and for this atom there is already one electron in the fourth shell. Between potassium and copper certain irregularities are apparent from inspection of Table 1.2.

Chemical properties of the atoms are related closely to the arrangement of the electrons in the outer shells. Atoms in which the electrons form closed shells or, in some cases, closed sub-shells are inactive chemically, and their valences are zero. This is true for the noble gases, helium, neon, argon, krypton. The two electrons of helium close the first shell. With neon the shell is closed. For argon the $3p$ sub-shell is closed; so also is the $4p$ sub-shell with krypton.

Atoms with shells which are only partially closed are active chemically, particularly those atoms which have just one electron in the outermost shell, as is the case for hydrogen, lithium, sodium, potassium; and particularly those atoms which lack one electron of having their shells closed, such as fluorine, chlorine, bromine. It is to be expected that two atoms, one from each of the two groups, will combine readily to form a molecule; an illustration is sodium chloride.

The foregoing description, which assumes that the atom is a minute mechanical system composed of electrical particles, energy, and angular momentum, is only approximately correct. Its chief virtue lies in the fact that this concept leads readily to a pictorial representation. The wave-mechanical conception of an atom leads directly to the correct evaluation of the stationary states of energy which an atom may occupy, and to the correct values of angular momentum. This is accomplished, in this new theory, without injecting special restrictions of any kind, whereas the older theories required the assistance of special rules designed to give the correct answer in a particular case. The wave-mechanical atom, however, cannot be represented accurately by any pictorial scheme.

As an electron moves through an electric field the total energy E of the electron remains constant, so that any change in the potential energy V is balanced by an equal and opposite change in kinetic energy K (and momentum m_0v). When the electron approaches a positive charge the potential energy declines while the kinetic energy and momentum increase.

An electron in a stationary state of energy is represented, in wave mechanics, by a system of standing or stationary waves. A standing or stationary wave is composed of two progressive or running waves of like amplitude, frequency, and wave length, which move past each other with equal but oppositely directed velocities. At certain points in space the running waves cancel each other at all instants, giving nodes in the standing wave; halfway between the nodes are antinodes in the standing wave. The amplitude ψ of the standing wave satisfies the following equation,

$$\frac{\partial^2\psi}{\partial x^2} + \frac{\partial^2\psi}{\partial y^2} + \frac{\partial^2\psi}{\partial z^2} + \frac{4\pi^2\nu^2}{u^2}\psi = 0 \tag{II.2}$$

in which ν is the frequency, and u is the velocity, of the running waves comprising the standing wave. The wave length of the running waves is $\lambda_0 = u/\nu$. If the wave length is now made dependent upon the momentum of the electron by the equation

$$\lambda_0 = \frac{h}{m_0 v} \tag{II.3}$$

the standing wave equation is related to the momentum and kinetic energy of the electron.

Since

$$K = \tfrac{1}{2}m_0v^2 = E - V \tag{II.4}$$

$$v = \sqrt{\frac{2(E - V)}{m}} \tag{II.5}$$

and

$$\lambda_0 = \frac{h}{\sqrt{2m(E - V)}} = \frac{u}{\nu} \tag{II.6}$$

The equation for the amplitude of the standing wave then becomes

$$\frac{\partial^2\psi}{\partial x^2} + \frac{\partial^2\psi}{\partial y^2} + \frac{\partial^2\psi}{\partial z^2} + \frac{8\pi^2 m}{h^2}(E - V)\psi = 0 \tag{II.7}$$

This is called the Schrödinger equation.

To apply this equation to a particular case, it is necessary to substitute for V the potential energy function of the electron. To illustrate, the potential energy of an electron with charge $-e$ at a distance r from a charge $+Q$ is equal to $V = -\frac{Qe}{4\pi\epsilon r} = \frac{1}{4\pi\epsilon}\frac{-Qe}{(x^2 + y^2 + z^2)^{1/2}}$. The Schrödinger equation possesses satisfactory solutions for only certain particular values of total energy E, which prove to be the only stationary states of energy which the electron may have. This type of approach gives the correct values for the energy states of the atom, but unfortunately it is not adapted to pictorial representation.

APPENDIX III

HISTORICAL OUTLINE

Until only a few years ago, the study of physical theory was divided into half a dozen or more distinct fields—physics, chemistry, mechanics, electricity, magnetism, light, etc.—each based on its own group of experimental facts and hypotheses. Today the distinctions between these fields are superficial, for the physicist has obtained a deeper insight into physical processes and has a broader conception of the connections underlying physical phenomena. The various fields of study are still retained for reasons of convenience despite the development of the electron theory, which attempts to correlate all physical phenomena and express the relations in terms of laws that apply to all fields of physical theory.

The electron theory rests upon the recognition of the electrical constitution of matter, and many of the basic relations of this theory have been obtained from studies of electricity and magnetism. It is appropriate then to consider a factual background against which the more recent developments may stand in sharp relief.

The first knowledge of these matters was obtained by the Greeks, who observed that rubbed amber attracts light particles (600 B.C.) and that certain iron ores attract pieces of iron (425 B.C.). Apparently the Greeks made no use of this information, but fourteen hundred years later (A.D. 1000) European navigators were using suspended magnets or compasses for guidance of their ships. About two hundred and fifty years later, we find Peter Peregrinus (1269) shaping a magnet out of lodestone into the form of a sphere. By placing a needle at various points on the sphere and drawing lines tangent to the needle, Peregrinus found these lines converging like meridians toward two points on the sphere's surface, which he called poles.

The observations of Peregrinus were extended by William Gilbert, physician to Queen Elizabeth. In his book *De Magnete*, published in 1600, he explains that he formed a spherical magnet of lodestone, drew meridians upon it which converged at the poles, and noted that it was a counterpart of the earth. Gilbert floated two magnets and was able to show that the poles differ from each other and that like poles repel and unlike attract each other.

Gilbert is known mainly for his work on magnetism, but he contributed a great deal to the knowledge of the behavior of electrified bodies by showing that many other bodies are like amber in that they too have the ability, when rubbed, to attract light bodies to themselves. Gilbert called these bodies *electrics* after *elektron*, which is the Greek word for amber. Certain other bodies which failed to exhibit the property of amber he called *nonelectrics*. In 1640, von Guericke, the mayor of Magdeburg, discovered that electrified bodies may exhibit repulsion as well as attraction.

The true character of Gilbert's nonelectrics was not understood until Stephen Gray (1729) succeeded in transmitting the "Electrick Vertue" from a glass rod to another body placed at some distance by connecting the two bodies with a nonelectric. In modern terminology, electrics are insulators and nonelectrics are conductors of electricity. Gray's experiments yielded one other result of importance. By electrifying two wooden cubes, one hollow and the other solid, he showed that they produced the same effects and concluded that the electrification is identified with the surfaces and not the volumes.

The two hundred years that followed Gray's experiments were extremely productive, both in new discoveries and in the development of significant theory correlating the observations. The first quarter of this period was devoted to the further accumulation of experimental information.

Du Fay's work (1733), which was in many respects merely a repetition of Gray's experiments, is significant in that he recognized and stated his belief in the existence of two kinds of electricity, vitreous and resinous, or + and −, as suggested later by Franklin. Du Fay noticed further that like electricities repel whereas unlike kinds attract each other, and he showed that conductors as well as insulators may, with proper handling, be made to acquire electric charge.

Von Kleist and von Musschenbroek (1745 and 1746) independently discovered that electricity from a charged sphere could be stored in a bottle if it contained copper wire or mercury. Shortly Gralatt improved the apparatus by placing metal on the outside as well as in. This soon became known as the Leyden jar and is identical with the condenser of the present day. The Leyden jar permitted a much larger concentration of electricity and greatly intensified the spark of discharge, creating a very considerable interest.

Franklin's famous kite experiment was undertaken in 1752. He had previously noted the facility with which a metallic point will discharge an electrified conductor, and he proposed and carried out an experiment to "draw off the fire from the cloud." Franklin's work brought him world-wide fame and no doubt helped to reduce the feeling of superstitious awe held by many with respect to lightning, though it brought denunciation from many pulpits for interfering with the wrath of God.

Up to Franklin's time, studies in magnetism and electrostatics had been entirely *qualitative*, but in 1785 these studies were placed on a *quantitative* basis by Coulomb's work with the torsion balance. Coulomb showed by direct experiment that electric charges (also magnetic poles) exert forces on each other that vary as the inverse square of the distance. Michel and Priestley had earlier inferred the inverse square law, but the quantitative study of electricity and magnetism must be dated from Coulomb's experiments. This step was particularly significant in that it made possible the application of the mathematics developed after Newton's enunciation (in about 1680) of the inverse square law as applied to gravitational phenomena.

During the latter part of the eighteenth century a great deal of attention was paid to Galvani's experiments on "animal electricity." Galvani found that a muscle of a frog's leg touched by a piece of metal twitched violently every time an electrostatic machine placed nearby was discharged. But presently he discovered that the effect could be produced merely by touching the muscle nerve at two points with a piece of metal. Opinion was divided as to the real source of the effect; some thought with Galvani that the effect was electrical in nature, much like the discharge of a Leyden jaw, others that he had discovered a new fluid of nerve origin. But Volta believed the real cause lay in the metals used.

In 1800 Volta found the means for greatly increasing the strength of the effects. He made the first battery or "voltaic pile" by placing disks of copper, zinc, and moistened pasteboard in series and found that, upon closing the circuit, this arrangement gave a *continuous* electrical effect, whereas the Leyden jar had to be re-electrified after each discharge. Thus the nineteenth century opened with a clear recognition of the existence of the electric current and possessed a method for its generation.

Later in the same year Nicholson and Carlisle connected a voltaic pile to platinum wires immersed in water and showed the decomposition of water into hydrogen and oxygen. During the next few years, Humphry Davy, Grothuss, and Berzelius sought and discovered numerous relations between electricity and chemistry. In particular, Davy passed a current through certain salts and soon discovered potassium and sodium. Berzelius is known especially for his ionic theory that every compound is composed of two parts with opposite electrifications and that chemical combination results when oppositely charged ions unite, giving a stable compound in which the electricities are neutralized.

Meanwhile Poisson (1812) gave impetus to the application of mathematics by his development of the concept of the potential as applied to electric and magnetic fields separately. Shortly after this, observations were made clearly connecting the hitherto unrelated facts of electricity and magnetism. Believing intuitively in this relation, Oersted in 1820 more or less accidentally noticed the deflection of a compass placed near an electric current and discovered the converse, the deflection of an electric current by a magnet.

Spurred on by this discovery, Ampère almost immediately devised simple experiments which demonstrated the force relations between two circuits carrying currents, and showed that the magnetic effects of a current and those of a permanent magnet are similar in nature. Fifty years later James Clerk Maxwell, appreciating the significance of Ampère's work, referred to him as the "Newton of Electricity."

The years 1821–25 resulted in three more important contributions: the invention of the first direct-current motor by Faraday (1821), who produced continuous motion of a wire carrying current about a magnetic pole; the first demonstration of current induction by Arago (1824), who showed that a disk of copper revolving beneath a compass placed coaxially causes the later to rotate; and the construction of the first electromagnet by Sturgeon (1825).

The year 1827 saw the formulation of Ohm's law ($V = IR$), which was the first statement of the existence of a quantitative relation between the electric charge movement or current flow, the strength of the battery (voltage or potential), and the reaction coefficient of the circuit itself (resistance). The results of Ohm's experiments relating to the conditions which govern the presence and quantity of current and of Ampère's investigations of its magnetic and force effects form the foundations for the applications of electricity and magnetism to direct-current engineering.

Far-reaching as these results proved to be, we have still to consider the much more significant contributions of Michael Faraday, who even now stands out as the greatest experimental scientist. Hundreds of investigations are described in his *Experimental Researches in Electricity*. The most important perhaps are those in which he set out to find, and did discover (1831), that an electric current changing in time induces an electromotive force in nearby conductors properly placed. This discovery made possible present-day alternating-current applications, though fifty years passed before any use was developed.

We are indebted to Faraday for the next great stride. Before his time experimenters focused their attention on what we might call the end effects—the electricity and the magnetism which cause certain phenomena and the phenomena themselves. Faraday intuitively felt that the intervening space or substance subjected to electric or magnetic influence was the region that would repay careful study. Experimenting with condensers, he found that their capacities were altered by the substitution of other materials for air, and he devised a most convenient system of tabulating substances according to their specific inductive capacities or dielectric constants.

Very soon Sir William Thomson (later Lord Kelvin) began his distinguished career as a physicist, which resulted in valuable contributions to the study and clarification of ideas concerning electricity. He early rescued from oblivion an essay on the potential function of mathematics written by George Green in 1828, and by giving a simple interpretation of the concept of potential developed one of the most powerful theoretical tools of applied mathematics.

Kelvin did more than any other to emphasize the importance of accurate measurement of physical observations, thus striking the keynote of modern physical methods. Measurement calls for units of reference. Kelvin was largely responsible for devising the CGS electrostatic and electromagnetic systems of units based on the point electric charge and point magnetic pole respectively. In 1856 Weber and Kohlrausch found that the ratio between the electrostatic and electromagnetic units of measurement was 3×10^{10}, a number very close to the velocity of light, measured in centimeters per second. The explanation of this result was not made until Maxwell's work appeared.

Toward the middle of the nineteenth century, scientists came to realize more and more clearly that all physical phenomena have a common underlying physical quantity—energy. The facts and ideas concerning energy and its transformations are summed up in the theory of thermodynamics, so called because the main outlines were obtained through the study of heat phenomena. Joule showed in 1845 that the expenditure of a definite amount of mechanical energy always results in the production of a definite number of heat units. In 1847 Helmholtz wrote a memorable paper on the "conservation of energy."

Five years later Kelvin and Clausius developed the foundations of thermodynamics, summed up in the well-known first and second laws of thermodynamics, which may be stated rather loosely to be:

First Law. Accompanying any physical change, energy may be transformed from one state to another but without any change in its total amount. (This is the law of the conservation of energy.)

Second Law. Whenever energy is transformed from one state to another a portion appears in the form of heat. (This is the law of the degradation of energy. According to this view, heat is the lowest form of energy, to which in time all energy will degenerate.)

Thus there was developed the "doctrine of energy," which underlies modern theories.

The foundations of the "classical theory" had been laid, and it was time for one to arise who could gather together the isolated facts of experiment and could weld the parts into a unified structure by the use of mathematical tools. James Clerk Maxwell (1873) performed this task so thoroughly that for many years it was felt that there remained little more to be done. Maxwell's contribution was more than a mere coordination of the results obtained by others. By developing Faraday's suggestion that the region surrounding the electricity or current is of great importance Maxwell was led to make as assumption that has since proved to be a forward step of paramount importance, namely, that even in free space an electric field changing in time must be regarded as a "displacement electric current." The incorporation of this assumption into the statement of the previous theory led him to a set of relations now known as Maxwell's equations, which assert that electric and magnetic fields, if changing in time, generate each other and by their combined action propagate energy as waves out into space with the velocity of light. Thereafter it was but a short step to the assertion that light is nothing but an electromagnetic wave phenomenon. Common enough today, the idea was a veritable bombshell to physicists of the nineteenth century.

Such revolutionary ideas usually receive slow credence. But fifteen years later (1888) Hertz was able to verify Maxwell's prediction as to the electromagnetic wave character of light by discharging an induction coil through a sphere gap which emitted waves out into space. These waves, reaching the other end of Hertz's laboratory, caused a series of sparks to jump across a spark gap placed in a metallic ring. By 1901, Marconi, using these waves, was successfully transmitting messages by wireless. Thus Maxwell's work must ever remain one of the great contributions to science—a triumph for theoretical investigation.

Many scientists naïvely believed that the main outlines of science had been firmly established by the development and verification of the classical theory, but they were rudely shaken from their complacence by the advent of a new series of experiments, which resulted in the discovery of X rays by Röntgen in 1895, of new types of emanations from salts by Becquerel in 1896, of radium by Mme. Curie in 1898. These discoveries together with many new facts acquired through the study of the conduction of electricity through gases at low pressures indicated the existence of phenomena that could not be explained by the classical theory. Maxwell's contribution was not overthrown, however; it must be regarded as incomplete rather than incorrect. He treated the behavior of large groups of atoms, and developed laws which describe the group action under certain conditions. Such laws must be regarded as the "statistical averages" of the operations of the group composed of many individuals and cannot be applied to the individuals themselves. New methods of study and new points of view were found to be necessary for consideration of the individual particle, and since the advent of the twentieth century many physicists have made this subject the focus of their research. They have gathered many new experimental facts and have developed new theories to account for them, and their work has produced a picture of the atom which shows that it is composed of the basic electrical elements, positive and negative charges.

As long ago as the Grecian period, philosophers speculated upon the possibility that all substances might be divisible into ultimate units or atoms. Not, however, until the development of chemistry do we find the concept of the atom based upon an experimental rather than a speculative foundation. Beginning with the superstitious searchings of the alchemists, centuries of painstaking study and research passed before the supernatural was eliminated and the culmination of the chemical approach to the mystery of the atom was given in the Periodic Table of the Elements by Mendeleef in 1870. After arranging the elements according to their relative atomic weights, Mendeleef formed the list of atoms into seven groups in each of which similar chemical and physical properties are found. Experiments in chemistry, though extremely valuable, yield information about the interactions between atoms rather than knowledge of the inner electrical nature of the atom. The latter problem has been approached only by the methods of the modern atomic physicist.

As early as 1833, Faraday was forced to suggest, though reluctantly, that electricity might be atomic in nature. He was led to this assumption by his studies in electrolysis, which showed him that a definite quantity of electricity passing through an electrolyte deposits out of solution a definite quantity of the material forming the electrolyte. Further investigations in this field brought many new facts to light, but an even more fruitful source of knowledge was found in the study of the passage of electricity through gases at low pressures.

Though Watson began work along this line in 1752, no striking results were obtained until 1878, when Crookes, aided by Geissler's new type of mercury-air pump giving a higher vacuum, produced his very striking display of electrical discharges through gas at a low pressure, which are called "cathode rays." But it is really to Sir J. J. Thomson that we are most indebted for our knowledge of the "cathode ray." The end result of this series of study and experiment (1894) was the definite assurance that the "cathode rays" are particles of negative electricity (called electrons) which possess mass and therefore inertia, which travel along paths that are bent in the presence of either electric or magnetic fields, and which travel with velocities that may approach the speed of light. Thomson's experiments gave in addition the ratio of the electric charge to the mass of the particles. In 1917, Millikan had completed the work of measuring the charge of these particles so that it then became possible to evaluate the mass accurately.

Sir Ernest Rutherford (1899) began a series of investigations of radioactive elements which were known to disintegrate, and with the help of others showed that the emanations are of three kinds: α-particles, relatively slow moving and positively charged, later shown to be helium nuclei or aggregates of two positive units and two neutrons; β-particles (electrons), swift moving and negative; and γ-rays, which were shown to be energy emanations similar to X rays, but shorter in wave length, moving with the speed of light. These emanations proved to be the same no matter what element served as their source. In the face of this evidence it would be hard to deny that all atoms are composed of three elemental units: the positive charge, the electron or negative charge, and neutrons bound together by forces resulting from the energy of their contiguous positions and motions.

Since the opening of the twentieth century, physical research in both experimental and theoretical fields has leaped forward at a pace so rapid that it has been difficult to assimilate and coordinate the results as swiftly as they have appeared. The number of workers as well as organizations interested in research has increased tremendously in the last few years. Many of the great industrial concerns have opened departments of research and provided sums of money for investigation so that great promise may be held out for an even more productive future.

Early in the century Planck found that Maxwell's classical relations failed to explain the experimental facts of radiant energy. Hitherto it had been assumed that light or heat is propagated through space in the form of continuous waves. Planck's hypothesis, forming the basis of the quantum theory, assumed that radiant energy is emitted from the originating source in "chunks" or "quanta." The elementary quantum is not a fixed amount of energy but follows the rule that Planck's constant (h) multiplied by the frequency of the emitted wave gives the size of the unit. Since frequency is the reciprocal of time, h is of the order of energy-time or action, a concept which has proved to be of a more fundamental nature than energy. Planck's constant h has been shown to be equal to 6.624×10^{-34} joule-second. Here then is the foundation for the scientists' belief that action as well as electricity and matter is atomic in nature. But there are difficulties, for even today the quantum idea has not been able to explain the fact of the interference of light, which is most adequately satisfied by the old wave theory arising from Maxwell's relations. Further studies on specific heats at low temperatures and on photoelectricity are apparently consistent only with a quantum hypothesis. We may expect the quantum idea to remain even though the details may be altered.

This century has been the age of startling developments, for in 1905 Einstein brought forward his special relativity theory, which proved to be very upsetting because he asserted that the ideas of space and time coming down from Newton's day could not be entirely true but could be at best only good approximations. He asserted that space and time could not be absolute and independent but must be curiously intermixed so that any physical event can be described only in terms of the relative constant motion of the observer. To an observer standing upon the earth viewing a train moving relatively to him with a constant velocity, lengths must appear shortened and time move more slowly than to a second observer placed aboard the train. Velocities may not be added directly, nor is mass a constant quantity. A mass m_0 as seen by an observer traveling with it, appears larger to another observer moving with a constant relative velocity v, in this way

$$m_v = \frac{m_0}{\sqrt{1 - \left(\frac{v}{c}\right)^2}} \tag{III.1}$$

where c is the velocity of light in free space. At the velocity of light, the mass m_v would be infinite. In this theory, mass becomes interchangeable with energy through the relationships:

$$U_0 = m_0c^2 \tag{III.2}$$

$$U_v = m_vc^2 = \frac{m_0c^2}{\sqrt{1 - \left(\frac{v}{c}\right)^2}} \tag{III.3}$$

so that infinite energy would result from a mass moving with the velocity of light. This is of course a limiting case never quite reached. Before this time, Kaufman and Bucherer experimented with β-electrons ejected by radioactive substances, which travel with a velocity of about 98 per cent of that of light, and found that their ratio of charge to mass varied with the velocity. Strangely enough, if the charge is assumed constant, the variation in mass which they found is that given above by Einstein.

Ten years later (1915) Einstein extended his ideas to cover accelerated relative motion and developed the general relativity theory, in which gravitation appears as a consequence of the distortion of space-time in the neighborhood of material bodies. In the period between Newton and Einstein, gravitation had successfully defied every attempt to correlate it with other natural phenomena.

To an increasing extent throughout this century physicists have been developing spectrum analysis into the most powerful laboratory tool yet discovered for prying into the mystery of the constitution of the atom. Side by side, checking each other at every step, the speculations of the mathematical physicist and the results of laboratory investigation largely given by spectral studies are forcing the atomic riddle of today to become the commonplace of tomorrow.

As the evidence increased, it was natural that attempts should be made to construct an atom model which would give a true and usable image of the atom itself. In 1913 Rutherford and Bohr proposed the planetary atom model, in which electrons rotate in orbits about a nuclear sun. This model is described in Appendix II. Unfortunately the atom model, while easy to visualize, fails to predict the correct spectral wave lengths, and a wave-mechanical type of atom has taken its place, at least where accurate calculations of atomic magnitudes are required. The wave-mechanical atom was developed originally by De Broglie and Schrödinger in 1926.

Since then many workers have used their methods so that present-day knowledge concerning the electronic shells of the atom is fairly complete.

Index

CATALOGUE OF DOVER BOOKS

PHYSICS

General physics

FOUNDATIONS OF PHYSICS, R. B. Lindsay & H. Margenau. Excellent bridge between semi-popular works & technical treatises. A discussion of methods of physical description, construction of theory; valuable for physicist with elementary calculus who is interested in ideas that give meaning to data, tools of modern physics. Contents include symbolism, mathematical equations; space & time foundations of mechanics; probability; physics & continua; electron theory; special & general relativity; quantum mechanics; causality. "Thorough and yet not overdetailed. Unreservedly recommended," NATURE (London). Unabridged, corrected edition. List of recommended readings. 35 illustrations. xi + 537pp. 5⅜ x 8.
S377 Paperbound **$2.75**

FUNDAMENTAL FORMULAS OF PHYSICS, ed. by D. H. Menzel. Highly useful, fully inexpensive reference and study text, ranging from simple to highly sophisticated operations. Mathematics integrated into text—each chapter stands as short textbook of field represented. Vol. 1: Statistics, Physical Constants, Special Theory of Relativity, Hydrodynamics, Aerodynamics, Boundary Value Problems in Math. Physics; Viscosity, Electromagnetic Theory, etc. Vol. 2: Sound, Acoustics, Geometrical Optics, Electron Optics, High-Energy Phenomena, Magnetism, Biophysics, much more. Index. Total of 800pp. 5⅜ x 8. Vol. 1 S595 Paperbound **$2.00**
Vol. 2 S596 Paperbound **$2.00**

MATHEMATICAL PHYSICS, D. H. Menzel. Thorough one-volume treatment of the mathematical techniques vital for classic mechanics, electromagnetic theory, quantum theory, and relativity. Written by the Harvard Professor of Astrophysics for junior, senior, and graduate courses, it gives clear explanations of all those aspects of function theory, vectors, matrices, dyadics, tensors, partial differential equations, etc., necessary for the understanding of the various physical theories. Electron theory, relativity, and other topics seldom presented appear here in considerable detail. Scores of definitions, conversion factors, dimensional constants, etc. "More detailed than normal for an advanced text . . . excellent set of sections on Dyadics, Matrices, and Tensors," JOURNAL OF THE FRANKLIN INSTITUTE. Index. 193 problems, with answers. x + 412pp. 5⅜ x 8. S56 Paperbound **$2.00**

THE SCIENTIFIC PAPERS OF J. WILLARD GIBBS. All the published papers of America's outstanding theoretical scientist (except for "Statistical Mechanics" and "Vector Analysis"). Vol I (thermodynamics) contains one of the most brilliant of all 19th-century scientific papers—the 300-page "On the Equilibrium of Heterogeneous Substances," which founded the science of physical chemistry, and clearly stated a number of highly important natural laws for the first time; 8 other papers complete the first volume. Vol II includes 2 papers on dynamics, 8 on vector analysis and multiple algebra, 5 on the electromagnetic theory of light, and 6 miscellaneous papers. Biographical sketch by H. A. Bumstead. Total of xxxvi + 718pp. 5⅝ x 8⅜.
S721 Vol I Paperbound **$2.50**
S722 Vol II Paperbound **$2.00**
The set **$4.50**

BASIC THEORIES OF PHYSICS, Peter Gabriel Bergmann. Two-volume set which presents a critical examination of important topics in the major subdivisions of classical and modern physics. The first volume is concerned with classical mechanics and electrodynamics: mechanics of mass points, analytical mechanics, matter in bulk, electrostatics and magnetostatics, electromagnetic interaction, the field waves, special relativity, and waves. The second volume (Heat and Quanta) contains discussions of the kinetic hypothesis, physics and statistics, stationary ensembles, laws of thermodynamics, early quantum theories, atomic spectra, probability waves, quantization in wave mechanics, approximation methods, and abstract quantum theory. A valuable supplement to any thorough course or text.
Heat and Quanta: Index. 8 figures. x + 300pp. 5⅜ x 8½. S968 Paperbound **$2.00**
Mechanics and Electrodynamics: Index. 14 figures. vii + 280pp. 5⅜ x 8½.
S969 Paperbound **$1.75**

THEORETICAL PHYSICS, A. S. Kompaneyets. One of the very few thorough studies of the subject in this price range. Provides advanced students with a comprehensive theoretical background. Especially strong on recent experimentation and developments in quantum theory. Contents: Mechanics (Generalized Coordinates, Lagrange's Equation, Collision of Particles, etc.), Electrodynamics (Vector Analysis, Maxwell's equations, Transmission of Signals, Theory of Relativity, etc.), Quantum Mechanics (the Inadequacy of Classical Mechanics, the Wave Equation, Motion in a Central Field, Quantum Theory of Radiation, Quantum Theories of Dispersion and Scattering, etc.), and Statistical Physics (Equilibrium Distribution of Molecules in an Ideal Gas, Boltzmann statistics, Bose and Fermi Distribution, Thermodynamic Quantities, etc.). Revised to 1961. Translated by George Yankovsky, authorized by Kompaneyets. 137 exercises. 56 figures. 529pp. 5⅜ x 8½. S972 Paperbound **$2.50**

ANALYTICAL AND CANONICAL FORMALISM IN PHYSICS, André Mercier. A survey, in one volume, of the variational principles (the key principles—in mathematical form—from which the basic laws of any one branch of physics can be derived) of the several branches of physical theory, together with an examination of the relationships among them. Contents: the Lagrangian Formalism, Lagrangian Densities, Canonical Formalism, Canonical Form of Electrodynamics, Hamiltonian Densities, Transformations, and Canonical Form with Vanishing Jacobian Determinant. Numerous examples and exercises. For advanced students, teachers, etc. 6 figures. Index. viii + 222pp. 5⅜ x 8½. S1077 Paperbound **$1.75**

Acoustics, optics, electricity and magnetism, electromagnetics, magnetohydrodynamics

THE THEORY OF SOUND, Lord Rayleigh. Most vibrating systems likely to be encountered in practice can be tackled successfully by the methods set forth by the great Nobel laureate, Lord Rayleigh. Complete coverage of experimental, mathematical aspects of sound theory. Partial contents: Harmonic motions, vibrating systems in general, lateral vibrations of bars, curved plates or shells, applications of Laplace's functions to acoustical problems, fluid friction, plane vortex-sheet, vibrations of solid bodies, etc. This is the first inexpensive edition of this great reference and study work. Bibliography. Historical introduction by R. B. Lindsay. Total of 1040pp. 97 figures. 5⅜ x 8.
S292, S293, Two volume set, paperbound, **$4.70**

THE DYNAMICAL THEORY OF SOUND, H. Lamb. Comprehensive mathematical treatment of the physical aspects of sound, covering the theory of vibrations, the general theory of sound, and the equations of motion of strings, bars, membranes, pipes, and resonators. Includes chapters on plane, spherical, and simple harmonic waves, and the Helmholtz Theory of Audition. Complete and self-contained development for student and specialist; all fundamental differential equations solved completely. Specific mathematical details for such important phenomena as harmonics, normal modes, forced vibrations of strings, theory of reed pipes, etc. Index. Bibliography. 86 diagrams. viii + 307pp. 5⅜ x 8. S655 Paperbound **$1.50**

WAVE PROPAGATION IN PERIODIC STRUCTURES, L. Brillouin. A general method and application to different problems: pure physics, such as scattering of X-rays of crystals, thermal vibration in crystal lattices, electronic motion in metals; and also problems of electrical engineering. Partial contents: elastic waves in 1-dimensional lattices of point masses. Propagation of waves along 1-dimensional lattices. Energy flow. 2 dimensional, 3 dimensional lattices. Mathieu's equation. Matrices and propagation of waves along an electric line. Continuous electric lines. 131 illustrations. Bibliography. Index. xii + 253pp. 5⅜ x 8.
S34 Paperbound **$2.00**

THEORY OF VIBRATIONS, N. W. McLachlan. Based on an exceptionally successful graduate course given at Brown University, this discusses linear systems having 1 degree of freedom, forced vibrations of simple linear systems, vibration of flexible strings, transverse vibrations of bars and tubes, transverse vibration of circular plate, sound waves of finite amplitude, etc. Index. 99 diagrams. 160pp. 5⅜ x 8. S190 Paperbound **$1.35**

LIGHT: PRINCIPLES AND EXPERIMENTS, George S. Monk. Covers theory, experimentation, and research. Intended for students with some background in general physics and elementary calculus. Three main divisions: 1) Eight chapters on geometrical optics—fundamental concepts (the ray and its optical length, Fermat's principle, etc.), laws of image formation, apertures in optical systems, photometry, optical instruments etc.; 2) 9 chapters on physical optics—interference, diffraction, polarization, spectra, the Rayleigh refractometer, the wave theory of light, etc.; 3) 23 instructive experiments based directly on the theoretical text. "Probably the best intermediate textbook on light in the English language. Certainly, it is the best book which includes both geometrical and physical optics," J. Rud Nielson, PHYSICS FORUM. Revised edition. 102 problems and answers. 12 appendices. 6 tables. Index. 270 illustrations. xi +489pp. 5⅜ x 8½. S341 Paperbound **$2.50**

PHOTOMETRY, John W. T. Walsh. The best treatment of both "bench" and "illumination" photometry in English by one of Britain's foremost experts in the field (President of the International Commission on Illumination). Limited to those matters, theoretical and practical, which affect the measurement of light flux, candlepower, illumination, etc., and excludes treatment of the use to which such measurements may be put after they have been made. Chapters on Radiation, The Eye and Vision, Photo-Electric Cells, The Principles of Photometry, The Measurement of Luminous Intensity, Colorimetry, Spectrophotometry, Stellar Photometry, The Photometric Laboratory, etc. Third revised (1958) edition. 281 illustrations. 10 appendices. xxiv + 544pp. 5½ x 9¼. S319 Clothbound **$10.00**

EXPERIMENTAL SPECTROSCOPY, R. A. Sawyer. Clear discussion of prism and grating spectrographs and the techniques of their use in research, with emphasis on those principles and techniques that are fundamental to practically all uses of spectroscopic equipment. Beginning with a brief history of spectroscopy, the author covers such topics as light sources, spectroscopic apparatus, prism spectroscopes and graphs, diffraction grating, the photographic process, determination of wave length, spectral intensity, infrared spectroscopy, spectrochemical analysis, etc. This revised edition contains new material on the production of replica gratings, solar spectroscopy from rockets, new standard of wave length, etc. Index. Bibliography. 111 illustrations. x + 358pp. 5⅜ x 8½. S1045 Paperbound **$2.25**

FUNDAMENTALS OF ELECTRICITY AND MAGNETISM, L. B. Loeb. For students of physics, chemistry, or engineering who want an introduction to electricity and magnetism on a higher level and in more detail than general elementary physics texts provide. Only elementary differential and integral calculus is assumed. Physical laws developed logically, from magnetism to electric currents, Ohm's law, electrolysis, and on to static electricity, induction, etc. Covers an unusual amount of material; one third of book on modern material: solution of wave equation, photoelectric and thermionic effects, etc. Complete statement of the various electrical systems of units and interrelations. 2 Indexes. 75 pages of problems with answers stated. Over 300 figures and diagrams. xix +669pp. 5⅜ x 8. S745 Paperbound **$2.75**

MATHEMATICAL ANALYSIS OF ELECTRICAL AND OPTICAL WAVE-MOTION, Harry Bateman. Written by one of this century's most distinguished mathematical physicists, this is a practical introduction to those developments of Maxwell's electromagnetic theory which are directly connected with the solution of the partial differential equation of wave motion. Methods of solving wave-equation, polar-cylindrical coordinates, diffraction, transformation of coordinates, homogeneous solutions, electromagnetic fields with moving singularities, etc. Index. 168pp. 5⅜ x 8. S14 Paperbound **$1.75**

PRINCIPLES OF PHYSICAL OPTICS, Ernst Mach. This classical examination of the propagation of light, color, polarization, etc. offers an historical and philosophical treatment that has never been surpassed for breadth and easy readability. Contents: Rectilinear propagation of light. Reflection, refraction. Early knowledge of vision. Dioptrics. Composition of light. Theory of color and dispersion. Periodicity. Theory of interference. Polarization. Mathematical representation of properties of light. Propagation of waves, etc. 279 illustrations, 10 portraits. Appendix. Indexes. 324pp. 5⅜ x 8. S178 Paperbound **$2.00**

THE THEORY OF OPTICS, Paul Drude. One of finest fundamental texts in physical optics, classic offers thorough coverage, complete mathematical treatment of basic ideas. Includes fullest treatment of application of thermodynamics to optics; sine law in formation of images, transparent crystals, magnetically active substances, velocity of light, apertures, effects depending upon them, polarization, optical instruments, etc. Introduction by A. A. Michelson. Index. 110 illus. 567pp. 5⅜ x 8. S532 Paperbound **$2.45**

ELECTRICAL THEORY ON THE GIORGI SYSTEM, P. Cornelius. A new clarification of the fundamental concepts of electricity and magnetism, advocating the convenient m.k.s. system of units that is steadily gaining followers in the sciences. Illustrating the use and effectiveness of his terminology with numerous applications to concrete technical problems, the author here expounds the famous Giorgi system of electrical physics. His lucid presentation and well-reasoned, cogent argument for the universal adoption of this system form one of the finest pieces of scientific exposition in recent years. 28 figures. Index. Conversion tables for translating earlier data into modern units. Translated from 3rd Dutch edition by L. J. Jolley. x + 187pp. 5½ x 8¾. S909 Clothbound **$6.00**

ELECTRIC WAVES: BEING RESEARCHES ON THE PROPAGATION OF ELECTRIC ACTION WITH FINITE VELOCITY THROUGH SPACE, Heinrich Hertz. This classic work brings together the original papers in which Hertz—Helmholtz's protegé and one of the most brilliant figures in 19th-century research—probed the existence of electromagnetic waves and showed experimentally that their velocity equalled that of light, research that helped lay the groundwork for the development of radio, television, telephone, telegraph, and other modern technological marvels. Unabridged republication of original edition. Authorized translation by D. E. Jones. Preface by Lord Kelvin. Index of names. 40 illustrations. xvii + 278pp. 5⅜ x 8½. S57 Paperbound **$1.75**

PIEZOELECTRICITY: AN INTRODUCTION TO THE THEORY AND APPLICATIONS OF ELECTROMECHANICAL PHENOMENA IN CRYSTALS, Walter G. Cady. This is the most complete and systematic coverage of this important field in print—now regarded as something of scientific classic. This republication, revised and corrected by Prof. Cady—one of the foremost contributors in this area—contains a sketch of recent progress and new material on Ferroelectrics. Time Standards, etc. The first 7 chapters deal with fundamental theory of crystal electricity. 5 important chapters cover basic concepts of piezoelectricity, including comparisons of various competing theories in the field. Also discussed: piezoelectric resonators (theory, methods of manufacture, influences of air-gaps, etc.); the piezo oscillator; the properties, history, and observations relating to Rochelle salt; ferroelectric crystals; miscellaneous applications of piezoelectricity; pyroelectricity; etc. "A great work," W. A. Wooster, NATURE. Revised (1963) and corrected edition. New preface by Prof. Cady. 2 Appendices. Indices. Illustrations. 62 tables. Bibliography. Problems. Total of 1 + 822pp. 5⅜ x 8½.
S1094 Vol. I Paperbound **$2.50**
S1095 Vol. II Paperbound **$2.50**
Two volume set Paperbound **$5.00**

MAGNETISM AND VERY LOW TEMPERATURES, H. B. G. Casimir. A basic work in the literature of low temperature physics. Presents a concise survey of fundamental theoretical principles, and also points out promising lines of investigation. Contents: Classical Theory and Experimental Methods, Quantum Theory of Paramagnetism, Experiments on Adiabatic Demagnetization. Theoretical Discussion of Paramagnetism at Very Low Temperatures, Some Experimental Results, Relaxation Phenomena. Index. 89-item bibliography. ix + 95pp. 5⅜ x 8. S943 Paperbound **$1.25**

SELECTED PAPERS ON NEW TECHNIQUES FOR ENERGY CONVERSION: THERMOELECTRIC METHODS; THERMIONIC; PHOTOVOLTAIC AND ELECTRICAL EFFECTS; FUSION, Edited by Sumner N. Levine. Brings together in one volume the most important papers (1954-1961) in modern energy technology. Included among the 37 papers are general and qualitative descriptions of the field as a whole, indicating promising lines of research. Also: 15 papers on thermoelectric methods, 7 on thermionic, 5 on photovoltaic, 4 on electrochemical effect, and 2 on controlled fusion research. Among the contributors are: Joffe, Maria Telkes, Herold, Herring, Douglas, Jaumot, Post, Austin, Wilson, Pfann, Rappaport, Morehouse, Domenicali, Moss, Bowers, Harman, Von Doenhoef. Preface and introduction by the editor. Bibliographies. xxviii + 451pp. 6⅛ x 9¼. S37 Paperbound **$3.00**

SUPERFLUIDS: MACROSCOPIC THEORY OF SUPERCONDUCTIVITY, Vol. I, Fritz London. The major work by one of the founders and great theoreticians of modern quantum physics. Consolidates the researches that led to the present understanding of the nature of superconductivity. Prof. London here reveals that quantum mechanics is operative on the macroscopic plane as well as the submolecular level. Contents: Properties of Superconductors and Their Thermodynamical Correlation; Electrodynamics of the Pure Superconducting State; Relation between Current and Field; Measurements of the Penetration Depth; Non-Viscous Flow vs. Superconductivity; Micro-waves in Superconductors; Reality of the Domain Structure; and many other related topics. A new epilogue by M. J. Buckingham discusses developments in the field up to 1960. Corrected and expanded edition. An appreciation of the author's life and work by L. W. Nordheim. Biography by Edith London. Bibliography of his publications. 45 figures. 2 Indices. xviii + 173pp. 5⅝ x 8⅜. S44 Paperbound **$1.45**

SELECTED PAPERS ON PHYSICAL PROCESSES IN IONIZED PLASMAS, Edited by Donald H. Menzel, Director, Harvard College Observatory. 30 important papers relating to the study of highly ionized gases or plasmas selected by a foremost contributor in the field, with the assistance of Dr. L. H. Aller. The essays include 18 on the physical processes in gaseous nebulae, covering problems of radiation and radiative transfer, the Balmer decrement, electron temperatives, spectrophotometry, etc. 10 papers deal with the interpretation of nebular spectra, by Bohm, Van Vleck, Aller, Minkowski, etc. There is also a discussion of the intensities of "forbidden" spectral lines by George Shortley and a paper concerning the theory of hydrogenic spectra by Menzel and Pekeris. Other contributors: Goldberg, Hebb, Baker, Bowen, Ufford, Liller, etc. viii + 374pp. 6⅛ x 9¼. S60 Paperbound **$2.95**

THE ELECTROMAGNETIC FIELD, Max Mason & Warren Weaver. Used constantly by graduate engineers. Vector methods exclusively: detailed treatment of electrostatics, expansion methods, with tables converting any quantity into absolute electromagnetic, absolute electrostatic, practical units. Discrete charges, ponderable bodies, Maxwell field equations, etc. Introduction. Indexes. 416pp. 5⅜ x 8. S185 Paperbound **$2.00**

THEORY OF ELECTRONS AND ITS APPLICATION TO THE PHENOMENA OF LIGHT AND RADIANT HEAT, H. Lorentz. Lectures delivered at Columbia University by Nobel laureate Lorentz. Unabridged, they form a historical coverage of the theory of free electrons, motion, absorption of heat, Zeeman effect, propagation of light in molecular bodies, inverse Zeeman effect, optical phenomena in moving bodies, etc. 109 pages of notes explain the more advanced sections. Index. 9 figures. 352pp. 5⅜ x 8. S173 Paperbound **$1.85**

FUNDAMENTAL ELECTROMAGNETIC THEORY, Ronold P. King, Professor Applied Physics, Harvard University. Original and valuable introduction to electromagnetic theory and to circuit theory from the standpoint of electromagnetic theory. Contents: Mathematical Description of Matter—stationary and nonstationary states; Mathematical Description of Space and of Simple Media—Field Equations, Integral Forms of Field Equations, Electromagnetic Force, etc.; Transformation of Field and Force Equations; Electromagnetic Waves in Unbounded Regions; Skin Effect and Internal Impedance—in a solid cylindrical conductor, etc.; and Electrical Circuits—Analytical Foundations, Near-zone and quasi-near zone circuits, Balanced two-wire and four-wire transmission lines. Revised and enlarged version. New preface by the author. 5 appendices (Differential operators: Vector Formulas and Identities, etc.). Problems. Indexes. Bibliography. xvi + 580pp. 5⅜ x 8½. S1023 Paperbound **$2.75**

Hydrodynamics

A TREATISE ON HYDRODYNAMICS, A. B. Basset. Favorite text on hydrodynamics for 2 generations of physicists, hydrodynamical engineers, oceanographers, ship designers, etc. Clear enough for the beginning student, and thorough source for graduate students and engineers on the work of d'Alembert, Euler, Laplace, Lagrange, Poisson, Green, Clebsch, Stokes, Cauchy, Helmholtz, J. J. Thomson, Love, Hicks, Greenhill, Besant, Lamb, etc. Great amount of documentation on entire theory of classical hydrodynamics. Vol I: theory of motion of frictionless liquids, vortex, and cyclic irrotational motion, etc. 132 exercises. Bibliography. 3 Appendixes. xii + 264pp. Vol II: motion in viscous liquids, harmonic analysis, theory of tides, etc. 112 exercises, Bibliography. 4 Appendixes. xv + 328pp. Two volume set. 5⅜ x 8.
S724 Vol I Paperbound **$1.75**
S725 Vol II Paperbound **$1.75**
The set **$3.50**

HYDRODYNAMICS, Horace Lamb. Internationally famous complete coverage of standard reference work on dynamics of liquids & gases. Fundamental theorems, equations, methods, solutions, background, for classical hydrodynamics. Chapters include Equations of Motion, Integration of Equations in Special Gases, Irrotational Motion, Motion of Liquid in 2 Dimensions, Motion of Solids through Liquid-Dynamical Theory, Vortex Motion, Tidal Waves, Surface Waves, Waves of Expansion, Viscosity, Rotating Masses of liquids. Excellently planned, arranged; clear, lucid presentation. 6th enlarged, revised edition. Index. Over 900 footnotes, mostly bibliographical. 119 figures. xv + 738pp. 6⅛ x 9¼. S256 Paperbound **$3.75**

HYDRODYNAMICS, H. Dryden, F. Murnaghan, Harry Bateman. Published by the National Research Council in 1932 this enormous volume offers a complete coverage of classical hydrodynamics. Encyclopedic in quality. Partial contents: physics of fluids, motion, turbulent flow, compressible fluids, motion in 1, 2, 3 dimensions; viscous fluids rotating, laminar motion, resistance of motion through viscous fluid, eddy viscosity, hydraulic flow in channels of various shapes, discharge of gases, flow past obstacles, etc. Bibliography of over 2,900 items. Indexes. 23 figures. 634pp. 5⅜ x 8. S303 Paperbound **$2.75**

Mechanics, dynamics, thermodynamics, elasticity

MECHANICS, J. P. Den Hartog. Already a classic among introductory texts, the M.I.T. professor's lively and discursive presentation is equally valuable as a beginner's text, an engineering student's refresher, or a practicing engineer's reference. Emphasis in this highly readable text is on illuminating fundamental principles and showing how they are embodied in a great number of real engineering and design problems: trusses, loaded cables, beams, jacks, hoists, etc. Provides advanced material on relative motion and gyroscopes not usual in introductory texts. "Very thoroughly recommended to all those anxious to improve their real understanding of the principles of mechanics." MECHANICAL WORLD. Index. List of equations. 334 problems, all with answers. Over 550 diagrams and drawings. ix + 462pp. 5⅜ x 8. S754 Paperbound **$2.00**

THEORETICAL MECHANICS: AN INTRODUCTION TO MATHEMATICAL PHYSICS, J. S. Ames, F. D. Murnaghan. A mathematically rigorous development of theoretical mechanics for the advanced student, with constant practical applications. Used in hundreds of advanced courses. An unusually thorough coverage of gyroscopic and baryscopic material, detailed analyses of the Coriolis acceleration, applications of Lagrange's equations, motion of the double pendulum, Hamilton-Jacobi partial differential equations, group velocity and dispersion, etc. Special relativity is also included. 159 problems. 44 figures. ix + 462pp. 5⅜ x 8. S461 Paperbound **$2.25**

THEORETICAL MECHANICS: STATICS AND THE DYNAMICS OF A PARTICLE, W. D. MacMillan. Used for over 3 decades as a self-contained and extremely comprehensive advanced undergraduate text in mathematical physics, physics, astronomy, and deeper foundations of engineering. Early sections require only a knowledge of geometry; later, a working knowledge of calculus. Hundreds of basic problems, including projectiles to the moon, escape velocity, harmonic motion, ballistics, falling bodies, transmission of power, stress and strain, elasticity, astronomical problems. 340 practice problems plus many fully worked out examples make it possible to test and extend principles developed in the text. 200 figures. xvii + 430pp. 5⅜ x 8. S467 Paperbound **$2.00**

THEORETICAL MECHANICS: THE THEORY OF THE POTENTIAL, W. D. MacMillan. A comprehensive, well balanced presentation of potential theory, serving both as an introduction and a reference work with regard to specific problems, for physicists and mathematicians. No prior knowledge of integral relations is assumed, and all mathematical material is developed as it becomes necessary. Includes: Attraction of Finite Bodies; Newtonian Potential Function; Vector Fields, Green and Gauss Theorems; Attractions of Surfaces and Lines; Surface Distribution of Matter; Two-Layer Surfaces; Spherical Harmonics; Ellipsoidal Harmonics; etc. "The great number of particular cases . . . should make the book valuable to geophysicists and others actively engaged in practical applications of the potential theory," Review of Scientific Instruments. Index. Bibliography. xiii + 469pp. 5⅜ x 8. S486 Paperbound **$2.50**

THEORETICAL MECHANICS: DYNAMICS OF RIGID BODIES, W. D. MacMillan. Theory of dynamics of a rigid body is developed, using both the geometrical and analytical methods of instruction. Begins with exposition of algebra of vectors, it goes through momentum principles, motion in space, use of differential equations and infinite series to solve more sophisticated dynamics problems. Partial contents: moments of inertia, systems of free particles, motion parallel to a fixed plane, rolling motion, method of periodic solutions, much more. 82 figs. 199 problems. Bibliography. Indexes. xii + 476pp. 5⅜ x 8. S641 Paperbound **$2.50**

MATHEMATICAL FOUNDATIONS OF STATISTICAL MECHANICS, A. I. Khinchin. Offering a precise and rigorous formulation of problems, this book supplies a thorough and up-to-date exposition. It provides analytical tools needed to replace cumbersome concepts, and furnishes for the first time a logical step-by-step introduction to the subject. Partial contents: geometry & kinematics of the phase space, ergodic problem, reduction to theory of probability, application of central limit problem, ideal monatomic gas, foundation of thermo-dynamics, dispersion and distribution of sum functions. Key to notations. Index. viii + 179pp. 5⅜ x 8. S147 Paperbound **$1.50**

ELEMENTARY PRINCIPLES IN STATISTICAL MECHANICS, J. W. Gibbs. Last work of the great Yale mathematical physicist, still one of the most fundamental treatments available for advanced students and workers in the field. Covers the basic principle of conservation of probability of phase, theory of errors in the calculated phases of a system, the contributions of Clausius, Maxwell, Boltzmann, and Gibbs himself, and much more. Includes valuable comparison of statistical mechanics with thermodynamics: Carnot's cycle, mechanical definitions of entropy, etc. xvi + 208pp. 5⅜ x 8. S707 Paperbound **$1.45**

PRINCIPLES OF MECHANICS AND DYNAMICS, Sir William Thomson (Lord Kelvin) and Peter Guthrie Tait. The principles and theories of fundamental branches of classical physics explained by two of the greatest physicists of all time. A broad survey of mechanics, with material on hydrodynamics, elasticity, potential theory, and what is now standard mechanics Thorough and detailed coverage, with many examples, derivations, and topics not included in more recent studies. Only a knowledge of calculus is needed to work through this book. Vol. I (Preliminary): Kinematics; Dynamical Laws and Principles; Experience (observation, experimentation, formation of hypotheses, scientific method); Measures and Instruments; Continuous Calculating Machines. Vol. II (Abstract Dynamics): Statics of a Particle—Attraction; Statics of Solids and Fluids. Formerly Titled "Treatise on Natural Philosophy." Unabridged reprint of revised edition. Index. 168 diagrams. Total of xlii + 1035pp. 5⅜ x 8½.
Vol. I: S966 Paperbound **$2.35**
Vol. II: S967 Paperbound **$2.35**
Two volume Set Paperbound **$4.70**

INVESTIGATIONS ON THE THEORY OF THE BROWNIAN MOVEMENT, Albert Einstein. Reprints from rare European journals. 5 basic papers, including the Elementary Theory of the Brownian Movement, written at the request of Lorentz to provide a simple explanation. Translated by A. D. Cowper. Annotated, edited by R. Fürth. 33pp. of notes elucidate, give history of previous investigations. Author, subject indexes. 62 footnotes. 124pp. 5⅜ x 8.
S304 Paperbound **$1.25**

MECHANICS VIA THE CALCULUS, P. W. Norris, W. S. Legge. Covers almost everything, from linear motion to vector analysis: equations determining motion, linear methods, compounding of simple harmonic motions, Newton's laws of motion, Hooke's law, the simple pendulum, motion of a particle in 1 plane, centers of gravity, virtual work, friction, kinetic energy of rotating bodies, equilibrium of strings, hydrostatics, sheering stresses, elasticity, etc. 550 problems. 3rd revised edition. xii + 367pp. 6 x 9. S207 Clothbound **$4.95**

THE DYNAMICS OF PARTICLES AND OF RIGID, ELASTIC, AND FLUID BODIES; BEING LECTURES ON MATHEMATICAL PHYSICS, A. G. Webster. The reissuing of this classic fills the need for a comprehensive work on dynamics. A wide range of topics is covered in unusually great depth, applying ordinary and partial differential equations. Part I considers laws of motion and methods applicable to systems of all sorts; oscillation, resonance, cyclic systems, etc. Part 2 is a detailed study of the dynamics of rigid bodies. Part 3 introduces the theory of potential; stress and strain, Newtonian potential functions, gyrostatics, wave and vortex motion, etc. Further contents: Kinematics of a point; Lagrange's equations; Hamilton's principle; Systems of vectors; Statics and dynamics of deformable bodies; much more, not easily found together in one volume. Unabridged reprinting of 2nd edition. 20 pages of notes on differential equations and the higher analysis. 203 illustrations. Selected bibliography. Index. xi + 588pp. 5⅜ x 8. S522 Paperbound **$2.45**

A TREATISE ON DYNAMICS OF A PARTICLE, E. J. Routh. Elementary text on dynamics for beginning mathematics or physics student. Unusually detailed treatment from elementary definitions to motion in 3 dimensions, emphasizing concrete aspects. Much unique material important in recent applications. Covers impulsive forces, rectilinear and constrained motion in 2 dimensions, harmonic and parabolic motion, degrees of freedom, closed orbits, the conical pendulum, the principle of least action, Jacobi's method, and much more. Index. 559 problems, many fully worked out, incorporated into text. xiii + 418pp. 5⅜ x 8.
S696 Paperbound **$2.25**

DYNAMICS OF A SYSTEM OF RIGID BODIES (Elementary Section), E. J. Routh. Revised 7th edition of this standard reference. This volume covers the dynamical principles of the subject, and its more elementary applications: finding moments of inertia by integration, foci of inertia, d'Alembert's principle, impulsive forces, motion in 2 and 3 dimensions, Lagrange's equations, relative indicatrix, Euler's theorem, large tautochronous motions, etc. Index. 55 figures. Scores of problems. xv + 443pp. 5⅜ x 8. S664 Paperbound **$2.50**

DYNAMICS OF A SYSTEM OF RIGID BODIES (Advanced Section), E. J. Routh. Revised 6th edition of a classic reference aid. Much of its material remains unique. Partial contents: moving axes, relative motion, oscillations about equilibrium, motion. Motion of a body under no forces, any forces. Nature of motion given by linear equations and conditions of stability. Free, forced vibrations, constants of integration, calculus of finite differences, variations, precession and nutation, motion of the moon, motion of string, chain, membranes. 64 figures. 498pp. 5⅜ x 8. S229 Paperbound **$2.45**

DYNAMICAL THEORY OF GASES, James Jeans. Divided into mathematical and physical chapters for the convenience of those not expert in mathematics, this volume discusses the mathematical theory of gas in a steady state, thermodynamics, Boltzmann and Maxwell, kinetic theory, quantum theory, exponentials, etc. 4th enlarged edition, with new material on quantum theory, quantum dynamics, etc. Indexes. 28 figures. 444pp. 6⅛ x 9¼.
S136 Paperbound **$2.65**

THE THEORY OF HEAT RADIATION, Max Planck. A pioneering work in thermodynamics, providing basis for most later work, Nobel laureate Planck writes on Deductions from Electrodynamics and Thermodynamics, Entropy and Probability, Irreversible Radiation Processes, etc. Starts with simple experimental laws of optics, advances to problems of spectral distribution of energy and irreversibility. Bibliography. 7 illustrations, xiv + 224pp. 5⅜ x 8.
S546 Paperbound **$1.75**

FOUNDATIONS OF POTENTIAL THEORY, O. D. Kellogg. Based on courses given at Harvard this is suitable for both advanced and beginning mathematicians. Proofs are rigorous, and much material not generally avaliable elsewhere is included. Partial contents: forces of gravity, fields of force, divergence theorem, properties of Newtonian potentials at points of free space, potentials as solutions of Laplace's equations, harmonic functions, electrostatics, electric images, logarithmic potential, etc. One of Grundlehren Series. ix + 384pp. 5⅜ x 8.
S144 Paperbound **$1.98**

THERMODYNAMICS, Enrico Fermi. Unabridged reproduction of 1937 edition. Elementary in treatment; remarkable for clarity, organization. Requires no knowledge of advanced math beyond calculus, only familiarity with fundamentals of thermometry, calorimetry. Partial Contents: Thermodynamic systems; First & Second laws of thermodynamics; Entropy; Thermodynamic potentials: phase rule, reversible electric cell; Gaseous reactions: van't Hoff reaction box, principle of LeChatelier; Thermodynamics of dilute solutions: osmotic & vapor pressures, boiling & freezing points; Entropy constant. Index. 25 problems. 24 illustrations. x + 160pp. 5⅜ x 8. S361 Paperbound **$1.75**

THE THERMODYNAMICS OF ELECTRICAL PHENOMENA IN METALS and A CONDENSED COLLECTION OF THERMODYNAMIC FORMULAS, P. W. Bridgman. Major work by the Nobel Prizewinner: stimulating conceptual introduction to aspects of the electron theory of metals, giving an intuitive understanding of fundamental relationships concealed by the formal systems of Onsager and others. Elementary mathematical formulations show clearly the fundamental thermodynamical relationships of the electric field, and a complete phenomenological theory of metals is created. This is the work in which Bridgman announced his famous "thermomotive force" and his distinction between "driving" and "working" electromotive force. We have added in this Dover edition the author's long unavailable tables of thermodynamic formulas, extremely valuable for the speed of reference they allow. Two works bound as one. Index. 33 figures. Bibliography. xviii + 256pp. 5⅜ x 8. S723 Paperbound **$1.65**

TREATISE ON THERMODYNAMICS, Max Planck. Based on Planck's original papers this offers a uniform point of view for the entire field and has been used as an introduction for students who have studied elementary chemistry, physics, and calculus. Rejecting the earlier approaches of Helmholtz and Maxwell, the author makes no assumptions regarding the nature of heat, but begins with a few empirical facts, and from these deduces new physical and chemical laws. 3rd English edition of this standard text by a Nobel laureate. xvi + 297pp. 5⅜ x 8. S219 Paperbound **$1.75**

THE MATHEMATICAL THEORY OF ELASTICITY, A. E. H. Love. A wealth of practical illustration combined with thorough discussion of fundamentals—theory, application, special problems and solutions. Partial Contents: Analysis of Strain & Stress, Elasticity of Solid Bodies, Elasticity of Crystals, Vibration of Spheres, Cylinders, Propagation of Waves in Elastic Solid Media, Torsion, Theory of Continuous Beams, Plates. Rigorous treatment of Volterra's theory of dislocations, 2-dimensional elastic systems, other topics of modern interest. "For years the standard treatise on elasticity," AMERICAN MATHEMATICAL MONTHLY. 4th revised edition. Index. 76 figures. xviii + 643pp. 6⅛ x 9¼. S174 Paperbound **$3.25**

STRESS WAVES IN SOLIDS, H. Kolsky, Professor of Applied Physics, Brown University. The most readable survey of the theoretical core of current knowledge about the propagation of waves in solids, fully correlated with experimental research. Contents: Part I—Elastic Waves: propagation in an extended plastic medium, propagation in bounded elastic media, experimental investigations with elastic materials. Part II—Stress Waves in Imperfectly Elastic Media: internal friction, experimental investigations of dynamic elastic properties, plastic waves and shock waves, fractures produced by stress waves. List of symbols. Appendix. Supplemented bibliography. 3 full-page plates. 46 figures. x + 213pp. 5⅜ x 8½.
S1098 Paperbound **$1.75**

Relativity, quantum theory, atomic and nuclear physics

SPACE TIME MATTER, Hermann Weyl. "The standard treatise on the general theory of relativity" (Nature), written by a world-renowned scientist, provides a deep clear discussion of the logical coherence of the general theory, with introduction to all the mathematical tools needed: Maxwell, analytical geometry, non-Euclidean geometry, tensor calculus, etc. Basis is classical space-time, before absorption of relativity. Partial contents: Euclidean space, mathematical form, metrical continuum, relativity of time and space, general theory. 15 diagrams. Bibliography. New preface for this edition. xviii + 330pp. 5⅜ x 8.
S267 Paperbound **$2.00**

ATOMIC SPECTRA AND ATOMIC STRUCTURE, G. Herzberg. Excellent general survey for chemists, physicists specializing in other fields. Partial contents: simplest line spectra and elements of atomic theory, building-up principle and periodic system of elements, hyperfine structure of spectral lines, some experiments and applications. Bibilography. 80 figures. Index. xii + 257pp. 5⅜ x 8. S115 Paperbound **$2.00**

THE PRINCIPLE OF RELATIVITY, A. Einstein, H. Lorentz, H. Minkowski, H. Weyl. These are the 11 basic papers that founded the general and special theories of relativity, all translated into English. Two papers by Lorentz on the Michelson experiment, electromagnetic phenomena. Minkowski's SPACE & TIME, and Weyl's GRAVITATION & ELECTRICITY. 7 epochmaking papers by Einstein: ELECTROMAGNETICS OF MOVING BODIES, INFLUENCE OF GRAVITATION IN PROPAGATION OF LIGHT, COSMOLOGICAL CONSIDERATIONS, GENERAL THEORY, and 3 others. 7 diagrams. Special notes by A. Sommerfeld. 224pp. 5⅜ x 8.
S81 Paperbound **$1.75**

EINSTEIN'S THEORY OF RELATIVITY, Max Born. Revised edition prepared with the collaboration of Gunther Leibfried and Walter Biem. Steering a middle course between superficial popularizations and complex analyses, a Nobel laureate explains Einstein's theories clearly and with special insight. Easily followed by the layman with a knowledge of high school mathematics, the book has been thoroughly revised and extended to modernize those sections of the well-known original edition which are now out of date. After a comprehensive review of classical physics, Born's discussion of special and general theories of relativity covers such topics as simultaneity, kinematics, Einstein's mechanics and dynamics, relativity of arbitrary motions, the geometry of curved surfaces, the space-time continuum, and many others. Index. Illustrations, vii + 376pp. 5⅜ x 8. S769 Paperbound **$2.00**

ATOMS, MOLECULES AND QUANTA, Arthur E. Ruark and Harold C. Urey. Revised (1963) and corrected edition of a work that has been a favorite with physics students and teachers for more than 30 years. No other work offers the same combination of atomic structure and molecular physics and of experiment and theory. The first 14 chapters deal with the origins and major experimental data of quantum theory and with the development of conceptions of atomic and molecular structure prior to the new mechanics. These sections provide a thorough introduction to atomic and molecular theory, and are presented lucidly and as simply as possible. The six subsequent chapters are devoted to the laws and basic ideas of quantum mechanics: Wave Mechanics, Hydrogenic Atoms in Wave Mechanics, Matrix Mechanics, General Theory of Quantum Dynamics, etc. For advanced college and graduate students in physics. Revised, corrected republication of original edition, with supplementary notes by the authors. New preface by the authors. 9 appendices. General reference list. Indices. 228 figures. 71 tables. Bibliographical material in notes, etc. Total of xxiii + 810pp. 5⅜ x 8⅜.
S1106 Vol. I Paperbound **$2.50**
S1107 Vol. II Paperbound **$2.50**
Two volume set Paperbound **$5.00**

WAVE MECHANICS AND ITS APPLICATIONS, N. F. Mott and I. N. Sneddon. A comprehensive introduction to the theory of quantum mechanics; not a rigorous mathematical exposition it progresses, instead, in accordance with the physical problems considered. Many topics difficult to find at the elementary level are discussed in this book. Includes such matters as: the wave nature of matter, the wave equation of Schrödinger, the concept of stationary states, properties of the wave functions, effect of a magnetic field on the energy levels of atoms, electronic spin, two-body problem, theory of solids, cohesive forces in ionic crystals, collision problems, interaction of radiation with matter, relativistic quantum mechanics, etc. All are treated both physically and mathematically. 68 illustrations. 11 tables. Indexes. xii + 393pp. 5⅜ x 8½. S1070 Paperbound **$2.25**

BASIC METHODS IN TRANSFER PROBLEMS, V. Kourganoff, Professor of Astrophysics, U. of Paris. A coherent digest of all the known methods which can be used for approximate or exact solutions of transfer problems. All methods demonstrated on one particular problem —Milne's problem for a plane parallel medium. Three main sections: fundamental concepts (the radiation field and its interaction with matter, the absorption and emission coefficients, etc.); different methods by which transfer problems can be attacked; and a more general problem—the non-grey case of Milne's problem. Much new material, drawing upon declassified atomic energy reports and data from the USSR. Entirely understandable to the student with a reasonable knowledge of analysis. Unabridged, revised reprinting. New preface by the author. Index. Bibliography. 2 appendices. xv + 281pp. 5⅜ x 8½.
S1074 Paperbound **$2.00**

PRINCIPLES OF QUANTUM MECHANICS, W. V. Houston. Enables student with working knowledge of elementary mathematical physics to develop facility in use of quantum mechanics, understand published work in field. Formulates quantum mechanics in terms of Schroedinger's wave mechanics. Studies evidence for quantum theory, for inadequacy of classical mechanics, 2 postulates of quantum mechanics; numerous important, fruitful applications of quantum mechanics in spectroscopy, collision problems, electrons in solids; other topics. "One of the most rewarding features . . . is the interlacing of problems with text," Amer. J. of Physics. Corrected edition. 21 illus. Index. 296pp. 5⅜ x 8. S524 Paperbound **$2.00**

PHYSICAL PRINCIPLES OF THE QUANTUM THEORY, Werner Heisenberg. A Nobel laureate discusses quantum theory; Heisenberg's own work, Compton, Schroedinger, Wilson, Einstein, many others. Written for physicists, chemists who are not specialists in quantum theory, only elementary formulae are considered in the text; there is a mathematical appendix for specialists. Profound without sacrifice of clarity. Translated by C. Eckart, F. Hoyt. 18 figures. 192pp. 5⅜ x 8. S113 Paperbound **$1.25**

ENGINEERING AND TECHNOLOGY

General and mathematical

ENGINEERING MATHEMATICS, Kenneth S. Miller. A text for graduate students of engineering to strengthen their mathematical background in differential equations, etc. Mathematical steps very explicitly indicated. Contents: Determinants and Matrices, Integrals, Linear Differential Equations, Fourier Series and Integrals, Laplace Transform, Network Theory, Random Function . . . all vital requisites for advanced modern engineering studies. Unabridged republication. Appendices: Borel Sets; Riemann-Stieltjes Integral; Fourier Series and Integrals. Index. References at Chapter Ends. xii + 417pp. 6 x 8½. S1121 Paperbound **$2.00**

MATHEMATICAL ENGINEERING ANALYSIS, Rufus Oldenburger. A book designed to assist the research engineer and scientist in making the transition from physical engineering situations to the corresponding mathematics. Scores of common practical situations found in all major fields of physics are supplied with their correct mathematical formulations—applications to automobile springs and shock absorbers, clocks, throttle torque of diesel engines, resistance networks, capacitors, transmission lines, microphones, neon tubes, gasoline engines, refrigeration cycles, etc. Each section reviews basic principles of underlying various fields: mechanics of rigid bodies, electricity and magnetism, heat, elasticity, fluid mechanics, and aerodynamics. Comprehensive and eminently useful. Index. 169 problems, answers. 200 photos and diagrams. xiv + 426pp. 5⅜ x 8½. S919 Paperbound **$2.50**

MATHEMATICS OF MODERN ENGINEERING, E. G. Keller and R. E. Doherty. Written for the Advanced Course in Engineering of the General Electric Corporation, deals with the engineering use of determinants, tensors, the Heaviside operational calculus, dyadics, the calculus of variations, etc. Presents underlying principles fully, but purpose is to teach engineers to deal with modern engineering problems, and emphasis is on the perennial engineering attack of set-up and solve. Indexes. Over 185 figures and tables. Hundreds of exercises, problems, and worked-out examples. References. Two volume set. Total of xxxiii + 623pp. 5⅜ x 8.
S734 Vol I Paperbound **$1.85**
S735 Vol II Paperbound **$1.85**
The set **$3.70**

MATHEMATICAL METHODS FOR SCIENTISTS AND ENGINEERS, L. P. Smith. For scientists and engineers, as well as advanced math students. Full investigation of methods and practical description of conditions under which each should be used. Elements of real functions, differential and integral calculus, space geometry, theory of residues, vector and tensor analysis, series of Bessel functions, etc. Each method illustrated by completely-worked-out examples, mostly from scientific literature. 368 graded unsolved problems. 100 diagrams. x + 453pp. 5⅝ x 8⅜. S220 Paperbound **$2.00**

THEORY OF FUNCTIONS AS APPLIED TO ENGINEERING PROBLEMS, edited by R. Rothe, F. Ollendorff, and K. Pohlhausen. A series of lectures given at the Berlin Institute of Technology that shows the specific applications of function theory in electrical and allied fields of engineering. Six lectures provide the elements of function theory in a simple and practical form, covering complex quantities and variables, integration in the complex plane, residue theorems, etc. Then 5 lectures show the exact uses of this powerful mathematical tool, with full discussions of problem methods. Index. Bibliography. 108 figures. x + 189pp. 5⅜ x 8.
S733 Paperbound **$1.35**

Aerodynamics and hydrodynamics

AIRPLANE STRUCTURAL ANALYSIS AND DESIGN, E. E. Sechler and L. G. Dunn. Systematic authoritative book which summarizes a large amount of theoretical and experimental work on structural analysis and design. Strong on classical subsonic material still basic to much aeronautic design . . . remains a highly useful source of information. Covers such areas as layout of the airplane, applied and design loads, stress-strain relationships for stable structures, truss and frame analysis, the problem of instability, the ultimate strength of stiffened flat sheet, analysis of cylindrical structures, wings and control surfaces, fuselage analysis, engine mounts, landing gears, etc. Originally published as part of the CALCIT Aeronautical Series. 256 Illustrations. 47 study problems. Indexes. xi + 420pp. 5⅜ x 8½.
S1043 Paperbound **$2.25**

FUNDAMENTALS OF HYDRO- AND AEROMECHANICS, L. Prandtl and O. G. Tietjens. The well-known standard work based upon Prandtl's lectures at Goettingen. Wherever possible hydrodynamics theory is referred to practical considerations in hydraulics, with the view of unifying theory and experience. Presentation is extremely clear and though primarily physical, mathematical proofs are rigorous and use vector analysis to a considerable extent. An Engineering Society Monograph, 1934. 186 figures. Index. xvi + 270pp. 5⅜ x 8.
S374 Paperbound **$1.85**

FLUID MECHANICS FOR HYDRAULIC ENGINEERS, H. Rouse. Standard work that gives a coherent picture of fluid mechanics from the point of view of the hydraulic engineer. Based on courses given to civil and mechanical engineering students at Columbia and the California Institute of Technology, this work covers every basic principle, method, equation, or theory of interest to the hydraulic engineer. Much of the material, diagrams, charts, etc., in this self-contained text are not duplicated elsewhere. Covers irrotational motion, conformal mapping, problems in laminar motion, fluid turbulence, flow around immersed bodies, transportation of sediment, general charcteristics of wave phenomena, gravity waves in open channels, etc. Index. Appendix of physical properties of common fluids. Frontispiece + 245 figures and photographs. xvi + 422pp. 5⅜ x 8. S729 Paperbound **$2.25**

WATERHAMMER ANALYSIS, John Parmakian. Valuable exposition of the graphical method of solving waterhammer problems by Assistant Chief Designing Engineer, U.S. Bureau of Reclamation. Discussions of rigid and elastic water column theory, velocity of waterhammer waves, theory of graphical waterhammer analysis for gate operation, closings, openings, rapid and slow movements, etc., waterhammer in pump discharge caused by power failure, waterhammer analysis for compound pipes, and numerous related problems. "With a concise and lucid style, clear printing, adequate bibliography and graphs for approximate solutions at the project stage, it fills a vacant place in waterhammer literature," WATER POWER. 43 problems. Bibliography. Index. 113 illustrations. xiv + 161pp. 5⅜ x 8½.
S1061 Paperbound **$1.65**

AERODYNAMIC THEORY: A GENERAL REVIEW OF PROGRESS, William F. Durand, editor-in-chief. A monumental joint effort by the world's leading authorities prepared under a grant of the Guggenheim Fund for the Promotion of Aeronautics. Intended to provide the student and aeronautic designer with the theoretical and experimental background of aeronautics. Never equalled for breadth, depth, reliability. Contains discussions of special mathematical topics not usually taught in the engineering or technical courses. Also: an extended two-part treatise on Fluid Mechanics, discussions of aerodynamics of perfect fluids, analyses of experiments with wind tunnels, applied airfoil theory, the non-lifting system of the airplane, the air propeller, hydrodynamics of boats and floats, the aerodynamics of cooling, etc. Contributing experts include Munk, Giacomelli, Prandtl, Toussaint, Von Karman, Klemperer, among others. Unabridged republication. 6 volumes bound as 3. Total of 1,012 figures, 12 plates. Total of 2,186pp. Bibliographies. Notes. Indices. 5⅜ x 8.
S328-S330 Clothbound, The Set **$17.50**

APPLIED HYDRO- AND AEROMECHANICS, L. Prandtl and O. G. Tietjens. Presents, for the most part, methods which will be valuable to engineers. Covers flow in pipes, boundary layers, airfoil theory, entry conditions, turbulent flow in pipes, and the boundary layer, determining drag from measurements of pressure and velocity, etc. "Will be welcomed by all students of aerodynamics," NATURE. Unabridged, unaltered. An Engineering Society Monograph, 1934. Index. 226 figures, 28 photographic plates illustrating flow patterns. xvi + 311pp. 5⅜ x 8.
S375 Paperbound **$1.85**

SUPERSONIC AERODYNAMICS, E. R. C. Miles. Valuable theoretical introduction to the supersonic domain, with emphasis on mathematical tools and principles, for practicing aerodynamicists and advanced students in aeronautical engineering. Covers fundamental theory, divergence theorem and principles of circulation, compressible flow and Helmholtz laws, the Prandtl-Busemann graphic method for 2-dimensional flow, oblique shock waves, the Taylor-Maccoll method for cones in supersonic flow, the Chaplygin method for 2-dimensional flow, etc. Problems range from practical engineering problems to development of theoretical results. "Rendered outstanding by the unprecedented scope of its contents . . . has undoubtedly filled a vital gap," AERONAUTICAL ENGINEERING REVIEW. Index. 173 problems, answers. 106 diagrams. 7 tables. xii + 255pp. 5⅜ x 8. S214 Paperbound **$1.45**

HYDRAULIC TRANSIENTS, G. R. Rich. The best text in hydraulics ever printed in English . . . by one of America's foremost engineers (former Chief Design Engineer for T.V.A.). Provides a transition from the basic differential equations of hydraulic transient theory to the arithmetic intergration computation required by practicing engineers. Sections cover Water Hammer, Turbine Speed Regulation, Stability of Governing, Water-Hammer Pressures in Pump Discharge Lines, The Differential and Restricted Orifice Surge Tanks, The Normalized Surge Tank Charts of Calame and Gaden, Navigation Locks, Surges in Power Canals—Tidal Harmonics, etc. Revised and enlarged. Author's prefaces. Index. xiv + 409pp. 5⅜ x 8½.
S116 Paperbound **$2.50**

HYDRAULICS AND ITS APPLICATIONS, A. H. Gibson. Excellent comprehensive textbook for the student and thorough practical manual for the professional worker, a work of great stature in its area. Half the book is devoted to theory and half to applications and practical problems met in the field. Covers modes of motion of a fluid, critical velocity, viscous flow, eddy formation, Bernoulli's theorem, flow in converging passages, vortex motion, form of effluent streams, notches and weirs, skin friction, losses at valves and elbows, siphons, erosion of channels, jet propulsion, waves of oscillation, and over 100 similar topics. Final chapters (nearly 400 pages) cover more than 100 kinds of hydraulic machinery: Pelton wheel, speed regulators, the hydraulic ram, surge tanks, the scoop wheel, the Venturi meter, etc. A special chapter treats methods of testing theoretical hypotheses: scale models of rivers, tidal estuaries, siphon spillways, etc. 5th revised and enlarged (1952) edition. Index. Appendix. 427 photographs and diagrams. 95 examples, answers. xv + 813pp. 6 x 9.
S791 Clothbound **$8.00**

FLUID MECHANICS THROUGH WORKED EXAMPLES, D. R. L. Smith and J. Houghton. Advanced text covering principles and applications to practical situations. Each chapter begins with concise summaries of fundamental ideas. 163 fully worked out examples applying principles outlined in the text. 275 other problems, with answers. Contents: The Pressure of Liquids on Surfaces; Floating Bodies; Flow Under Constant Head in Pipes; Circulation; Vorticity; The Potential Function; Laminar Flow and Lubrication; Impact of Jets; Hydraulic Turbines; Centrifugal and Reciprocating Pumps; Compressible Fluids; and many other items. Total of 438 examples. 250 line illustrations. 340pp. Index. 6 x 8⅞. S981 Clothbound **$6.00**

THEORY OF SHIP MOTIONS, S. N. Blagoveshchensky. The only detailed text in English in a rapidly developing branch of engineering and physics, it is the work of one of the world's foremost authorities—Blagoveshchensky of Leningrad Shipbuilding Institute. A senior-level treatment written primarily for engineering students, but also of great importance to naval architects, designers, contractors, researchers in hydrodynamics, and other students. No mathematics beyond ordinary differential equations is required for understanding the text. Translated by T. & L. Strelkoff, under editorship of Louis Landweber, Iowa Institute of Hydraulic Research, under auspices of Office of Naval Research. Bibliography. Index. 231 diagrams and illustrations. Total of 649pp. 5⅜ x 8½. Vol. I: S234 Paperbound **$2.00**
Vol. II: S235 Paperbound **$2.00**

THEORY OF FLIGHT, Richard von Mises. Remains almost unsurpassed as balanced, well-written account of fundamental fluid dynamics, and situations in which air compressibility effects are unimportant. Stressing equally theory and practice, avoiding formidable mathematical structure, it conveys a full understanding of physical phenomena and mathematical concepts. Contains perhaps the best introduction to general theory of stability. "Outstanding," Scientific, Medical, and Technical Books. New introduction by K. H. Hohenemser. Bibliographical, historical notes. Index. 408 illustrations. xvi + 620pp. 5⅜ x 8⅜. S541 Paperbound **$3.50**

THEORY OF WING SECTIONS, I. H. Abbott, A. E. von Doenhoff. Concise compilation of subsonic aerodynamic characteristics of modern NASA wing sections, with description of their geometry, associated theory. Primarily reference work for engineers, students, it gives methods, data for using wing-section data to predict characteristics. Particularly valuable: chapters on thin wings, airfoils; complete summary of NACA's experimental observations, system of construction families of airfoils. 350pp. of tables on Basic Thickness Forms, Mean Lines, Airfoil Ordinates, Aerodynamic Characteristics of Wing Sections. Index. Bibliography. 191 illustrations. Appendix. 705pp. 5⅜ x 8. S558 Paperbound **$3.25**

WEIGHT-STRENGTH ANALYSIS OF AIRCRAFT STRUCTURES, F. R. Shanley. Scientifically sound methods of analyzing and predicting the structural weight of aircraft and missiles. Deals directly with forces and the distances over which they must be transmitted, making it possible to develop methods by which the minimum structural weight can be determined for any material and conditions of loading. Weight equations for wing and fuselage structures. Includes author's original papers on inelastic buckling and creep buckling. "Particularly successful in presenting his analytical methods for investigating various optimum design principles," AERONAUTICAL ENGINEERING REVIEW. Enlarged bibliography. Index. 199 figures. xiv + 404pp. 5⅝ x 8⅜. S660 Paperbound **$2.50**

Electricity

TWO-DIMENSIONAL FIELDS IN ELECTRICAL ENGINEERING, L. V. Bewley. A useful selection of typical engineering problems of interest to practicing electrical engineers. Introduces senior students to the methods and procedures of mathematical physics. Discusses theory of functions of a complex variable, two-dimensional fields of flow, general theorems of mathematical physics and their applications, conformal mapping or transformation, method of images, freehand flux plotting, etc. New preface by the author. Appendix by W. F. Kiltner. Index. Bibliography at chapter ends. xiv + 204pp. 5⅜ x 8½. S1118 Paperbound **$1.50**

FLUX LINKAGES AND ELECTROMAGNETIC INDUCTION, L. V. Bewley. A brief, clear book which shows proper uses and corrects misconceptions of Faraday's law of electromagnetic induction in specific problems. Contents: Circuits, Turns, and Flux Linkages; Substitution of Circuits; Electromagnetic Induction; General Criteria for Electromagnetic Induction; Applications and Paradoxes; Theorem of Constant Flux Linkages. New Section: Rectangular Coil in a Varying Uniform Medium. Valuable supplement to class texts for engineering students. Corrected, enlarged edition. New preface. Bibliography in notes. 49 figures. xi + 106pp. 5⅜ x 8. S1103 Paperbound **$1.25**

INDUCTANCE CALCULATIONS: WORKING FORMULAS AND TABLES, Frederick W. Grover. An invaluable book to everyone in electrical engineering. Provides simple single formulas to cover all the more important cases of inductance. The approach involves only those parameters that naturally enter into each situation, while extensive tables are given to permit easy interpolations. Will save the engineer and student countless hours and enable them to obtain accurate answers with minimal effort. Corrected republication of 1946 edition. 58 tables. 97 completely worked out examples. 66 figures. xiv + 286pp. 5⅜ x 8½.
S974 Paperbound **$1.85**

GASEOUS CONDUCTORS: THEORY AND ENGINEERING APPLICATIONS, J. D. Cobine. An indispensable text and reference to gaseous conduction phenomena, with the engineering viewpoint prevailing throughout. Studies the kinetic theory of gases, ionization, emission phenomena; gas breakdown, spark characteristics, glow, and discharges; engineering applications in circuit interrupters, rectifiers, light sources, etc. Separate detailed treatment of high pressure arcs (Suits); low pressure arcs (Langmuir and Tonks). Much more. "Well organized, clear, straightforward," Tonks, Review of Scientific Instruments. Index. Bibliography. 83 practice problems. 7 appendices. Over 600 figures. 58 tables. xx + 606pp. 5⅜ x 8. S442 Paperbound **$3.25**

INTRODUCTION TO THE STATISTICAL DYNAMICS OF AUTOMATIC CONTROL SYSTEMS, V. V. Solodovnikov. First English publication of text-reference covering important branch of automatic control systems—random signals; in its original edition, this was the first comprehensive treatment. Examines frequency characteristics, transfer functions, stationary random processes, determination of minimum mean-squared error, of transfer function for a finite period of observation, much more. Translation edited by J. B. Thomas, L. A. Zadeh. Index. Bibliography. Appendix. xxii + 308pp. 5⅜ x 8. S420 Paperbound **$2.25**

TENSORS FOR CIRCUITS, Gabriel Kron. A boldly original method of analyzing engineering problems, at center of sharp discussion since first introduced, now definitely proved useful in such areas as electrical and structural networks on automatic computers. Encompasses a great variety of specific problems by means of a relatively few symbolic equations. "Power and flexibility . . . becoming more widely recognized," Nature. Formerly "A Short Course in Tensor Analysis." New introduction by B. Hoffmann. Index. Over 800 diagrams. xix + 250pp. 5⅜ x 8. S534 Paperbound **$2.00**

SELECTED PAPERS ON SEMICONDUCTOR MICROWAVE ELECTRONICS, edited by Sumner N. Levine and Richard R. Kurzrok. An invaluable collection of important papers dealing with one of the most remarkable devolopments in solid-state electronics—the use of the **p-n** junction to achieve amplification and frequency conversion of microwave frequencies. Contents: General Survey (3 introductory papers by W. E. Danielson, R. N. Hall, and M. Tenzer); General Theory of Nonlinear Elements (3 articles by A. van der Ziel, H. E. Rowe, and Manley and Rowe); Device Fabrication and Characterization (3 pieces by Bakanowski, Cranna, and Uhlir, by McCotter, Walker and Fortini, and by S. T. Eng); Parametric Amplifiers and Frequency Multipliers (13 articles by Uhlir, Heffner and Wade, Matthaei, P. K. Tien, van der Ziel, Engelbrecht, Currie and Gould, Uenohara, Leeson and Weinreb, and others); and Tunnel Diodes (4 papers by L. Esaki, H. S. Sommers, Jr., M. E. Hines, and Yariv and Cook). Introduction. 295 Figures. xiii + 286pp. 6½ x 9¼. S1126 Paperbound **$2.25**

THE PRINCIPLES OF ELECTROMAGNETISM APPLIED TO ELECTRICAL MACHINES, B. Hague. A concise, but complete, summary of the basic principles of the magnetic field and its applications, with particular reference to the kind of phenomena which occur in electrical machines. Part I: General Theory—magnetic field of a current, electromagnetic field passing from air to iron, mechanical forces on linear conductors, etc. Part II: Application of theory to the solution of electromechanical problems—the magnetic field and mechanical forces in non-salient pole machinery, the field within slots and between salient poles, and the work of Rogowski, Roth, and Strutt. Formery titled "Electromagnetic Problems in Electrical Engineering." 2 appendices. Index. Bibliography in notes. 115 figures. xiv + 359pp. 5⅜ x 8½. S246 Paperbound **$2.25**

Mechanical engineering

DESIGN AND USE OF INSTRUMENTS AND ACCURATE MECHANISM, T. N. Whitehead. For the instrument designer, engineer; how to combine necessary mathematical abstractions with independent observation of actual facts. Partial contents: instruments & their parts, theory of errors, systematic errors, probability, short period errors, erratic errors, design precision, kinematic, semikinematic design, stiffness, planning of an instrument, human factor, etc. Index. 85 photos, diagrams. xii + 288pp. 5⅜ x 8. S270 Paperbound **$2.00**

A TREATISE ON GYROSTATICS AND ROTATIONAL MOTION: THEORY AND APPLICATIONS, Andrew Gray. Most detailed, thorough book in English, generally considered definitive study. Many problems of all sorts in full detail, or step-by-step summary. Classical problems of Bour, Lottner, etc.; later ones of great physical interest. Vibrating systems of gyrostats, earth as a top, calculation of path of axis of a top by elliptic integrals, motion of unsymmetrical top, much more. Index. 160 illus. 550pp. 5⅜ x 8. S589 Paperbound **$2.75**

MECHANICS OF THE GYROSCOPE, THE DYNAMICS OF ROTATION, R. F. Deimel, Professor of Mechanical Engineering at Stevens Institute of Technology. Elementary general treatment of dynamics of rotation, with special application of gyroscopic phenomena. No knowledge of vectors needed. Velocity of a moving curve, acceleration to a point, general equations of motion, gyroscopic horizon, free gyro, motion of discs, the damped gyro, 103 similar topics. Exercises. 75 figures. 208pp. 5⅜ x 8. S66 Paperbound **$1.75**

STRENGTH OF MATERIALS, J. P. Den Hartog. Distinguished text prepared for M.I.T. course, ideal as introduction, refresher, reference, or self-study text. Full clear treatment of elementary material (tension, torsion, bending, compound stresses, deflection of beams, etc.), plus much advanced material on engineering methods of great practical value: full treatment of the Mohr circle, lucid elementary discussions of the theory of the center of shear and the "Myosotis" method of calculating beam deflections, reinforced concrete, plastic deformations, photoelasticity, etc. In all sections, both general principles and concrete applications are given. Index. 186 figures (160 others in problem section). 350 problems, all with answers. List of formulas. viii + 323pp. 5⅜ x 8. S755 Paperbound **$2.00**

PHOTOELASTICITY: PRINCIPLES AND METHODS, H. T. Jessop, F. C. Harris. For the engineer, for specific problems of stress analysis. Latest time-saving methods of checking calculations in 2-dimensional design problems, new techniques for stresses in 3 dimensions, and lucid description of optical systems used in practical photoelasticity. Useful suggestions and hints based on on-the-job experience included. Partial contents: strained and stress-strain relations, circular disc under thrust along diameter, rectangular block with square hole under vertical thrust, simply supported rectangular beam under central concentrated load, etc. Theory held to minimum, no advanced mathematical training needed. Index. 164 illustrations. viii + 184pp. 6⅛ x 9¼. S720 Paperbound **$2.00**

APPLIED ELASTICITY, J. Prescott. Provides the engineer with the theory of elasticity usually lacking in books on strength of materials, yet concentrates on those portions useful for immediate application. Develops every important type of elasticity problem from theoretical principles. Covers analysis of stress, relations between stress and strain, the empirical basis of elasticity, thin rods under tension or thrust, Saint Venant's theory, transverse oscillations of thin rods, stability of thin plates, cylinders with thin walls, vibrations of rotating disks, elastic bodies in contact, etc. "Excellent and important contribution to the subject, not merely in the old matter which he has presented in new and refreshing form, but also in the many original investigations here published for the first time," NATURE. Index. 3 Appendixes. vi + 672pp. 5⅜ x 8. S726 Paperbound **$3.25**

APPLIED MECHANICS FOR ENGINEERS, Sir Charles Inglis, F.R.S. A representative survey of the many and varied engineering questions which can be answered by statics and dynamics. The author, one of first and foremost adherents of "structural dynamics," presents distinctive illustrative examples and clear, concise statement of principles—directing the discussion at methodology and specific problems. Covers fundamental principles of rigid-body statics, graphic solutions of static problems, theory of taut wires, stresses in frameworks, particle dynamics, kinematics, simple harmonic motion and harmonic analysis, two-dimensional rigid dynamics, etc. 437 illustrations. xii + 404pp. 5⅜ x 8½. S1119 Paperbound **$2.00**

THEORY OF MACHINES THROUGH WORKED EXAMPLES, G. H. Ryder. Practical mechanical engineering textbook for graduates and advanced undergraduates, as well as a good reference work for practicing engineers. Partial contents: Mechanisms, Velocity and Acceleration (including discussion of Klein's Construction for Piston Acceleration), Cams, Geometry of Gears, Clutches and Bearings, Belt and Rope Drives, Brakes, Inertia Forces and Couples, General Dynamical Problems, Gyroscopes, Linear and Angular Vibrations, Torsional Vibrations, Transverse Vibrations and Whirling Speeds (Chapters on vibrations considerably enlarged from previous editions). Over 300 problems, many fully worked out. Index. 195 line illustrations. Revised and enlarged edition. viii + 280pp. 5⅝ x 8¾. S980 Clothbound **$5.00**

THE KINEMATICS OF MACHINERY: OUTLINES OF A THEORY OF MACHINES, Franz Reuleaux. The classic work in the kinematics of machinery. The present thinking about the subject has all been shaped in great measure by the fundamental principles stated here by Reuleaux almost 90 years ago. While some details have naturally been superseded, his basic viewpoint has endured; hence, the book is still an excellent text for basic courses in kinematics and a standard reference work for active workers in the field. Covers such topics as: the nature of the machine problem, phoronomic propositions, pairs of elements, incomplete kinematic chains, kinematic notation and analysis, analyses of chamber-crank trains, chamber-wheel trains, constructive elements of machinery, complete machines, etc., with main focus on controlled movement in mechanisms. Unabridged republication of original edition, translated by Alexander B. Kennedy. New introduction for this edition by E. S. Ferguson. Index. 451 illustrations. xxiv + 622pp. 5⅜ x 8½. S1124 Paperbound **$3.00**

ANALYTICAL MECHANICS OF GEARS, Earle Buckingham. Provides a solid foundation upon which logical design practices and design data can be constructed. Originally arising out of investigations of the ASME Special Research Committee on Worm Gears and the Strength of Gears, the book covers conjugate gear-tooth action, the nature of the contact, and resulting gear-tooth profiles of: spur, internal, helical, spiral, worm, bevel, and hypoid or skew bevel gears. Also: frictional heat of operation and its dissipation, friction losses, etc., dynamic loads in operation, and related matters. Familiarity with this book is still regarded as a necessary prerequisite to work in modern gear manufacturing. 263 figures. 103 tables. Index. x + 546pp. 5⅜ x 8½. S1073 Paperbound **$2.75**

Optical design, lighting

THE SCIENTIFIC BASIS OF ILLUMINATING ENGINEERING, Parry Moon, Professor of Electrical Engineering, M.I.T. Basic, comprehensive study. Complete coverage of the fundamental theoretical principles together with the elements of design, vision, and color with which the lighting engineer must be familiar. Valuable as a text as well as a reference source to the practicing engineer. Partial contents: Spectroradiometric Curve, Luminous Flux, Radiation from Gaseous-Conduction Sources, Radiation from Incandescent Sources, Incandescent Lamps, Measurement of Light, Illumination from Point Sources and Surface Sources, Elements of Lighting Design. 7 Appendices. Unabridged and corrected republication, with additions. New preface containing conversion tables of radiometric and photometric concepts. Index. 707-item bibliography. 92-item bibliography of author's articles. 183 problems. xxiii + 608pp. 5⅜ x 8½. S242 Paperbound **$2.85**

OPTICS AND OPTICAL INSTRUMENTS: AN INTRODUCTION WITH SPECIAL REFERENCE TO PRACTICAL APPLICATIONS, B. K. Johnson. An invaluable guide to basic practical applications of optical principles, which shows how to set up inexpensive working models of each of the four main types of optical instruments—telescopes, microscopes, photographic lenses, optical projecting systems. Explains in detail the most important experiments for determining their accuracy, resolving power, angular field of view, amounts of aberration, all other necessary facts about the instruments. Formerly "Practical Optics." Index. 234 diagrams. Appendix. 224pp. 5⅜ x 8. S642 Paperbound **$1.75**

APPLIED OPTICS AND OPTICAL DESIGN, A. E. Conrady. With publication of vol. 2, standard work for designers in optics is now complete for first time. Only work of its kind in English; only detailed work for practical designer and self-taught. Requires, for bulk of work, no math above trig. Step-by-step exposition, from fundamental concepts of geometrical, physical optics, to systematic study, design, of almost all types of optical systems. Vol. 1: all ordinary ray-tracing methods; primary aberrations; necessary higher aberration for design of telescopes, low-power microscopes, photographic equipment. Vol. 2: (Completed from author's notes by R. Kingslake, Dir. Optical Design, Eastman Kodak.) Special attention to high-power microscope, anastigmatic photographic objectives. "An indispensable work," J., Optical Soc. of Amer. "As a practical guide this book has no rival," Transactions, Optical Soc. Index. Bibliography. 193 diagrams. 852pp. 6⅛ x 9¼. Vol. 1 S366 Paperbound **$3.50**
Vol. 2 S612 Paperbound **$2.95**

Miscellaneous

THE MEASUREMENT OF POWER SPECTRA FROM THE POINT OF VIEW OF COMMUNICATIONS ENGINEERING, R. B. Blackman, J. W. Tukey. This pathfinding work, reprinted from the "Bell System Technical Journal," explains various ways of getting practically useful answers in the measurement of power spectra, using results from both transmission theory and the theory of statistical estimation. Treats: Autocovariance Functions and Power Spectra; Direct Analog Computation; Distortion, Noise, Heterodyne Filtering and Pre-whitening; Aliasing; Rejection Filtering and Separation; Smoothing and Decimation Procedures; Very Low Frequencies; Transversal Filtering; much more. An appendix reviews fundamental Fourier techniques. Index of notation. Glossary of terms. 24 figures. XII tables. Bibliography. General index. 192pp. 5⅜ x 8. S507 Paperbound **$1.85**

CALCULUS REFRESHER FOR TECHNICAL MEN, A. Albert Klaf. This book is unique in English as a refresher for engineers, technicians, students who either wish to brush up their calculus or to clear up uncertainties. It is not an ordinary text, but an examination of most important aspects of integral and differential calculus in terms of the 756 questions most likely to occur to the technical reader. The first part of this book covers simple differential calculus, with constants, variables, functions, increments, derivatives, differentiation, logarithms, curvature of curves, and similar topics. The second part covers fundamental ideas of integration, inspection, substitution, transformation, reduction, areas and volumes, mean value, successive and partial integration, double and triple integration. Practical aspects are stressed rather than theoretical. A 50-page section illustrates the application of calculus to specific problems of civil and nautical engineering, electricity, stress and strain, elasticity, industrial engineering, and similar fields.—756 questions answered. 566 problems, mostly answered. 36 pages of useful constants, formulae for ready reference. Index. v + 431pp. 5⅜ x 8. T370 Paperbound **$2.00**

METHODS IN EXTERIOR BALLISTICS, Forest Ray Moulton. Probably the best introduction to the mathematics of projectile motion. The ballistics theories propounded were coordinated with extensive proving ground and wind tunnel experiments conducted by the author and others for the U.S. Army. Broad in scope and clear in exposition, it gives the beginnings of the theory used for modern-day projectile, long-range missile, and satellite motion. Six main divisions: Differential Equations of Translatory Motion of a projectile; Gravity and the Resistance Function; Numerical Solution of Differential Equations; Theory of Differential Variations; Validity of Method of Numerical Integration; and Motion of a Rotating Projectile. Formerly titled: "New Methods in Exterior Ballistics." Index. 38 diagrams. viii + 259pp. 5⅜ x 8½. S232 Paperbound **$1.75**

LOUD SPEAKERS: THEORY, PERFORMANCE, TESTING AND DESIGN, N. W. McLachlan. Most comprehensive coverage of theory, practice of loud speaker design, testing; classic reference, study manual in field. First 12 chapters deal with theory, for readers mainly concerned with math. aspects; last 7 chapters will interest reader concerned with testing, design. Partial contents: principles of sound propagation, fluid pressure on vibrators, theory of moving-coil principle, transients, driving mechanisms, response curves, design of horn type moving coil speakers, electrostatic speakers, much more. Appendix. Bibliography. Index. 165 illustrations, charts. 411pp. 5⅜ x 8. S588 Paperbound **$2.25**

MICROWAVE TRANSMISSION, J. C. Slater. First text dealing exclusively with microwaves, brings together points of view of field, circuit theory, for graduate student in physics, electrical engineering, microwave technician. Offers valuable point of view not in most later studies. Uses Maxwell's equations to study electromagnetic field, important in this area. Partial contents: infinite line with distributed parameters, impedance of terminated line, plane waves, reflections, wave guides, coaxial line, composite transmission lines, impedance matching, etc. Introduction. Index. 76 illus. 319pp. 5⅜ x 8. S564 Paperbound **$1.50**

MICROWAVE TRANSMISSION DESIGN DATA, T. Moreno. Originally classified, now rewritten and enlarged (14 new chapters) for public release under auspices of Sperry Corp. Material of immediate value or reference use to radio engineers, systems designers, applied physicists, etc. Ordinary transmission line theory; attenuation; capacity; parameters of coaxial lines; higher modes; flexible cables; obstacles, discontinuities, and injunctions; tunable wave guide impedance transformers; effects of temperature and humidity; much more. "Enough theoretical discussion is included to allow use of data without previous background," Electronics. 324 circuit diagrams, figures, etc. Tables of dielectrics, flexible cable, etc., data. Index. ix + 248pp. 5⅜ x 8. S459 Paperbound **$1.65**

RAYLEIGH'S PRINCIPLE AND ITS APPLICATIONS TO ENGINEERING, G. Temple & W. Bickley. Rayleigh's principle developed to provide upper and lower estimates of true value of fundamental period of a vibrating system, or condition of stability of elastic systems. Illustrative examples; rigorous proofs in special chapters. Partial contents: Energy method of discussing vibrations, stability. Perturbation theory, whirling of uniform shafts. Criteria of elastic stability. Application of energy method. Vibrating systems. Proof, accuracy, successive approximations, application of Rayleigh's principle. Synthetic theorems. Numerical, graphical methods. Equilibrium configurations, Ritz's method. Bibliography. Index. 22 figures. ix + 156pp. 5⅜ x 8. S307 Paperbound **$1.85**

ELASTICITY, PLASTICITY AND STRUCTURE OF MATTER, R. Houwink. Standard treatise on rheological aspects of different technically important solids such as crystals, resins, textiles, rubber, clay, many others. Investigates general laws for deformations; determines divergences from these laws for certain substances. Covers general physical and mathematical aspects of plasticity, elasticity, viscosity. Detailed examination of deformations, internal structure of matter in relation to elastic and plastic behavior, formation of solid matter from a fluid, conditions for elastic and plastic behavior of matter. Treats glass, asphalt, gutta percha, balata, proteins, baker's dough, lacquers, sulphur, others. 2nd revised, enlarged edition. Extensive revised bibliography in over 500 footnotes. Index. Table of symbols. 214 figures. xviii + 368pp. 6 x 9¼. S385 Paperbound **$2.45**

THE SCHWARZ-CHRISTOFFEL TRANSFORMATION AND ITS APPLICATIONS: A SIMPLE EXPOSITION, Miles Walker. An important book for engineers showing how this valuable tool can be employed in practical situations. Very careful, clear presentation covering numerous concrete engineering problems. Includes a thorough account of conjugate functions for engineers—useful for the beginner and for review. Applications to such problems as: Stream-lines round a corner, electric conductor in air-gap, dynamo slots, magnetized poles, much more. Formerly "Conjugate Functions for Engineers." Preface. 92 figures, several tables. Index. ix + 116pp. 5⅜ x 8½. S1149 Paperbound **$1.25**

THE LAWS OF THOUGHT, George Boole. This book founded symbolic logic some hundred years ago. It is the 1st significant attempt to apply logic to all aspects of human endeavour. Partial contents: derivation of laws, signs & laws, interpretations, eliminations, conditions of a perfect method, analysis, Aristotelian logic, probability, and similar topics. xviii + 424pp. 5⅜ x 8. S28 Paperbound **$2.00**

SCIENCE AND METHOD, Henri Poincaré. Procedure of scientific discovery, methodology, experiment, idea-germination—the intellectual processes by which discoveries come into being. Most significant and most interesting aspects of development, application of ideas. Chapters cover selection of facts, chance, mathematical reasoning, mathematics, and logic; Whitehead, Russell, Cantor; the new mechanics, etc. 288pp. 5⅜ x 8. S222 Paperbound **$1.35**

FAMOUS BRIDGES OF THE WORLD, D. B. Steinman. An up-to-the-minute revised edition of a book that explains the fascinating drama of how the world's great bridges came to be built. The author, designer of the famed Mackinac bridge, discusses bridges from all periods and all parts of the world, explaining their various types of construction, and describing the problems their builders faced. Although primarily for youngsters, this cannot fail to interest readers of all ages. 48 illustrations in the text. 23 photographs. 99pp. 6⅛ x 9¼. T161 Paperbound **$1.00**

Technological, historical

A DIDEROT PICTORIAL ENCYCLOPEDIA OF TRADES AND INDUSTRY, Manufacturing and the Technical Arts in Plates Selected from "L'Encyclopédie ou Dictionnaire Raisonné des Sciences, des Arts, et des Métiers" of Denis Diderot. Edited with text by C. Gillispie. This first modern selection of plates from the high point of 18th century French engraving is a storehouse of valuable technological information to the historian of arts and science. Over 2000 illustrations on 485 full-page plates, most of them original size, show the trades and industries of a fascinating era in such great detail that the processes and shops might very well be reconstructed from them. The plates teem with life, with men, women, and children performing all of the thousands of operations necessary to the trades before and during the early stages of the industrial revolution. Plates are in sequence, and show general operations, closeups of difficult operations, and details of complex machinery. Such important and interesting trades and industries are illustrated as sowing, harvesting, bee-keeping, cheesemaking, operating windmills, milling flour, charcoal burning, tobacco processing, indigo, fishing, arts of war, salt extraction, mining, smelting, casting iron, steel, extracting mercury, zinc, sulphur, copper, etc., slating, tinning, silverplating, gilding, making gunpowder, cannons, bells, shoeing horses, tanning, papermaking, printing, dyeing, and more than 40 other categories. Professor Gillispie, of Princeton, supplies a full commentary on all the plates, identifying operations, tools, processes, etc. This material, presented in a lively and lucid fashion, is of great interest to the reader interested in history of science and technology. Heavy library cloth. 920pp. 9 x 12. T421 Two volume set **$18.50**

CHARLES BABBAGE AND HIS CALCULATING ENGINES, edited by P. Morrison and E. Morrison. Babbage, leading 19th century pioneer in mathematical machines and herald of modern operational research, was the true father of Harvard's relay computer Mark I. His Difference Engine and Analytical Engine were the first machines in the field. This volume contains a valuable introduction on his life and work; major excerpts from his autobiography, revealing his eccentric and unusual personality; and extensive selections from "Babbage's Calculating Engines," a compilation of hard-to-find journal articles by Babbage, the Countess of Lovelace, L. F. Menabrea, and Dionysius Lardner. 8 illustrations, Appendix of miscellaneous papers. Index. Bibliography. xxxviii + 400pp. 5⅜ x 8. T12 Paperbound **$2.00**

HISTORY OF HYDRAULICS, Hunter Rouse and Simon Ince. First history of hydraulics and hydrodynamics available in English. Presented in readable, non-mathematical form, the text is made especially easy to follow by the many supplementary photographs, diagrams, drawings, etc. Covers the great discoveries and developments from Archimedes and Galileo to modern giants—von Mises, Prandtl, von Karman, etc. Interesting browsing for the specialist; excellent introduction for teachers and students. Discusses such milestones as the two-piston pump of Ctesibius, the aqueducts of Frontius, the anticipations of da Vinci, Stevin and the first book on hydrodynamics, experimental hydraulics of the 18th century, the 19th-century expansion of practical hydraulics and classical and applied hydrodynamics, the rise of fluid mechanics in our time, etc. 200 illustrations. Bibliographies. Index. xii + 270pp. 5¾ x 8.
S1131 Paperbound **$2.00**

BRIDGES AND THEIR BUILDERS, David Steinman and Sara Ruth Watson. Engineers, historians, everyone who has ever been fascinated by great spans will find this book an endless source of information and interest. Dr. Steinman, recipient of the Louis Levy medal, was one of the great bridge architects and engineers of all time, and his analysis of the great bridges of history is both authoritative and easily followed. Greek and Roman bridges, medieval bridges, Oriental bridges, modern works such as the Brooklyn Bridge and the Golden Gate Bridge, and many others are described in terms of history, constructional principles, artistry, and function. All in all this book is the most comprehensive and accurate semipopular history of bridges in print in English. New, greatly revised, enlarged edition. 23 photographs, 26 line drawings. Index. xvii + 401pp. 5⅜ x 8. T431 Paperbound **$2.00**

Prices subject to change without notice.

Dover publishes books on art, music, philosophy, literature, languages, history, social sciences, psychology, handcrafts, orientalia, puzzles and entertainments, chess, pets and gardens, books explaining science, intermediate and higher mathematics, mathematical physics, engineering, biological sciences, earth sciences, classics of science, etc. Write to:

Dept. catrr.
Dover Publications, Inc.
180 Varick Street, N.Y. 14, N.Y.